S. Selberherr, H. Stippel, and E. Strasser (eds.)

Simulation of Semiconductor Devices and Processes

Vol. 5

Springer-Verlag Wien GmbH

Univ.-Prof. Dipl.-Ing. Dr. Siegfried Selberherr
Dipl.-Ing. Hannes Stippel
Dipl.-Ing. Ernst Strasser
Institut für Mikroelektronik
Technische Universität Wien, Austria

© 1993 Springer-Verlag Wien
Originally published by Springer Wien New York in 1993
Softcover reprint of the hardcover 1st edition 1993

Typesetting: Camera ready by authors

Printed on acid-free paper

With 530 Figures

ISBN 978-3-7091-7372-5   ISBN 978-3-7091-6657-4 (eBook)
DOI 10.1007/978-3-7091-6657-4

Preface

The "Fifth International Conference on Simulation of Semiconductor Devices and Processes" (SISDEP 93) continues a series of conferences which was initiated in 1984 by K. Board and D. R. J. Owen at the University College of Wales, Swansea, where it took place a second time in 1986. Its organization was succeeded by G. Baccarani and M. Rudan at the University of Bologna in 1988, and W. Fichtner and D. Aemmer at the Federal Institute of Technology in Zürich in 1991. This year the conference is held at the Technical University of Vienna, Austria, September 7 – 9, 1993.

This conference shall provide an international forum for the presentation of outstanding research and development results in the area of numerical process and device simulation. The miniaturization of today's semiconductor devices, the usage of new materials and advanced process steps in the development of new semiconductor technologies suggests the design of new computer programs. This trend towards more complex structures and increasingly sophisticated processes demands advanced simulators, such as fully three-dimensional tools for almost arbitrarily complicated geometries. With the increasing need for better models and improved understanding of physical effects, the Conference on Simulation of Semiconductor Devices and Processes brings together the simulation community and the process- and device engineers who need reliable numerical simulation tools for characterization, prediction, and development.

The conference committee of SISDEP 93 has prepared an excellent program with three invited papers, eighty papers for oral presentation and forty posters, selected from a total of two hundred and twenty-seven abstracts. Their distribution reflects the international nature of the conference: 27 from Germany, 16 from the USA, 14 from France, 10 from Austria, 6 from Italy and Switzerland each, 5 from Canada, England and Japan, 4 from The Netherlands, 3 from Australia and Spain, 2 from Belgium, the Czech Republic, Korea and Lithuania and 1 from Chile, China, Finland, Hungary, Ireland, Poland, Russia, Slovakia, Sweden, Taiwan and Ukraine. SISDEP 93 covers the following topics: process simulation and equipment modeling, device modeling and simulation of complex structures, device simulation and parameter extraction for circuit models, integration of process, device and circuit simulation, practical applications of simulation, algorithms and software.

The proceedings are printed from direct lithographs of the authors' manuscripts. The editors are not responsible for any inaccuracies, comments or opinions expressed in the papers.

We would like to express our sincere appreciation to the authors for their high quality contributions, their cooperation and effort. Besides we would like to thank the members of the conference committee for carrying out the paper selection work with care and competence.

Siegfried Selberherr
Hannes Stippel
Ernst Strasser

Institute for Microelectronics
Vienna, September 1993

Supporting Organizations

Austria Mikro Systeme International AG
Bundesministerium für Wissenschaft und Forschung
Der Bürgermeister der Stadt Wien
Digital Equipment Corporation, Austria
Die Erste Österreichische Spar-Casse – Bank
Erwin-Schrödinger Gesellschaft für Mikrowissenschaften
Forschungsförderungsfonds für die gewerbliche Wirtschaft
IEEE Austria Section
IEEE Electron Devices Society
IEEE Region 8
Der Landeshauptmann für Niederösterreich
Kammer der gewerblichen Wirtschaft für Wien
Siemens AG Österreich
Siemens AG, Zentralabteilung Forschung und Entwicklung
Siemens Entwicklungszentrum für Mikroelektronik Ges.m.b.H.
Technische Universität Wien
Textilmaschinenfabrik Dr.Ernst Fehrer AG
Österreichische Elektrizitätswirtschafts-AG
Vereinigung Österreichischer Industrieller

Conference Committee

G. Baccarani	Università di Bologna	ITALY
K. De Meyer	IMEC	BELGIUM
W. Fichtner	ETH Zürich	SWITZERLAND
M. Fukuma	NEC	JAPAN
H. Jacobs	Siemens	GERMANY
S. Laux	IBM	USA
C. Lombardi	SGS-Thompson	ITALY
P. Mole	BNR Europe	UNITED KINGDOM
M. Orlowski	Motorola	USA
A. Poncet	CNET/CNS	FRANCE
H. Ryssel	Universität Erlangen-Nürnberg	GERMANY
W. Schilders	Philips	THE NETHERLANDS
S. Selberherr	Technische Universität Wien	AUSTRIA
T. Toyabe	Hitachi	JAPAN
H. Van der Vorst	Rijksuniversiteit Utrecht	THE NETHERLANDS

Table of Contents

Challenges to Achieving Accurate Three-Dimensional Process Simulation

M. E. Law

Florida Integrated Research in Silicon Technology, Department of Electrical
Engineering, University of Florida
339 Larsen Hall, Gainsville, FL 32611, USA

Abstract

This paper will identify the reasons three-dimensional process simulators are not
widely available, when three-dimensional device simulators are widely available.
There appear to be four major obstacles; metrology, models, numerics, and
structural barriers. Each of these will be discussed and possible solutions will
be provided.

1. Introduction

The TCAD community generally recognizes that sub micron MOS and bipolar devices
require three-dimensional simulation. In addition, DRAM structures and latchup phenomena are three-dimensional problems. For these reasons, a great deal of work has
been performed on three-dimensional device simulation. In fact, three-dimensional
device simulators are available from both university and commercial sources. Doping
and material information for these device simulators is usually developed from either analytic models or two-dimensional process simulation. Three-dimensional process simulation would be useful not only for analysis of three-dimensional structures,
but also in providing accurate doping and material information for use with three-
dimensional device modeling. The accuracy of a device simulation can be no better
than the structural input — "garbage in, garbage out." Yet despite the apparent
need for three-dimensional process simulation, there are no widely available process
simulators, and little activity is focused on their development. Why?

There have been two traditional branches to process modeling. The first major area
is in the simulation of surface evolution and lithography. This activity is topologically
two-dimensional, but geometrically three-dimensional. In other words, simulators exist to evolve a two-dimensional surface plane in a three-dimensional space. The other
activity is bulk simulation, which focuses on material growth and dopant diffusion.
This activity requires the solution of a complex, nonlinear system of equations in the
materials of interest. For three-dimensional simulation, it requires three-dimensional
grid and structure.

For surface simulation, the state-of-the-art is very good. Excellent results have been
attained in modeling the surface evolution. For example, the work of the Berkeley
group on surface evolution during resist development requires advanced computational

geometry to evolve the surface without kinks, folds, and loops [1, 2]. An engineering workstation is sufficient for these types of simulations. In addition to simulation, verification of the simulator is also required. For surfaces, this can also be quite advanced. Slinkmann et al., have recently used a scanning force probe microscope to profile the surface of DRAM trench technology [3]. This technique and others allow new models to be developed and characterized effectively. Consequently, these simulators offer three-dimensional capability currently, and will not be discussed further.

The real problem lies with bulk simulation. Bulk three-dimensional process simulation is not currently available. There are three main reasons for this lack of capability. The first major area is measurement techniques. It is not difficult to measure the I-V characteristics of a transistor, but it is costly and less reliable to measure a doping profile. The second reason is the numerical cost. Three-dimensional process simulation is thought to be an expensive computation [4]. The final reason is the difficulty in performing process modeling work in universities and company labs.

2. Challenges

2.1. Metrology

One of the most difficult challenges facing one, two, or three-dimensional bulk process simulation is in obtaining verification of models. Measurement techniques used to investigate doping profiles are not very accurate and for the most part are inherently one dimensional. This limits our ability to verify the predictions of the models in more than one dimension, and limits the scalability of simpler empirical models. This is further complicated by the fact that all of the available analysis techniques are expensive and difficult to perform.

There are three main techniques available to measure dopant profiles. The first is Secondary Ion Mass Spectroscopy (SIMS) which is a destructive technique that makes an etch pit using ion sputtering and examines the backscattered ions. This technique can accurately measure the chemical concentration of the dopant, but can not determine thin dopant layer concentrations very well due to the broadening of the peak. SIMS detection limits place the lower bound on concentration at $10^{16} cm^{-3}$, and measurement of doping levels below this is usually not accurate. Because the raw SIMS data is a particle count, the concentration values must be scaled by a dose calibration. This can also result in some uncertainty in the profiles. This technique only gives resolution in the direction of the etch, and therefore is one dimensional only.

Spreading Resistance Profiling (SRP) measures the resistance as a function of depth. SRP measurement systems extract the mobile carrier concentration from the resistance by using mobility model as a function of concentration. The mobile carrier concentration can then be related to the doping concentration. There are several major problems with this technique that limit its accuracy. The first problem is the mobility model, particularly at high concentrations when damage and precipitates can affect the scattering, but are not included in the mobility model. The second effect is carrier spilling and depletion layers that alter the carrier concentration from the doping concentration. These two effects limit SRP to characterization of moderately doped profiles that are not ultra shallow. Related to SRP is capacitance-voltage profiling since it also measures carrier concentrations. This technique can be accurately coupled to Poisson solvers to obtain doping profiles, but is limited to lightly doped layers so that they can be depleted from a surface voltage.

The final measurement technique is junction staining. This technique allows chemical delineation of the junction based on the doping type, and is the only technique that can be used to examine doping in two dimensions. However, it gives only a single data point.— the junction depth. This limits its ability for verification. Some clever structures have been used with junction staining to obtain two-dimensional information [5].

There has been research work on multidimensional measurement techniques. Work has been performed using multiple stains, which allow multiple concentrations to be resolved [6]. This approach makes use of the two-dimensional nature of staining to add additional data beyond just a junction position. Subrahmanyan used this technique to verify two-dimensional simulations of doping profiles near mask edges [7]. Goodwin-Johannson has investigated using multiple angled SIMS beams to reconstruct two-dimensional doping profiles. This work shows promise, but is limited by the resolution of the SIMS spot. A large spot produces averaging over a wide lateral area [8]. Finally, electrical information can be used to deconvolve two-dimensional profiles, *e.g.*, Khalil and Faricelli [9], but is limited often times by uncertainties in other structural data. This limitation in measurement technology severely hampers development of accurate multi-dimensional simulators. Since it is impossible to verify two-dimensional doping profiles, why bother to compute them in three-dimensions?

2.2. Models

In device simulation, there are many levels of simulation complexity available. There are drift-diffusion simulators, energy balance codes, full hydrodynamic solvers, and finally Monte Carlo techniques that use a variety of approximations to the band and scattering problems. This richness of complexity is not available in process simulation. There have been essentially two approaches to bulk simulation; a phenomenological approach, best exemplified by the decision tree simulator PREDICT [10], and a point-defect-based approach, as demonstrated by SUPREM-IV [11, 12]. The phenomenological approach would be the natural tack to take for three-dimensional bulk simulation since it requires less CPU time, but development of these models is delayed by the lack of multidimensional measurement techniques of doping profiles.

This leaves full point-defect-based simulators, which offer the promise of accurate multidimensional simulation due to their complete encapsulation of the full physics of dopant diffusion. These simulations rely on accurate calculation of point defect concentrations that are then used to compute the dopant diffusion. Although conceptually it promises great accuracy, there are two main difficulties. The first is parameterization and second is computation time.

Figure 1 shows the interstitial diffusivity in silicon as a function of temperature and experimentalist. This large scatter in the data does not inspire confidence in the predictive ability of the simulator. Although this experimental scatter can be explained [5], it does point out one of the problems with the point-defect-based simulators. Both point defect types, interstitials and vacancies, can not be measured directly and can only be measured indirectly through their effect on processes, *e.g.*, diffusion. This leads to circular reasoning — there must be excess interstitials because phosphorus diffuses faster, and the reason for the faster diffusion of phosphorus is that there are excess interstitials present. This circular reasoning loop is difficult to break, and has resulted in a great deal of controversy about the point defect behavior in silicon.

Finally, there are a large number of equations to be solved in point-defect-based simulators. Because the approach has always been to include as much physics as possible in the simulation to obtain accuracy, it is often necessary to solve 3–5 partial differential equations to obtain the solution for a single dopant profile. This adds to the burden placed on the numerics of the simulation.

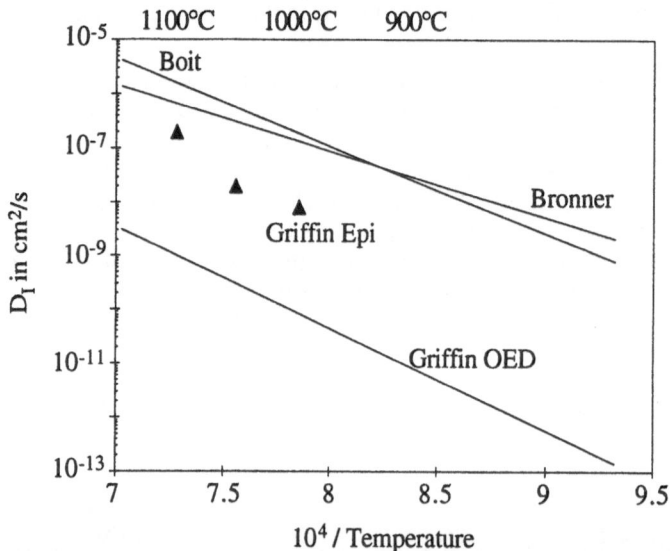

Figure 1: Experimental investigations of the interstitial diffusivity as a function of temperature [5, 13–15]. Different results are obtained under different experimental conditions.

2.3. Numerics

There are two major challenges numerically to be solved before three-dimensional process simulation can become a reality. The first involves the linear algebra and the second involves grid generation and adaptation. Both issues must be addressed, but the second is more the province of process simulation than device simulation.

Linear algebra is complicated due to the immense problem size involved. A minimum of roughly one hundred grid points is required to represent a dopant profile to 1there will be approximately one million nodes. As stated earlier, point-defect-based simulators require three solution variables to solve for a single dopant. The resulting linear sparse system will require approximately three million unknowns. Since solution of linear systems requires approximate solution time on the order of the number of unknowns to the 3/2 power, the computation time required for three-dimensional process simulation is very large, even for super computers. This situation is similar to that for device modeling, and many of the same iterative techniques can be exploited. SUPREM-IV uses a conjugate gradient technique with a block preconditioner to achieve speeds up around a factor of two for two-dimensional problems [12].

The second problem concerns three-dimensional grid generation and adaptation. Generation is a problem that has been under study for device simulation [16–18]. Many of these techniques can be adapted for use with process simulation, however, it is important to consider refinement techniques for process simulation since the profile changes during the simulation. The initial grid may not be adequate for the final structure. Lin et al., have developed error estimators for diffusion simulation that may help with this problem [19].

The main challenge for grid generation is the adaptation that must occur due to the moving oxide/silicide interface. Oxidation and silicidation are natural companion processes to dopant diffusion simulation, and therefore the grid must be adapted to conform to the growing layer driven by surface reactions. This represents a key problem for 2D simulation reliability, as a large number of the problems with current versions of SUPREM-IV come from the grid adaptation algorithm. Some encouraging 2D results have been presented [20, 21], but further work will be required for three-dimensional simulation.

2.4. Cultural Barriers

The simplest answer to the lack of three-dimensional process simulation is that there are few research labs that are actively pursuing this goal, particularly when compared to the amount of activity in device simulation. Lack of work in the field necessarily produces slower progress. At this conference, for example, 15 of the 80 papers selected for oral presentation concern bulk process simulation. I don't believe that this ratio is atypical for most process and device simulation conferences, and is even lower at the IEDM conference. The paper ratio accurately reflects the distribution of world wide effort on process modeling as compared to device, applications, and numerical research. Why is there less research activity in the area of process simulation than device simulation?

The first reason is facilities. Obviously, industrial electronics firms have access to advanced fabrication facilities. However, these facilities are not designed to build process modeling test cases, but instead focus on advanced device structures. The lack of dedicated facilities for experimental work impedes progress. Another barrier is the perceived difficulty in obtaining results. Rather than using resources to characterize and model a particular process, a series of shotgun experiments is run to optimize the structure. In the short term, this approach is probably effective in reducing costs and development time. In the long run, however, little is learned that can be reused for the next generation of technology. Only the largest companies, to date, have been able to afford to invest in process modeling.

On the other hand, most university research labs do not have extensive fabrication facilities and can not build structures at all. This is in sharp contrast to the situation for device modeling, where transistors are available for measurement from many corporate sources. Fabrication facilities are beyond the means of most universities, and it is increasingly difficult to support those universities that do have facilities. A large fraction of the university based process modeling done in the last decade has come from Stanford and Duke, both of which have large, expensive fabrication facilities. This lack of facilities for experimental work translates into a lack of progress.

Finally, the sheer complexity of the problem discourages researchers. In addition to the diffusion simulation, the flow equations must be treated for oxide and silicide growth in three-dimensions, and accurate calculation of doping profiles from implantation must be obtained. The wide variety of knowledge and disciplines makes it difficult to form an effective process simulation group.

2.5. Opportunity — Software Engineering

Figure 2 illustrates the organization of most of the core of process simulation tools. For a complete tool, user interface and post processing tools are required, but these will be ignored in this discussion. There are three major components; grid, physics, and linear algebra. The grid and linear algebra are directly related to the challenges discussed in section 2.3, and the physics relates to the challenges in section 2.2.

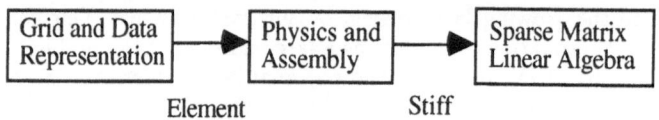

Figure 2: Basic organization of the core of process simulation tools.

These code sections have fairly well defined interfaces that make it possible to organize and write them independently. Object-oriented languages are a natural development framework for this data encapsulation. The code components communicate through elements and stiffness matrices. The physics component needs to be able to assemble the Jacobian for the required PDE's for each element in the mesh. In two-dimensional simulation, the element would be a triangle and for three-dimensional simulation the elements could be either tetrahedra or bricks. The physics code produces a small, dense Jacobian matrix usually referred to as the stiffness matrix for historical reasons. The dense stiffness matrices can then be copied or summed into the matrix component solver one by one to assemble the large sparse matrix.

If a code is structured in this fashion, it is possible for a diverse community to be actively developing the final software package. Mathematicians can work on sparse matrix techniques and insert their code with advanced process models and physics. New grid generation schemes can be implemented at the other end of the scale, without affecting the remaining code components. This has the potential for accelerating development of a three-dimensional process simulator, because each of the major challenge areas can be addressed cooperatively by a wide range of groups in both academic and industrial environments. This in turn tends to lower the structural barriers present to development of a three-dimensional process simulator.

Finally, three-dimensional process simulation will be a computationally expensive undertaking. Intelligent systems need to be developed to manage the simulation of an entire process flow and device structure. The entire structure does not need to be simulated by a three-dimensional code, and many parts can be handled adequately by a one- or two-dimensional code. Such a system is under development that builds 3D device structures from mask information and one-, two-, and three-dimensional process simulation [22]. Such a tool will be required to perform efficient simulation of full device structures.

3. Conclusions

There are four major challenges to obtaining accurate three-dimensional process simulators, metrology, models, numerics, and structural. Metrology is the major stumbling block, because it is very difficult to measure two-dimensional profiles, much less three. Model development, particularly phenomenological models that would be more computationally efficient, are slowed due to the lack of sophisticated measurement techniques. These two factors contribute to lack of motivation for development, since without accurate, verifiable models three-dimensional process simulators "garbage in, garbage out."

Numerical problems also provide a major hurdle to development. The first problem is efficient inversion of the large sparse linear system. This problem is similar to the problem faced by device simulation in three-dimensions and can be attacked by some of the same methods. The other problem is grid generation and adaptation. Adaptation is required so that diffusion and oxidation/silicidation problems can be solved at the same time. There has been little work on this topic for three-dimensional simulation.

Finally, there is widespread lack of effort in part due to lack of facilities and apparent benefit for development of three-dimensional process simulation. Overcoming the first three hurdles will help make three-dimensional process simulation more attractive. A possible solution to some of these problems is dividing the program into a set of well-defined interfaces between major modules, so that co-development at several locations is possible. This would allow greater leverage to be applied to the problem, and lower the individual cost of development.

Acknowledgements

I would like to thank the National Science Foundation, SEMATECH, IBM and Silvaco for support of my process modeling research. I would also like to thank Kevin S. Jones and Heemyong Park for critical reading of this paper.

References

[1] E.W. Scheckler and A.R. Neureuther, Numerical Analysis of Process and Devices Workshop, Seattle, p. 9, 1992.

[2] J.J. Helmsen, E.W. Scheckler, A.R. Neureuther and C.H. Sequin, Numerical Analysis of Process and Devices Workshop, Seattle, p. 3, 1992.

[3] J. Slinkmann, Private Communication.

[4] R.W. Dutton and M.R. Pinto, Proceedings of the IEEE, 74 (12), 1986.

[5] P.B. Griffin and J.D. Plummer, International Electron Devices Meeting, Los Angeles, p. 522, 1986.

[6] R. Subrahmanyan, H.Z. Massoud and R.B. Fair, Appl. Phys. Lett., 52, p. 2145, 1988.

[7] R. Subrahmanyan, H.Z. Massoud and R.B. Fair, J. Electrochem. Soc., 135, p. 1573, 1990.

[8] S. Goodwin-Johannson, R. Subrahmanyan, C.E. Floyd and H.Z. Massoud, IEEE Trans. CAD, 8(4), p. 323, 1989.

[9] N. Khalil and J. Faricelli, Simulation of Semiconductor Devices and Processes, Vienna, 1993.

[10] R.B. Fair and J.E. Rose, International Conference on Computer Aided Design, Santa Clara, 1987.

[11] M.E. Law, C.S. Rafferty and R.W. Dutton, International Electron Devices Meeting, Los Angeles, p. 518, 1986.

[12] M.E. Law and R.W. Dutton, IEEE Trans. on CAD, 7(2), p. 181, 1988.

[13] C. Boit, F. Lau and R. Sittig, Applied Physics A, 50, p. 197, 1990.

[14] G.B. Bronner and J.D. Plummer, J. Appl. Phys., 61(12), p. 5286, 1987.

[15] P.B. Griffin, P.M. Fahey, J.D. Plummer and R.W. Dutton, Appl. Phys. Lett, 47(3), p. 319, 1985.

[16] J. Fuhrmann and K. Gärtner, Simulation of Semiconductor Devices and Processes, Vienna, 1993.

[17] N. Hitschfeld and W. Fichtner, Simulation of Semiconductor Devices and Processes, Vienna, 1993.

[18] N. Hitschfeld, P. Conti and W. Fichtner, Simulation of Semiconductor Devices and Processes, p. 165, 1991.

[19] C.C. Lin, M.E. Law and R.E. Lowther, IEEE Trans. CAD, 12(10), 1993.

[20] Z.H. Sahul, R.W. Dutton and M. Noell, Simulation of Semiconductor Devices and Processes, Vienna, 1993.

[21] K. Smith and R.E. Bank, NUPAD, Seattle, p. 187, 1992.

[22] C. Hegarty, T. Feudel, N. Hitschfeld, R. Ryter, N. Strecker, M. Westermann and W. Fichtner, Process Physics and Modeling in Semiconductor Technology Symposium, Honolulu, p. 565, 1993.

SIMULATION OF SEMICONDUCTOR DEVICES AND PROCESSES Vol. 5
Edited by S. Selberherr, H. Stippel, E. Strasser – September 1993

9

Modeling Nano-Structure Devices

K. Hess[††] and L. F. Register[‡]

[‡]Beckman Institute for Advanced Science and Technology and Coordinated Science
Laboratory, University of Illinois at Urbana-Champaigne
Urbana, IL 61801, USA
[†]Department of Electrical and Computer Engineering,
University of Illinois at Urbana-Champaigne
Urbana, IL 61801, USA

Abstract

Fundamental problems of and approaches to modeling nanostructure devices
are reviewed. First the requirements for modeling charge transport in classical
and nanostructure devices are compared and contrasted. Then the quantum
mechanical concepts of transmission probabilities and eigen energies in nanostructures are related back to the classical concepts of resistance and capacitance,
respectively. Next a small illustrative sampling of numerical approaches to calculation of the quantum mechanical properties of nanostructures is presented.
Finally examples are given of how such theoretical concepts and numerical
methods can be applied to modeling existing and future devices.

1. Introduction

In conventional semiconductor devices, most quantum transport effects can be treated
indirectly. The effects of the rapidly varying crystal potential on electron transport
can be modeled via the concepts of effective masses, energy gaps, and the positively
charged quasi-particle holes. On the macroscopic level, charge transport can then
be modeled using the concepts of classical mechanics, aided sometimes by the Pauli
exclusion principle and the Born approximation for scattering problems. Only the
possibility of electron-hole recombination across the energy gap and the corresponding
non-conservation of particle numbers bears witness to the deeper quantum mechanical
character of the problem.

Of course, size quantization effects do appear in some conventional devices. The
conduction channel of a MOSFET extends only a few nanometers from the Si-SiO$_2$
interface into the silicon and, thus, the charge carrier motion perpendicular to the
interface is quantized. However, charge transport over the oxide barrier is usually
negligible, and the carriers can be treated as quasi two-dimensional particles in the
plane of the conduction channel and still be modeled semi-classically. How far the
dimensions of the conduction channel can be shrunk in this latter plane before quantization leads to prominent effects is currently the subject of speculation.

In addition to quantization effects, the current mainstays of semiconductor device
technology, electron and hole gases and the formation of homojunctions by doping,
also present barriers to shrinkage as the desired step from the vacuum to the dense

solid has not been achieved entirely after the invention of the point contact transistor by Bardeen and Brattain in 1947 [1]. On the nano-scale, doping is inhomogeneous and majority carriers and particularly minority carriers form dilute gases in conventional devices [2]. If shrinkage is to substantially continue, the homojunction and these dilute gasses must yield to heterojunctions and Fermi-liquids. The increased use of heterojunctions in silicon technology and the interesting science surrounding super-lattices and most recently nanostructures and mesoscopic systems indicate motion in this direction.

Nanostructures and mesoscopic systems currently provide an interesting playground for applications oriented science [3, 4]. From an engineering point of view, the mechanism that will make nanostructure devices tick and permit integration on a gigantic scale has yet to be found. In our opinion, however, nanostructure devices will involve regions of significantly unbroken carrier phase extending over many atomic distances, *mesoscopic systems or structures*, alternating with regions where phase coherence is broken and excess carrier energy is dissipated, *reservoirs*, such as contacts and interconnects. Within these mesoscopic structures current will be carried by charge carrier transport within, over and (via tunneling) through heterojunction barriers. Further, these devices will exhibit problems of multi-scale; nanostructure regions, governed by quantum mechanical properties, will be connected to much larger regions, where classical transport models are appropriate. These envisioned transport mechanisms already are evidenced in existing devices containing nanostructures such as quantum well lasers and the now familiar resonant tunneling diode and, therefore, set the stage for the modeling and simulation tools that must be developed.

2. Basic physical theories

Quantum transport theory encompasses a wide range of phenomena including super-conductivity, quantum Hall effect, tunneling, non-parabolic and multi-band energy band structures, carrier-carrier interactions and carrier-phonon interactions. Nevertheless, it is often possible to translate the quantum theoretical problem of carrier transport in nanostructures into conventional engineering concepts such as of resistance, capacitance and inductance. Below, only resistance and capacitance are addressed due to spatial limitations. Quantum inductance is discussed extensively in Ref. [5].

The Landauer-Büttiker theory allows direct calculation of charge currents and interpretation of resistance in terms of quantum mechanical transmission coefficients [6, 7, 8]. The Landauer-Büttiker theory, as well as the Bardeen transfer Hamiltonian approach [9], view the mesoscopic (phase coherent) region on the basis of scattering theory, i.e. as simply an object that scatters the charge carrier's wave function. A simple classical analogy to this approach is the use of thermionic emission theory within a drift-diffusion simulator where transport over a heterojunction barrier from one reservoir to another is characterized by the classical transmission coefficient, a step function in energy. Extending this latter method to include quantum mechanical transmission coefficients, as for a tunneling diode, gives for the electron current density perpendicular to the barrier(s) J_{12} [10, 11],

$$J_{12} = 2\frac{e}{(2\pi)^3\hbar}\int_{-\infty}^{\infty}dk_{\parallel}\int_{E_{c,max}}^{\infty}dE_{\perp}\ T(E_{\perp}, k_{\parallel})[f_1(E) - f_2(E)]. \tag{1}$$

Here, e is the (negative) electron charge, the leading factor of 2 accounts for spin degeneracy, T is the transmission probability, $f_1(E)$ and $f_2(E)$ are the occupation probabilities for the electrons as a function of total energy in their respective reservoirs, and $E_{c,\text{max}}$ is the greater of the two minimum conduction band energies of the reservoirs. The Landauer-Büttiker theory generalizes this approach to arbitrary mesoscopic geometries and multiple reservoirs. Specifically, in the linear response regime, the current I_{12} flowing from one reservoir to a second via some mesoscopic structure (perhaps in addition to the current flow to and from other reservoirs) is given by [7, 8]

$$I_{12} = 2\frac{e^2}{h} \sum_{n_1,n_2} T_{n_1 n_2} \frac{\mu_1 - \mu_2}{e}. \tag{2}$$

Here, μ_1 and μ_2 are the respective chemical potentials of the two reservoirs, and the $T_{n_1 n_2}$ are the transmission probabilities between the states of the two reservoirs at the Fermi-surface energy. Because $(\mu_1 - \mu_2)/e$ represents the voltage drop V between the reservoirs, the conductance can be identified in terms of the transmission coefficients as $2(e^2/h) \sum_{n_1,n_2} T_{n_1 n_2}$. This formula, Eq. (2), is readily generalized beyond the linear response regime and for high temperatures; however, in either, case the transmission coefficients do not incorporate dissipative processes within the mesoscopic region. Single electron transmission coefficients can be generalized to include dissipative processes, but the significance of this generalized concept becomes unclear when Pauli exclusion effects must be considered simultaneously [12, 13]. Examples of familiar device geometries where the Landauer-Büttiker approach may be applicable include the transition region between the source contact and the quasi-two-dimensional channel of a MOSFET, and the region about an undoped quantum well within the heavily doped classical transport regions in a semiconductor laser. This approach has also been extremely successful in explaining a large amount of experimental results for nanostructures including conductance steps for transport through ultrasmall constrictions [3].

As transmission coefficients are converted to resistance via Landauer-Büttiker theory, the energy spectra of nanostructure systems can be converted to capacitance. The capacitance C of an isolated island or dot of conducting material is defined as the ratio of the stored charge Q in the dot to the voltage V of the dot [14], or, in a form more conducive to nanostructure applications, $C \equiv e^2(N/\mu)$ where N is the total number of electrons within the dot and μ is the chemical potential of the dot. Similarly, the differential capacitance, $C_d \equiv \Delta Q/\Delta V$, is given by [15],

$$C_d(N) \equiv \frac{e^2}{\mu(N+1) - \mu(N)}, \tag{3}$$

where the chemical potential $\mu(N)$ is given by,

$$\mu(N) = E(N) - E(N-1). \tag{4}$$

Here, $E(N)$ is the total energy of the N-electron system. In metallic systems, capacitance and differential capacitance are effectively independent of N [14], even in many structures exhibiting coulomb blockade effects where μ changes in discrete steps on the scale of the thermal energy with the addition of single electrons to the dot [3, 4, 16]. However, in semiconductor nanostructures with few electrons, capacitance does depend on N due to the electrostatic interactions among the electrons, size quantization contributions to the energies $E(N)$ and Pauli exclusion effects [15, 17, 18].

The interpretation of conductance in terms of electron transmission probabilities, via Landauer-Büttiker theory, and capacitance in terms of multi-electron energies, reduces the problem of calculating nanostructure device parameters to one of solving the Schrödinger equation for one or more charge carriers within regions of phase coherence. Numerical methods for accomplishing this latter goal are considered next.

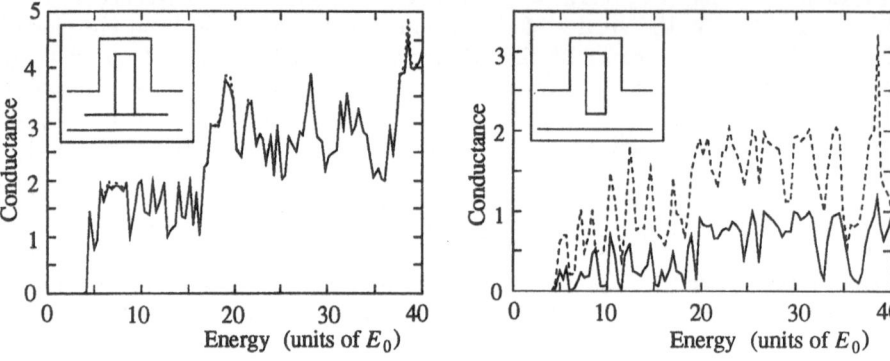

Figure 1: Conductance (in units of $2e^2/h$) as a function of energy (in units of the transverse ground-state energy of of the leads) for two similar mesoscopic structures, shown in the inserts. The exact conductances are indicated by solid lines, while the dashed lines are the simple (classical) sum of the conductances of the upper and lower channels in the absence of the other. (After Ref. [19], ©1992 The American Physical Society.)

3. Numerical methods

There are many numerical approaches to calculating the required transmission coefficients and energy spectra for charge carriers in nano-scale systems. Here, however, we have room to discuss only a few. A common feature of two of the methods discussed is that each attempts to model a many particle problem: one, electron transport in the presence of phonons and, the other, the effects of coulomb and exchange interactions within many electron systems.

One computational efficient method for simulating electron transport in multi-dimensional mesoscopic structures employs a mode matching technique [19]. The analyzed structures are assumed to be defined by hard walls (i.e., potential walls of infinite height) and separable into rectangular sections. Within each of these sections at a given energy, the wave function solution of the time-independent Schrödinger equation can be written as the sum of the products of propagating and evanescent modes in the longitudinal direction and standing waves in the transverse direction. At each interface, enforcing the continuity of the wave function and its normal derivative produces a linear system of equations with a sparse coefficient matrix that can be efficiently solved. Because of the efficiency of the method but restrictions on device geometries, this method is best suited for addressing fundamental questions of quantum transport such as how do conductances add in mesoscopic structures, as in the calculations of conductance, via Eq. (2), in the examples of Fig. 1.

A more computationally intensive but more flexible approach to simulating transmission through multidimensional mesoscopic structures is based on solution of the time-dependent Schödinger equation [20]. This method features robust open boundary conditions and an arbitrarily variable potential function that allows simulation of the transient through steady-state transport in a broad range of mesoscopic structures, and has recently been updated to allow calculation of week dissipative coupling to phonons when Pauli exclusion effects are not critical [21]. Specifically, if the Schrödinger equation for an electron–few-oscillator (few-phonon-mode) system,

$$i\hbar\frac{\partial}{\partial t}\Psi_{n_1,\ldots,n_K}(\vec{r},t) = \left[-\frac{\hbar^2}{2m^*}\nabla_{\vec{r}}^2 + V(\vec{r}) + \sum_{k=1}^{K}\hbar\omega_k\left(n_k+\frac{1}{2}\right)\right]\Psi_{n_1,\ldots,n_K}(\vec{r},t)$$

$$(5)$$

$$+\sum_{k=1}^{K}\left(\frac{\hbar}{2\omega_k K}\right)^{\frac{1}{2}}M_k(\vec{r})\left[\sqrt{n_k}\,\Psi_{n_1,\ldots,n_k-1,\ldots,n_K}(\vec{r},t)+\sqrt{n_k+1}\,\Psi_{n_1,\ldots,n_k+1,\ldots,n_K}(\vec{r},t)\right],$$

is solved numerically, quantum transport with first order accurate phonon scattering can be treated within the limits of the single electron picture. Here, $V(\vec{r})$ is the applied potential seen by the electron within the mesoscopic structure, and the n_k are the discretized coordinates of the few K oscillators corresponding to their uncoupled eigenstates. The coupling functions $M_k(\vec{r})$ are obtained stochastically as,

$$M_k(\vec{r}) = \sum_{\vec{q}}\lambda_{\vec{q}}M_{\vec{q}}(\vec{r}),$$

$$(6)$$

where the $M_q(\vec{r})$ are the coupling functions for the true electron-phonon system for phonon modes $\vec{q}$, and the λ_s are random numbers $(\langle\lambda_{\vec{q}}\rangle = 0, \langle\lambda_{\vec{q}}\lambda_{\vec{q}'}\rangle = \delta_{\vec{q},\vec{q}'})$ such that the $M_k(\vec{r})$ have the spatial correlation function characteristic of the original system's coupling functions,

$$\langle M_k(\vec{r})M_k(\vec{r}')\rangle = \sum_{\vec{q}}M_{\vec{q}}(\vec{r})M_{\vec{q}}(\vec{r}').$$

$$(7)$$

Because the correlation function on the right-hand-side of Eq. (7) is the common factor of all first order calculations in the electron-phonon coupling (as, for example, in "golden rule" calculations of scattering rates), the identity of Eq. (7) assures, on average, the equivalence of the true multiphonon system and the few-oscillator system of Eq. (5) to first order. As the number of samples K is increased, the sampling error is reduced and second and higher order artifacts of using a small oscillator system are reduced at least as $1/K$, while some true higher order process are retained. Figure 2 shows simulation results for emission of 50meV polar optical phonons by an approximately 50meV electron incident through a quasi-one-dimensional quantum wire into a quasi-two-dimensional semiinfinite plane, each 50Å thick in the direction perpendicular to the plane of the simulation. The ground-state energy of an electron within the wire is approximately 40meV such that real (as opposed to virtual) phonon emission is only possible within the semiinfinite plane. Twenty samples of $M_k(\vec{r})$ were used, and much of the fluctuation in Fig 2(d) is due to sampling error. However, the calculated probabilities of reflection (inherently without phonon emission), transmission through the simulation region without phonon emission, and transmission with phonon emission are 14.6%, 64.5% and 20.9%, respectively, all to within approximately ±3% RMS deviation.

Calculation of the energy spectrum and, in turn, the capacitance for several electrons in a small island of material, requires balancing a desire for rigor with computational resources. One approach that appears well suited to the task is the local density functional formalism [15]. For this method, the Schrödinger equation that must be self-consistently solved for each electron i in the presence of the others j is,

$$-\frac{\hbar^2}{2m^*}\nabla^2\psi_i(\vec{r}_i) + [V_b(\vec{r}_i) + V_c(\vec{r}_i) + V_{ex}(\vec{r}_i) + V_{corr}(\vec{r}_i)]\psi_i(\vec{r}_i) = E\psi(\vec{r}_i).$$

$$(8)$$

Here, $\psi_i(\vec{r}_i)$ is the wave function for the ith electron, $V_b(\vec{r}_i)$ is the built in potential of the dot of material and any applied potential, and V_c is the coulomb interaction term among the electrons given by,

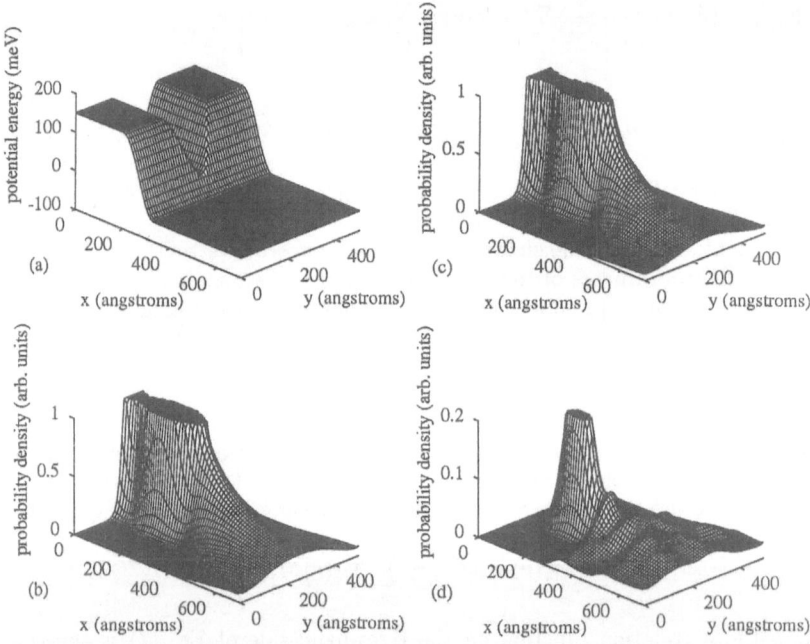

Figure 2: (a) Potential energy function and (b) probability density function for an uncoupled electron. (c) Probability density for the corresponding coupled electron-phonon system in the ground-state of the uncoupled phonon system (prior to phonon emission) and (d) the probability density in an excited state of the phonon system (after phonon emission). The peaks in the probability density have been cutoff to allow viewing of the regions of lower probability. (After Ref. [21].)

$$V_c(\vec{r}_i) = \sum_{j=1}^{N} \frac{1}{4\pi\epsilon_0\epsilon_r} \int_S \frac{e^2|\psi_i(\vec{r}_j)|^2}{|\vec{r}_i - \vec{r}_j|} d\vec{r}_j \,, \tag{9}$$

where N is the total number of electrons in the dot and ϵ_0 and ϵ_r are the absolute and relative dielectric constants, respectively. Nearby metallic electrodes (gates) can also be modeled via the method of images. $V_{ex}(\vec{r}_i)$ and $V_{corr}(\vec{r}_i)$ are the exchange and correlation terms, respectively. These latter terms have received extensive attention in the recent literature, and the easiest representations to include for a thin layer of charge in numerical calculations are the polynomial expressions from the theory of Tanatar and Ceperley [22]. Figure 3 shows the calculated differential capacitance, Eq. (3), vs. electron number for an isolated small quasi-two-dimensional box with hard walls and a quasi-parabolic confinement potential due to a uniformly distributed positive charge of $100|e|$.

4. Current and future device applications

The computational problems encountered in numerical simulation of quantum well lasers are illustrative of those that can be expected in general nanostructure simula-

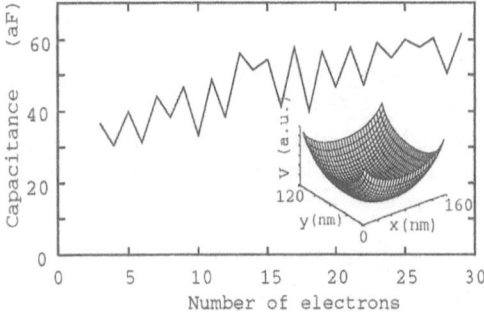

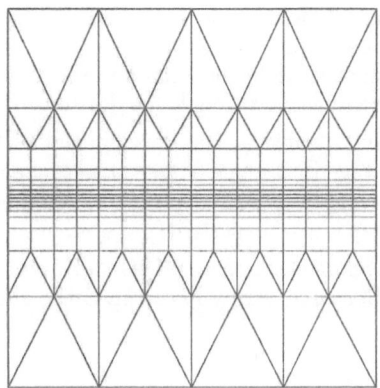

Figure 3: Differential capacitance vs. electron number for an isolated 160nm×120nm quasi-two-dimensional box with hard walls and a quasi-parabolic confinement potential due to a uniformly distributed charge of 100|e|. The inset shows the self-consistent potential for 10 electrons. (After Ref. [15].)

Figure 4: Grid used for numerical simulation of charge transport within a semiconductor quantum well laser. (After Ref. [23].)

tions. For a laser, the width of the quantum well region(s) relative to the entire laser thickness is as an inch to the height of the Sears Tower in Chicago and, thus, a grid system is required that provides suitable resolution both within and outside the quantum well region, as shown in the neighborhood of the quantum well in Fig. 4 [23]. The explosion of grid points near the well clearly demonstrates the multi-scale nature of the problem. Further, outside of the quantum well, simple classical drift-diffusion theory provides a good model of charge transport, while quantum mechanical calculations are required to obtain the energy spectrum and density of states for the charge carriers within the well to model photon emission. Thus, the question is one of how to connect these regions, and a straightforward solution answer lies in the spirit of the Landauer-Büttiker and Bardeen transfer Hamiltonian approaches. To a large extent, the entire quantum well region can be treated as a scattering (recombination) center within the drift diffusion simulator. Carriers within and above the well then can communicate with each other by a multipoint approach, as in the case of thermionic emission [23]. A rather complex understanding of laser switching, phonon emission and carrier-carrier interactions is still required to model carrier capture in the quantum well, but at least for quasi-two-dimensional quantum wells such information is available. For quantum dot lasers, if indeed carriers can be collected efficiently by quantum dots, an approach such as that of Eq. (5) will be required to model the process.

One particularly intriguing device concept of the future is the single-electron transistor [3]. The basic component of such a device will be a dot of material in which the chemical potential changes discretely on the scale of the thermal energy with the number of electrons. For metallic dots, the total energies can be calculated from $E(N) = (Ne)^2/2C$ where the capacitance C is effectively a geometry dependent constant only. However, in semiconductor nanostructures with few electrons, calculation of the the multi-electron energies will require an approach such as that of Eqs. (8) and (9). These energies can then be translated into chemical potentials and capacitances via Eqs. (3) and (4). If the dot is loosely coupled to an external reservoir via a mesoscopic barrier or other structure, a current will flow according to the differences in chemical potentials as described by Eq. (2). However, because the chemical potential of the dot changes discretely with the number of electrons, coulomb exclusion and blockade effects can prevent or control current flow as evidenced by the periodic conductance oscillations reported in many publications [3, 4, 16]. A gate (perhaps another dot) then can be used to control the energy spectrum of the dot and, thus, conduction through the dot to provide transistor action.

In conclusion, the development of theory and computational resources is now at a stage that much of nanostructure electronics can be understood and simulated. However, deep seated problems remain with regard to the multi-scale and many-body aspects of nanostructure systems.

Acknowledgement: This work was supported by the United States Office of Naval Research, Army Research Office and National Science Foundation (NCCE).

References

[1] J. Bardeen and W. H. Brattain, *Phys. Rev.* **74**, 230 (1948).

[2] A carrier density of $10^{18}/cm^3$ translates to one carrier per 1,000 cubic nanometers.

[3] *Mesoscopic Phenomena in Solids*, ed. by B. L. Altshuler, P. A. Lee and R. A. Webb (North-Holland, Amsterdam, 1991). (Vol. 30 of *Modern Problems in Condensed Matter Sciences*, ed. by V. M. Agranovich and A. A. Maradudin.)

[4] *Nanostructures and Mesoscopic Systems*, ed. by W. P. Kirk and M. A. Reed (Academic Press, Boston, 1992).

[5] J. R. Tucker and M. J. Feldman, *Reviews of Modern Physics* **57**, 1055 (1985).

[6] R. Landauer, *IBM J. Res. and Develop.* **1**, 233 (1957).

[7] M. Büttiker, *Phys. Rev. Lett.* **57**, 1761 (1986).

[8] A. D. Stone and A. Szafer, *IBM J. Res. Develop.* **32**, 384 (1988).

[9] J. Bardeen , *Phys. Rev. Lett.* **6**, 57 (1961).

[10] C. B. Duke in *Solid State Physics*, ed. by F. Seitz, D. Turnbull and H. Ehrenreich (Academic Press, New York, 1969) Vol. 10, pp 24–32.

[11] Because electrons most easily exhibit quantum mechanical behavior in semiconductors, they are usually referred to here. However, the theoretical and numerical methods discussed here apply equally well to holes.

[12] F. Sols, *Annals of Physics* **214**, 386 (1992).

[13] F. Sols, unpublished.

[14] See, for example, P. Lorrain and D. R. Corson, *Electromagnetic Fields and Waves* (W. H. Freeman and Company, San Francisco, 1970), p76.

[15] M. Macucci, K. Hess and G. J. Iafrate, unpublished.

[16] *Condensed Matter: Special Issue on Single Charge Tunneling* **85**, ed. by H. Grabert, 319 (1991).

[17] D. V. Averin, A. N. Korotkov, and K. K. Likharev, *Phys. Rev. B* **44**, 6199 (1991).

[18] Y. Meir, N. S. Wingreen, and P. A. Lee, *Phys. Rev. Lett.* **66** 3048 (1991).

[19] M. Macucci and K. Hess, *Phys. Rev. B* bf 46, 15357 (1992).

[20] L. F. Register, U. Ravaioli and K. Hess, *J. Appl. Phys.* **69**, 7153 (1991). [Erratum: **71**, 1555 (1992).]

[21] L. F. Register and K. Hess, unpublished.

[22] B. Tanatar and D. M. Ceperley, *Phys. Rev. B* **39**, 5005 (1989).

[23] M. Grupen and K. Hess, unpublished.

About Boltzmann Equations for Transport Modeling in Semiconductors

F. Poupaud

Laboratoire J.A. Dieudonné, URA 168 du CNRS Université de Nice
Parc Valrose, F-06108 Nice Cédex, FRANCE

Abstract

We present some results on the mathematical analysis of kinetic equations for modelling transport processes in semiconductors. We focus our attention on the connection between the kinetic models and the fluids ones based on drift-diffusion or hydrodynamic equations. Asymptotic analysis gives hydrodynamic coefficients in terms of microscopic quantities and allows to derive accurate boundary conditions.

1. Introduction

Transport modelling in modern submicronic devices requires a kinetic or quantum description. It explains the increase of interest in numerical simulation and mathematical analysis of the Boltzmann equations for semiconductors.

Perhaps one of the major issue for modelling is to couple simulations on the three levels : quantum, kinetic and fluids, depending on the regions of the device. Doing this requires a very precise description of the mathematical connection between this three levels. We present here some result on this topic.

2. Existence theory for kinetic models and connection with quantum physics.

One of the most general kinetic models is based on the following Boltzmann equation, [1] :

$$\partial_t f + v(k).\nabla_x f - \frac{q}{\hbar} E.\nabla_k f = Q(f), \tag{1}$$

where the distribution function f depends on time t, position x and wave vector k. The electric field is denoted by E and satisfies the Poisson equation. Therefore it depends on the distribution f through the concentration n :

$$n(t,x) = \int f(t,x,k) \frac{dk}{4\pi^3}. \tag{2}$$

Therefore the Boltzmann equation is nonlinear. Moreover if no degeneracy assumption is made, the collision operator reads :

$$Q(f) = \int s(k', k)f(k)(1 - f(k')) - s(k, k')f(k')(1 - f(k)) \, dk'. \tag{3}$$

Thus it is also nonlinear. Above, s is the scattering rate. From the principle of detailed balance, it follows that :

$$s(k', k)M(k) = s(k, k')M(k') \tag{4}$$

where M is the normalized Maxwellian :

$$M(k) = C \exp(-\frac{\varepsilon(k)}{k_B T}), \int M(k) \, dk = 1. \tag{5}$$

Here $\varepsilon(k)$ is the energy of carriers, k_B is the Boltzmann constant and T the temperature of the crystal. The velocity $v(k)$ depends on the energy band through the relation

$$v(k) = \frac{1}{\hbar} \nabla_k \varepsilon(k). \tag{6}$$

Recent progress in non linear analysis, due mainly to Diperna and Lions [2], allow to obtain a complete existence theory of (1), for boundary value problems [7], [8], as well as for the Cauchy problem [3]. More interesting from a physical point of view is the determination of entropies related to (1). Indeed we have

Proposition 1 *Let κ be any increasing function, then for every distribution f*

$$\int Q(f)\kappa(g) \, dk \leq 0, \; with \; g = \frac{f}{M(1 - f)}. \tag{7}$$

Define

$$H(\lambda, k) = \int_0^\lambda \kappa(\frac{s}{M(1 - s)}) \, ds, \tag{8}$$

then the functions

$$S(t, x) = -\int H(f(t, x, k), k) \, dk \tag{9}$$

are entropies for (1).

This proposition follows from straightforward algebra using (4), see [3]. One consequence is the determination of equilibrium distributions for Q which are nothing but Fermi-Dirac distributions

$$F = \frac{1}{1 + \exp(\frac{\varepsilon(k) - q\mu}{k_B T})}. \tag{10}$$

Concerning the connection between quantum physics and Boltzmann equations, it is well known that it can be provided by tracking wave packets. However, especially for coupling numerical methods, one wish to have a more precise derivation of Boltzmann equations from Bloch equations. With P.A. Markowitch and N.J. Mauser [10], we have introduced a Wigner function approach which is well adapted to describe periodic materials. Namely if ψ is a wave function the related Wigner series is

$$w(x, k) = \sum_\mu e^{ik\mu}\psi(x + \mu/2)\bar{\psi}(x - \mu/2) \tag{11}$$

where μ belong to the lattice. Then it can be shown that the Wigner function w tends to the distribution function f when the scaled Planck constant and the scaled characteristic length of the lattice vanish. Moreover macroscopic quantities are easily computed from the Wigner function w.

3. Fluid approximations

One way to analyse the connection between the Boltzmann equation (1) and fluid models is to use asymptotic analysis. Thus a scaled version of (1) is

$$\alpha^2 \, \partial_t f + \alpha \, (v(k).\nabla_x f - E.\nabla_k f) = Q(f), \tag{12}$$

where α is a relative value of the mean free path. Then we have to find the limit of the distribution f as α vanishes. In a very general framework, with F. Golse we have proved in [9] that the function f tends to a Fermi-Dirac distribution F (10). The chemical potential μ solves the drift diffusion equation

$$\partial_t n + div j = 0, \; n = n(\mu), \; j = \Pi(\mu).\nabla_x(\mu - V), \tag{13}$$

where the positive definite matrix μ is entirely determined by the scattering croos section s. The proof relies on entropies estimates based on Proposition 1 and on mean compactness results of [2]. Using parabolic band approximation and relaxation time model, it is possible to compute explicit values of the coefficients in terms of Fermi integrals, [6]. Of course if a non degeneracy assumption is made, we recover the Schockley equation.

Interesting for applications is the analysis of boundary conditions. In first approximation a Dirichlet boundary condition is derived for the concentration. But using such a condition, we neglect a boundary layer. As a concequence non physical discontinuities of the concentration at the boundaries appear in numerical simulations. This boundary layer can be taken into account if the Dirichlet boundary condition $n = n_0$ is replaced by a Robin boundary condition

$$n - \frac{l}{D} j.\nu = n_0 \tag{14}$$

where D is a diffusion coefficient, ν the outward normal and l the so-called extrapolation length. This length is also determined by the microscopic scattering mechanisms, see [11].

It seems much more difficult to derive hydrodynamic equations from the Boltzmann equation without ad-hoc assumptions on the distribution function. An attempt is this direction is to introduce an other scaling of the Boltzmann equation which leads to

$$\alpha(\partial_t f + v(k).\nabla_x f) - E.\nabla_k f = Q(f). \tag{15}$$

When α tends to 0, the function f tends to $nF(E,k)$ where $F(E,k)$ is the homogeneous steady state

$$E.\nabla_k F = Q(F), \int F \, dk = 1. \tag{16}$$

At a first order of approximation, n solves the Ohm law, see [5]. The second order of approximation can only formally be performed. It leads to extended drift-diffuion models [5] or to hydrodynamic systems where the pressure tensor has non isotropic components in the direction of the electric field, [4].

References

[1] P.A. Markowitch, C.A. Ringhoffer, C. Schmeiser, Semiconductor Equations, Springer, 1990.

[2] R.J. Diperna, P.L. Lions, *Global weak solutions of Vlasov Maxwell systems*, Com. Pure Appl. Math. **XLII** (1989) p 729-757.

[3] F. Poupaud, *On a system of non linear Boltzmann equation of semiconductor physics*, SIAM J. on Appl. Math. **50** n°6 (1990) p 1593-1606.

[4] F. Poupaud, *Derivation of hydrodynamic models for semiconductors from the Boltzmann equation*, Appl. Math. Lett. **4** n°1 (1991) p 75-79.

[5] F. Poupaud, *Runaway phenomena and fluid approximation under high fields in semiconductor kinetic theory*, Z. Angew. Math. Mech. **72** (1992) p 359-372.

[6] F. Poupaud, C. Schmeiser, *Charge Transport in Semiconductors with Degeneracy Effects*, Math. Meth. in Appl. Sci. **14** (1991) p 301-318.

[7] F. Poupaud, *Boundary value problems for the stationary Vlasov-Maxwell system*, Forum Math. **4** (1992) p 499-527.

[8] A. Nouri, F. Poupaud, *Boundary value problems for the stationary Vlasov-Maxwell-Boltzmann system under Fermi-Dirac statistics*, submitted to SIAM J. Math. Anal.

[9] F. Golse, F. Poupaud, *Limite fluide des équations de Boltzmann des semiconducteurs pour une statistique de Fermi-Dirac*, J. on Asympt. Analysis **6** (1992) p 135-160.

[10] P.A. Markowich, N.J. Mauser, F. Poupaud, *A Wigner Function Approach to (Semi)classical Limits : Electrons in a Periodic Potential*, submitted to J. of Math. Phys.

[11] F. Poupaud, A. Yamnahakki, *Conditions limites du second ordre pour l'équation de dérive diffusion des semiconducteurs*, Internal Report 330 (1993), Lab. J.A. Dieudonné, Univ. de Nice.

SIMULATION OF SEMICONDUCTOR DEVICES AND PROCESSES Vol. 5
Edited by S. Selberherr, H. Stippel, E. Strasser – September 1993

21

Applied TCAD in Mega-Bits Memory Design

H. Masuda, H. Pimingstorfer[†], H. Sato, K. Tsuneno, K. Ichikawa, H. Tobe,
H. Miyazawa, M. Nakamura, K. Kajigaya, O. Tsuchiya, and T. Matsumoto

Hitachi, Ltd.
2326 Imai, Ome-shi, JAPAN
[†]Institute for Microelectronics, TU Vienna
Gußhausstraße 27-29, A-1040 Wien, AUSTRIA

Abstract

This paper describes a methodology of TCAD application in VLSI design and
development. Simulation–based circuit model parameter generation for chip
design purpose is one of the key topics in TCAD. Several critical phenomena,
such as CMOS latchup etc., were also analyzed to verify feasibility and performance of the memory process. Two months of TCAD analysis were required,
in which twelve sets of MOS model parameters were generated by VISTA with
the computational cost of six hours on six CPUs of SGI-IRIS machines.

1. Basic Data for TCAD Design

For submicron devices, the shallow junction formation is one of the essential processes which determine the device performance. To verify the impurity profile of
ion–implanted and annealed diffusion layer, we used SIMS measurement for B, P,
and As impurities. Over a hundred samples with various doping conditions have
been analyzed. Enhanced diffusion model parameters with annealing temperature
and dose have been extracted. Figure 1 shows an example of a Phosphorous im-
plantation/diffusion profile which exhibits dose dependent TED (Transient Enhanced
Diffusion) phenomena [1]. The two–dimensional profile of the Phosphorous diffusion was also verified as shown in Fig. 2 [2]. The result shows an isotropic nature
of TED for Phosphorous. In circuit design, threshold control and driving current
of the device are two major characteristics which determine chip performance and
yield. We found an anomalous degradation of submicron MOS driving current based
on a study of intrinsic drain current, as shown in Fig. 3 [3]. A simple model which
describes carrier–velocity–saturation has been developed to clarify the phenomena.
Experimental $I_{ds} - L$ curves in submicron NMOS and the proposed model allow us
prediction of effective drain current of submicron NMOS as shown in Fig. 4.

2. Analysis of Critical Phenomena

In CMOS memory process, a couple of critical phenomena have to be evaluated.
Simulation works on CMOS latchup immunity [4], memory cell alpha–particle induced
soft error [5], Si crystal defects formation due to LOCOS process related mechanical
stress [6] and electric–field enhanced SRH recombination [7] have been conducted.

Qualitative (semi–quantitative) studies predicted a relative process margin with help of extensive experimental data in conventional processes. One example of the analysis is the CMOS latchup immunity being enhanced in shallow wells with retrograde structure, as shown in Fig. 5.

3. Generation of Circuit Model Parameters

In VLSI memory development, concurrent engineering, i.e. simultaneous work on chip/circuit design and fabrication process does appear in many cases. Our TCAD methodology to meet this needs is shown below:

1. Initial process debugging by 2D process analysis
   When the process designer proposed the initial process flow, fabrication steps have been analyzed using 2D process simulators with calibrated parameters (cf. Section 1) with a focus on ion implantation and annealing. A couple of mis-settings in impurity dose were found by simulation. Also optimum conditions in ion implantation dose and energy were determined by the initial TCAD verification.

2. Worst–case device definitions
   In circuit design, worst case analysis has been used to ensure operation even in the worst process variation extreme. In our case, gate length definition, gate–oxide thickness variation, and channel dose variation were chosen as process variations to be considered in circuit simulation. Based on this design strategy, $I - V$ characteristics of six device structures for both N&PMOS, have been defined for circuit design and analysis.

3. Generation of $I - V$ data
   The TCAD system VISTA [8] conducts 2D process/device simulations for the twelve devices of N&PMOS's. Computed I_{ds} data in bias conditions of V_{bb}, V_{gs} and V_{ds} are arranged into a pre–determined format for parameter extraction. Six hours in CPU time were required to compute twelve sets of $I - V$ data (2448 bias points total) on six SGI-IRIS CPUs in parallel.

4. Calibration of generated $I - V$ data
   Since some simulation models are assumed to remain uncertain as well as unknown process variations in fabrication line, an experimental $I_{ds} - L$ curve of a similar submicron NMOS (see Section 1) is used to calibrate the generated $I - V$ data. A simple I_{ds} normalization (scaling) at maximum drain current proved to be accurate for submicron NMOS as shown in Fig. 6.

5. MOS model parameter extraction
   The calibrated $I - V$ data are formatted as input deck for the MOS model [9] parameter extraction program. Four examples of the extracted MOS model are shown in Fig. 7. The averaged RMS error of the twelve devices obtained in this work is less than 1%.

4. Conclusions

A TCAD application in VLSI memory design has been conducted. Simulation–based circuit model parameter generation for chip design purposes was performed based on a TCAD database. Critical phenomena such as CMOS latchup were analyzed to verify feasibility and performance of the memory process. Two months of TCAD analysis were required, in which twelve sets of MOS model parameters were generated by VISTA with the computational cost of six hours on six CPUs of SGI-IRIS machines.

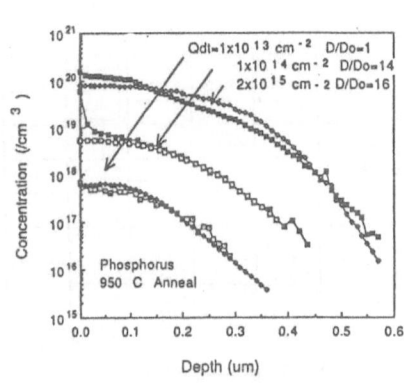

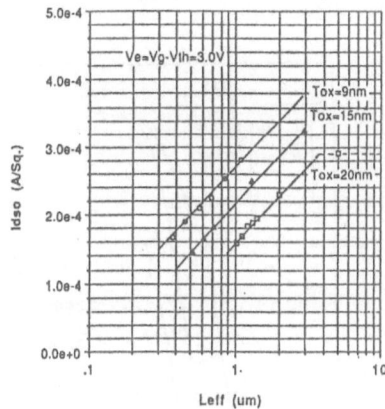

Fig.1 Determination of Phosphorous Implant./diffusion parameters at 950C furnes annealing.

Fig. 3 Anomalous degradation of drain current (of unit area: W/L=1) found in submicron NMOS measurements.

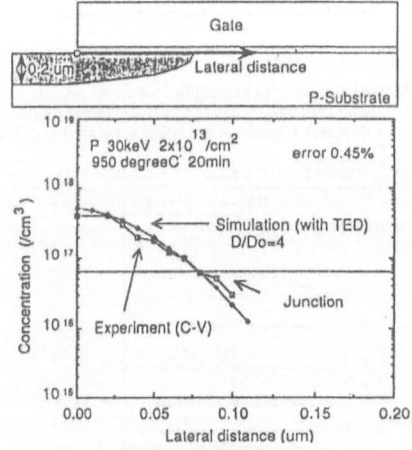

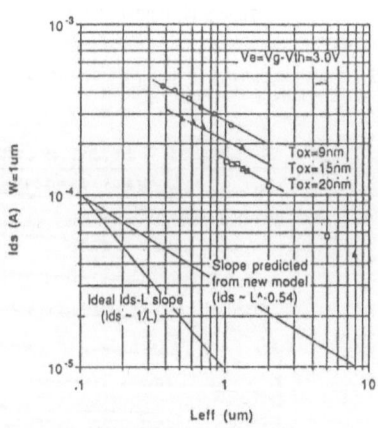

Fig.2 Experimental verification of 2D diffusion of phosphorous with TED (Transient Enhanced Diffusion) effects.

Fig.4 Empirical Ids-L curves and comparison with proposed model which describes velocity-saturation effect.

References

[1] H. Sato et al., VPAD'93 Tech. Digest, pp.86–87, May 1993

[2] K. Kubota et al., ICMTS'91 Tech. Digest, March 1991

[3] K. Tsuneno et al., VPAD'93 Tech. Digest, pp.152–153, May 1993

[4] D. J. Sleeter et al., IEEE Trans. ED, 39, pp.2592–2599, 1992

[5] H. Masuda et al., IEDM Tech. Digest, pp.496–499, 1985

[6] N. Saito et al., IEDM Tech. Digest, pp.695–698, 1989

[7] A. Schenk, Solid State Electron., 35, pp.1585–1596, 1992

[8] S. Halama et al., Technology CAD Systems Workshop, Vienna, Sept. 1993

[9] H. Masuda et al., IEEE Trans. CAD, 10, pp.161–170, 1991

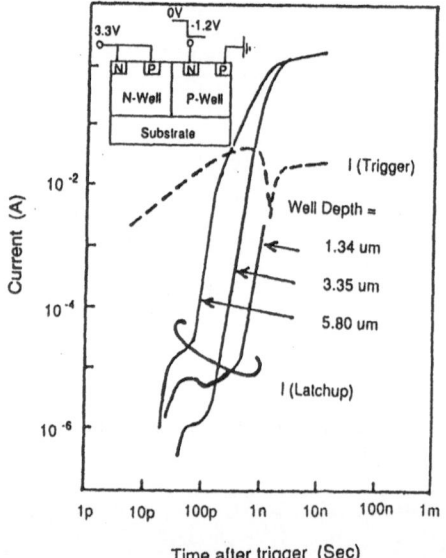

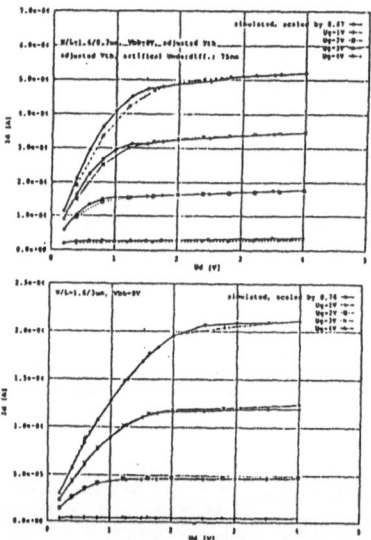

Fig.5 2D latchup simulation as a parameter of depth profiles in N- & P-wells.

Fig.6 Evaluations of scaling technique to adjust simulated Id-Vd curves to experimental one. (dashed: experiments, solid: simulated & scaled)

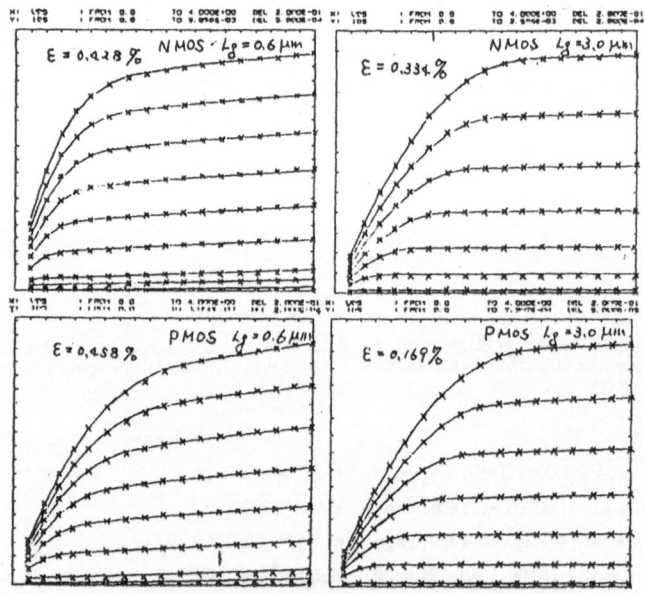

Fig. 7 Circuit MOS models for 0.6um and 3.0um N & PMOS's, that are generated using I-V process/device analysis and automatic parameter extraction. (symbol: simulated & scaled data, solid line: models)

Process Flow Representation within the VISTA Framework

Ch. Pichler and S. Selberherr

Institute for Microelectronics, TU Vienna
Gußhausstraße 27-29, A-1040 Wien, AUSTRIA

Abstract

The execution of multi-step simulation sequences involving a number of independent simulation tools is taken care of by the VISTA simulation flow control module which allows for the definition of a simulation task by means of the simulation flow description, a representation of the process flow using symbolic names to call simulation tools. Large process flows can be modeled by using predefined sequences, where the calculation results of all intermediate simulation steps remain available for analysis at a later time. Data level integration is based on the PIF data exchange format which is used to create a wafer state description containing all wafer geometry and material data after each simulation step.

1. Introduction

Due to long fabrication times in a wafer fab the simulation of semiconductor fabrication processes and device behavior are an indispensable aid for experimenting with device designs and process parameters. Technology CAD (TCAD) has successfully used process simulation tools to model production process steps during the design stages of a device. A wide variety of such tools exist, each being more or less specialized to perform a specific task. In order to integrate simulation tools and to present them to the user in a uniform way, TCAD frameworks have been developed, e.g. [1], [2], [3]. Drawing on experience gained from these undertakings, we took on devising and implementing a *simulation flow control* module for the Viennese Integrated System for Technology–CAD Applications (VISTA) [4], [5], putting special emphasis on an open concept which allows for the integration of arbitrary simulators, and on a simple, extendible representation of simulation sequences, enabling the process engineer to quickly apply changes to the process design, investigate the results, and optimize device performance.

2. Multistep Process Simulation

Consider a set of process simulation tools such as PROMIS [6] or SAMPLE [7], each capable of performing the numerical simulation of one or more VLSI fabrication process steps. Typically, the user invokes a simulation by specifying the values of the tool's input parameters via an input deck, i.e., a text file containing directives for setting these parameters, which is read by the tool upon execution. Additionally, initial geometry and dopant distribution data have to be provided in a format readable by the tool; after the successful completion of a simulation run, the output data are written to a file.

Should the user intend to simulate a series of process steps, the appropriate simulation tools have to be called sequentially, with the output data of a tool run being used as the input data for the next one. Therefore, a tool must be able to understand the data generated by its predecessor in the process sequence, i.e. the tools have to share a common data format or be able to translate from a foreign format to their own. Calling the tools one after the other in a UNIX environment is usually accomplished by a shell script which generates an input deck and calls a simulator for each simulation step. As the number of input parameters for a single tool is of the order of ten to hundred, modifying the script in the case of a change of a parameter value or of the process sequence is a tedious and error-prone task.

If we look at a process simulation sequence from different levels of abstraction we are able to identify a number of problem domains. At the *data level*, we have to ensure that the results generated by a tool are understood by the next one. At the *tool control level*, the simulator has to be provided with a set of key values encoding the parameters for a particular process step. Depending on the tool, various methods have to be provided for passing the process settings, e.g., generating an input deck. At the *task level*, we want to specify our simulation intent as intuitively as possible, concentrating on design parameters rather than on tool invocation subtleties. If we are not satisfied with the final outcome of a simulation sequence and want to modify our design, it should not be necessary to completely rewrite the process flow description, i.e., predefined process sequences should remain unchanged even if minor alterations are to be made. Using such *process modules* previously written to build up larger process flows greatly simplifies the process flow design.

Offering the user a comfortable means for automatically executing a large series of simulation steps reduces the cycle time in the design iteration loop. The following section presents our approach towards this goal, where we concentrate on the aspects of a generic tool integration concept.

3.   Tool Integration and Process Flow Representation within VISTA

The Viennese Integrated System for Technology–CAD Applications (VISTA) uses the Profile Interchange Format (PIF) [8] as a common data format to exchange wafer data between simulation tools. Simulators which do not adhere to this policy have to get a wrapping program which establishes a PIF interface. Tool integration and task definition issues with respect to process flow simulation are subsumed under the VISTA *simulation flow control* module. Essentially, this module consists of three parts, the XLISP binding functions for the simulation tools, the tool control module, and the task control module.

In order to make executable modules available to the TCAD shell, they have to have a representation in the shell's extension language, XLISP. These *binding functions* enable the user to invoke a simulation step just like any other XLISP function. All calling details for the execution of a simulation module are hidden from the user, and so are all data manipulations necessary to feed wafer data into the simulator and to get the calculation's results back. All parameter values, file names, and other control data for a tool run are passed as LISP *key* arguments to avoid errors due to a wrong argument order. These control data are translated into the simulator's control argument format, e.g., an input file is generated, or a command line is synthesized, or simulator specific PIF objects are added to the PIF input file. While the PIF defines a syntax for wafer description, it does not enforce a semantically strict representation of wafer data. In generating an input data file, ambiguities brought about by this semantic liberty are resolved by the binding function. For instance, we have to insure that all symbolic names used in a PIF file to describe material types are understood by the tool. If a material is not recognized, its name is replaced by an appropriate alias. Similarly, simulators do not agree upon geometry orientations (clockwise or counter clockwise), or they might require the presence of certain PIF objects which a preceding tool has not supported. In order to establish a standard data interface for plugging in simulation tools, the XLISP binding combines wafer data from before and after the simulation run to generate the current *wafer state*, i.e., a complete description of the wafer geometry and impurity concentrations reflecting the current state of the wafer.

Simulators usually require a large number of input parameters which alter some aspects of the computation and are not changed anymore once a tool is tuned to optimally satisfy the user's simulation requests. The *tool control* part passes a complete set of arguments to the simulator, merging user defined values and default values so that one is only obliged to enter those parameters one's concerned about. By switching at run-time between directories with the files containing the default values for the tools, the simulator behavior may be modified without affecting user settings.

In this context, we call a *task* any sequence of computations carried out on related data, e.g. a series of process simulation steps working on a common wafer, or an iteration loop which performs the same sequence of calculations on a set of initial geometries which might be automatically generated from a prototype geometry description, and the like. The *task control* module is the only one interacting directly with the user. A task is defined by writing a *simulation flow description* in LISP syntax, where symbolic names are used to specifiy a call to one of the *registered* simulation tools or to execute some control commands for data handling during the execution of a task. For instance, if the user wants to call a predefined process sequence, the process reference keyword PROCESS followed by the name of the file containig the process sequence is used. An optional override mechanism allows any parameter in any subprocess to be modified. The process reference and parameter override mechanisms work recursively.

The following example shows a small part of a wafer fab run traveller as it appears in the simulation flow description.

```
(
   (start-with :phys-pif-infile "InitGeom.pbf")
   (monte-carlo-implant :elem "BORON" :dose  1e13 :energy  30.)
   (anneal :temp 900 :time (35 "min"))
   (isotropic-deposition :time 225. :material ("SiO2" 0.0015))
   (anisotropic-etch :time 68. :material-default (0. 0.0001)
                     :material ("SiO2" 0 0.005))
   (monte-carlo-implant :elem "BORON" :dose 1e15 :energy 45.)
   (anneal :temp 900 :time (20 "min"))
)
```

The sequence shown above defines the process steps necessary to simulate the fabrication of an LDD (lightly-doped drain) structure of a p-channel MOS transistor. The PIF file InitGeom.pbf contains a PIF model of the wafer to be processed, basically a chunk of silicon partially covered by a nitride layer defining the gate location.

4. Conclusions

The simulation flow control module provides a comfortable means for defining, executing and modifying multistep simulation tasks. Existing process sequences can be easily modified to optimize device characteristics, large process flows can be built up from process modules. The XLISP tool binding functions establish an interface which allows for the integration of a large class of simulation tools. These tools are available to the user as plug-in modules for his simulation tasks. If the user chooses so, all intermediate calculation results remain available for analysis at a later time, simplifying error recovery as well as a detailed examination of process steps.

Acknowledgements

Our work is significantly supported by Digital Equipment Corporation at Hudson, USA; and Siemens Corporation at Munich, Germany.

References

[1] D. S. Boning, *Semiconductor Process Design: Representations, Tools, and Methodologies*, PhD Thesis, Massachusetts Institute of Technology, January 1991.

[2] A. S. Wong, *Technology Computer-Aided Design Frameworks and the PROSE Implementation*, PhD Thesis, University of California, Berkeley, 1992.

[3] E. W. Scheckler et al., *A Utility-Based Integrated System for Process Simulation*, IEEE Trans. Comp. Aided Design, Vol. 11, No. 7, pp. 911-920, 1992.

[4] H. Pimingstorfer et al., *A Technology CAD Shell*, SISDEP IV, pp. 409-416, 1991.

[5] S. Halama et al., *Consistent User Interface and Task Level Architecture of a TCAD System*, NUPAD IV, pp. 237-242, 1992.

[6] G. Hobler et al., *RTA-Simulation with the 2D Process Simulator PROMIS*, NUPAD III, pp. 13-14, 1990

[7] W. G. Oldham et al., *A General Simulator for VLSI Lithography and Etching Processes: Part II-Application to Deposition and Etching*, IEEE Trans. Electron Devices, Vol. ED-27, No. 8, pp. 1455-1459, 1980.

[8] St. G. Duvall, *An Interchange Format for Process and Device Simulation*, IEEE Trans. Comp. Aided Design, Vol. 7 No. 7 pp. 741-754, 1988.

SIMULATION OF SEMICONDUCTOR DEVICES AND PROCESSES Vol. 5
Edited by S. Selberherr, H. Stippel, E. Strasser – September 1993

29

A Powerful TCAD System Including Advanced RSM Techniques for Various Engineering Optimization Problems

R. Cartuyvels, R. Booth[†], S. Kubicek, L. Dupas, and K. M. De Meyer

IMEC
Kapeldreef 75, B-3001 Leuven, BELGIUM
[†]AT&T
1247 South Cedar Crest Blvd., Allentown, PA 18103, USA

Abstract

This paper presents the NORMAN/DEBORA TCAD system developed at IMEC to design and optimize sub-micron IC technology using process and device simulators. The versatility of the TCAD system will be shown for two important problems encountered in IC technology design and optimization.

1. Introduction

TCAD systems have become indispensable tools to investigate various engineering problems in the field of IC technology. Process design and optimization problems, simulator calibration and sensitivity analysis across several design levels (process → device → circuit) can be handled with an increasing efficiency using the automated coupling features a TCAD system provides. Although simulation provides a cheap way to experiment, it still remains quite CPU-time intensive, even if the TCAD system creates a fully automated environment. We have adopted the concepts of the Response Surface Methodology (RSM), which have been successfully used in various experimental fields [1], to efficiently plan a number of simulation experiments using the TCAD system to schedule and sequence them automatically. For these purposes, we developed an open TCAD system (NORMAN) including a novel design of experiments concept (Target Oriented Design), a unique parameter transformation technique to improve the fitting accuracy of the response models and a non-linearly constrained optimizer (DEBORA). We'll discuss its use in a few of the previously mentioned engineering problems.

2. Process design and optimization

Designing a process involves determining the settings of a high number of control variables such that a number of constraints upon device or circuit characteristics are met (figure 1). Optimizing the process is improving the process design such that an objective is optimized. Several objectives can be defined : minimize the process sensitivity w.r.t. to disturbances upon variables (e.g. an implantation dose), minimize the characteristic's target deviations, ... The mathematical formulation for minimizing the process sensitivity is shown in figure 2. $R_j(\vec{x})$ denotes response j (e.g. a device characteristic), L_j and H_j are user defined boundary specifications for response j; l_i and h_i are bounds upon x_i (design variable i), $FS(\vec{x})$ is the process sensitivity function containing weighted contributions for each response sensitivity. The weights (w_j) allow the process engineer to investigate inevitable trade-off situations. The overall objective function to be optimized is a weighted combination of the target deviation sum of squares objective function and the previously defined process sensitivity objective function. We developed an optimizer (DEBORA) to solve these types of problems. The responses are modeled by truncated Taylor series based upon the results of a chosen experimental design. We developed the TOD concept [2] which allows to compute higher order models in a huge number of variables restricting the model accuracy to the region of interest and consuming a minimum number of experiments. TOD based models are used to identify and rank a number of solution subspaces in the initial design space and to screen out less significant variables. The ranking is based on the sensitivity function. The subspace with the lowest sensitivity will be further investigated using a CCF design after screening out less significant variables. Using the NORMAN/DEBORA TCAD system, we managed to find an optimal process design for our 0.5 μm process in 3 days. We looked at 10 responses which are listed in figure 1. We applied the RSM technique using a TOD in 15 process variables and performed an optimization to minimize the process sensitivity as well as the process target deviations. The optimal solution is compared with an initial solution in figure 4 for 3 responses (threshold voltage for nmos and pmos, leakage current pmos). The process sensitivity is reflected in the spread of the distribution due to normal distributions on the process variables. The target is reflected in the mean of the response distribution. Note that the overall target deviation as well as the overall process sensitivity is reduced.

In addition, we developed a new transformation technique to improve the model accuracy of models based on CCF designs. Each input variable x_i is transformed to $z_i(x_i, \alpha_i, \beta_i)$ (figure 3) by solving iteratively for α_i and β_i in the characteristic ratio equation. The result (figure 5) shows the residual plot, experimented over a 2-D grid, for the original model and the transformed model of a nmos threshold voltage as a function of well dose and well drive-in temperature. The transformation indicates a square root dependence for the well dose agreeing with the physical model for the doping level dependence of the threshold voltage.

3. Simulator calibration

The goal is to tune model coefficients of the simulator such that the difference between experimental and simulated results is minimized. Using the RSM approach, the difference function $\Phi(\vec{x}_s(a_1, \ldots, a_k), \vec{x}_e)$, which is modeled based on the results of an experimental design in the model coefficients (a_i), is minimized ($\vec{x}_e$ is the experimental characteristic and $\vec{x}_s$ is the simulated one). This is shown for the threshold voltage of the nmos and pmos in figure 6 (dashed line is before calibration, solid line after calibration and triangles are measurements) where the difference function to be minimized is expressed as :

$$\Phi(\vec{x}_s(a_1, a_2, a_3), \vec{x}_e) = \sum_{i=n,p} (Vt_i(\vec{x}_s(a_1, a_2, a_3)) - Vt_i(\vec{x}_e))^2$$

$\vec{x}_s(a_1, a_2, a_3)$ is the simulated $I_{ds} - V_{gs}$ characteristic with a_1 being the boron segregation coefficient, a_2 being a pre-exponential coefficient of positive vacancy states (boron diffusion) and a_3 being a coefficient for modelling neutral vacancy states (phosphorus diffusion).

References

[1] R.H. Myers, A.I. Khuri, W.H. Carter, jr. *Response Surface Methodology : 1966-1988*, Technometrics Vol.31, no.2,pp 137-157,1989.

[2] R. Cartuyvels, R. Booth, L. Dupas, K. De Meyer. *Process Technology Optimization Using An Integrated Process and Device Simulation Sequencing System*, ESSDERC '92, Leuven.

response	lower bound	upper bound	target	unit
threshold voltage nmos	0.6	0.7	0.65	V
body effect nmos	*	1.0	$\ll$	$\sqrt{V}$
leakage current nmos	*	10^{-10}	$\ll$	A
max. current nmos	0.001	*	$\gg$	A
max. chan. doping dens. nmos	*	5.E17	$\ll$	cm^{-3}
threshold voltage pmos	-0.7	-0.6	-0.65	V
body effect pmos (abs. value)	*	0.8	$\ll$	$\sqrt{V}$
leakage current pmos (abs. value)	*	10^{-10}	$\ll$	A
max. current pmos (abs. value)	0.001	*	$\gg$	A
max. chan. doping dens. pmos	*	5.E17	$\ll$	cm^{-3}

Figure 1: response specifications ; '*' means 'no bound' ; '$\ll$' means 'as low as possible' ; '$\gg$' means 'as high as possible'

$$\min FS(x_1, \ldots, x_n)$$

subject to

$$L_j \le R_j(\vec{x}) \le H_j \quad j=1..m$$
$$l_i \le x_i \le h_i \quad i=1..n$$

where

$$FS(\vec{x}) = \sum_{j=1}^{m} w_j FS_j(\vec{x})$$

$$FS_j(\vec{x}) = \sum_{i=1}^{n} \left(\frac{\partial R_j(\vec{x})}{\partial x_i} \right)^2$$

$$\vec{x} = (x_1, \ldots, x_n)$$

Figure 2: process sensitivity optimization

original model
$$F(\vec{x}) = a_0 + a_1 x_1 + \ldots + a_n x_n + a_{1.2} x_1 x_2 + \ldots + a_{n-1.n} x_{n-1} x_n + a_{1.1} x_1^2 + \ldots + a_{n.n} x_n^2$$

transformed model
$$T(\vec{z}) = b_0 + b_1 z_1 + \ldots + b_n z_n + b_{1.2} z_1 z_2 + \ldots + b_{n-1.n} z_{n-1} z_n$$

with

$$z_i(x_i, \alpha_i, \beta_i) = \frac{[(x_i + \beta^2 + 1)^\alpha - 1]}{\alpha}$$

Figure 3: model transformations (Box-Cox)

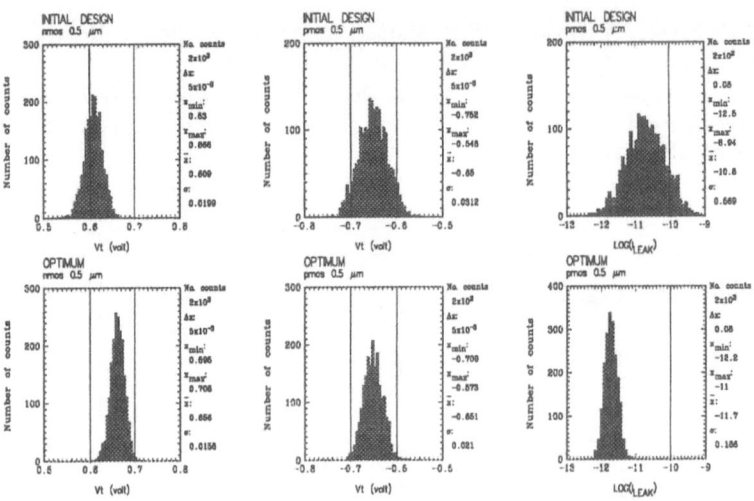

Figure 4: histogram plots of an initial (up) and optimal (down) design

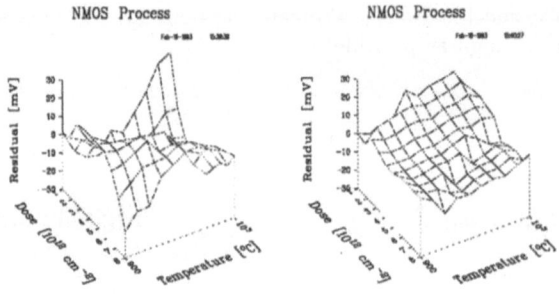

Figure 5: residual for a Vt response (left = original, right = transformed)

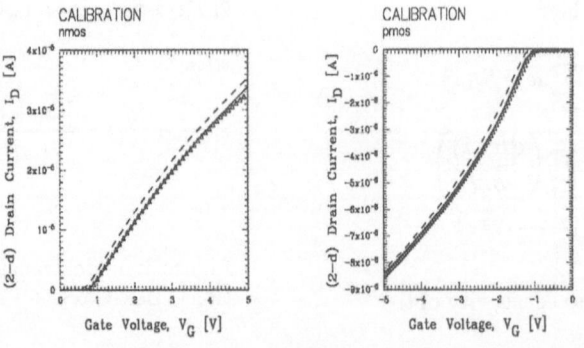

Figure 6: process calibration for accurate threshold voltage prediction

SIMULATION OF SEMICONDUCTOR DEVICES AND PROCESSES Vol. 5
Edited by S. Selberherr, H. Stippel, E. Strasser – September 1993

33

Modeling of VLSI MOSFET Characteristics Using Neural Networks

Ph. Lindorfer and C. Bulucea

National Semiconductor Corporation
2900 Semiconductor Dr., Santa Clara, CA 95052, USA

Abstract

Neural modeling of transistor current-voltage characteristics is explored as a possible solution to the complexity and accuracy problems currently encountered with analytical representations of VLSI devices. The neural modeling methodology is discussed along with first results obtained for a 0.8 μm CMOS process. The drain and substrate current-voltage characteristics of an n-channel MOSFET device are modeled over a drain current range of 10 orders of magnitude, from deep subthreshold to high-current operation.

1. Introduction

The compact MOSFET models for circuit simulation are customarily built around the parabolic approximation derived in the classical long-channel theory. While this approach is adequate for channel lengths larger than 2 μm, it tends to lead to excessively complicated parameter definitions when modified for submicron devices. Moreover, the good match between models and experimental data reported for particular processes fails to yield similar results on different processes. This explains the quite respectable size of the compact modeling literature, as well as the continuous demand for more accurate modeling coming from circuit designers.

To cope with the increasing complexity of the state-of-the-art VLSI scenario, we propose the more pragmatic approach of representing the transistor as a computer-simulated neural network, the *connection strengths* (or *weights*) of which are computed by *training* the network on measured or simulated data. The neural networks methodology is essentially used here as a convenient multi-variable, multi-function interpolation tool, rather than as a predictive tool, as usually reported.

The basic terminology of neural networks, typed in *italics*, is assumed to be known, and the readers not yet exposed to it are referred to the introductory book [1] and to the representative selection of the classical neurocomputing papers [2].

2. Modeling Methodology with Neural Networks

The principal steps of developing the neural network model of an electronic device are described in the flowchart of Fig. 1. The *training* and *testing* data are obtained from electrical measurements, or from numerical simulations, at a given input parameter matrix. Since neural modeling does not involve any physics-based validation, it is important that the data is checked for sanity by 2-D or 3-D plotting.

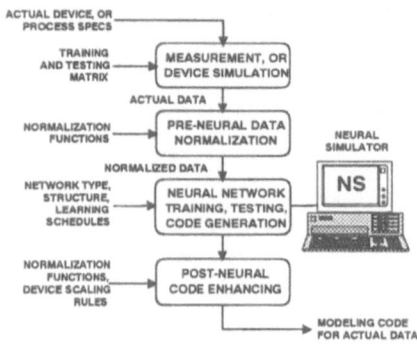

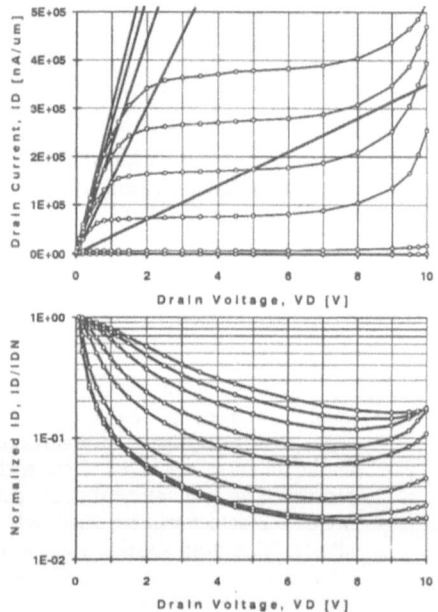

Fig. 1 - Simplified flowchart of the device modeling methodology based on neural networks.

Fig. 2 - Drain I-V characteristics used for training: (a) actual (curves with circles), and normalization (straight lines) data at $V_G = 0, 1, 2, 3, 4$, and 5 V, and (b) normalized data at $V_G = 0, 0.4, 0.8, 1, 1.5, 2, 3, 4$, and 5 V.

The neural network algorithms assume that the input and output values of the modeled system are *scaled* to finite linear ranges, such as from -1.0 to +1.0, that are compatible with the *transfer functions* of the *processing elements* (or *neurons*) involved, and do not cause their *saturation*. This poses a prime difficulty to neural modeling of semiconductor devices, where the output currents vary by many orders of magnitude. Simple linear presentation of the actual data results, at best, in modeling the two uppermost decades of the output quantities. Conversely, simple logarithmic presentation compresses excessively the upper ranges, losing modeling accuracy there. To cope with this difficulty, adequate normalization functions have to be devised for each output quantity, such that the data are uniformly compressed to a linear range.

Following normalization, the *training data* are fed into the neural simulator, together with the previously determined *network type, structure* and *learning schedules*. The neural simulation software repeatedly presents the data to the network and adjusts the *connection weights* using error minimization algorithms [3]. When a limit or a predetermined error for the output quantities is reached, the connection weights are set to their final values and, if tested satisfactorily, the network is able to model the device. A computer code is automatically generated for the normalized data, which is externally enhanced for denormalization and inclusion of device scaling rules.

3. Data Preparation for Neural Network Training

The *training* data used for this study were generated using MINIMOS and VISTA [4], a Technology Computer-Aided Design environment, both from the Technical University of Vienna, using the process characteristics of the company's $0.8 \mu m/5V$ CMOS process. The goal was to model the DC characteristics of the drain and substrate currents as functions of the gate and drain voltages, such as to continuously cover the full gate voltage range, from subthreshold to high-current operation (0 to 5 V), and the drain voltage range from zero to incipient avalanche multiplication (0 to 10 V). The substrate was biased to source potential.

Fig. 2a illustrates the normalization process for the drain current, where the linear region I-V characteristics at each gate voltage, $I_{DN} = g_{ch}(V_G) \cdot V_D$ are used as normalization functions, with $g_{ch}(V_G)$ determined at $V_D = 0.1$ V. To simplify *training*, a look-up table representation of the normalization function was preferred to an analytical function. The normalized drain current I_D/I_{DN} used for *training*, plotted in Fig. 2b, has a reduced dynamic range, and varies monotonically with the gate voltage. The network was actually presented the logarithm of this ratio. A similar normalization scheme was used for the substrate current, where the substrate current determined at $V_D = 10$ V was used as a normalization function, also in a look-up table representation as a function of V_G. A complete *training* set in this case consisted of 10 ordered data subsets at constant gate voltages, each having 22 drain voltage points.

4. Neural Network Architecture, Parameters, Results

With the ultimate goal of modeling for VLSI design, the networks considered were extremely simple, in order to minimize the calculation time of circuit simulators. Different network architectures and learning strategies were tried, all of which were of the *back-propagation* type. The PC/386 version of the NeuralWare Professional II/Plus [3] software was chosen, from several commercially available packages, due to its multiple options of neural paradigms and monitoring devices. The best results were obtained with networks having two *hidden layers*, the typical architecture being

2 * input (*linear*) $\Rightarrow$ 4...8 * hidden1 (*tanh*) $\Rightarrow$ 3...4 * hidden2 (*tanh*) $\Rightarrow$ 2 * output (*tanh*),

where the transfer function of the *processing element* is specified in parentheses.

The *Delta-Rule learning* was generally preferred, with data presented sequentially.

Fig. 3 shows a typical well behaving network around the beginning of the training process. The *Summation Value* instrument averages its values over a displaying *epoch* equal to the number of sets (220) in the training file. The *RMS Error* instrument averages the root mean square error of the outputs over the same period. The arrows on the error curve mark the *transition points* of the learning schedule, where the *learning* coefficients are changed to avoid *processing element saturation*. The final value of the *linear correlation coefficient* of the desired and actual outputs, printed on the vertical axis of the *confusion matrix* (see Fig. 3), is 0.9989. High values of this coefficient, in the range reported here are required for accurate modeling.

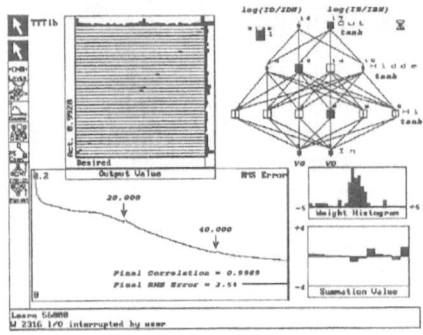

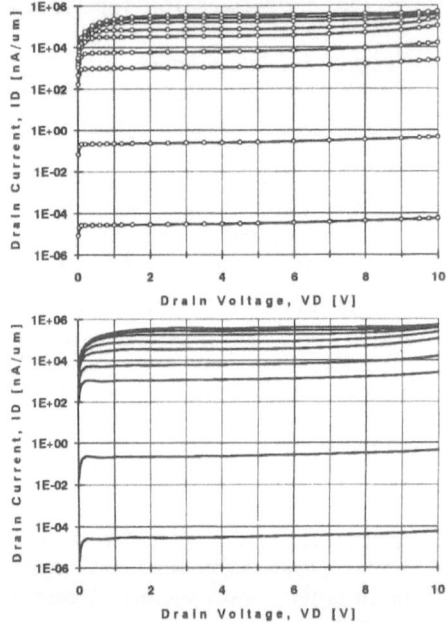

Fig. 3 - Typical NeuralWare Professional II/Plus screen of a well behaving network, around the beginning of the training.

Fig. 4 - Characteristics of a VLSI n-channel MOSFET: (a) real, and (b) modeled by a $2 \Rightarrow 6 \Rightarrow 4 \Rightarrow 2$ network. $V_G = 0, 0.4, 0.8, 1.0, 1.5, 2, 3, 4, 5$ V.

Fig. 4 illustrates the results of this modeling experience, for the simple network represented in Fig. 3, comparing the real (a) and modeled (b) characteristics of the drain current over the full range of gate and drain voltages, covering ten orders of magnitude. The network yielded similar results for the substrate current. The modeling accuracy is quite surprising considering the relative simplicity of the network used.

5. Conclusions

The neural representation is shown to be a practical and universally applicable modeling tool for accurate simulation of submicron VLSI devices. The vast variety of options available for neural modeling, ranging from fundamentally different network architectures and normalization functions, to details of training algorithms and schedules invite further communication exchange.

References

[1] Jeannette Lawrence, *Introduction to Neural Networks and Expert Systems*, California Scientific Software, Nevada City CA, 1992.

[2] James A. Anderson and Edward Rosenfeld, *Neurocomputing: Foundations of Research*, MIT Press, Cambridge MA, 1990.

[3] NeuralWare Staff, *Neural Computing - NeuralWorks Professional II /PLUS and NeuralWorks Explorer*, Software User's Manual, NeuralWare Inc., Pittsburgh PA, 1991.

[4] S. Halama et al., *Consistent User Interface and Task Level Architecture of a TCAD System*, Proceedings NUPAD IV, Seattle WA, 1992.

SIMULATION OF SEMICONDUCTOR DEVICES AND PROCESSES Vol. 5
Edited by S. Selberherr, H. Stippel, E. Strasser – September 1993

Coupling a Statistical Process-Device Simulator with a Circuit Layout Extractor for a Realistic Circuit Simulation of VLSI Circuits

W. Kuźmicz, W. Denisiuk, J. Gempel, Z. Jaworski, M. Niewczas, A. Pfitzner,
E. Piwowarska, W. Pleskazc, and A. Wojtasik

Institute for Microelectronics and Optoelectronics, Warsaw University of Technology
ul. Koszykowa 75, PL-00662 Warszawa, POLAND

Abstract

This paper discusses methodology of statistical simulation of an IC design which includes disturbances described by independent random variables, spatially correlated random disturbances and deterministic process parameters distribution on a wafer. The method of coupling of a process/device simulator with a circuit extractor is proposed. Practical example of an operational amplifier design optimization is presented.

1. Introduction

The performance and parametric yield of integrated circuits often suffer from process-related variations of parameters of IC components. Ordinary simulation tools are not sufficient to predict the IC performance when process-related parameter variations are of primary concern. A method of statistical simulation has been proposed [1] which accounts for random global and independent local variations of process parameters in a hierarchical way [2]. A statistical process/device simulator implementing this simple model of process-related variations [3] can provide good approximation for the variations of parameters of single devices. If statistical process/device simulation is coupled with circuit simulation, useful statistical information can be obtained for circuits which are not very sensitive to local variations of device parameters. However, experience [4, 5] shows that for precise analog circuits and for large digital VLSI chips spatial dependencies of deterministic variations and correlations of random variations cannot be ignored. For realistic statistical simulation a more complex model of process-related variations is necessary. Such a model should be able to handle: (i) global variations, (ii) local variations described by deterministic functions of device position on the wafer, (iii) local variations described by independent random variables, and (iv) local variations described by spatially dependent random variables. In this work we propose a methodology of statistical simulation which accounts for such variations and describe practical implementation of this methodology.

2. The problem of coupling CAD tools

The basic idea of statistical process/device/circuit simulation is to couple four tools: a process simulator, a device simulator, a circuit extractor and a circuit simulator, and use them in a Monte Carlo loop. However, in practice coupling an extractor with a process and device simulator is not straightforward. If local variations and their spatial dependencies are to be accounted for, the process simulation should be performed separately for every component before device simulation. Moreover, the circuit netlist and all the geometrical information (i.e. locations, shapes and sizes of the components) have to be extracted from the layout by a circuit extractor and passed to the device simulator. This requires an enormous amount of data to be transferred from process simulator and from extractor to device simulator (e.g. descriptions of all the shapes and one- or two-dimensional doping profiles at hundreds of locations). To avoid this we split the device simulation task into two: modeling of passive components is performed within the circuit extractor and the rest of the device modeling is integrated with process simulation. More precisely, we base our solution on the following assumptions:

Assumption 1. For the purpose of this work all the components can be divided into two classes: *process-dependent components* and *geometry-dependent components*. The first class includes components which are geometrically simple but whose parameters depend on fine details of the results of the manufacturing process. A good example is a MOS transistor. Its parameters depend strongly on the doping profiles in the channel and source/drain regions but the mask geometry can be described by just two numbers: channel width and length. The second class includes components of complex geometry but simple dependence on the manufacturing process. An example is parasitic resistance of a polysilicon region of complex shape. Its dependence on processing parameters can be characterized by a single number - the sheet resistance. To minimize the data flow between the simulation tools, it is convenient to combine the modeling of process-dependent components with the process simulation and modeling of geometry-dependent components with circuit extraction.

Assumption 2. Performance of well designed integrated circuits may depend critically on *local* variations and spatial dependencies of parameters of a set of components called *critical components*, but not on *local* variations of parameters of *parasitic* passive components. In a general case the set of critical components may include active devices and possibly some passive components, e.g. capacitors in a SC filter, but in current implementation MOS transistors are the only critical components available. We assume that all critical components are sufficiently small so that all the physical parameters such as oxide thickness, doping, resistivity etc. can be considered uniform within the component area.

These assumptions allow to organize the simulation in the way shown in Fig. 1.

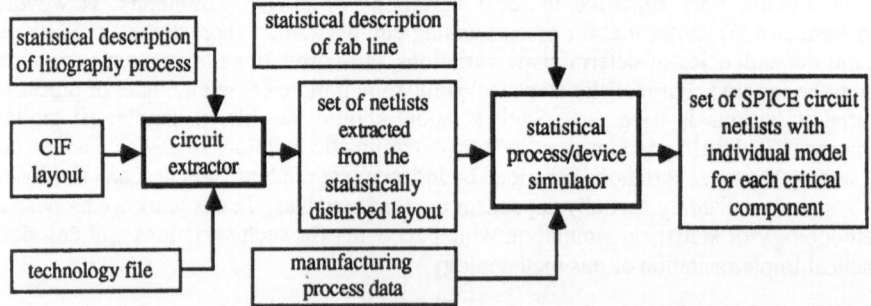

Fig.1. Generation of a set of netlists representing circuits affected by process variations and layout disturbances.

First, the extractor simulates layout disturbances and performs extraction creating a netlist which also includes the (x,y) coordinates of all the critical components. Then the process and device simulator adds a separate device model for each critical component on the netlist. To generate these models, full process simulation and device modeling is repeated for every critical component, and the positions of components are taken into account when disturbed process parameters are generated. The outcome of modeling is a data file for a circuit simulator. Such file enables for electrical simulation which accounts for local and global process parameter variations and lithography related disturbances for all critical components. Having such a statistical sample of netlists one can make a number of circuit simulations and estimate the range of values for electrical parameters of the design (e.g. delay time, input offset voltage, gain, slew-rate, linearity).

3. Special features of the programs

To implement the methodology described above, we developed two programs: EXCESS II - a layout extractor which can handle layouts with lithography-related deformations, and SYPRUS - a statistical process/device simulator which combines numerical efficiency with accuracy of process simulation and MOS device modeling. Both programs can simulate global random variations, local uncorrelated variations, local spatially correlated variations and wafer-wide deterministic distribution of process parameters and layout geometry. The circuit extractor EXCESS II [6] takes a CIF file as input, uses appropriate technology data to simulate layout deformations and then carries out circuit extraction. The extractor can process shapes described by polygons with arbitrary angles what is important if one wants to extract a circuit from the layout affected by arbitrary deformations. The device modeling in EXCESS II includes computations of W/L ratios for the transistors with process-disturbed geometry as well as modeling of RC parasitic devices [7]. The output consists of a set of netlists representing a sample of fabricated chips. Each netlist contains information about global coordinates of each critical component within the wafer. The process/device simulator SYPRUS generates disturbed process parameters and performs the process simulation and device modeling for each critical device from each netlist (e.g. a SPICE model for each MOS transistor) and outputs the netlist in SPICE format. All important geometrical dependencies such as dependence of threshold voltage on transistor channel length are taken into account. As a result, a set of netlists is produced which includes devices with models corresponding to variations of such parameters as etching rate, mask alignment, oxidation time, temperatures etc.

4. Example: optimization of an operational amplifier design

An example of application of our approach is shown in fig. 2. The sensitivity of input offset voltage of a CMOS operational amplifier to the layout of the input stage was investigated. Since the amplifier was to be used as a building block in an artificial neural network chip, the goal was to make the whole design as compact as possible while maintaining low offset input voltage. Two versions of input stage layout were considered: the common centroid layout (Fig.2a) and the simple layout (Fig. 2b). The process data were typical for an industrial n-well CMOS process. All types of variations were included. The positions of simulated circuits are shown in Fig.3. The results of simulation for 180 simulated circuits were as follows. For both layouts the mean value of the input offset voltage was almost the same (0.037 mV), but for the common centroid layout the standard deviation was 3.2 mV while for the simple layout the standard deviation was significantly larger: 5.0 mV. As expected, the common centroid layout yields smaller input offset voltage due to partial compensation of local variations. These results may help to decide whether for a particular application the input offset voltage of the simple layout is acceptable or the common centroid layout must be used.

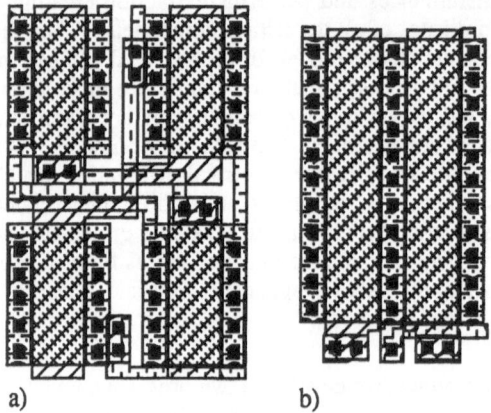
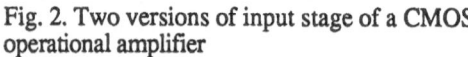
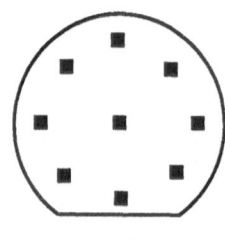

a)   b)

Fig. 2. Two versions of input stage of a CMOS   Fig. 3. Positions of simulated
operational amplifier   circuits on a wafer

5. Conclusions

For realistic prediction of performance and yield of integrated circuits sensitive to local variations of process parameters and layout disturbances it is necessary to include spatial dependencies of these variations. A method of coupling of circuit extractor with process/device simulator and circuit simulator has been proposed. Special properties of the programs and the method of coupling allow to include spatial dependencies of process variations in the statistical simulation of the circuit's behavior. An example shows how such a simulation may help to find an optimal design when there is a tradeoff between circuit size and performance.

Acknowledgment: Critical remarks of W.Maly are gratefully acknowledged.

This work has been supported by following KBN grants: 800329101 and 800359101.

References:

[1] W.Maly, A.J.Strojwas, "Statistical simulation of the IC manufacturing process", IEEE Trans. Computer-Aided Design, vol. CAD-1, pp.120-131, 1982.
[2] W.Maly, "Modeling of Random Phenomena in CAD of IC - A Basic Theoretical Consideration", Proceedings of ISCAS '85, pp. 427-430, 1985.
[3] S.R.Nassif, A.J.Strojwas, S.W.Director, "FABRICS II: A Statistically Based IC Fabrication Process Simulator", IEEE Trans. Computer-Aided Design, vol. CAD-3, pp.20-46, 1984.
[4] C. Michael, M. Ismail, "Statistical Modeling of Device Mismatch for Analog MOS Integrated Circuits", IEEE J. of Solid-State Circuits, vol. 27, pp. 154-166, 1992.
[5] J.K.Kibarian, A.J.Strojwas, "Using Spatial Information to Analyze Correlations Between Test Structure Data", IEEE Trans. on Semiconductor Manufacturing, vol. 4, pp. 219-225, 1991.
[6] An early version of EXCESS was described in: D.Korzec, A.Wojtasik, M.Syrzycki, E.Piwowarska, W.Pleskacz, W. Kuzmicz and W. Maly "Device and Parasitic Oriented Circuit Extractor", Proc. IEEE Int Conference on Computer Design ICCD '87, pp. 430-433, New York, USA, 1987
[7] M. Niewczas and A. Wojtasik, "Modeling of VLSI RC Parasitics Based on the Network Reduction Algorithm", submitted for publication in IEEE Trans. on Computer Aided Design.
[8] W. Maly, "Computer-Aided Design for VLSI Circuit Manufacturability", Proceedings of the IEEE, vol. 78, pp. 356-392, 1990.

Simulation of Self-Heating Effects in a Power p-i-n Diode

K. Kells, S. Müller, G. Wachutka[†], and W. Fichtner

Integrated Systems Laboratory, ETH-Zürich
Gloriastraße 35, CH-8092 Zürich, SWITZERLAND
[†]Quantum Electronics Laboratory, ETH-Zürich
ETH-Hönggerberg, CH-8093 Zürich, SWITZERLAND

Abstract

To accurately predict the effects of self-heating in a power p-i-n diode, we have applied self-consistent device simulation using a thermodynamically rigorous electrothermal model [1] implemented in the device/circuit simulator *Simul* [2]. Results of steady-state and high-voltage turn-off simulations with external electrical and thermal circuit elements are presented comparing the isothermal and self-heating cases.

1. Introduction

Two time frames are important when dealing with non-isothermal effects in the operation of semiconductor devices. Over longer times, a rise in lattice temperature due to self-heating can change mobility and intrinsic density in the bulk and/or channel of the device. On much shorter time scales, local hot spots may produce electrical currents and heat flow due to high temperature gradients. Extreme local heating can lead to local device breakdown, which can result in device failure.

Non-isothermal effects are included through the thermodynamic model in three ways. First, we introduce to the well-known current equations the gradient of the local lattice temperature T as an additional current driving force:

$$\vec{J_n} = -q\mu_n n(\nabla\phi_n + P_n\nabla T) \quad \text{and} \quad \vec{J_p} = -q\mu_p p(\nabla\phi_p + P_p\nabla T), \tag{1}$$

where P_n and P_p are the absolute thermoelectric powers. Since the quasi-Fermi potentials ϕ_n and ϕ_p also depend on temperature, $\nabla\phi_n$ and $\nabla\phi_p$ include temperature gradients which must not be neglected when extending the drift-diffusion model to non-isothermal conditions. Then, in addition to the Poisson and current continuity equations, we solve self-consistently the following time-dependent equation for the temperature:

$$c\frac{\partial T}{\partial t} = \nabla \cdot \kappa\nabla T + H \tag{2}$$

where κ is the thermal conductivity, c is the heat capacity, and H is the heat generation. Finally, the temperature dependence of the bulk parameters such as the intrinsic number, the mobility, and the thermoelectric powers is taken into account.

2. Dynamical equation and model considerations in *Simul*

The heating term H in Eqn. 2 is the full transient heat expression as given in [1] with the exception that the terms involving $\partial n/\partial t$ and $\partial p/\partial t$ are neglected as in [3], since estimations show that this term is negligible compared with the other heating terms up to $\partial n/\partial t$ or $\partial p/\partial t \approx 10^{21} \mathrm{cm}^{-3} \mathrm{s}^{-1}$. The upper bound of $\partial n/\partial t$ or $\partial p/\partial t$ in the p-i-n diode simulations was estimated to be $10^{14}\,\mathrm{cm}^{-3}\mathrm{s}^{-1}$.

Solution of the dynamical equations is carried out through decoupled iteration between the heat flow equation and the coupled system of electrical equations [4, 5].

The two most important temperature-dependent models in the p-i-n diode simulation are the effective intrinsic density and the mobility. Our mobility model takes into account doping dependence (μ_{doping}) and carrier-carrier scattering (μ_{cc}) combined using the Matthiessen rule [2]:

$$\mu_{\mathrm{doping}} = \mu_{\mathrm{min1}} \cdot \exp\left(\frac{-P_c}{N_i}\right) + \frac{\mu_L\,(T/300)^{-\varsigma} - \mu_{\mathrm{min2}}}{1 + (N_i/C_r)^\alpha} - \frac{\mu_1}{1 + (C_s/N_i)^2} \tag{3}$$

$$\mu_{\mathrm{cc}} = \frac{D\,(T/300)^{3/2}}{\sqrt{(n\,p)}} \left[\ln\left(1 + F(T/300)^2(pn)^{-1/3}\right)\right]^{-1}. \tag{4}$$

The dominating temperature effect is a decrease in mobility with rising temperature due to the explicit temperature dependence in Eqn. 3.

Our effective intrinsic density model is by default that of Slotboom [6]. In the steady state, we also investigated the models given in [7,

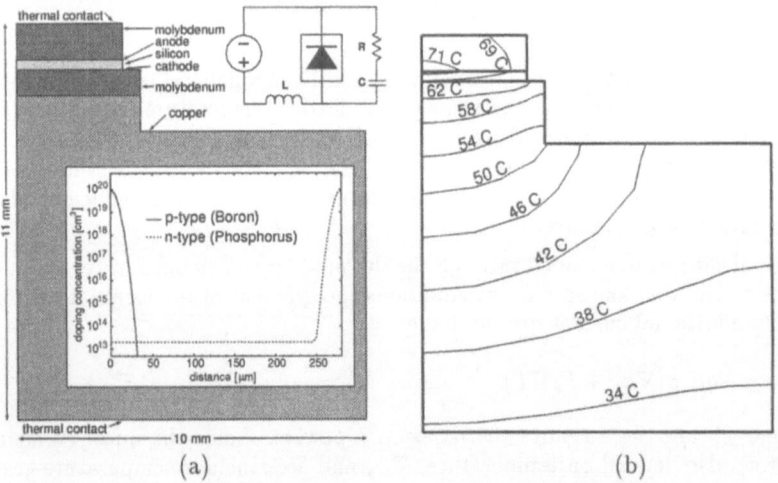

Figure 1: p-i-n diode geometry, doping profile, and external circuit (a) and steady-state temperature distribution (b) in long-lifetime diode forward biased to $200\,\mathrm{A/cm}^2$.

3. Results and Discussion

The simulation device consists of a $280\,\mu\mathrm{m}$ thick and $3.0\,\mathrm{mm}$ wide silicon p-i-n diode sandwiched between two molybdenum plates and mounted on a copper heatsink. The simulation domain and the impurity concentration profile of the diode are shown in Fig. 1.a. Two different carrier lifetimes are used for simulation; the "long-lifetime diode" has lifetimes $\tau_n = 28\,\mu\text{s}$ and $\tau_p = 9.33\,\mu\text{s}$, while the "short-lifetime diode" has lifetimes $\tau_n = 2.8\,\mu\text{s}$ and $\tau_p = 0.933\,\mu\text{s}$.

We assume that heat generated in the active area of the device flows to the surroundings through thermally resistive thermal contacts at the bottom of the heatsink ($R_{\text{th}} = 0.3\,\text{K}/\text{W}$) and along the top of the upper molybdenum plate ($R_{\text{th}} = 10\,\text{K}/\text{W}$) which are tied to $T = 300\,\text{K}$. Homogeneous Neumann boundary conditions (zero heat flux) are chosen for all other device surfaces.

The results of the steady-state simulation are shown in Fig. 2.a. At lower currents, the increase of the intrinsic density with temperature acts to increase current, while the decrease of the mobility with temperature dominates at higher currents and temperatures.

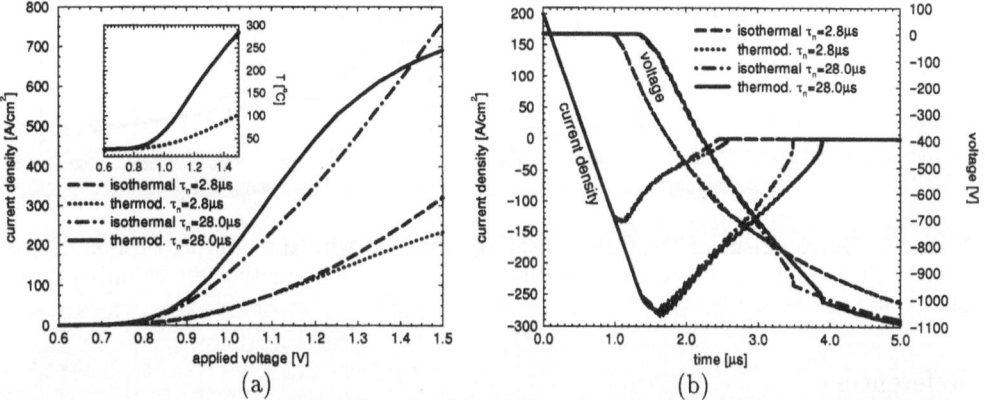

Figure 2: Steady-state current and maximum temperature vs. voltage (a) and comparison of isothermal and thermodynamic transient turn-off behavior (b) for both diodes.

For the transient simulation, both p-i-n diodes are turned off to a final blocking voltage of $-1\,\text{kV}$ from steady-state forward bias conditions of $200\,\text{A}/\text{cm}^2$. The electrical circuit used for the turn-off simulations is shown in Fig. 1.a. The values of R, L, and C for both diodes are $4.47\,\Omega$, $10\,\mu\text{H}$, and $2\,\mu\text{F}$, respectively, giving rise to a di/dt during turn-off of $-100\,\text{A}/\mu\text{s}$ using a total electrical contact area of $0.3\,\text{cm}^2$. The voltage supply switches linearly from the forward-bias voltage to $-1\,\text{kV}$ in $10\,\text{ns}$.

Fig. 2.b. shows the turn-off current as a function of time for both devices comparing the isothermal and thermodynamic case. The turn-off current in the thermodynamic simulations approaches zero more slowly than in the isothermal case, which is especially visible in the long-lifetime diode. This result agrees with the measured and simulated results published in [9].

The contributions of the individual heat generation mechanisms to the total heat in the long-lifetime diode integrated over the entire device are depicted in Fig. 3.a. In the initial state ($t = 0\,\text{s}$), the dominant heating mechanism is recombination heat at the p+-i and i-n+ junctions, in line with results presented in [10]. Another important term is the Peltier heat, which is also generated at the junctions. Since the sign of the Peltier heat is dependent on the direction of current flow, it is interesting to note that this heat is negative at $t = 0\,\text{s}$, but becomes positive during turn-off when the current reverses direction.

Joule heat generated by hole current is the dominating heating mechanism during turn-off, followed in importance by electron Joule heat and Peltier heat. After the diode has reached the blocking voltage, current due to impact ionization leads to small electron and hole Joule heat contributions. The current and consequently the Joule heating decrease with decreasing device temperature, until the final thermal and electrical turn-off state is attained after approximately 5 seconds.

Shown in Fig. 3.b. is the temperature in the long-lifetime diode measured along the vertical axis along the left of the simulation domain for various times during the turn-off simulation. A temperature peak of about 85 °C is reached at about 4 μs into the turn-off simulation.

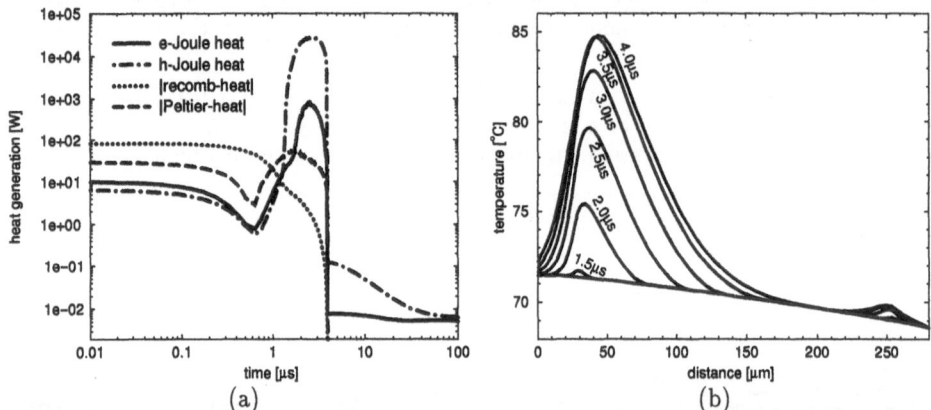

Figure 3: Development of the various heating terms (a) with time during turn-off of the long-lifetime diode and lattice temperature (b) across the long-lifetime diode during turn-off.

References

[1] G. Wachutka, "Rigorous thermodynamic treatment of heat generation and conduction in semiconductor device modeling," *IEEE Trans.*, vol. CAD-9, pp. 1141–1149, 1990.

[2] IIS, *Simul Manual.* Integrated Systems Laboratory, ETH Zurich, Switzerland, 1.0 (alpha) ed., 1992.

[3] P. Wolbert, *Modeling and Simulation of Semiconductor Devices in TRENDY.* PhD thesis, University Twente, 1991. publ. by PROSA, P.O. Box 8091, 7550 KB Hengelo, Netherlands.

[4] V. Alwin, D. Navon, and L. Turgeon, "Time-dependent carrier flow in a transistor structure under nonisothermal conditions," *IEEE Trans. Elec. Dev.*, vol. ED-24, pp. 1297–1304, 1977.

[5] P. Gough, P. Walker, and K. Wright, "Electrothermal simulation of power semiconductor devices," in *Proc. ISPSD*, pp. 89–94, 1991.

[6] J. W. Slotboom and H. C. de Graaff, "Bandgap narrowing in silicon bipolar transistors," *IEEE Trans. Elec. Dev.*, vol. ED-24, no. 8, pp. 1123–25, 1977.

[7] J. del Alamo, S. Swirhun, and R. M. Swanson, "Measuring and modeling minority carrier transport in heavily doped Silicon," *Solid-State Electronics*, vol. 28, no. 1, pp. 47–54, 1985.

[8] D. B. M. Klaassen, J. W. Slotboom, and H. C. de Graaff, "Unified apparent bandgap narrowing in n- and p-type Silicon," *Solid-State Electronics*, vol. 35, no. 2, pp. 125–29, 1992.

[9] R. Kraus, T. Türkes, and H. Mattausch, "Modelling the self-heating of power devices," in *Proc. ISPSD*, pp. 124–129, 1992.

[10] D. Kakati, S. Ramanan, and V. Ramamurthy, "Computer-aided electrothermal analysis of a semiconductor device," in *Proc. of NASECODE IV Conf.*, pp. 326–331, 1985.

Impact of Cell Geometries and Electrothermal Effects on IGBT Latch-Up in 2D-Simulation

H. Brunner, Y. C. Gerstenmaier, and H.-J. Mattausch

Corporate Research and Development, Siemens AG
Otto-Hahn-Ring 6, D-81739 München, GERMANY

Abstract

The influence of the emitter cell geometry of an insulated gate bipolar transistor (IGBT) on the forward behaviour is investigated by electrothermal device simulation. For comparison isothermal device simulation is used at different temperatures. The latch-up behaviour of cellular cell and stripe designs is calculated. As a result the stripe design possesses a higher latch-up ruggedness. However the impact of the cell geometry on IGBT latch-up is affected by electrothermal effects. The variation of the collector-emitter current with the lattice temperature distribution at latch-up is different for stripe and cell designs. The temperature dependence of the junction voltage and the hole current proportion to the total current are electrothermal effects which influence IGBT latch-up.

1. Introduction

The IGBT shown in Fig. 1 is a conductivity modulated MOSFET with a p doped emitter region. The conductivity modulation occures due to the injection of minority carriers into the n^- drift region. This device can be operated at high current densities even when designed to support high blocking voltages. However, the IGBT possesses a parasitic four-layer thyristor structure (n^+ source-p base-n^- base-p drain). When the parasitic thyristor is turned on, the collector current cannot be controlled further by the insulated gate and the device is often damaged. This mode is called IGBT latch-up. When the hole current which flows underneath the n^+ emitter region produces a forward voltage drop which exceeds the n^+ source to p base junction voltage, electrons are injected from the n^+ emitter into the p base region and turn on the parasitic thyristor. The investigation of IGBT latch-up is an electrothermal problem, due to the temperature dependence of the junction voltage, and other physical parameters. It has been measured, that the latching current decreases with increasing temperature [1]. The simulated structure in Fig. 1 possesses a n^- base designed for a blocking capability of 600V. The thermal boundary condition in the case of nonisothermal simulation is represented by a thermal resistance to a heat sink held at 300K which is fixed to the drain electrode. The case of a stripe cell versus a circular cell design with rotational symmetry is considered. A plane view of the emitter geometries is depicted in Fig. 2 , the shaded area indicates the minority-carrier collector area which is surrounded by the n^+ source.

For the numerical investigation of the latch-up behaviour with self heating effects electrothermal simulation is essential [2]. The device simulator MEDICI [3] is used which includes a fully coupled simultaneous solution of Poisson's, carrier continuity and heat flow equations . For steady state solutions the heat flow equation takes the form

$$H = -\nabla(\lambda(T)\nabla T) \tag{1}$$

where H is the heat generation determined by the total current density, electric field and the recombination rate. λ is the thermal conductivity of the material. The electron and hole current densities

$$\mathbf{J_n} = qn\mu_n\mathbf{E} + qD_n\nabla n + q\frac{D_n}{T}n\nabla T \tag{2}$$

$$\mathbf{J_p} = qp\mu_p\mathbf{E} - qD_p\nabla p - q\frac{D_p}{T}p\nabla T \tag{3}$$

include an additional current component with the temperature gradient as a driving force. Isothermal simulation is performed by solving the electrical semiconductor equations at constant temperature.

2. Results and Discussion

The n-channel inversion layer in the p base underneath the gate oxide which makes forward conduction possible, is supported by a gate voltage of 20V. The typical IGBT forward characteristics is the exponential rising of the collector-emitter current J_{CE} in the low voltage region and the MOSFET-like current saturation or the latch-up. Static IGBT forward characteristcs performed with electrothermal simulation are illustrated in Fig. 3. In the case of the circular cell latch-up occures at V_{CE}=7.3V and J_{CE}=1148A/cm^2 and the maximal lattice temperature amounts to 377K in the inversion layer of the p base region. For the stripe cell latch-up arises at a higer voltage and current density (V_{CE}=10.1V and J_{CE}=1360A/cm^2) with a maximal lattice temperature of 445K . Due to the smaller minority-carrier collector area of the circular cell (see Fig. 2) hole current crowding occures and increases the lateral voltage drop of the hole current beneath the n$^+$ source. The stripe cell structure has much less current crowding due to the larger minority-carrier collector area so that the necessary voltage drop for IGBT latch-up arises at a larger collector-emitter current. In Fig. 6 the hole current flow lines, the electron concentration and the internal lattice temperature distributions in the device are shown at the bias point A of the stripe cell. The electron concentration indicates the beginning of electron injection into the p base, which initiates the turn on of the parasitic thyristor. The results of isothermal simulation at the constant lattice temperatures 300K and 400K are shown in Fig. 4-5. At 300K the collector-emitter voltage of the circular cell amounts to 80V when latch-up occures and the stripe cell represents at this temperature a latch-up safe device until the blocking voltage is reached. At temperature of 400K (Fig. 5) the circular cell latches at V_{CE}=13.1V and J_{CE}=1516A/cm^2 the stripe cell at V_{CE}=315V. The built-in voltage of the n$^+$ source p-well junction decreases with temperature and is the dominating thermal effect in the isothermal calculations. Considering the circular cell, in the non-isothermal case with T_{max}=377K latch-up occures at a lower collector-emitter current compared to the isothermal case with T_{max}=400K. In this comparison the variation of the hole current proportion with non-isothermal conditions plays an important role. The cross sections of hole currents (Fig. 7) and temperature distributions (Fig. 8) at constant collector-emitter current show the relationship of these physical quantities for different thermal conditions. The difference of the hole currents at the isothermal temperature distributions 300K and 400K are relatively small. The highest hole current occures (solid line) at non-isothermal temperature distribution

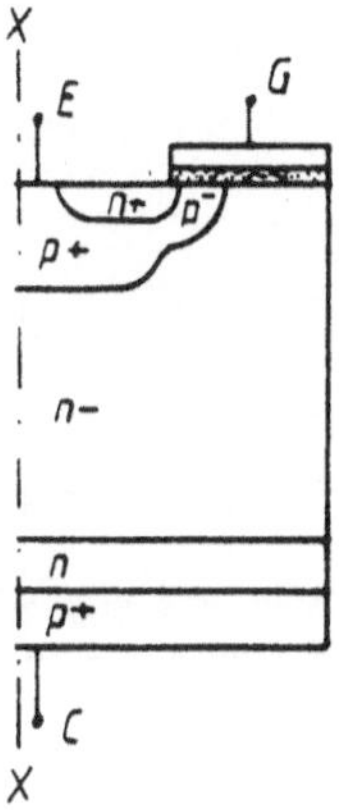

Figure 1: IGBT structure

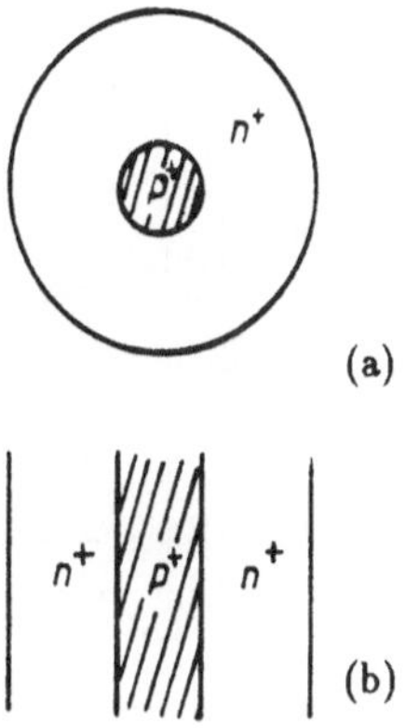

Figure 2: circular (a)
and stripe (b) cell

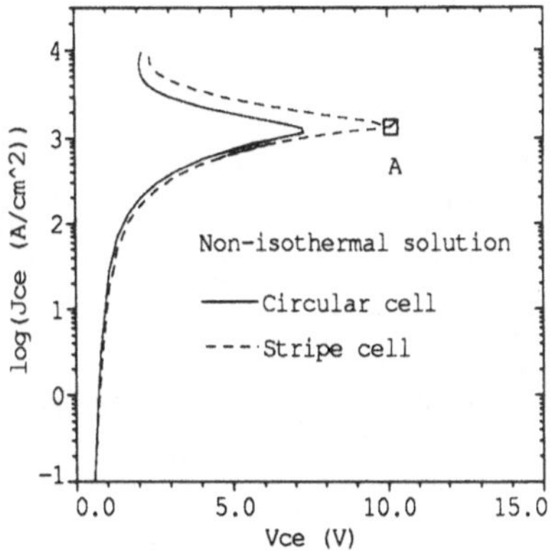

Figure 3: non-isothermal forward
characteristics

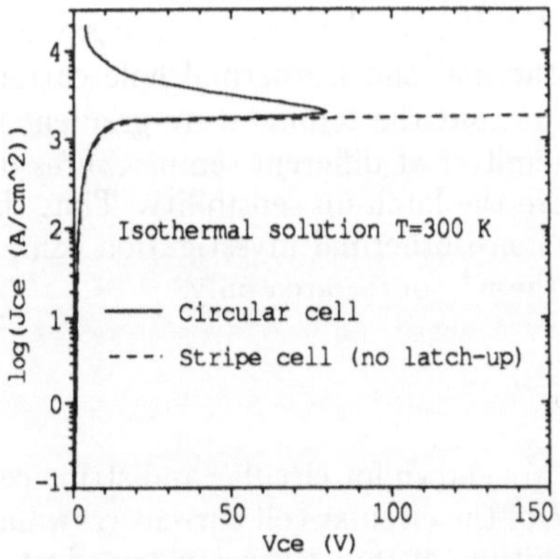

Figure 4: isothermal forward characteristics

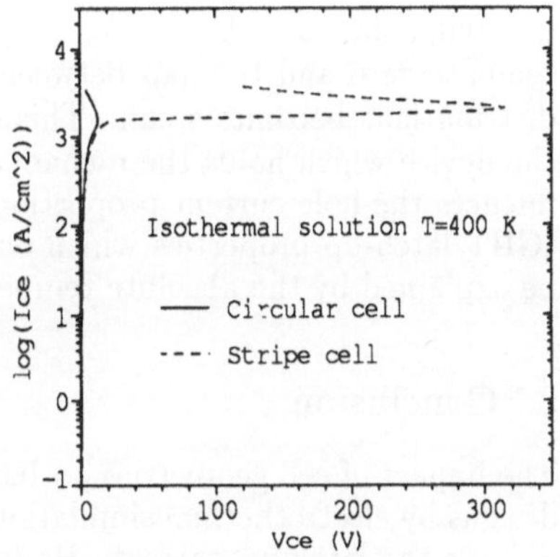

Figure 5: isothermal forward characteristics

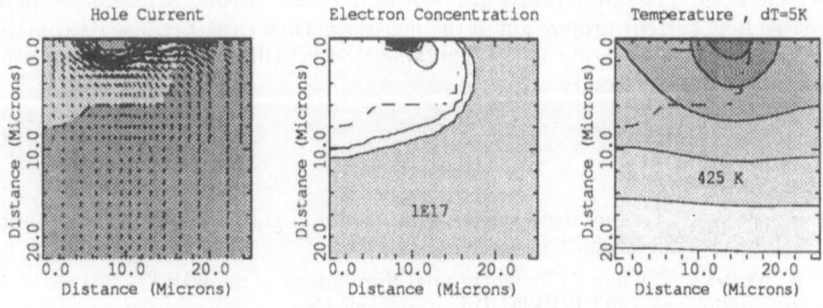

Figure 6:   hole current flow lines, electron concentration, temperatur contours
(T_{max}=445K, ΔT=5K) for stripe design at bias point A (darker color → higher value)

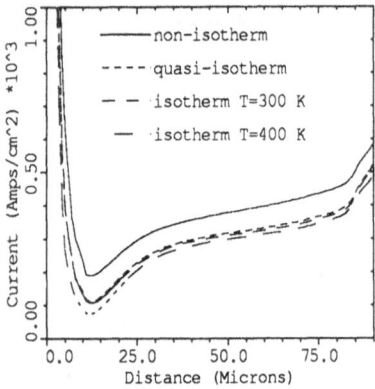

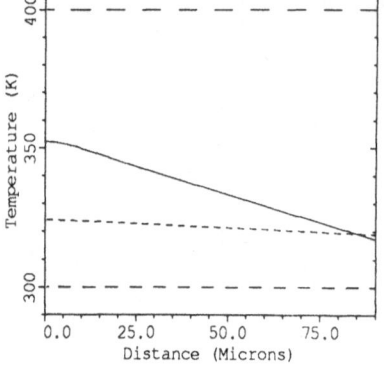

Figure 7: cross sections of hole currents at J_{CE}=1000A/cm² for circular cell at x-x axis

Figure 8: cross sections of temperatures at J_{CE}=1000A/cm² for circular cell at x-x axis

in the IGBT. The quasi-isothermal case is computed with a very large thermal conductivity, chosen as $\lambda = 10^5 \lambda_{Si}$ ($\lambda_{Si} = 1.4\,\text{W/cmK}$). Hence, the temperature gradient tends to zero and the gap between the non-isothermal and isothermal hole current distributions becomes small. This example shows that the temperature gradient in the device which holds the n-channel and the p-emitter at different temperatures influences the hole current proportion and therefore the latch-up sensibility. Thus the IGBT latch-up properties which occures in the non-isothermal investigation cannot be explained by the absolute temperature near the n⁺ source area only.

3. Conclusion

The impact of cell geometries on IGBT latch-up are shown for circular and stripe cell designs by electrothermal simulation. In the case of the circular cell current crowding reduces the latch up current. However the magnitude of this cell geometry effect is clearly influenced by temperature. The difference between the latch-up currents varies with the temperature. The temperature dependence of the junction voltage is the dominating effect at isothermal conditions. In the non-isothermal case the variation of the hole current proportion with temperature gradient plays an important role for IGBT latch-up. The temperature gradient term of Eq. (2) and (3) cannot explain this increased hole current proportion in the non-isothermal case, because its contribution is to small and of opposite sign. This question is still unsettled, therefore further investigations are necessary.

References

[1] B.J.Baliga, "Temperature behavior of insulated gate transistor characteristics," Solid State Electron., 28, pp.289-297 (1985)

[2] V. Axelrad, R.Klein "Electrothermal Simulation of an IGBT" Proceedings of 1992 ISPSD, Tokyo, pp.158-159

[3] Technology Modeling Associates, Inc., Palo Alto, California, USA. MEDICI users's manual, Jan 1992.

SIMULATION OF SEMICONDUCTOR DEVICES AND PROCESSES Vol. 5
Edited by S. Selberherr, H. Stippel, E. Strasser – September 1993

On the Influence of Thermal Diffusion and Heat Flux on Bipolar Device and Circuit Performance

M. Stecher, B. Meinerzhagen, I. Bork, and W. L. Engl

Institut für Theoretische Elektrotechnik, University of Aachen
Kopernikusstraße 16, D-52074 Aachen, GERMANY

Abstract

For device models which consider energy transport the modeling of the flux components due to spatially inhomogeneous carrier temperatures is still a controversial issue. In this paper the influence of these flux components on device and circuit performance is evaluated by the example of a state of the art bipolar technology using mixed level 2D-device/circuit simulation.

1. Introduction

The derivation of hydrodynamic (HD) models from Boltzmann's transport equation (e.g. [1]) depends on a number of approximations (e.g. relaxation time approximation (RTA)) which are of limited validity. Consequently, there is still some controversy concerning the details of the HD models. One of the most controversial issues is the modeling of the flux components driven by the gradients of the carrier temperatures, namely the thermal diffusion (TD) components of the current densities and the heat flux (HF) components (nonconvective parts) of the energy flux densities, respectively. Recently it has been shown that the application of the macroscopic RTA leads to an incorrect modeling of HF and TD under certain conditions [2]. So far the influence of these components on HD device modeling results has been monitored based on unipolar, 1D N^+NN^+ test structures [3],[4], [6].

In this paper, for the first time the influence of TD and HF is discussed based on realistic state of the art bipolar devices and circuits. A related study on the influence of energy transport related effects in general has been published previously [7].

2. The Influence of HF and TD on Device Characteristics

As a test device a realistic, down scaled 2D NPN bipolar transistor structure with a base width of $50nm$ has been selected whose doping profiles (figure 1)) have been optimized to achieve minimal ECL gate delay. All results presented in this paper have been achieved with the simulator GALENE III and the generalized HD model (GHDM) [1]. The GHDM is consistent to an advanced Monte Carlo Model and it has been shown that it reproduces the results of the MC model within short channel MOSFET's very well [5].

For the purpose of this paper the TD and HF components in the GHDM have been modified, so that both components can be independently scaled down using two constant factors f_{td} and f_{hf}, respectively. The modified expressions for the electron current density J and the electron energy flux density S are given below:

$$J = -\frac{q}{m^*}\left\{\tau_i^* qn\nabla\Psi - \tau_i kT^*\nabla n - f_{td}\,\tau_i n\nabla kT^*\right\}$$

$$S = -\frac{5}{2q}kT^*\tau_s^*\tau_i^{*-1}\left\{J + f_{hf}\,\frac{q}{m^*}\tau_i n\nabla kT^*\right\}$$

All symbols have their usual meaning as defined in [1]. In order to demonstrate the influence on modeling results caused by either a reduction of TD or HF, three different HD models are compared in this paper: the original GHDM, a modified GHDM with reduced TD ($f_{td} = 0.3$) and a modified GHDM with reduced HF ($f_{hf} = 0.5$). Since it turns out that for all relevant bias conditions holes remain close to thermal equilibrium inside the emitter and base regions, modeling differences for holes are negligible so that in this paper the modeling differences for electrons are predominately demonstrated. The modeling results on the device level are summarized in figures 2-5, respectively. Figure 2 shows the influence on collector current. It can be seen that the early voltage

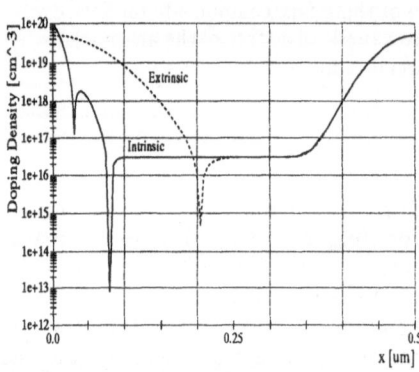

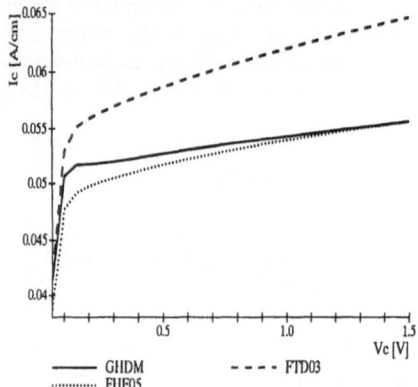

Figure 1: Doping profiles of the bipolar transistor

Figure 2: $I_C(V_{CE})$ characteristic for $V_{BE} = 0.8V$

decreases for reduced TD as well as for reduced HF. On the other hand a significant influence on collector current can only be observed by reducing TD. This influence of TD can be understood by analyzing the directions of the two electron diffusion current components within the base region near the emitter. In this region the TD component and the diffusion component driven by the gradient of electron density have opposite directions. Consequently electron injection into the base is increased by a reduced TD component implying a higher collector current. Modeling differences for electron temperature are only minor at least for higher V_{CE} as can be seen in fig. 3. Fig. 4 shows that drift velocity increases at the beginning of the high velocity zone (base collector junction) and decreases at the end of this zone (beginning of the buried layer) when TD is reduced. This is in good agreement with results achieved for N^+NN^+ structures [6]. On the other hand reducing HF has only a minor effect on drift velocity. This is again consistent with [6] because for N^+NN^+ structures reducing HF reduces drift velocity only within the region of the so called "spurious" velocity overshoot peak while only minor drift velocity changes are observed otherwise and within the bipolar structure of this paper no "spurious" velocity overshoot peak exists. Fig. 5 shows that common base f_t is increased by reducing TD while reducing HF has no influence. The increase of f_t is a direct consequence of the increased current gain for reduced TD.

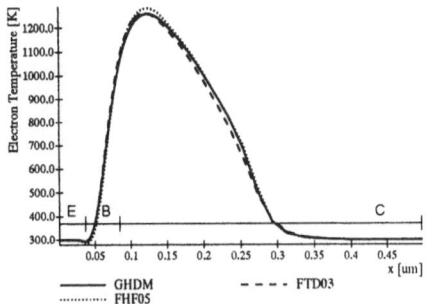

Figure 3: Electron temperature within the intrinsic transistor at $V_{BE} = 0.8V, V_{CE} = 1.4V$

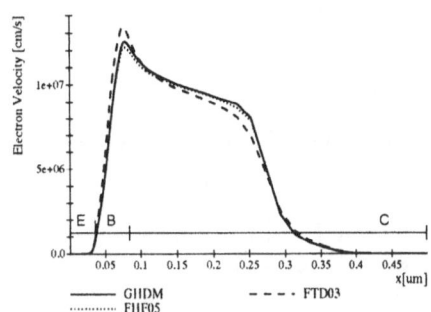

Figure 4: Electron velocity within the intrinsic transistor at $V_{BE} = 0.8V, V_{CE} = 1.4V$

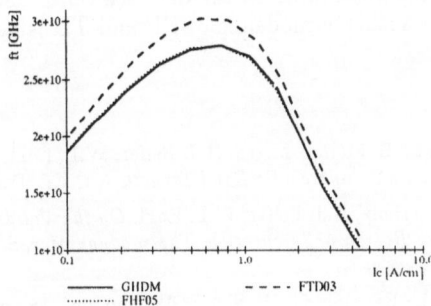

Figure 5: Common base f_T at $V_{CB} = 1V$

3. The Influence of HF and TD on ECL Gate Delay

The consequences of reducing HF and TD for the circuit level are difficult to estimate based on device level results. To overcome this situation the mixed level device/circuit simulation system GALENE III/ CEDUSA has been applied for studying the effects on the circuit level without introducing simplifying assumptions on the device level like in the case of compact models. The ECL inverter circuit analyzed by this simulator system is given in fig. 6. The circuit simulator CEDUSA allows to perform the GALENE III device simulations for the four BJT's within this circuit in parallel, which decreases the required CPU-time drastically [8]. Fig. 7 shows the transient response functions V_{nout} at the inverter output after applying the ramp V_{inp} at the inverter input with the inverter being in steady state before. It can be seen that the differences in the output signals are only minor even in the case of reduced TD.

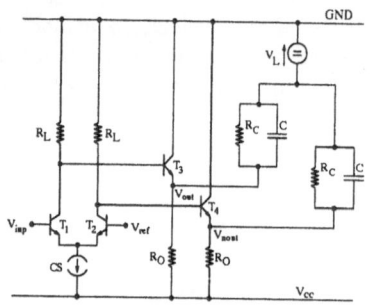

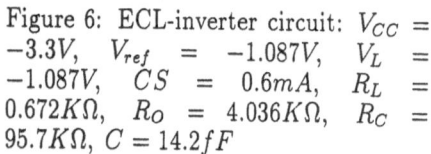

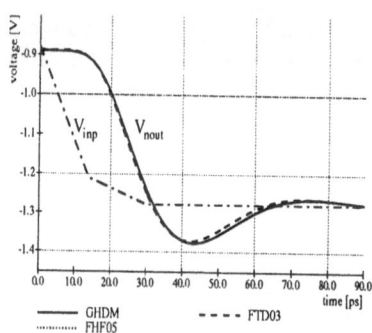

Figure 6: ECL-inverter circuit: $V_{CC} = -3.3V$, $V_{ref} = -1.087V$, $V_L = -1.087V$, $CS = 0.6mA$, $R_L = 0.672K\Omega$, $R_O = 4.036K\Omega$, $R_C = 95.7K\Omega$, $C = 14.2fF$

Figure 7: Transient response functions V_{nout} to the applied voltage ramp V_{inp}

4. Conclusion

Though a modified modeling of HF and TD introduces visible modeling differences on the device level, only minor differences can be observed on the circuit level even for devices which are scaled down somewhat beyond state of the art. This implies that presently and in near future an accurate modeling of TD and HF is not necessary for the optimization of ECL circuit performance. However, as soon as further scaling causes the effects already observable on the device level to become significant on the circuit level as well, an accurate modeling of HF and TD is needed.

References

[1] R. Thoma, A. Emunds, B. Meinerzhagen, H.J. Peifer, W.L. Engl, *Hydrodynamic Equations for Semiconductors with Nonparabolic Band Structure*, IEEE Trans. on ED, p. 1343, 1991

[2] R. Thoma, K.P. Westerholz, H.J. Peifer, W.L. Engl, *On the Validity of the Relaxation Time Approximation in the Region of Irreversible Thermodynamics*, Semicond. Sci. Technol., p. B328, 1992

[3] D. Chen, E.C. Kan, U. Ravaioli, C.-W. Shu, R.W. Dutton, *An Improved Energy Transport Model including Non-Parabolicity and Non-Maxwellian Distribution Effects*, IEEE Elec. Dev. Letters, p. 26, 1992

[4] A. Gnudi, F. Odeh, M. Rudan, *Investigation of Non-Local Transport Phenomena in Small Semiconductor Devices*, Europ. Trans. on Telecom. and rel. Technol., p. 307, 1990

[5] H.J. Peifer, B. Meinerzhagen, R. Thoma, W.L. Engl, "*Evaluation of Impact Ionization Modeling in the Farmework of Hydrodynamic Equations*, IEDM Tech. Dig., p. 131, Washington, 1991

[6] B. Meinerzhagen, H.J. Peifer, R. Thoma, W.L. Engl, *On the Consistency of the Hydrodynamic and the Monte Carlo Models*, Proceedings of the Workshop on Computational Electronics, University of Illinois at Urbana-Champaign, p.7, 1992

[7] M. Stecher, B. Meinerzhagen, I. Bork, J.M.J. Krücken, P. Maas, W. L. Engl, *Influence of Energy Transport Related Effects on NPN BJT Device Performance and ECL Gate Delay Analyzed by 2D Parallel Mixed Level Device Circuit Simulation*, published at VPAD, Nara, 1993

[8] B. Meinerzhagen, J.M.J. Krücken, K.H. Bach, M. Stecher, W.L. Engl, *A Modular Approach to Parallel Mixed Level Device/Circuit Simulation*, VPAD Tech. Dig., Kawasaki, p. 170, 1990

2-D Electrothermal Simulation and Failure Analysis of GTO Turn-off with Complete Chopper Circuit Parasitics

Y. C. Gerstenmaier and H. Brunner

Corporate Research and Development, Siemens AG
Otto-Hahn-Ring 6, D-81739 München, GERMANY

Abstract

2-D electrothermal simulations of GTO-thyristor turn-off process including a complete chopper circuit and parasitic stray inductances will be presented. Turn-off failure is investigated for a single GTO cell and on a wafer scale using a homogeneous wafer, which has in parallel a one segment GTO representing a local perturbation on the wafer by a slightly different doping or carrier lifetime profile. During the spike voltage extreme power densities may result in the perturbed segment due to current filamentation. In high power GTO turn-off two destruction mechanisms have to be dealt with: 1.current-filamentation during the spike voltage period and 2.dynamic avalanche during the tail-phase.

1. Simulation of Single GTO Cell and Wafer Scale GTO

In this paper* for the first time 2-D electrothermal simulations of GTO-thyristor turn-off process including a complete chopper circuit and parasitic stray inductances will be presented. The device simulator MEDICI [1] solves the complete semiconductor equations (Poisson and continuity equations) together with the heat flow equation for the dynamic development of the lattice temperature within a general external network. Additionally thermal resistors and capacitors can be included in order to allow for the cooling of the device by the packaging.

In Fig.1 the simulated circuit is shown with RCD-protection (snubber) circuit, inductive load, clamp and gate-drive circuit. Fig.2 displays the simulated GTO half cell with anode short-structure and Fig.3 shows a simulation result for a 4.5kV/3000A GTO, that reveals the characteristics of measured results like storage time, spike voltage U_{DSP}, peak off-state voltage U_{DM} and tail-current behaviour.

At the beginning of the turn-off process the GTO has a homogeneous temperature of $300°K$. The maximal local temperature of $316°K$ is attained during the spike voltage at the center junction in the middle under the cathode contact. As was already observed in the isothermal simulations of [2] the power density at that instant is very high, but according to our result not sufficient to cause a dramatic or dangerous temperature increase. In Fig.4 the current density to be turned off is ten times larger. Now temperatures around the silicon melting point of $1700°K$ arise and would cause thermal destruction. The turn-off failure becomes apparent by the sudden anode-voltage breakdown in the tail-phase. Despite the negative gate-voltage and current

*This work was supported by **eupec**, a company of AEG and Siemens

the device is not able to sustain the blocking voltage, because of the local loss of blocking capability at the center junction (Fig.5 and Fig.6). This is due to the fact that the temperature at the hot spot exceeds the intrinsic temperature of $550°K$ considerably.

Turn-off failure in the tail-phase has already been observed in isothermal 1-D simulation ([3] and references cited there) and 2-D simulation [4] and is due to dynamic avalanche generation, i.e. the steepening of the electric field gradient in the n-base by hole transport from the anode. In [4] turn-off failure was solely attributed to this. The measurements of [5] were done without snubber cicuit and clamp, so that there was no spike voltage. They revealed that a temperature maximum occurs just after the anode current fall time.

On a wafer scale many individual GTO-Segments (> 2000) in parallel contribute to current transport. Due to inhomogenities between different cells, current redistribution during turn-off takes place [6] (independently of dynamic avalanche) and may destroy those segments, which carry the most heavy load. For a special circuit without stray inductances and snubber circuit (no spike voltage) it was shown in [6], that with a gate current as high as the load current or higher the current filamentation can be damped but for usual gate drives this is not possible. [7] presented an analytical model in order to demonstrate, that during dynamic avalanche current filamentation may take place even for completely homogeneous devices.

The spike voltage, which arises during fall time, may cause destruction due to current filamentation. The current filaments in this case are not caused by dynamic avalanche as in [7] but, as we shall see, by slight inhomogeneities in the lateral dopant or carrier lifetime distribution over the device area.

Because of the excessive number of grid-points it is not possible to simulate a whole GTO wafer. Therefore in Fig.7 the result is shown for a homogeneous 2.5kV/2.5kA GTO (half cell scaled to full wafer size of $20cm^2$), which has in parallel a one segment GTO ($0.024cm^2$ area) representing a local perturbation on the wafer by a slightly lower p-base doping of 10%. Fig.8 shows that precisely at the time of the spike voltage the one segment GTO takes over a maximum current 100 times larger than its on-state value. Since the maximum current of the small GTO segment exactly matches the time of the spike voltage, extreme power densities occur, which will lead to destruction. Somewhat lower current peaks are observed in the tail-period by dynamic avalanche, when the anode voltage is high. If avalanche generation in the parallel one segment GTO is turned off a smooth behaviour of the tail current emerges, whereas the current peak at spike voltage time is nearly the same. Therefore destruction at spike voltage time occurs independent of dynamic avalanche.

By suitable design the current redistribution can be reduced. In [8] similar isothermal simulations were carried out for 1-D devices in parallel; however without snubber circuit and spike voltage. Even though the p-base doping difference amounted to 28% the current redistribution between the cells was not dramatic.

2. Conclusion

The essential meaning of the spike voltage for GTO turn-off failure is revealed. Two destruction mechanisms have to be dealt with in high power GTO turn-off: 1. current-filamentation during the spike voltage period and 2. dynamic avalanche during the tail-phase. Both failure mechanisms can be investigated in numerical simulation with respect to design variations.

References

[1] Technology Modeling Associates. Inc. Palo Alto, California, USA, MEDICI user's manual. March 1992.

[2] M. Turowski, A. Napieralski, J.M. Dorkel, Proceedings EPE-MADEP Firenze 1991, p.0-363 - 0-368.

[3] Y.C. Gerstenmaier, ISPS'92 Proceedings, 1992, Czech Technical University in Prague, p.49.

[4] I. Omura and A. Nakagawa, Proceedings of 1992 International Symposium on Power Semiconductor Devices & ICs, Tokyo, pp.112-117.

[5] H. Ohashi and A. Nakagawa, IEEE IEDM-81 Tech. Dig., p.414, 1981.

[6] K. Lilja and H. Grüning, IEEE PESC Dig., p.398, 1990.

[7] G.K. Wachutka, IEEE Trans. on Electron Devices, Vol.38, p.1516, 1991.

[8] C.M. Johnson and P.R.Palmer, IEEE Trans. on Power Electronics, Vol.6, p.308, 1991.

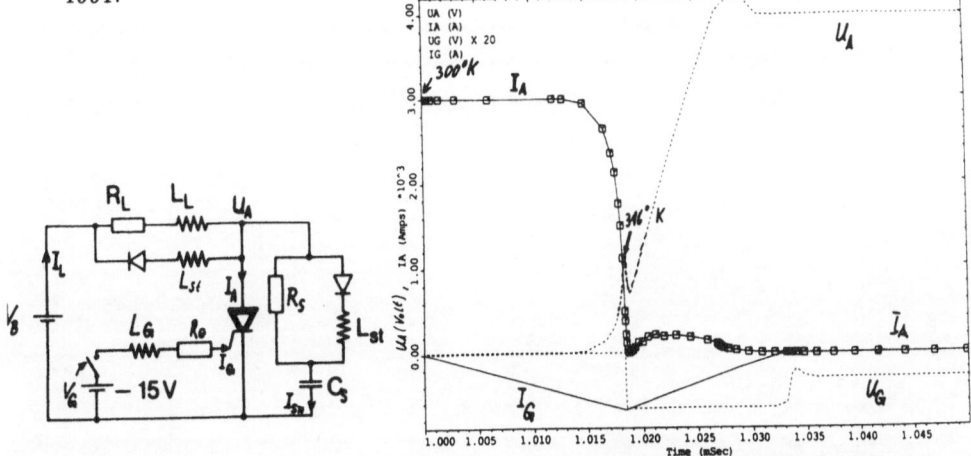

Figure 1: Simulated chopper-circuit for GTO turn-off.

Figure 3: 2-D electrothermal MEDICI simulation of GTO turn-off for circuit of Fig.1. Highest local temperature is marked at the spike voltage.

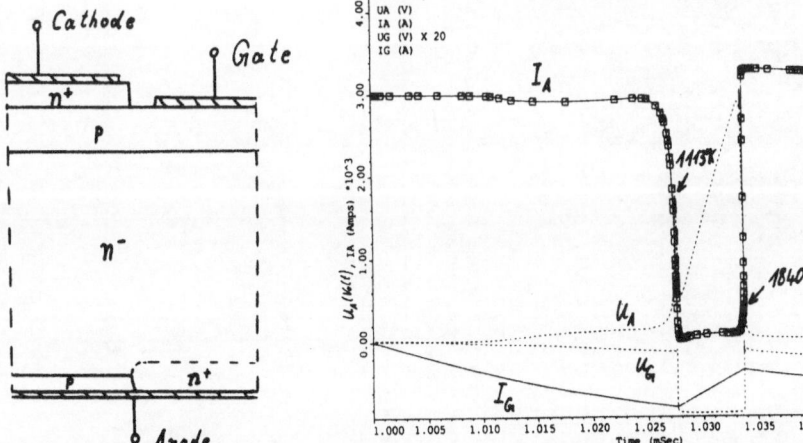

Figure 2: Simulated GTO half cell with anode short structure.

Figure 4: Destructive failure of GTO turn-off in 2-D simulation at ten times the current density of Fig.3.

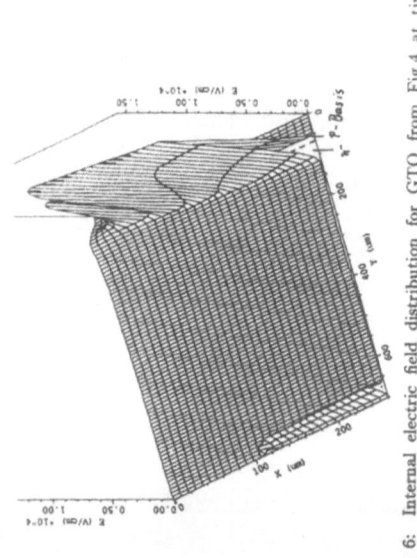

Figure 6: Internal electric field distribution for GTO from Fig. 4 at time t=1.2msec. At the hot spot the center pn-junction loses its blocking capability.

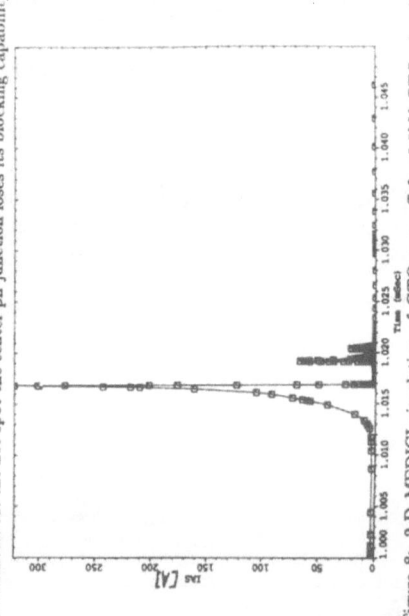

Figure 8: 2-D MEDICI simulation of GTO turn-off for 2.5kV GTO. Anode current of parallel one GTO-segment with 0.024cm² displayed (Fig. 7). The current maximum at spike voltage time is more than 100 times the on state current of 3A.

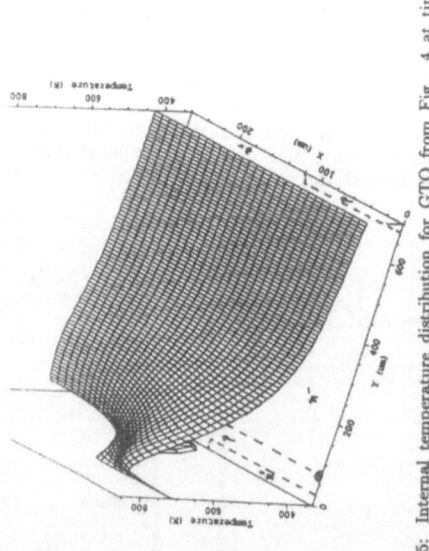

Figure 5: Internal temperature distribution for GTO from Fig. 4 at time t=1.2msec. At the spot the intrinsic temperature of the n^--doping of 550°K is exceeded.

Figure 7: 2-D MEDICI simulation of GTO turn-off for 2.5kV GTO of 20cm² area which has in parallel a one segment GTO with 10% lower p-base doping.

3D Thermal/Electrical Simulation of Breakdown in a BJT Using a Circuit Simulator and a Layout-to-Circuit Extraction Tool

B. H. Krabbenborg, H. C. de Graaff, A. J. Mouthan, H. Boezen[†], A. Bosma, and C. Tekin

MESA Research Institute, University of Twente
P.O.Box 217, NL-7500 AE Enschede, THE NETHERLANDS
[†]Philips Semiconductors
Gerstenweg 2, NL-6543 AE Nijmegen, THE NETHERLANDS

Abstract

A method for the generation of circuit models for fast thermal-electrical simulation of 3D device structures with a circuit simulator is proposed. It has been used for simulation of the influence of layout parameters on the Safe Operating Area of a BJT and to study the mechanisms that start breakdown processes. For a thermally instable switch-on behaviour of a BJT, a comparison with measurements has been made.

1. Introduction

In this paper we present a tool for 3D thermal-electrical simulation of devices using a circuit simulator. It can be used by circuit designers for optimizing the thermal-electrical characteristics of (power) devices by modifying layout parameters. Coupled thermal-electrical circuits are generated by a layout-to-circuit extractor using a device description based on the layout and the technology used [1]. A comparison with measurement results is done for thermal runaway in a BJT. An application is shown for optimization of the Safe Operating Area (SOA) of a BJT. The advantages of this simulation method compared to the alternative, thermal-electrical device simulation [2] (using the semiconductor transport equations), are discussed. Besides device optimization, this simulation method is used to obtain insight in the coupled thermal-electrical mechanisms that lead to thermal instability and breakdown.

2. Simulation method and circuit generation

The method for simulation of thermal-electrical device behaviour with a circuit simulator has been described in [3,4]. Our circuit model generation is based on the device layout (masks), technological process parameters (e.g., sheet resistance), and the chip dimensions. The circuit representing the device is built up with layers. Each layer represents a distinct part of the device (e.g., an emitter region of a BJT), and is built on a lateral triangular mesh with vertical elements located on the nodes (for the connection with other layers), and lateral elements on the edges (connecting nodes in a layer). The layers are defined by using logical functions regarding the mask information and technology description. The latter determines what type of circuit elements will be generated between the nodes in a layer or as a connection between different layers. Figure 1 shows a BJT built up with layers. For the discretisation two grids are used, one for the area in which the time dependent heat flow equation is solved, and one for the area in which the electrical model equations are solved. The meshes coincide in the area where the coupling between thermal and electrical circuit occurs by means of power dissipation and temperature. The transistor model that has been used in the simulations, is an Ebers-Moll model, extended with avalanche multi-plication and high injection effects [3]. The thermal network consists of resistances and capacitances. The thermal conductance was assumed constant here.

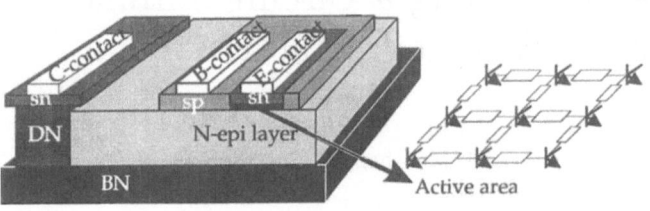

Figure 1:   BJT built up with layers. The active area is represented with 9 transistors in this case.

3. Verification, comparison with measurements

The switch-on behaviour of a BJT with five emitters (of 36x100 [μm²] each) has been used for a comparison of simulation and measurement results. The layout of this structure is shown in figure 2. Only the left half of the structure was simulated because of symmetry. A constant total emitter current I_E=100 [mA] and a constant base-collector voltage V_{BC}=55 [V] were applied at t=0 [s]. The five emitter potentials were all kept at the same level (error < 0.1 [mV]).

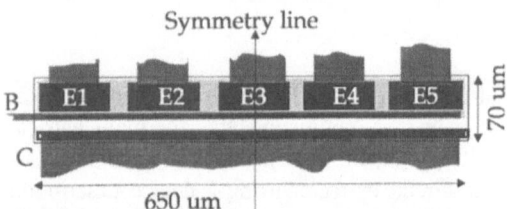

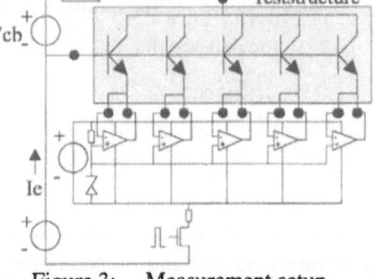

Figure 2:   Layout of test structure with 5 emitters.

Figure 3:   Measurement setup

This was done using a control circuit as shown in figure 3. Because of temperature gradients built up in the device, the current redistributes over the five emitters when time increases. Figure 4 shows the measured and the simulated emitter currents versus time. The measurement results show a difference between the currents of

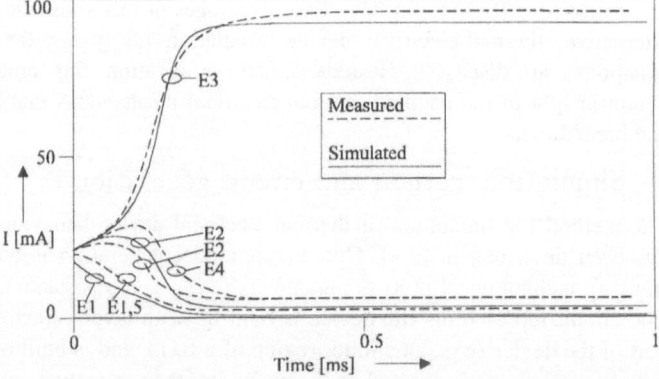

Figure 4:   Simulated and measured emitter currents. E1-E5: Emitter 1 to 5.

emitter 2 and 4. This is due to an unintended asymmetry in the device structure (and not because of offset in the control circuit). The central emitter conducts most of the total current after 0.2 ms. A good agreement between measurement and simulation results is obtained. Figure 5 shows the simulated collector current distribution and the temperature distribution in the emitter area after 0.1 and 1 [ms]. The base current crowding effect causes the high current densities at the emitter edges in figure 5A. The effect of the temperature distribution at t=0.1 [ms] (figure 5B) is visible in the current distribution (figure 5A); in y-direction the highest current occurs in the center. The maximum temperature rise in the transistor is about 260 [K] at t=1 [ms] (figure 5D).

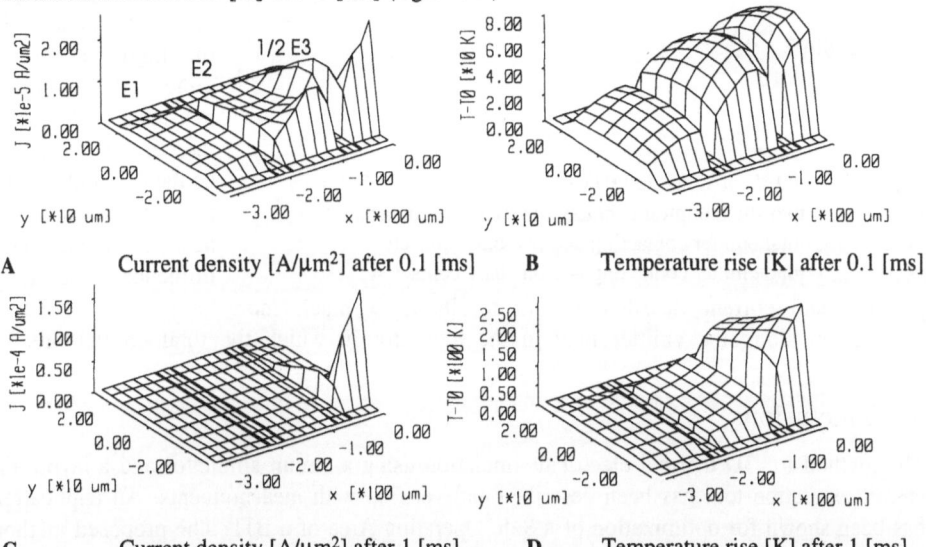

A   Current density [A/μm^2] after 0.1 [ms]   B   Temperature rise [K] after 0.1 [ms]

C   Current density [A/μm^2] after 1 [ms]   D   Temperature rise [K] after 1 [ms]

Figure 5:   Current (A & C) and temperature (B & D) distribution for t=0.1 [ms] (A & B), and t=1 [ms] (C & D). Note: Only in the emitter areas the data were plotted.

4. Optimization of the SOA of a BJT

In a BJT power dissipation in combination with avalanche multiplication are the major factors determining the boundary of the Safe Operating Area (SOA). Both factors depend

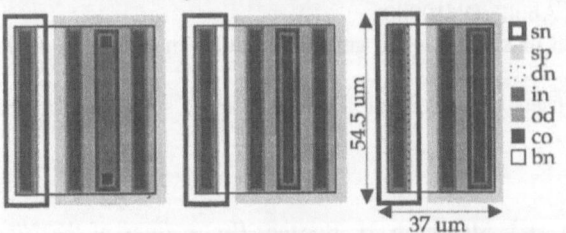

strongly on the transistor layout. Therefore a SOA can be optimized using layout modifications. The simulation results demonstrate the influence of an extra base contact and a modification in the emitter contacting on the SOA. Figures 6 and 7 show the layouts and the SOA's of the various transistors respectively.

Figure 6:   Layouts of simulated BJT's. Left: two small emitter contacts and two base contacts, middle: two base contacts, right: one base contact.

The extra base contact reduces the base resistance. This improves the SOA because it increases the snapback voltage with 4 [V] (10 %) in the high voltage ($V_{CE} > BV_{CE0}$), low current region where avalanche multiplication appears to be the dominant factor for the snapback point. In the high current, low voltage region the base resistance counteracts the positive temperature dependency of the collector current. The base resistance reduction also reduces this compensation, and

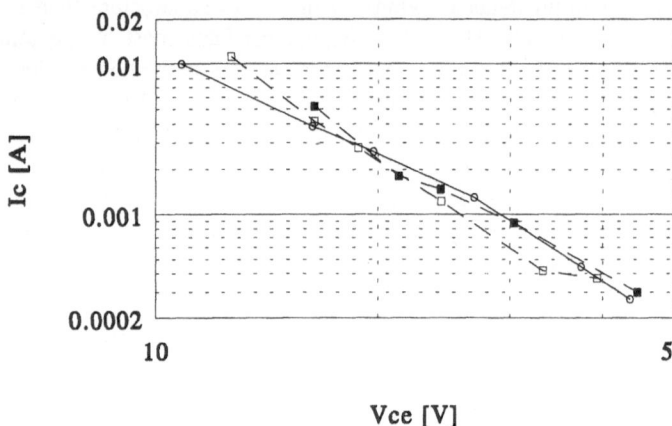

therefore a lower snapback voltage is obtained in this region. The smaller emitter contacts results in a higher emitter series resistance especially for the sections in the middle of an emitter area. In the high current regime the emitter resistance decreases the current in the non contacted emitter area and concentrates it near the contact. This influences the temperature and current distribution and results in a much more uniform temperature distribution along the emitter than in the transistor in which the total emitter area is contacted.

Figure 7:   SOA's of simulated BJT's.
■:   two small emitter contacts and two base contacts
○:   total emitter contacting and two base contacts
□:   total emitter contacting and one base contact.

5. Conclusions

The method for 3D thermal-electrical simulation using a circuit simulator and a layout-to-circuit extraction tool has been described and verified with measurements. An application has been shown for optimization of a Safe Operating Area of a BJT. The proposed method has a number of advantages compared to the alternative, 3D thermal-electrical device simulation (using the semiconductor transport equations):

1) It allows circuit designers to simulate the thermal-electrical device characteristics with the circuit simulator they are used to.
2) The cpu-time is orders of magnitude smaller. This makes it suitable for simulation of characteristics with many bias points. The average cpu time necessary for simulation of one bias point of the test structure in section 3 (containing 6925 circuit elements and 2074 nodes) was 125 [s] on an Apollo DN10000.
3) The layout-to-circuit extraction makes it easy to generate circuits and to study device behaviour for different layouts.
4) Because a circuit simulator is used, the generated 3D model can be simulated together with external components.

References

[1] B.H. Krabbenborg, A. Bosma, A.J. Mouthaan, H. Boezen, *Proceedings of the Ninth International Conference on the Numerical Analysis of Semiconductor devices and Integrated Circuits*, 6-9 April 1993, Coppermountain Colorado, USA, pp. 99-100.

[2] B.H. Krabbenborg, R. Beltman, Ph. Wolbert, A.J. Mouthaan, *Journal of Electrostatics*, Vol. 28, 1992, pp. 285-299.

[3] M. Latif, P.R. Bryant, *IEEE transactions on CAD*, Vol. CAD-1, No. 2, April 1982, pp. 94-101.

[4] P.C. Munro, F.-Q. Ye, *IEEE Journal of Solid State Circuits*, Vol. SC-26, No. 9, Sept. 1991, pp. 1321-1324.

Nonlocal Oxide Injection Models

K. P. Traar and A. v. Schwerin[†]

PSE, Siemens AG Austria
Gudrunstraße 11, A-1101 Wien, AUSTRIA
[†]Corporate Research and Development, Siemens AG
Otto-Hahn-Ring 6, D-81739 München, GERMANY

Abstract

Three models for hot carrier injection into the gate oxide layer of a MOSFET
are examined and compared with gate current measurement.

1. Introduction

In order to develop models for the simulation of hot carrier degradation of deep-sub-
μm MOSFET's, the oxide injection of hot carriers and the transport of the injected
carriers inside the gate oxide layer are investigated. Considering the sharply peaked
electric field near drain, it is clear that a model for the injection of hot carriers,
that relies purely on local values of electric field and carrier concentration, is not
appropriate. In section 2 we present three models to calculate the injected current.

To describe the oxide transport of injected carriers a 2-D continuity equation is solved
on the whole oxide bulk, resulting in oxide current densities, which allow to calculate
gate currents and are the necessary input for oxide trapping calculation.

The injection models and the oxide transport approach were implemented into the
device simulator MINIMOS. Gate current simulation results for all injection models
are compared with experimental data for the case of a 0.9 μm MOSFET with 10 nm
oxide thickness and purely As-doped drain junctions.

2. Models

Model I is based on the nonlocal ballistic lucky electron model introduced by Meinerzhagen for the calculation of oxide injection in MOSFETs [1]. In his work the oxide
injection current is given by

$$j_{inj}(x_0) = A\, n[x(d)]\, v_{sat} \cos(\theta) \exp\left(-\frac{d}{\lambda}\right) \exp\left(-\frac{d_{ii}}{\lambda_{ii}}\right) \tag{1}$$

where x_0 is the injection point, A is a constant, v_{sat} the saturation velocity, $x(d)$ that
point on the electric field line path ending at the interface point x_0 with $\Psi(x(d)) =
\Psi(x_0) - \Phi/q$ (for electrons), d the length of the path between $x(d)$ and x_0, λ the

inelastic mean free path, λ_{ii} the mean free path for impact ionization scattering events, d_{ii} the path length between x_0 and $x(d_{ii})$ with $\Psi(x(d_{ii})) - \Psi(x(d)) = \Phi_{ii}/q$,[1] and θ the angle between the electric field vector and the vector normal to the Si/SiO$_2$–interface.

The basic idea of Model II is to integrate contributions from all over the semiconductor to the injected current at the respective oxide interface point. The general form of this injection formula is given by

$$j_{inj}(x_0) = \frac{A}{2\lambda} \int_{-\pi/2}^{\pi/2} d\phi \, \cos\phi \int_0^\infty dr \, \exp\left(-\frac{r}{\lambda}\right) |j(r,\phi)| \, \exp\left(-\frac{\Phi - q\Delta\Psi}{qE_\parallel(r,\phi)\lambda'}\right) \quad (2)$$

where r is the distance between x_0 and the integration point (r, ϕ), and ϕ the angle between the connection line $x_0 - (r, \phi)$ and the interface normal. A is a proportionality constant , λ the inelastic mean free path, $j(r, \phi)$ the current density at the integration point in the semiconductor, $\Delta\Psi = \Psi(x_0) - \Psi(r, \phi)$ is used for the difference in electrostatic potential between integration and interface point. λ' is assumed to be smaller than λ, as it is used to estimate the "high energy temperature" $(k_B T_{high} := qE_\parallel\lambda')$.

In Model III the injected current is given by:

$$j_{inj}(x_0) \;=\; B \, v_{inj}^0 \, \lambda_{inj} \, \frac{\partial}{\partial n}\Big[\int_V \frac{d^2x'}{\lambda^2} \, G(x_0, x') \, n(x') \, \exp\left(-\frac{\Phi - q\,\Delta\Psi}{qF\lambda} \, s(F)\right)\Big] \quad (3)$$

where B is a constant, v_{inj}^0 is the particle velocitiy, λ_{inj} the mean free path at $\epsilon = \Phi$ respectively, the derivative is with respect to the interface normal, $k = \sqrt{3} \, (\lambda\lambda_{op})^{-1/2}$, and λ and λ_{op} are the total and the optical mean free path, $n(x')$ is the particle density, F is the driving force and s is a parameter to determine the high energy temperature of the distribution by solving a transcendental equation [2]. The propagator $G(x_0, x')$ essentially varies exponentially with $-k \, |x_o - x'|$, its detailed form is given in [3]. Note that the expression in the square brackets in equation 3 is essentially the product of the density of states and the isotropic part of the distribution function f_0.

For the threshold energy Φ, barrier lowering is taken into account according to [4].

3. Results and Discussion

A distinct difference between the results of Model I and Model II is found for values of U_G below the applied drain voltage U_D (see fig. 1 and 2). The reason is that behind the pinch off, where the maximum lateral electric field is found, there is no electric field component driving electrons into the oxide, causing a sharper decline of the injected current in the ballistic lucky electron model (Model I).

A comparison of measurement and simulation for Model III can be found in figure 3. Figure 4 shows simulation results of Model II and Model III of the gate current at 5.5V drain voltage for the full gate voltage range, together with the experimental data. For low gate voltage, also "positive" (i. e. hole) gate currents are predicted (still below the measurement limit of the equipment used for the gate current measurement). The bias region and the order of magnitude of this gate current, which is due to hot hole injection, is in good qualitative agreement with literature data [5].

---

[1] Φ_{ii} is the threshold energy for impact ionization $\approx 1.5\,\mathrm{eV}$

In Figures 5a and 5b the distribution of injected current along the Si/SiO_2–interface is shown for $V_D = 5.5V, V_G = 6V$. The slight shift of the peak injection current relative to the peak of the lateral electric field is due to the non-local nature of the injection models used.

4. Acknowledgment

This work was partially funded by the ADEQUAT project (JESSI Project BT1B, ESPRIT Project 7236). One of us (K. P. T) acknowledges funding by the Austrian Forschungsfoerderungsfond under grant no Zl. 3 /8481.

References

[1] B. Meinerzhagen. Consistent gate and substrate current modeling based on energy transport and the lucky electron concept. In *IEDM Tech. Dig.*, p.504, 1988

[2] L. V. Keldysh. Concerning the theory of impact ionisation. *Sov. Phys. JETP*, 21(6), pp. 1135–1144, December 1965.

[3] W. Hänsch. *The Drift Diffusion Equation and Its Applications in MOSFET Modeling*. Computational Microelectronics. Springer Verlag, Wien and NewYork, 1991.

[4] T. H. Ning, C. M. Osburn, and H. N. Yu, *J. Appl. Phys.*, Vol. 48, p. 286 (1977).

[5] N. S. Saks, P. L. Heremans, L. van den Hove, H. E. Maes, R. F. DeKeersmaecker, and G. J. Declerck. Observation of hot-hole injection in nMOS transistor using a modified floating-gate technique. *IEEE Trans. Electron Devices*, vol. ED-32, pp. 691–699, 1985.

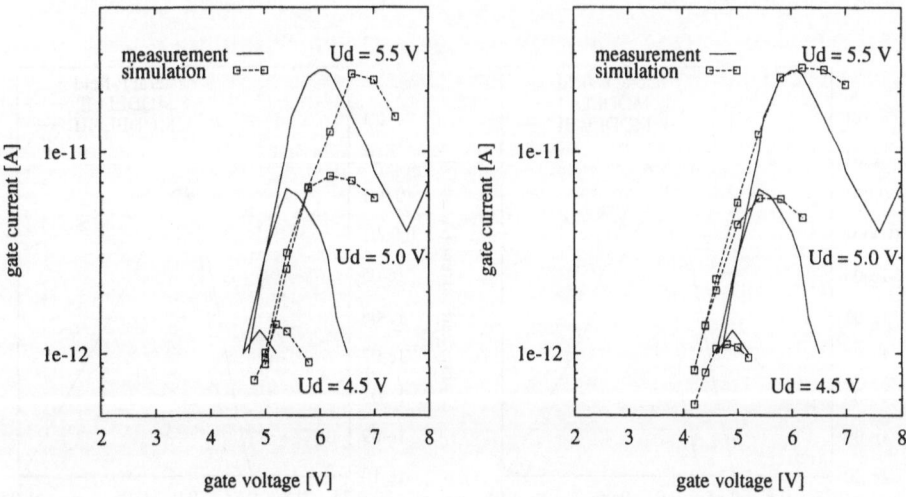

Figure 1: Gate current as a function of gate voltage for 3 different drain voltages U_D. Comparison of measurement (solid lines) and simulation using model I (dashed lines).

Figure 2: Gate current as a function of gate voltage for 3 different drain voltages U_D. Comparison of measurement (solid lines) and simulation using model II (dashed lines).

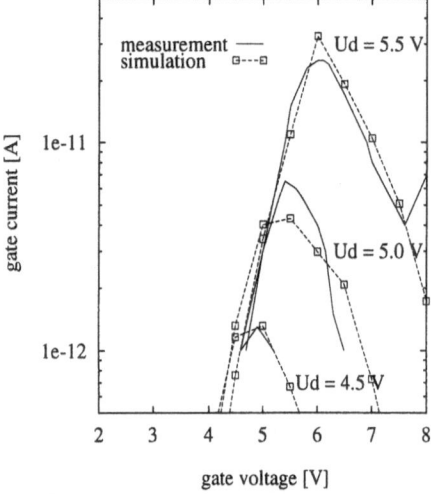

Figure 3: Gate current as a function of gate voltage for 3 different drain voltages U_D. Comparison of measurement (solid lines) and simulation using model III (dashed lines).

Figure 4: Gate current as a function of gate voltage for $U_D = 5.5V$. For $0 \leq U_G \leq 2$ the current is positive. Comparison of measurement (solid line) and simulation using model II (dashed line) and model III (dashed-dotted line).

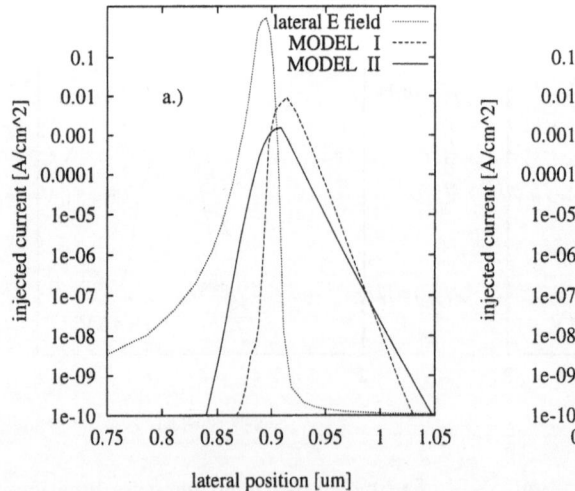

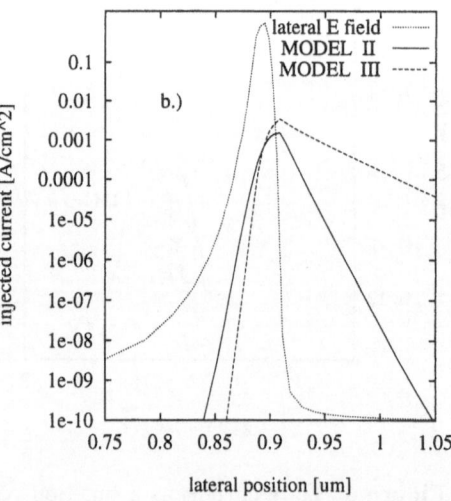

Figure 5: Injected current for the different models vs. lateral position for $U_G = 6V, U_D = 5.5V$. The gate edge is located at 0.9 μm. The electric field is plotted linearly with a maximum of $4.55 \cdot 10^5 V/cm^2$.

Electron Transport in Silicon Dioxide at Intermediate and High Electric Fields

M. Hackel, H. Kosina, and S. Selberherr

Institute for Microelectronics, TU Vienna
Gußhausstraße 27-29, A-1040 Wien, AUSTRIA

Abstract

A semiclassical Monte Carlo technique is employed to simulate the steady-state electron transport in silicon dioxide at intermediate and high electric fields. The electronic band-structure is modelled by a single parabolic, by a single nonparabolic as well as an isotropic four-band model. We find that the electronic behavior of silicon dioxide is mainly influenced by a single nonparabolic conduction-band.

1. Introduction

In the seventies and the early eighties the standard picture of electron transport in insulators assumed only one scattering process to be dominant, the longitudinal polar optical phonons due to the ionic character of Silicon dioxide (SiO_2) [1],[2],[3]. In the mid eighties new experimental data predicted breakdown fields higher than calculated in theoretical studies. Ridley [4] supposed a new scattering mechanism to prevent polar runaway, namely nonpolar acoustic phonons. Sparks et al. [5] derived a mathematical formulation of this process. Despite of the ionic character of the electronic bonds in alkali halides, nonpolar acoustic phonons stabilize the electronic distribution and increase the electrical breakdown-field. A theoretical investigation for SiO_2 was first made by Fischetti [6] and it was found that collisions between electrons and nonpolar acoustic phonons indeed have strong impact on the electronic behavior of oxides. The data of several experiments could be well reproduced [7].

2. The Physical Model

Insulators partially have ionic character, thus electrons lose energy to the lattice in form of polar longitudinal optical (LO) phonons. Electrons in SiO_2 strongly interact with two LO phonon modes via the polarization field of the ions ($\hbar\omega_{LO,1} = 0.153 \ eV$, $\hbar\omega_{LO,2} = 0.063 \ eV$). Electronic Bloch waves can be perturbed by displacement of an ion from its equilibrium position. This causes electrons to interact with the lattice via nonpolar acoustic and optical phonons. The nonpolar optical phonon ($\hbar\omega_{TO} = 0.132 \ eV$) is of less importance but allows band-to-band scattering [3],[4]. At low energies electrons can undergo nonpolar acoustic collisions. The scattering rate is approximated by an intraband elastic ansatz [8], whereas at high electron energies U-processes occur stabilizing the electronic transport in SiO_2 [6],[7].

3. The Transport Model

In this section we briefly describe our Monte Carlo algorithm to solve the Boltzmann transport equation (BTE). An excellent review on this method is given in [8]. The equation of motion and the duration of free flight are solved simultaneously by employing a Runge-Kutta algorithm instead of the usual self-scattering scheme. After performing a free flight the scattering process is randomly chosen according to the partial scattering rates. If one scattering process is selected the after-scattering state of the electron is calculated. Having evaluated the state of the scattered electron we perform another free flight till the maximum number of scattering events is reached.

4. Results

To model the band-structure in SiO_2 we implemented an isotropic four-band model with one nonparabolic (nonparabolicity α) and three parabolic bands [9]

$$\epsilon(1 + \alpha\epsilon) = \frac{\hbar^2 k^2}{2m^*} \quad \text{for } 0 \leq k \leq k_{max} \quad \text{band 1,} \tag{1}$$

$$\epsilon = \epsilon_0 \pm \frac{\hbar^2 k^2}{2m^*} \quad \text{for } 0 \leq k \leq k_{max} \quad \text{band 2, 3, 4.} \tag{2}$$

band	m^* $[m_e-]$	ϵ_1 $[eV]$	ϵ_2 $[eV]$	k_{max} $[nm^{-1}]$	multiplicity
1	0.50	0.00	5.52	11.54	6
2	1.34	5.52	9.31	11.54	6
3	1.05	7.00	9.00	7.42	12
4	1.05	9.00	11.00	7.42	12

Tab. 1: Parameters of the four-band model used

The parameters of our sperical band-structure are summarized in Tab. 1 and were extracted from band calculations of Chelikovsky and Schlüter [10]. The first two conduction-bands were appropriately fitted with respect to an equivalent volume of the first Brillouin zone. Bands three and four crudely account for the behavior of higher conduction-bands due to their parabolic character, but allow a qualitative estimation of their influence on electron transport in oxides.

In Fig. 1 our four-band model is shown. The density of states (Fig. 2) of one parabolic, one nonparabolic and the four-band model are compared, whereas the first bands have the same mass. It is clearly seen for intermediate and high energies that the density of states strongly differs. The main features of a realistic band-structure with two maxima at 5.5 eV and 9 eV and one minimum at 7 eV are well reproduced and strongly influence the electronic distribution at all energies.

Tab. 2 reports the band occupancy of the four-band model. At low electric fields only the first band contributes to electron transport in SiO_2. An occupation of the second band is detected for fields being larger than 2 MV/cm, whereas the third and fourth band are not occupied significantly. The main contribution to the electronic transport behavior in SiO_2 is affected by an appropriate selection of the first conduction-band.

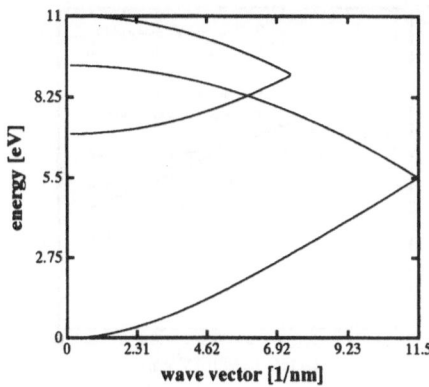 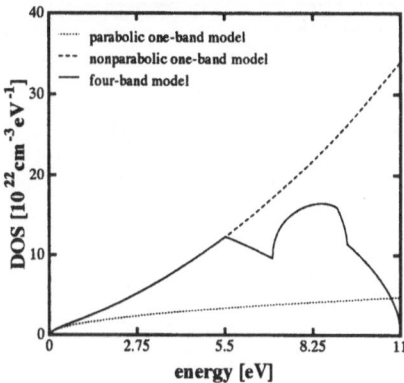

Fig. 1: Band-structure of the four-band model.

Fig. 2: DOS of the different band-structures.

$E \; [MV/cm]$	band 1 [%]	band 2 [%]	band 3 [%]	band 4 [%]
1	100.000	0.000	0.000	0.000
2	100.000	0.000	0.000	0.000
3	99.995	0.005	0.000	0.000
4	99.982	0.108	0.000	0.000
5	99.219	0.780	0.001	0.000
6	97.233	2.759	0.006	0.002
7	92.777	6.200	0.018	0.005
8	89.133	10.810	0.040	0.017
9	84.188	15.716	0.070	0.026
10	79.132	20.717	0.106	0.045

Tab. 2: Band occupancy of the four-band model.

Fig. 3 presents the total scattering rate for a temperature of 300 K and compares it with a single conduction-band. Nonpolar acoustic phonons set in at about 2.75 eV as the dominant scattering process (U-process). Compared with one-band models the different character of the four-band model again results in two maxima and one minimum. The discontinuity at the peak reflects the intravalley character of U-processes.

The dependence of the drift velocity of electronic carriers in SiO_2 versus electric field is plotted in Fig. 4. It increases till U-processes occur. A parabolic one-band model tends to lower velocities, whereas nonparabolicity increases the velocity. The split-off between the nonparabolic band-model and the four-band model is caused by a non-negligible occupancy of the second band at high electric fields. Our data are compared with the results of Fischetti [7]. In Fig. 5 the corresponding mobility in silicon dioxide is presented.

The average energy is plotted in Fig. 6. We observe that a single nonparabolic conduction-band and the four-band model do not exhibit any deviation, moreover, they almost demonstrate quantitative identical values. For data reported in literature [6],[7],[11] three different techniques have been employed to extract the energy as a function of the applied electric field, namely the carrier-separation technique, the electroluminiscence method and finally the vacuum-emission technique. We find a remarkable agreement with experimental data.

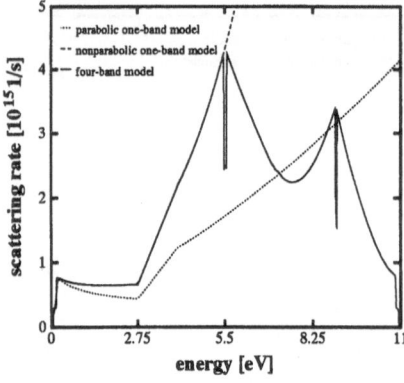

Fig. 3: Scattering rate of SiO$_2$.

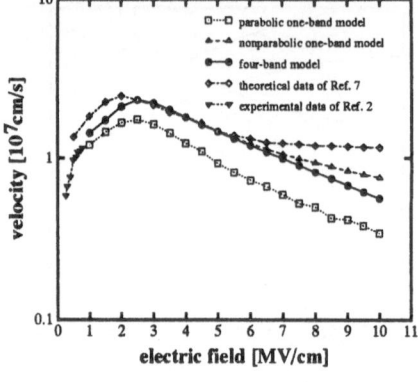

Fig. 4: Drift velocity in SiO$_2$.

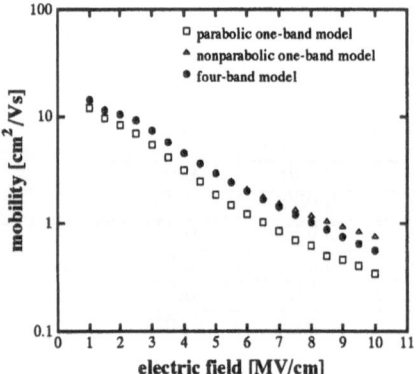

Fig. 5: Mobility in SiO$_2$.

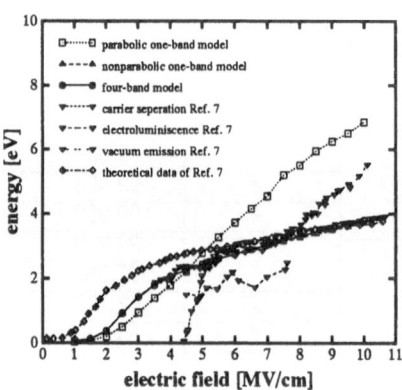

Fig. 6: Average energy in SiO$_2$.

Acknowledgment

The financial support of SIEMENS Munich, FRG is gratefully acknowledged.

References

[1] W. Porod and D.K. Ferry. *Physical Review Letters*, 54(11):1189–1191, 1985.

[2] H.-J. Fitting and J.-U. Friemann. *Physica status solidi (a)*, 69:349–358, 1985.

[3] H. Köstner Jr. and K. Hübner. *Physica status solidi (b)*, 118:293–301, 1983.

[4] B.K. Ridley. *Journal of Applied Physics*, 46(3):998–1007, 1975.

[5] M. Sparks, D.L. Mills, A.A. Maradudin, L.J. Sham, E. Loh Jr., and D.F. King. *Physical Review B*, 24(6):3519–3536, 1981.

[6] M.V. Fischetti. *Physical Review Letters*, 53(18):1755–1758, 1984.

[7] M.V. Fischetti, D.J. DiMaria, S.D. Brorson, T.N. Theis, and J.R. Kirtley. *Physical Review B*, 31(11):8124–8142, 1985.

[8] C. Jacoboni and L. Reggiani. *Review Modern Physics*, 55(3):645–705, 1983.

[9] R. Brunetti, C. Jacoboni, F. Venturi, E. Sangiorgi, and B. Ricco. *Solid-State Electronics*, 32(12):1663–1667, 1989.

[10] J.R. Chelikowsky and M. Schlüter. *Physical Review B*, 15(8):4020–4029, 1977.

[11] M.V. Fischetti, S.E. Laux, and D.J. DiMaria. *Applied Surface Science*, 39:578–596, 1989.

SIMULATION OF SEMICONDUCTOR DEVICES AND PROCESSES Vol. 5
Edited by S. Selberherr, H. Stippel, E. Strasser – September 1993

Efficient and Accurate Simulation of EEPROM Write Time and its Degradation Using MINIMOS

A. v. Schwerin, W. Bergner, and H. Jacobs

Corporate Research and Development, Siemens AG
Otto-Hahn-Ring 6, D-81739 München, GERMANY

Abstract

This paper demonstrates an efficient approach to flash-EEPROM write time simulation. MINIMOS is used to simulate the operation of the MOSFET equivalent device consisting of the Si substrate and the EEPROM's floating gate. The effective floating gate voltage is determined analytically using the coupling capacitances and the floating gate charge. With this method write time simulation is performed. Making use of selfconsistent trapping calculation, available in our version of MINIMOS, EEPROM write time degradation due to electron trapping in the gate oxide layer is estimated.

1. Introduction

Stacked double-poly flash EEPROM cells are widely used as devices for nonvolatile memories. Simulation of EEPROM tunneling erase has been performed by replacing the full device structure by a capacitor equivalent circuit [1], [2]. Full transient 2-D device simulation of flash EEPROM write and erase has been reported recently [3], [4]. In the latter case the full device structure is included in the device simulation, making use of charge boundary conditions for the floating gate. As electron injection is assumed to be constant within each time step, small time steps have to be chosen to achieve sufficient accuracy.

In this work we propose a hybrid approach using an equivalent capacitor model and results from the numerical simulation of the MOSFET structure formed by the floating gate and the silicon substrate. We use a modified version of MINIMOS [5] which includes elaborate models for non-local carrier injection and transport in the oxide. In spite of the substantially improved accuracy of gate current calculation, this approach proves to be computationally efficient. Moreover, our models in MINIMOS for oxide carrier trapping and trap generation allow for the selfconsistent calculation of the gate current decrease and the related write time degradation due to trapping of the injected electrons.

2. Models

To model an EEPROM device with MINIMOS (cf. Fig. 1) the equivalent floating gate (FG) voltage V_{FG} has to be determined. We consider the capacitor equivalent

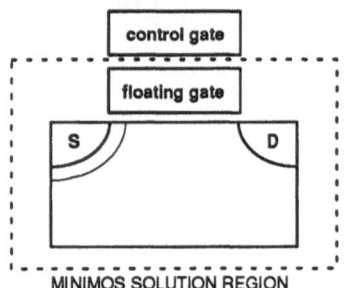

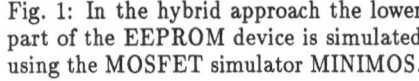

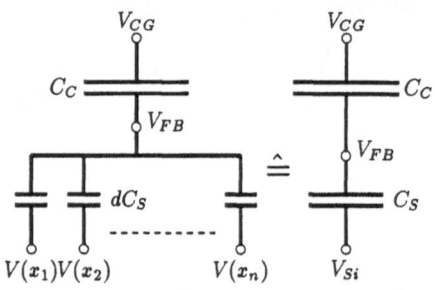

Fig. 1: In the hybrid approach the lower part of the EEPROM device is simulated using the MOSFET simulator MINIMOS.

Fig. 2: Capacitor equivalent circuit for EEPROM. C_C is the control gate capacitance, $dC_s = C_{ox} W \, dx$ represent the capacitance of an infinitesimal part of the channel.

circuit shown in Fig. 2. It can be shown that an effective voltage V_{Si} can be derived, yielding

$$V_{Si} = \frac{1}{L} \int_0^L \psi_s(x) \, dx + \phi_{ms} \tag{1}$$

representing the average of the surface potential ψ_s along the channel with ϕ_{ms} being the work function of the floating gate material. $V_{Si}(V_{FG})$ (for fixed drain voltage V_D) is extracted from MINIMOS simulations. Fig. 3 shows the result for V_{Si} for a $1\,\mu m$ device with $10\,nm$ oxide thickness at $V_D = 5\,V$. The relation between the FG-charge Q_{FG} and the terminal voltages is found by applying the charge conservation condition. This yields for fixed drain voltage:

$$- Q_{FG} = C_C(V_{CG} - V_{FG}) + C_S(V_{Si}(V_{FG}) - V_{FG}), \tag{2}$$

from which V_{FG} can be determined numerically. C_C is the capacitance between control and floating gate, V_{CG} the control gate (CG) voltage and C_S the capacitance between FG and silicon. During programming of the EEPROM the charge on the floating gate changes due to the gate current I_G:

$$\frac{dQ_{FG}}{dt} = I_G[V_{FG}(Q_{FG})]. \tag{3}$$

The solution of (3), using (2), yields the write time:

$$T_w = - \int_{V_{FG}^{ini}}^{V_{FG}^{end}} \frac{C_C + C_S(1 - \frac{dV_{Si}}{dV_{FG}})}{I_G(V_{FG})} \, dV_{FG}. \tag{4}$$

From this formula it is obvious that for write time calculation it suffices to know the gate current I_G and V_{Si} as a function of V_{FG} (which are taken from MINIMOS calculations). The integral in (4) is evaluated numerically. V_{FG}^{ini} and V_{FG}^{end} are calculated using (2) from $Q_{FG}^{ini} = 0$ and Q_{FG}^{end}, the latter being chosen from the CG threshold voltage target in the programmed state.

The calculation of the gate current I_G is done using a non-local injection formula [5]. The transport of the injected carriers through the gate oxide is described by a modified drift-diffusion equation. Its solution on the 2-D oxide region provides the current density j_{ox} at every location inside the oxide. Integration of j_{ox} along the gate contact results in I_G. Moreover, j_{ox} can be used together with trap rate equations to model gate oxide degradation due to electron trapping.

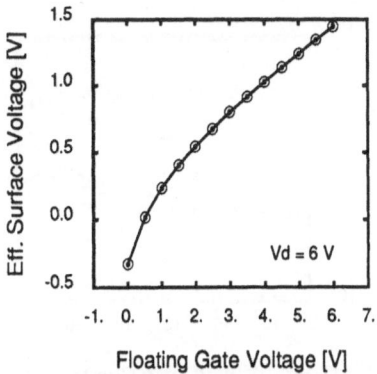

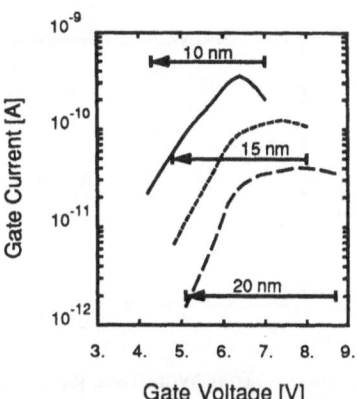

Fig. 3: Effective surface voltage V_{Si} as a function of the floating gate voltage applied in the MINIMOS calculations.

Fig. 4: Gate current as a function of gate voltage for different gate oxide thicknesses at $V_D = 6\,V$. The arrows indicate the range V_{FG}^{ini} to V_{FG}^{end} which correspond to $Q_{FG}^{ini} = 0$ and to the charge after EEPROM programming, Q_{FG}^{end}, respectively.

3. Application

In a first application example the model is used to quantify the expected improvement of flash EEPROM programming when the oxide thickness t_{ox} is reduced from 20 nm to 15 nm and to 10 nm. The channel doping was adjusted for each oxide thickness to guarantee the same control gate threshold voltage. The parameters of the injection model were taken from [5]. Fig. 4 shows the simulated gate current and the respective range of I_G-integration. The resulting write times for $V_D = 6\,V$ and $V_{CG} = 12\,V$ are 1460, 464, and 160 μs for t_{ox}=20, 15, and 10 nm, respectively.

In a second application the T_w-degradation due to electron trapping in the oxide between the Si-substrate and the floating gate is investigated. Due to electron trapping during the EEPROM write process the electric field in the Si-substrate and thus injection is changed; moreover the local electric field in the oxide layer and thus carrier transport through the oxide is modified. These effects cause a decrease of the EEPROM write current. Our model comprises the following trap mechanisms [6]:

- capture of electrons in bulk traps and oxide field dependent trap to band ionization:

$$\frac{dN^-}{dt} = j_{ox}\sigma \cdot (N_{tot} - N^-) - j_{ox}\beta(E_{ox}) \cdot N^- \qquad (5)$$

- oxide field dependent generation of electron traps:

$$\frac{dN_{tot}}{dt} = j_{ox}\gamma(E_{ox}) \qquad (6)$$

with N^- being the density of filled traps and N_{tot} the total trap density, σ the capture cross section, β the oxide field dependent trap-to-band ionization efficiency, and γ

the trap generation efficiency. The initial value for N_{tot} is the technology dependent intrinsic trap density. The system of differential equations (5, 6) is solved at every location inside the oxide. The simulation is done selfconsistently, i.e. the change in j_{ox} due to the build-up of oxide charge is taken into account by iterating the calculation of electron injection and electron trapping using appropriate time step control.

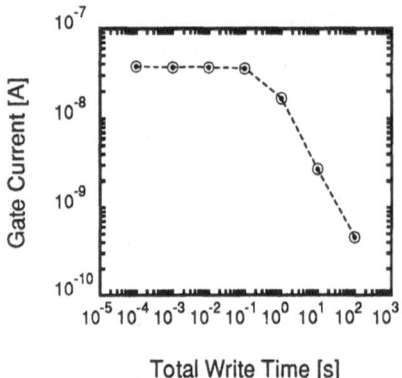

Total Write Time [s]   Write/Erase Cycles

Fig. 5: Degradation of gate current due to electron trapping in the gate oxide as a function of total write time, calculated selfconsistently with MINIMOS.

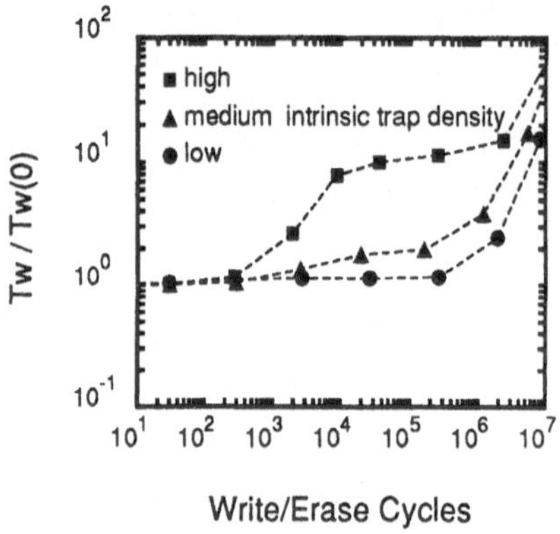

Fig. 6: Relative change of write time T_w as a function of the number of write/erase cycles for different intrinsic trap densities (MINIMOS calculation).

Fig. 5 shows the gate current as a function of accumulated write time. In Fig. 6 the change of EEPROM write time as a function of the number of write/erase cycles is demonstrated for three different values of the intrinsic trap density $N_{tot}(t = 0)$. The increase of T_w after $\approx 10^3$ write/erase cycles is due to electron capture in intrinsic traps, whereas after $\approx 2 \cdot 10^6$ cycles trap generation through hot electron injection becomes important.

4. Conclusion

This work shows that EEPROM write time simulation is efficiently done using a MOSFET simulator and a capacitor equivalent circuit. The gain in computational efficiency can be spent on more sophisticated gate current and carrier trapping models, which allow the investigation of EEPROM write time degradation.

Acknowledgement
This work was partially funded by the project ADEQUAT (JESSI Project BT1B, ESPRIT Project 7236).

References

[1] A. Bhattacharyya, *Solid-State Electron.* 27, p. 899, (1984).

[2] J. Suné et al., *IEDM Tech. Dig.*, 1991, p. 905.

[3] C. Fiegna et al., *IEEE Trans. Electr. Dev.*, ED-38, p. 603, (1991).

[4] S. Keeney et al., *IEEE Trans. Electr. Dev.*, ED-39, p. 2750, (1992).

[5] A. v. Schwerin et al., *IEDM Tech. Dig.* 1992, p. 543.

[6] Y. Nissan-Cohen et al., *J. Appl. Phys.*, 60, p. 2024, (1986).

Influence of Oxide-Damage on Degradation-Effects in Bipolar-Transistors

W. Bergner, B. Seidl, H. Wurzer, R. Mahnkopf, and H. Klose

Corporate Research and Development, Siemens AG
Otto-Hahn-Ring 6, D-81739 München, GERMANY

Abstract

To optimize sub-micron bipolar devices with respect to speed and stability, the precize physical mechanisms have to be understood and modeled, which describe the emitter-base breakdown itself, the charge injection into the emitter-base cap oxide and the resulting change of both forward and reverse charactersitics. The purpose of this paper is to analyze the physical effects involved, to model them and to show by comparison with measured data the accuracy of the calculated results.

1. Introduction

The device investigated is a bipolar transistor realized by a high performance 0.8 micron BICMOS process [1] (see Fig. 1 for a cross section). The 2-D doping profiles for this device were determined using the MIMAS simulation program [2] and verified by SIMS measurements in the vertical direction for both the active and inactive transistor regions. By using the device simulation program GALENE II [3], very good agreement between measured and calculated forward biased electrical characteristics was found, what indicates correct emitter/base doping profiles.

2. Stress process

A reverse bias applied between emitter and base causes leakage current due to the generation of carriers by band-to-band tunneling and impact ionization. Utilizing the device simulation program GALENE II this effect can be calculated selfconsistently [4]. A portion of the generated carriers is injected into the spacer oxide and damages the transistor. The model which describes this, weights the distance of the carriers to the oxide and their energy via the height of the oxide barrier and the thermal scattering length respectively, it has proven to be valid for MOS devices [5]:

$$j_{inj}(r_0) = \frac{A}{2\pi\lambda^2} \int_{Si} d^2r \, j(r) e^{-\frac{|r-r_0|}{\lambda}} \int_{\epsilon_b}^{\infty} f_{loc}(\epsilon, r) d\epsilon \qquad (1)$$

A is a normalization constant, λ the inelastic mean free path and $j(r)$ the current density in the silicon. ϵ_b measures the energy prependicular to the interface and

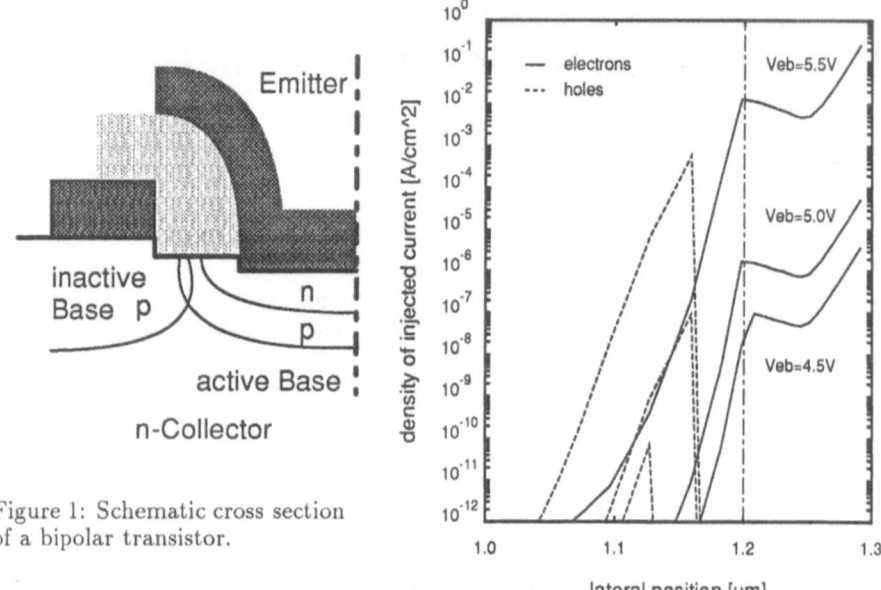

Figure 1: Schematic cross section
of a bipolar transistor.

Figure 2: Calculated distribution of injected electrons and holes along the
spacer for different stress voltages V_{eb}. The vertical line marks the base-
emitter junction.

relative to the oxide barrier. The distribution function f_{loc} is calculated as a first
approximation, from the local electric field. Figure 2 presents a calculation of this
electron and hole current density passing the silicon-silicondioxide interface. Due to
the field distribution under the spacer positive charge may be expected on the base
and negative charge on the emitter side of the p-n junction.

Next to that we extract the distribution of the damage created in the spacer and at
the spacer interface. The damage is then characterized by three quantities:

- The surface density of fixed charges q_{ss} represents the effective trapped charge
  in the spacer.

- The surface recombination velocity s_n and s_p results mainly from the density
  and capture cross section of midgap states at the interface.

- Interface traps created in the lower and upper part of the bandgap are modeled
  by a density increasing linearly between zero at midgap and d_d (d_a) at the
  valence (conduction) band edge. The charge trapped in these states is controlled
  by the level of the Fermi energy [6].

The absolute value of these quantities can be determined by comparing simulated
results with measurements [7], as shown in the following.

3. Degradation of reverse characteristics

We attribute the pronounced dependence of the change in reverse current on the
emitter-base voltage after stress to a modified field distribution due to injected charges,

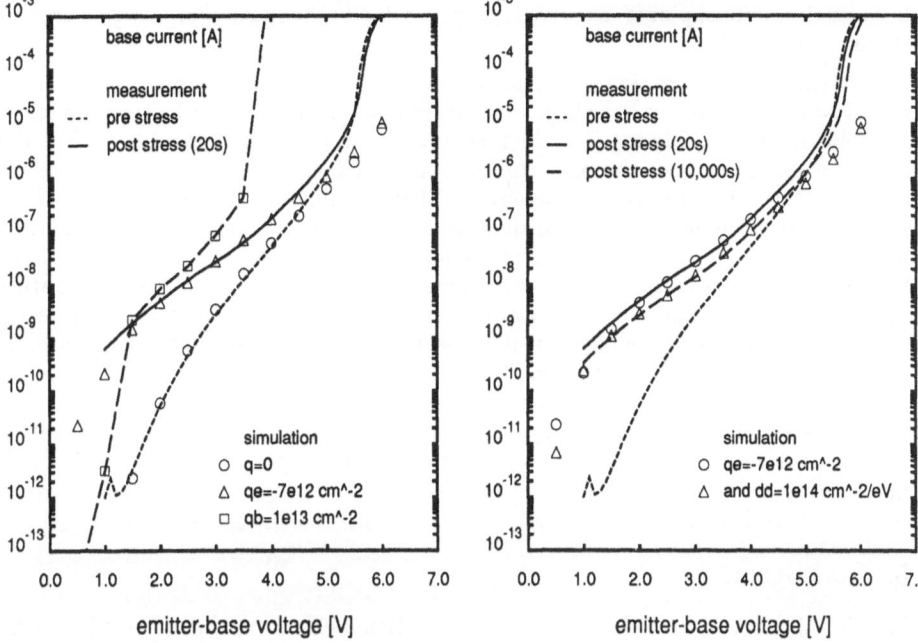

Figure 3: Comparison of measured and calculated reverse characteristics of a stressed device ($V_{eb} = 5.75V$, $t_{stress} = 30s$). Simulations carried out for no charge, a negative charge density at the emitter q_e and positive charge density at the base q_b.

Figure 4: Comparison of measured and calculated reverse characteristics of a stressed device after $t_{stress} = 10^4$s ($V_{eb} = 5.75V$). Simulations include negative charge at the emitter q_e and donator-like interface states with a density d_d in the bandgap.

which in turn influences tunneling and avalanche currents very sensitively. Thus, we investigated this effect on the bipolar transistor. Figure 3 shows simulations for different charge configurations. The assumption of a positive charge on the base side (q_b) enhances the leakage current both in the tunneling and the avalanche branch and therefore it is inadequate for a proper description of the reverse current degradation, which shows only a small shift in the avalanche branch. In contrast to this, a negative charge located at the emitter (q_e) increases only the tunneling generation. The cause is that with increasing reverse bias the space charge exceeds the amount of injected charge and reduces its effect on the breakdown characteristics to negligible values.

It has to be noted that the increase of the leakage current is a quite fast mechanism, the tunneling current reaches a maximum after about 30s stress ($V_{eb} = 5.75V$). Further stress will reduce the tunneling by a factor of 2 within 10^4s. We attribute this to the generation of donator-like interface states at the emitter side of the junction, as reported in literature [8]. Figure 4 shows the influence of positive charged states on the leackage current, the oxide charge is compensated partially.

4. Degradation of forward characteristics

Surface states generated at the interface by injected carriers have an impact not only on the charge distribution but also on the carrier life time (increased trap density) at the interface. Both effects can be observed by analyzing the forward characteristics, because the gummel plot is sensitive to traps located in the space charge region around the p-n junction.

To describe the degradation of the forward characteristics two different kinds of recombination mechanisms have been investigated: thermal and band-to-trap tunneling recombination. Considering just the first mechanism it was not possible to describe the nonideality factor of the base current below 0.6V emitter-base voltage correctly (open circles in Fig. 5). It is quite obvious that the simulation of the forward characteristics is not independent from the reverse condition. Therefore, we have to take into account the charge distribution at the interface, which was shown to be responsible for the degradation of the reverse current. This leads us to the curve marked with the open triangles. The remaining deviation of experiment and simulation can be explained by including the enhancement of recombination by tunneling (open squares). The occupation of traps in the bandgap by tunneling causes an increased capture cross section of the traps. This can be seen at low emitter-base forward voltage as long as tunneling between the trap level and the band edge is allowed. The introduction of this effect in the simulation gives good agreement between calculation and measured data.

5. Conclusions

Within this paper it was shown that the degradation effects on the forward base current due to emitter-base reverse biasing can be explained by band-to-trap tunneling. Degradation of the breakdown characteristics after stress on the other hand results from the increasing band-to-band tunneling generation due to negative charge on the emitter side of the base-emitter junction. Thus the understanding of tunneling mechanisms is therefore neccessary to explain the generation of hot carriers and their influence on both the forward and reverse base current characteristics.

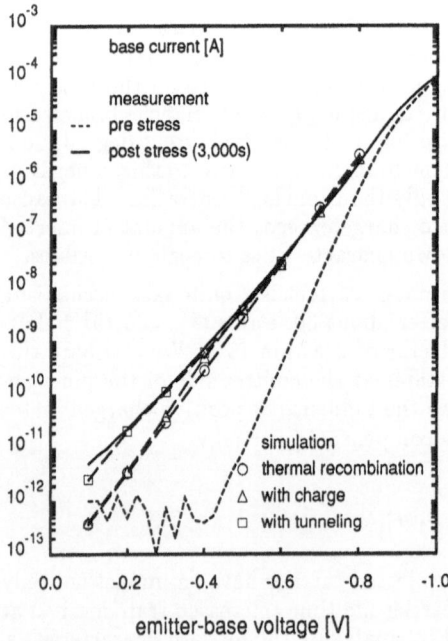

Figure 5: Comparison of measured and calculated forward characteristics of a stressed device ($V_{eb} = 5.75V$).

References

[1] H. Klose et al., Proc. SSDM 1990 (1990) 653.

[2] K. Steger and M. Paffrath, Trans. IEEE CDA, to appear.

[3] GALENE II User's Guide, RWTH Aachen, Germany.

[4] W. Bergner et al., Proc. ESSDERC 1992 (1992) 695.

[5] A.v. Schwerin, et al., IEDM Tech. Dig., 1992.

[6] A.v. Schwerin, et al., Trans. IEEE, ED-34, (1987) 2493.

[7] H. Wurzer, master's thesis, TU Munich, Germany (1992).

[8] D.A. Buchanan and D.J. DiMaria, J. Appl. Phys., 67, (1990) 7439.

SIMULATION OF SEMICONDUCTOR DEVICES AND PROCESSES Vol. 5   77
Edited by S. Selberherr, H. Stippel, E. Strasser – September 1993

The Application of Sparse Supernodal Factorization Algorithms for Structurally Symmetric Linear Systems in Semiconductor Device Simulation

A. Liegmann and W. Fichtner

Integrated Systems Laboratory, ETH-Zürich
Gloriastraße 35, CH-8092 Zürich, SWITZERLAND

1. Introduction

It is well known that the solution of sparse linear systems, generally expressed in the form $Ax = b$, is a core task of numerical simulation. In case of semiconductor device simulation the coefficient matrix A is unsymmetric, but structurally symmetric ([2]). The solution of linear systems can be achieved by iterative or direct methods. While iterative methods do not always lead to a solution due to matrix conditions, direct methods usually consume more time and memory.

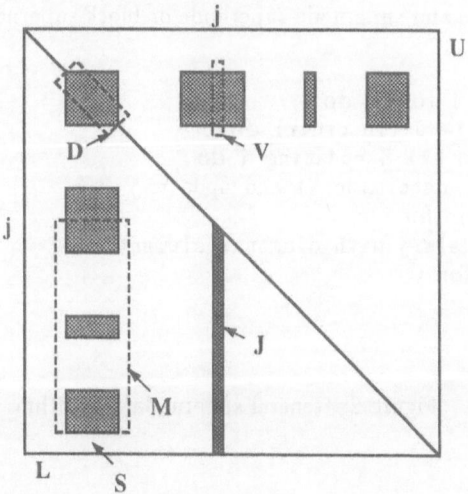

Figure 1: Updating column J by supernode S

2. Computational considerations

The problem of reducing the memory needed for a direct solver is closely related to the problem of minimizing the size of the LU decomposition as a result of reordering the coefficient matrix. In this respect the Minimum Degree reordering algorithm proved to be very successful for general sparse systems. This algorithm has been enhanced in terms of execution speed to what is referred to as the **Multiple Minimum Degree** algorithm ([5]).

On the other hand, speeding up a direct solver basically means a faster computation of the LU factorization. Extensive research in this area has lead to so-called *supernodal* techniques. The key concept of these techniques is what is nowadays referred to as a **supernode** [1]. During symbolic factorization, supernodes are identified as a set of consecutive columns in the factor L of the LU decomposition with the following structural properties. A supernode formed by, s say, adjacent columns consists of two blocks: a dense diagonal block of size $s \times s$ and a block of width s below the diagonal block where all columns share the same sparsity pattern. A sample supernode is depicted in Figure 1 denoted with the letter S. Computing column J involves the following steps: for all supernodes S updating column J determine vector $V1 = M*(D \cdot V)$ and then subtract it from the contents of J, i.e. $J = J - V1$ (see Fig. 1). The determination of vector V1 involves dense matrix-vector multiplications which run at vector speed on today's supercomputers. The subtraction step requires gather/scatter operations which are mostly hardware supported on many supercomputers. As a result, supernodal techniques make excellent use of the hardware features and thus are highly powerful.

The computational power of supernodal techniques applied to symmetric positive definite linear systems has been documented in papers like [1]. Since in semiconductor device simulation the linear systems are usually unsymmetric but structurally symmetric, we could apply supernodal techniques to both factors L (columns) and U (rows) simultaneously ([3]). Further enhancements have been added and more recently we implemented a whole collection of supernodal factorization algorithms which divide into supernode-node and supernode-supernode or block supernode methods ([4]).

```
for  J = 1 to  Ns do                              (1)
    for  j ∈ J (in order) do                      (2)
        for all K updating j do                   (3)
            determine V1 and update j
        end for
        scale j with diagonal element             (4)
    end for
end for
```

Figure 2: General supernodal algorithm

3. Supernode-node algorithms

Supernode-node updating describes a technique where only one column/row of the factors L and U is computed at a time, although the corresponding supernode might consist of several columns/rows. Figure 2 depicts the general framework for algorithms implementing this technique. A first glance at the algorithm already reveals the general form of supernode-node updating algorithms: a triple-nested for-loop (indicated with indices 1 through 3). The outermost loop runs over all supernodes J that were generated in the reordering and symbolic factorization steps. The next for-loop (2) goes one level deeper and scans over all nodes j of the current supernode J starting with the smallest index. The innermost loop (3) handles the contribution of all updating supernodes K to the current node j. Finally, column/row j has to be scaled by its diagonal element (4).

4. Block supernode algorithms

Block supernode factorization operates on groups of columns/rows or even a whole supernode at the same time instead of merely focusing on a single column/row. Doing so does not reduce the number of references to memory by any means, but by grouping them together memory fetch and store can be made more efficiently, i.e. using the same index map only once throughout a loop cycle ([6]).

On the other hand, supernode-supernode factorization increases storage overhead significantly, since the intermediate results for more than one column/row have to be kept and other data structures had to be added to support this technique. In our tests we have seen memory increase between at worst 6 to 20 times over our supernode-node implementations. Furthermore, the time necessary to do the set up and administration of these data structures cannot be neglected.

5. Benchmark

We present the timing results for a medium sized linear system stemming from a simulation of a MOSFET. The linear system has 12,000 unknowns and about 250,000 non-zero entries in the coefficient matrix. The benchmark was run on a Convex C220, a Cray-2, a Cray Y-MP, a NEC SX-3, and a Cray C98. The numbers shown in Table 1 are those of the best performing factorization algorithm in CPU seconds. For none of

	supernode-node	block supernode
Convex C220	12.93	18.93
Cray-2	3.38	3.48
Cray Y-MP	1.39	1.39
NEC SX-3	1.23	1.38
Cray C98	.73	.74

Table 1: Timing results for the best performing algorithm (seconds)

the machines used in the benchmark we found block supernode algorithms to perform better than the supernode-node algorithms. Mainly, there are two reasons for that:

- For all of the block supernode algorithms implemented we noticed a significant increase of scalar memory references. This increase is stemming from the additional data structure handling built into the block supernode algorithms. Obviously, this hurts especially on machines with scalar data caches like the Convex and the NEC. Here, block supernode methods loose performance by suffering from scalar data cache misses.

- Block supernode techniques are most effective when the supernodes contain many columns/rows, i.e. when the supernode partitioning consists of a small number of supernodes. A small supernode partitioning provides for bigger blocks during supernode update. In our test cases supernodes contain 5 to 6 columns/rows on average. Additionally, our factor columns/rows are very sparse (about 200 non-zero entries maximum) so that there are only a few cases during the factorization where we can exploit the potential of the block supernode algorithms.

6. Conclusion

In this paper we presented supernodal techniques suitable for structurally symmetric linear systems as they appear in semiconductor device simulation. Among these, supernode-node updating schemes perform best for this type of application. We have shown that block supernode methods cannot be exploited to their full potential which is due to the extreme sparsity of the linear systems and small supernode sizes.

7. Acknowledgement

The authors highly appreciate the support of Cray Research (Switzerland) for providing the original source code. Also, we thank the staff of the computer centers of the Swiss Institutes of Technology in Zurich and Lausanne as well as the Swiss Scientific Computing Center in Manno for providing us access to their supercomputers. Additionally, we are grateful to J.F. Bürgler and S. Müller (both from the Integrated System Laboratory) for providing the set of test cases. Our special thanks go to C. Pommerell (now AT&T Bell Labs) for a number of interesting discussions on the hallway of the laboratory, and to B.W. Peyton (ORNL) for his help on understanding the original code.

References

[1] C.C. Ashcraft, R.R. Grimes, J.G. Lewis, B.W. Peyton, and H.D. Simon. Progress in sparse matrix methods for large linear systems on vector supercomputers. *The International Journal of Supercomputer Applications*, 1(4):10–30, 1987.

[2] G. Heiser, C. Pommerell, J.Weis, and W. Fichtner. Three dimensional numerical semiconductor device simulation: Algorithms, architectures, results. *IEEE Transactions on Computer-Aided Design of Integrated Circuits*, 10(10):1218–1230, 1991.

[3] A. Liegmann. The application of supernodal techniques on the solution of structurally symmetric systems. Technical Report 92/5, Institut für Integrierte Systeme (ETH Zürich), 1992.

[4] A. Liegmann and W. Fichtner. The application of supernodal factorization algorithms for structurally symmetric linear systems in semiconductor device simulation. Technical Report 92/17, Institut für Integrierte Systeme (ETH Zürich), 1993. Submitted to *The International Journal of Supercomputer Applications*.

[5] J.W.H. Liu. Modification of the Minimum-Degree algorithm by multiple elimination. *ACM Transactions on Mathematical Software*, 11(2):141–153, 1985.

[6] E.G. Ng and B.W. Peyton. Block sparse Cholesky algorithms on advanced uniprocessor computers. Technical Report TM-11960, Oak Ridge National Laboratory, 1991.

Newton-GMRES Method for Coupled Nonlinear Systems Arising in Semiconductor Device Simulation

C. Simon[‡†], M. Sadkane[†], and S. Mottet[‡]

[‡]CNET Lannion B, France Telecom
Route de Trègastel, BP 40, F-22301 Lannion Cédex, FRANCE
[†]IRISA
Campus de Beaulieu, F-35042 Rennes Cédex, FRANCE

Abstract

We are interested in computing the solution of a system of coupled nonlinear PDE's which describes the electrical behaviour of semiconductor devices. This set of nonlinear equations is solved via a nonlinear version of the GMRES method [2]. This method consists in solving the linear system, that arises in Newton's method, by an iterative scheme, which constructs an orthonormal basis of a Krylov subspace, and minimizes the residual, over the current Krylov subspace. An advantage of this method over the classical ones is that the Jacobian is not stored and that little storage is required since the method restarts periodically whenever the size of the Krylov subspace reaches a maximum value fixed by the user.

1. Introduction

The classical steady state isothermal drift-diffusion conduction model in semiconductors is basically described by a set of three nonlinear PDE's (Poisson's equation, electron and hole continuity equations) :

$$div(\epsilon\nabla(\varphi)) = \rho \; ; \quad -\frac{1}{q}div(J_n) = -U \; ; \quad \frac{1}{q}div(J_p) = -U \qquad (1)$$

The symbols have their usual meaning of the semiconductor device theory. ϵ is the dielectric constant, φ the electrostatic potential, $U \equiv U(\varphi, \varphi_n, \varphi_p)$ the net recombination-generation rate where φ_n and φ_p are the electrochemical potentials of electron and hole. The expression of the charge density $\rho \equiv \rho(\varphi, \varphi_n, \varphi_p)$ is : $\rho = n - p - dop$, where dop is the residual doping level, $n \equiv n(\varphi, \varphi_n)$ and $p \equiv p(\varphi, \varphi_p)$ are the electron and hole free carrier densities. The drift-diffusion current densities are : $J_n = -q.n.\mu_n.\nabla(\varphi_n)$ and $J_p = -q.p.\mu_p.\nabla(\varphi_p)$ where $\mu_n \equiv \mu_n(\varphi, \varphi_n)$ and $\mu_p \equiv \mu_p(\varphi, \varphi_p)$ are the electron and hole mobilities.

The numerical solution of system (1) is carried out by discretizing the equations on a mesh. We have implemented a Flux Conservative Box Method (FCBM) scheme [6], which is a variant of the so called "Box Method". This transforms the system of PDE's (1) into a nonlinear algebraic system. We consider that the potentials φ,

φ_n and φ_p are the fundamental entities which characterize the device behaviour. So, this leads to a natural choice of unknowns $\varphi, \varphi_n, \varphi_p$, and the set of equations can be written as :

$$\begin{cases} \nabla. \left(\epsilon \nabla \varphi\right) - \rho = 0 \\ \nabla. \left(n \mu_n \nabla \varphi_n\right) + U = 0 \\ \nabla. \left(p \mu_p \nabla \varphi_p\right) - U = 0 \end{cases} \tag{2}$$

with appropriate boundary conditions. The domain of definition Ω is a meshed bounded domain belonging to $\mathbf{R}^l$ (l=1,2,3). The boundary of the domain, $\Gamma = \partial\Omega$, is divided into classes, each of them corresponding to a given boundary condition type (ohmic contact, Schottky contact, insulating boundary...).

For the sake of conciseness, the expressions of the coefficients in (2) are not reported. However, we can point out that, depending on the expressions used to describe the mobility law, the recombination-generation term (Shockley-Hall-Read, Auger, spontaneous band to band, impact ionization ...) and the statistics (Maxwell-Boltzmann or Fermi-Dirac) of the free carrier densities, this general model is adequate for a wide range of devices including silicon and III-V optoelectronic hetero-junction structures.

2. Newton-GMRES algorithm

We are interested in solving the nonlinear system (2) using a coupled method where the three unknowns φ, φ_n and φ_p are sought simultaneously. The system (2) can be written as a nonlinear mapping :

$$F(\Phi) = 0 \quad \text{where} \quad \Phi = (\varphi, \varphi_n, \varphi_p)^t \in \mathbf{R}^{3N}. \tag{3}$$

where N stands for the number of nodes of the mesh. Let us denote by $J(\Phi)$ the Jacobian matrix $\nabla F(\Phi)$ computed at vector Φ. Recall that Newton's method for solving the nonlinear equation (3) can be described as follows :
1. Choose an initial guess Φ_1 ;
2. **for** $k = 1, 2, \cdots$ **do**
    (i.) Solve the linear system : $J(\Phi_k)\delta_k = -F(\Phi_k)$;
    (ii.) Correct the approximate solution $\Phi_{k+1} = \Phi_k + \delta_k$;
end for

The Newton-GMRES method consists in solving the linear system $J(\Phi_k)\delta_k = -F(\Phi_k)$ in step (2.i.) using GMRES method [1, 2, 3], which builds iteratively an orthonormal basis of a Krylov subspace, whose dimension m is very small compared to the size of the matrix J, and minimizes the residual over it. Unlike the methods based on Gaussian elimination, this method only requires to compute, at each iteration k, the action of the Jacobian J on a vector. This is accomplished without explicit storage of the Jacobian matrix, using the approximation :

$$J(\Phi_k)\Psi_k \approx \frac{F(\Phi_k + \sigma\Psi_k) - F(\Phi_k)}{\sigma} \tag{4}$$

where σ is a small quantity chosen according to [4]. Moreover, in order to improve the conditioning of the linear system (2.i.), a simple scaling by row is used.

3. Numerical results

We have focused the study on two points which may have an impact on the quality of the method. The first one is the size m of the Krylov subspace, the second one is the parameter σ which appears in approximation (4).

- Concerning the size m of the Krylov subspace, the basic method consists in choosing a value m and then restarting the GMRES loop periodically every m iterations. This strategy is, in general, not recommended because it does not take care of the excessive storage and its numerical instability effect on the algorithm. On the other hand, if m is small enough, the GMRES algorithm restarts more frequently, and this results in a high execution time. We are still working on this point. So far, we have implemented an adaptive method as suggested in [3] and which may be described in the following : let us denote by RESL the norm of the linear residual given by GMRES, and RESN the norm of the nonlinear residual $||F(\Phi_k)||$ where Φ_k stands for the current approximate solution. We fixe a maximum size m_{max} and start with a small value, say $m_{min} \leq 5$, the size m of the Krylov subspace is then updated at each restart according to the following conditions :

  if $(RESN \leq 1.5 * RESL)$ **then** $m := min(m + k, m_{max})$ **endif**
  if $(RESN \geq 5 * RESL)$ **then** $m := max(m - k, m_{min})$ **endif**

  where k is a small integer,$(k \leq 5)$, by which the size m is either increased or decreased. Moreover, within the GMRES loop, we use two different criterions for the periodical restart : either the number of iterations has reached the maximum size m_{max} of the krylov subspace or $RESL < \epsilon$ where ϵ is a precision parameter depending on the machine precision ϵ_{mach} and the norm of the first non linear residual. Although the results obtained with this strategy are rather good, an improvement we are trying to find is a more general autoadaptive method which would adapt automatically the size m.

- Concerning the parameter σ in approximation (4), we have compared the choice suggested in [2], that is $\sigma = \sqrt{\epsilon_{mach}} \times ||F(\Phi_k)||$ (method A), with the adaptive one (method B) which has been successfully used in the context of CFD [4]. Both of the methods use the adaptive choice for the size m of the Krylov subspace. The cost of computing σ is equivalent to the cost of two residual evaluations. Consequently, the cost is minor when compared to the overall computation. The method could be improved because it seems not necessary to re-evaluate σ at each restart but rather every 5 or 10 restarts.

- The simulations have been performed, using the coupled method implemented in the 3D version of the simulator CARMES [5]. Two devices have been simulated. The first one is a classical Silicon pn junction with a $30 \times 30 \times 30$ rectangular mesh. Figure 1 shows the comparison between the static and the adaptive choice of the size of the Krylov subspace m. These results have been obtained under 5 Volts reverse bias (the size of the Jacobian matrix J is $3N = 81000$). We can observe that the adaptive method needs less iterations. The second device is a Buried Heterostructure (BH) used in a Semiconductor Optical Amplifier (SOA). It is mainly composed of a GaInAsP (p doped) active layer inserted into an InP (n doped) buffer layer and an InP (p doped) protection layer. The simulations are performed with a $20 \times 25 \times 30$ rectangular mesh (the size of the Jacobian matrix is $3N = 45000$), under .95 Volt direct bias. Figure 2 shows the comparison between method A and method B. Here again, the latter one needs less iterations than the former one.

4. Conclusion

We have investigated a Newton-GMRES method for solving nonlinear equations coming from semiconductor device simulation. We have shown that the method gives reasonable results if some of the parameters are chosen adequately. There are two important points left to be done. The first one concerns a further study about the adaptive choice of the Krylov subspace, and the second one concerns the finding of a preconditioner of the Jacobian matrix which would take into account some physical considerations.

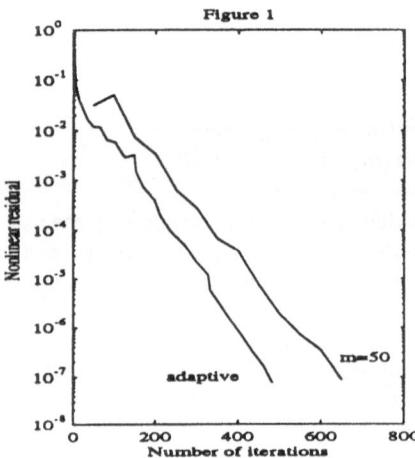
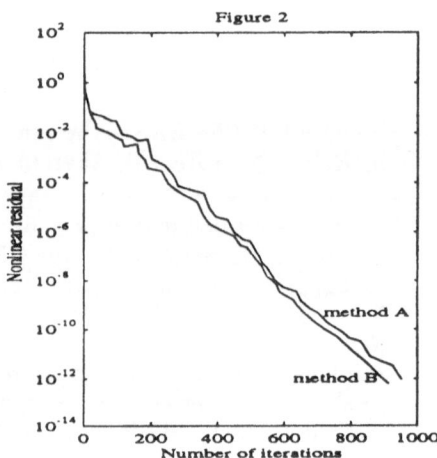

References

[1] Y. Saad and M. Schultz. GMRES: A Generalized Minimal Residual Algorithm for solving nonsymmetric linear systems. SIAM J. Sci. and Stat. Comp. Number 7. pp:856-869, 1986.

[2] P. Brown and Y. Saad. Hybrid Krylov methods for solving nonlinear systems of equations. SISSC. vol.11,no.3,pp 450-481,1990.

[3] T. Kerhoven and Y. Saad. On acceleration methods for coupled nonlinear elliptic. Numer. Math.60,525-548 (1992).

[4] Z. Johan. Data parallel finite element techniques for large scale computational fluid dynamics. PhD Thesis. Stanford University. July 1992.

[5] C. Simon, S. Mottet, J. E. Viallet. Autoadaptive Mesh Refinement. SISDEP91. Vol.4, pp225-233. W. Fichtner and D. Aemmer editors.

[6] C. Simon, S. Mottet. FCBM : Flux Conservative Box Method, a new discretization strategy of the semiconductor equations, in preparation.

Practical Use of a Hierarchical Linear Solver Concept for 3D MOS Device Simulation

O. Heinreichsberger, M. Thurner[†], and S. Selberherr

Institute for Microelectronics, TU Vienna
Gußhausstraße 27-29, A-1040 Wien, AUSTRIA
[†]Campus Based Engineering Center Vienna, Digital Equipment Corporation
Favoritenstraße 7, A-1040 Wien, AUSTRIA

Abstract

Three-dimensional device simulations of sidewall trench-isolated MOSFETs involve linear systems of equations that are extremely ill-conditioned if the geometric channel length exceeds $2\mu m$. Since conventional preconditioned iterative methods tend to fail in such cases we have implemented a robust iterative linear solver using a hierarchy of incomplete factorizations of the coefficient matrix.

1. Introduction

It is well-known that preconditioned iterative methods for the solution of linear systems are a key part of every three-dimensional semiconductor device simulator. Convergence speed, efficiency of implementation and robustness for very ill-conditioned matrices are preconditions for any linear system solver in a production environment. Recently the BiCGStab method as iterative routine [6] and the incomplete factorization method with threshold pivoting (ILUT) as preconditioner [4] have evolved as the most promising methods for solving even the worst conditioned linear systems in device simulation [1].
The condition numbers of matrices in 3D device simulators range from moderate values (e.g. the majority continuity equation in MOSFETs) to extremely ill-conditioned ones (e.g. the minority continuity equation for long channel MOSFETs). To cope with this variety of condition numbers, a hierarchical strategy (not to be confused with nested iterative methods) for combining different preconditioners has been proposed [1]. In this work we report our experience with a hierarchical linear system solver used in the three-dimensional part of MINIMOS 5.

2. Incomplete LU Factorizations

Incomplete LU factorizations of the nonsymmetric coefficient matrix are known to be the only viable preconditioning methods to solve the discrete continuity equations reliably. Other methods such as polynomial or hierarchical bases preconditioners are only applicable to relatively well-conditioned problems. There are various ways to factorize a sparse matrix incompletely: The standard method is to define some symbolic sparsity pattern for the incomplete factors and then computing matrix entries according to the pre-defined pattern. In the simplest case the original sparsity pattern of the matrix is used, denoted by ILU(0). For block diagonal matrices the sparsity pattern may be augmented easily by additional diagonals along the original sparsity pattern. If one additional diagonal is allocated along each existing diagonal we call this factorization ILU(1).

A more general class of ILU factorizations is obtained by keeping fill entries in the ILU factors only when they exceed some threshold, thus the sparsity pattern is defined during factorization. Clearly, for smaller values of the *drop threshold* parameter TOL a better ILU factorization with more fill-in can be created than for a larger TOL.

In this approach we precondition the BiCGStab method by two major *preconditioning levels* (Fig. 1):

- A vectorizable ILU preconditioner with positional dropping as the lowest hierarchy level,

- A parametric, non-vectorizable ILUT preconditioner in right position for higher levels.

In the first hierarchy we use either ILU(0) implemented in conjunction with the Eisenstat matrix-vector multiplication method to save about 30% of arithmetic work [2] or ILU(1). Both preconditioners are applied in split position.

The computation of the ILUT preconditioner for the second hierarchy level depends on two parameters: TOL which defines a drop tolerance weighted with the actual matrix rowsum for an individual matrix entry and FIL which determines the maximum number of matrix entries per row of an incomplete factor. TOL controls mainly the robustness of the preconditioner, FIL is necessary to limit memory usage, however, both parameters influence the robustness of the preconditioner.

To provide robustness for very ill-conditioned linear systems we use tolerances smaller than 10^{-5} and a fill-in of 20 beside the original sparsity pattern. This choice (in the two-dimensional parameter space) seems to yield an optimum in CPU time for our specific application (Fig. 2). An iteration limit of 400 for ILU(0) and 100 for ILU(1) is used as a switch between the hierarchies. For iteration termination we require the residual (left split preconditioned residual in case of ILU or true residual in case of ILUT) to be less than 10^{-8} to ensure convergence of the outer Gummel algorithm.

For ILUT factorization a row oriented compressed sparse matrix format is used. The factorization code is inherently sequential in nature and hence does not vectorize. Therefore the factorization must be implemented as efficient as possible. Efficiency can be exploited mainly by detecting unimportant matrix entries as early as possible and by an efficient sort algorithm to sort out the FIL largest elements in a row.

The package is written in FORTRAN 77 (iterative solver and preconditioner) and C (memory management and hierarchy sequencing) and has successfully been used on several platforms such as VAX/VMS, Alpha AXP/OpenVMS, DECstation/ULTRIX and HP9000/HP-UX.

The solver hierarchy is user configurable, if e.g. the user is equipped with a-priori knowledge on the conditioning of the problem the solver can easily be directed to automatically leap over less robust levels (in case of an ill-conditioned system). To our experience a two-level hierarchy (see Figure 1) is sufficient, hierarchies with more than one ILUT level have not proven advantageous in our simulations.

L_G (μm)	ILU(1)		ILUT(10^{-5})	
	ITMIN	CPU (s)	ITMIN	CPU (s)
1	190	3504	13	1770
2	350	5232	14	1882
3	749	6983	17	1880
4	805	8584	24	2181
5	1924	14725	26	2264
6	1046	8636	28	2184
7	2627	17226	29	1976
8	1804	13274	32	2251
9	2223	15946	35	2578

Table 1: Comparison of iteration counts and total simulation times (VAX 6420) for the first Gummel iteration using ILU(1) and ILUT(10^{-5}) preconditioned BiCGStab solvers for MOSFETs with various geometric channel lengths.

3. Solver performance

The sidewall trench isolation structure of the devices under consideration is given in Fig. 3 (width direction). The channel is limited in width direction by a shallow trench. Due to the small sidewall overlap a parasitic device is present with a smaller effective gate thickness thus lowering the threshold voltage [5]. The main doping profile as well as the field and sidewall doping profiles are constructed from two-dimensional slices. Device performance has been studied in dependence of gate length and width, sidewall overlap, field and sidewall doping.

Devices with different gate lengths (from $10\mu m$ down to $0.5\mu m$) have been simulated and their characteristics compared with measurements. We encountered serious convergence problems such as stagnation or iteration counts of several thousands for long channel devices ($L > 4\mu m$) at the beginning of Gummel's algorithm when using an ILUT(0) or ILUT(1) preconditioner. For CPU times and iteration counts of the minority continuity equation see Table 1. It is worth mentioning that this ill-conditioning is quite independent of biasing, specifically in the subthreshold region which was the domain of major interest during this investigation (see Fig. 4 for the short channel device showing the enhanced conductivity due to the large gate sidewall overlap).

In contrast to the ILU(0,1) preconditioner the CPU times when using the ILUT preconditioner are almost independent on the device channel lengths and much smaller. The iteration count in the ILUT level can be kept less than one hundred in all cases. Even for the largest problems the CPU time consumption is quite moderate provided that memory is large enough to keep the ILUT preconditioner in core. At least 32MB main memory is necessary for the preconditioner to solve problems involving grids with several 10^5 mesh points. The characteristic given in Fig. 4, containing 40 bias points has been simulated in 7 hours.

References

[1] M. Driessen, H.A.v.d. Vorst, *Bi-CGSTAB in Semiconductor Modeling*, Proc. SIS-DEP (1991).

[2] O. Heinreichsberger et al., *Fast Iterative Solution of Carrier Continuity Equations for Three-Dimensional Device Simulation*, SIAM J.Sci.Stat.Comp., Vol. 13, No. 1, pp. 289-306 (1992).

[3] C. Pommerell, *Solution of Large Unsymmetric Systems of Linear Equations*, Doctoral Thesis, ETH Z"urich (1992).

[4] Y. Saad, *Preconditioning Techniques for Nonsymmetric and Indefinite Linear Systems*, J.Comp. Appl.Math., Vol. 24, pp. 89-105 (1988).

[5] S.M. Sze, *High-Speed Semiconductor Devices*, John Wiley & Sons, p. 177 (1990).

[6] H.A.v.d. Vorst, *A Fast and Smoothly Converging Variant of Bi-CG for the Solution of Nonsymmetric Systems*, SIAM J.Sci.Stat.Comp., Vol. 13, No. 2, pp. 631-644 (1992).

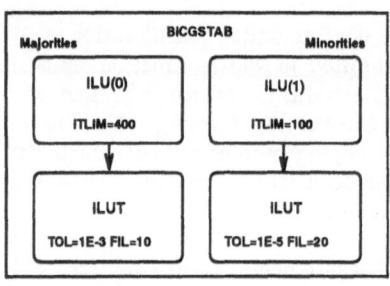

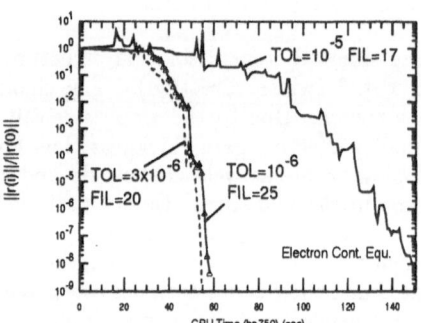

Fig. 1: Hierarchy Setup.

Fig. 2: Dependence on TOL.

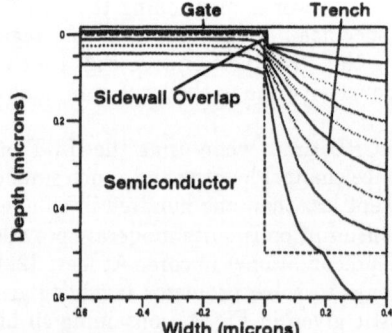

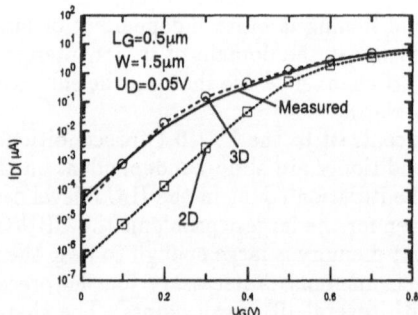

Fig. 3: Device geometry.

Fig. 4: Subtreshold characteristic.

SIMULATION OF SEMICONDUCTOR DEVICES AND PROCESSES Vol. 5
Edited by S. Selberherr, H. Stippel, E. Strasser – September 1993

Further Improvements in Nonsymmetric Hybrid Iterative Methods

H. A. Van der Vorst, D. R. Fokkema, and G. L. Sleijpen

Mathematical Institute, University of Utrecht
Budapestlaan 6, NL-3508 Utrecht, THE NETHERLANDS

Abstract

In the past few years new methods have been proposed that can be seen as
combinations of standard Krylov subspace methods, such as Bi–CG and GM-
RES. Such hybrid schemes include CGS, BiCGSTAB, QMRS, TFQMR, and
the nested GMRESR method. These methods have been successful in solving
relevant sparse nonsymmetric linear systems, but there is still a need for further
improvements. In this paper we will highlight some of the recent advancements
in the search for effective iterative solvers.

1. Bi-CGSTAB and variants

The residual $r_k = b - Ax_k$ in the Bi-Conjugate Gradient method, when applied to
$Ax = b$ with start x_0, can be written formally as $P_k(A)r_0$, where P_k is a k-degree
polynomial. These residuals are constructed with one operation with A and one with
A^T per iteration step. It was pointed out in [6] that with about the same amount of
computational effort one can construct residuals of the form $\tilde{r}_k = P_k^2(A)r_0$, which is
the basis for the CGS method.

In [7] it was shown that by a similar approach as for CGS, one can construct methods
for which r_k can be interpreted as $r_k = P_k(A)Q_k(A)r_0$, in which P_k is the polynomial
associated with BiCG and Q_k can be selected free under the condition that $Q_k(0) = 1$.
In [7] it was suggested to construct Q_k as the product of k linear factors $1 - \omega_j A$, where
ω_j was taken to minimize locally a residual. This approach leads to the BiCGSTAB
method. One weak point in BiCGSTAB is that we get break-down if an ω_j is equal to
zero. One may equally expect negative effects when ω_j is small. In fact, BiCGSTAB
can be viewed as the combined effect of BiCG and GMRES(1) steps. As soon as the
GMRES(1) part of the algorithm (nearly) stagnates, then the BiCG part in the next
iteration step cannot (or only poorly) be constructed.

Another dubious aspect of BiCGSTAB is that the factor Q_k has only real roots by
construction. It is well-known that optimal reduction polynomials for matrices with
complex eigenvalues may have complex roots as well.

This point of view was taken in [2] for the construction of the BiCGSTAB2 method.
In the odd-numbered iteration steps the Q-polynomial is expanded by a linear factor,
as in BiCGSTAB, but in the even-numbered steps this linear factor is discarded, and

factorize a sparse matrix incompletely: The standard method is to define some symbolic sparsity pattern for the incomplete factors and then computing matrix entries according to the pre-defined pattern. In the simplest case the original sparsity pattern of the matrix is used, denoted by ILU(0). For block diagonal matrices the sparsity pattern may be augmented easily by additional diagonals along the original sparsity pattern. If one additional diagonal is allocated along each existing diagonal we call this factorization ILU(1).

A more general class of ILU factorizations is obtained by keeping fill entries in the ILU factors only when they exceed some threshold, thus the sparsity pattern is defined during factorization. Clearly, for smaller values of the *drop threshold* parameter TOL a better ILU factorization with more fill-in can be created than for a larger TOL.

In this approach we precondition the BiCGStab method by two major *preconditioning levels* (Fig. 1):

- A vectorizable ILU preconditioner with positional dropping as the lowest hierarchy level,

- A parametric, non-vectorizable ILUT preconditioner in right position for higher levels.

In the first hierarchy we use either ILU(0) implemented in conjunction with the Eisenstat matrix-vector multiplication method to save about 30% of arithmetic work [2] or ILU(1). Both preconditioners are applied in split position.

The computation of the ILUT preconditioner for the second hierarchy level depends on two parameters: TOL which defines a drop tolerance weighted with the actual matrix rowsum for an individual matrix entry and FIL which determines the maximum number of matrix entries per row of an incomplete factor. TOL controls mainly the robustness of the preconditioner, FIL is necessary to limit memory usage, however, both parameters influence the robustness of the preconditioner.

To provide robustness for very ill-conditioned linear systems we use tolerances smaller than 10^{-5} and a fill-in of 20 beside the original sparsity pattern. This choice (in the two-dimensional parameter space) seems to yield an optimum in CPU time for our specific application (Fig. 2). An iteration limit of 400 for ILU(0) and 100 for ILU(1) is used as a switch between the hierarchies. For iteration termination we require the residual (left split preconditioned residual in case of ILU or true residual in case of ILUT) to be less than 10^{-8} to ensure convergence of the outer Gummel algorithm.

For ILUT factorization a row oriented compressed sparse matrix format is used. The factorization code is inherently sequential in nature and hence does not vectorize. Therefore the factorization must be implemented as efficient as possible. Efficiency can be exploited mainly by detecting unimportant matrix entries as early as possible and by an efficient sort algorithm to sort out the FIL largest elements in a row.

The package is written in FORTRAN 77 (iterative solver and preconditioner) and C (memory management and hierarchy sequencing) and has successfully been used on several platforms such as VAX/VMS, Alpha AXP/OpenVMS, DECstation/ULTRIX and HP9000/HP-UX.

The solver hierarchy is user configurable, if e.g. the user is equipped with a-priori knowledge on the conditioning of the problem the solver can easily be directed to automatically leap over less robust levels (in case of an ill-conditioned system). To our experience a two-level hierarchy (see Figure 1) is sufficient, hierarchies with more than one ILUT level have not proven advantageous in our simulations.

L_G (μm)	ILU(1)		ILUT(10^{-5})	
	ITMIN	CPU (s)	ITMIN	CPU (s)
1	190	3504	13	1770
2	350	5232	14	1882
3	749	6983	17	1880
4	805	8584	24	2181
5	1924	14725	26	2264
6	1046	8636	28	2184
7	2627	17226	29	1976
8	1804	13274	32	2251
9	2223	15946	35	2578

Table 1: Comparison of iteration counts and total simulation times (VAX 6420) for the first Gummel iteration using ILU(1) and ILUT(10^{-5}) preconditioned BiCGStab solvers for MOSFETs with various geometric channel lengths.

3. Solver performance

The sidewall trench isolation structure of the devices under consideration is given in Fig. 3 (width direction). The channel is limited in width direction by a shallow trench. Due to the small sidewall overlap a parasitic device is present with a smaller effective gate thickness thus lowering the threshold voltage [5]. The main doping profile as well as the field and sidewall doping profiles are constructed from two-dimensional slices. Device performance has been studied in dependence of gate length and width, sidewall overlap, field and sidewall doping.

Devices with different gate lengths (from $10\mu m$ down to $0.5\mu m$) have been simulated and their characteristics compared with measurements. We encountered serious convergence problems such as stagnation or iteration counts of several thousands for long channel devices ($L > 4\mu m$) at the beginning of Gummel's algorithm when using an ILUT(0) or ILUT(1) preconditioner. For CPU times and iteration counts of the minority continuity equation see Table 1. It is worth mentioning that this ill-conditioning is quite independent of biasing, specifically in the subthreshold region which was the domain of major interest during this investigation (see Fig. 4 for the short channel device showing the enhanced conductivity due to the large gate sidewall overlap).

In contrast to the ILU(0,1) preconditioner the CPU times when using the ILUT preconditioner are almost independent on the device channel lengths and much smaller. The iteration count in the ILUT level can be kept less than one hundred in all cases. Even for the largest problems the CPU time consumption is quite moderate provided that memory is large enough to keep the ILUT preconditioner in core. At least 32MB main memory is necessary for the preconditioner to solve problems involving grids with several 10^5 mesh points. The characteristic given in Fig. 4, containing 40 bias points has been simulated in 7 hours.

References

[1] M. Driessen, H.A.v.d. Vorst, *Bi-CGSTAB in Semiconductor Modeling*, Proc. SIS-DEP (1991).

[2] O. Heinreichsberger et al., *Fast Iterative Solution of Carrier Continuity Equations for Three-Dimensional Device Simulation*, SIAM J.Sci.Stat.Comp., Vol. 13, No. 1, pp. 289-306 (1992).

[3] C. Pommerell, *Solution of Large Unsymmetric Systems of Linear Equations*, Doctoral Thesis, ETH Z"urich (1992).

[4] Y. Saad, *Preconditioning Techniques for Nonsymmetric and Indefinite Linear Systems*, J.Comp. Appl.Math., Vol. 24, pp. 89-105 (1988).

[5] S.M. Sze, *High-Speed Semiconductor Devices*, John Wiley & Sons, p. 177 (1990).

[6] H.A.v.d. Vorst, *A Fast and Smoothly Converging Variant of Bi-CG for the Solution of Nonsymmetric Systems*, SIAM J.Sci.Stat.Comp., Vol. 13, No. 2, pp. 631-644 (1992).

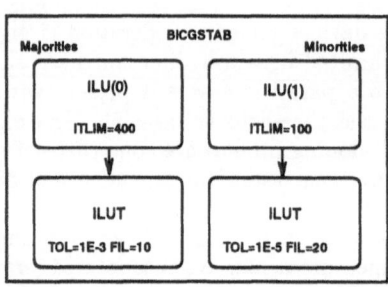

Fig. 1: Hierarchy Setup.

Fig. 2: Dependence on TOL.

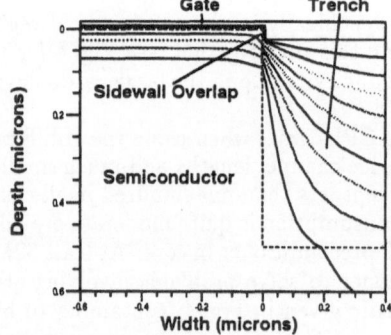

Fig. 3: Device geometry.

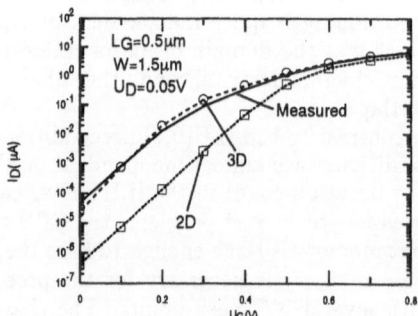

Fig. 4: Subtreshold characteristic.

Rigorous Microscopic Drift-Diffusion Theory and its Applications to Nanostructures

M. Orlowski

Advanced Products Research and Development Laboratory, Motorola Inc.
3501 Ed Bluestein Blvd., Austin, TX 78721, USA

Abstract

A recently proposed approach [1] to describe rigorously diffusion and diffusion-related reactions on arbitrary networks in terms of elementary jumps has been extented to include drift-diffusion capability. The method is based on combined density function approach to diffusion and on an adjacency matrix concept used in graph theory. The new method allows flexible and selective mixing of drift and diffusion on any two-dimensional domain of arbitrary outer and inner geometry. In particular, partial identity rank 4 tensors are introduced to allow preservation of results of previous operations not affected by the subsequent mechanisms. As an application it is shown that switching-off behavior of currents depends sensitively on the domain geometry.

1. Introduction to the Tensorial Diffusion Approach

The present approach is based on a tensorial formulation of the adjacency matrix in 2D proposed recently [1] to describe rigorously diffusion on arbitrary grain boundary networks. To calculate elementary jumps in 2D lattice with nodes (i, j) a four dimensional tensor $\mathcal{T}$ is defined with elements $\tau(i, j, m, n) = 1$ if the nodes (i, j) and (m, n) are connected with each other, and $\tau(i, j, m, n) = 0$ otherwise. Note that the connection is directional, i.e. the node (i, j) can be connected with the node (m, n) but not vice versa. That means that a particle can jump from (i, j) to (m, n) but not necessarily back. Clearly, this property allows to define drift along arbitrary axis and also any local emission and absorption mechanisms. For example, to describe diffusion allowing jumps to the nearest neighbors, only tensor elements which differ in only one of their coordinates by a unity will have the value one and all other zero: i.e., for example, $\tau(i, j, i + 1, j) = 1$. To evaluate the consecutive jumps in the entire domain a specific tensor product has to be constructed. This is accomplished by the following definition: $\mathcal{T} \bullet \mathcal{T} \equiv (\tau^{(2)}(i, j, m, n)) = \sum_{(k,l)} \tau(i, j, k, l) \cdot \tau(k, l, m, n)$. In order to evaluate the M-th tensor product, a tensor contraction $\hat{\mathcal{T}}^{(M)}_{(i,j)}$ with elements $\hat{\tau}^{(M)}_{(i,j)}(m, n)$ is defined. The contracted tensor gives the probability for the particle which started to diffuse at the node (i, j) to arrive after M jumps at nodes (m, n) $\forall$ m and n. To illustrate this, consider the diffusion of an implanted profile $g(x_i, y_j; t = 0)$ for a time $t = t^* \geq 0$. If f is the frequency of the diffusion coefficient then $M = f \cdot t^*$ and and the evolution of the dopant profile is given by $g(x_m, y_n; t = t^*) = \sum_{(i,j)} g(i, j; t = 0) \cdot \hat{\tau}^{(M)}_{(i,j)}(m, n)$.

2. Extension to Drift and Interactions

So far, the method [1] has been employed random walk diffusional jumps to the nearest neighboring nodes described by the tensor T_{sr} which we now characterize by an additional subscript sr, for short range. However, in the present formalism other mechanisms can be considered easily, since all mechanisms are defined by a specific distribution of the 1s and the 0s of the respective tensor elements. For example, long range diffusional jumps will be defined by a tensor T_{lr}, a drift of particles in a specific direction by the tensor T_d, absorbtion at certain regions by the tensor T_{ab}, and emission from some regions by the tensor T_{em}, and so on. To consider a composite dynamics of all the mechanisms invoked, we have simply to form a multiple product of the elementary mechanisms considered, followed by a power tensor of this product: $(T_{em}^{n5} \bullet T_{ab}^{n4} \bullet T_d^{n3} \bullet T_{lr}^{n2} \bullet T_{sr}^{n1})^M$, where the exponents ni $(i = 1, .., 5)$ denote the relative rapidity of the mechanism compared with the others, and M determines how often the whole cycle is supposed to take place, i.e., properly scaled M describes the process time.

The incorporation of drift is easily accomplished. Suppose the drift is along the x-axis (with the index i), then the only non-zero tensor elements are $\tau(i, j, m, j) = 1$ if $m = i + 1$ and $n = j$. Suppose the drift is along the xy-diagonal of the domain then the respective drift is a composite drift of lateral drift, $\tau_{lat}(i, j, m, n) = 1$ only if $m = i$ and $n = j + 1$ and zero otherwise, and by a subsequent vertical drift downwards, given by $\tau_{ver}(i, j, m, n) = 1$ only if $m = i + 1$ and $n = j$ and zero otherwise. The composite operation describing the drift downward the diagonal is accomplished by applying the tensor product $T_{drif} = T_{ver} \bullet T_{lat}$.

In many cases there might be parts of the domain which are not affected by a specific operation, for example by drift in a selected subdomain $\mathcal{D}_s$, or even on the same subdomain there might be mechanisms operating on different time scales. In this case the tensor reflecting the operation on the subdomain, $\mathcal{D}_s$, will have zero tensor elements outside this particular subdomain, and by necessity, zero tensor elements representing all other mechanisms. Because of these vanishing tensor elements, the results of all previous operations and existing distributions outside of the subdomain will not be preserved under the current operation. In order to preserve the results of previous operations in subdomains not considerered under current time step, the currently applied tensor has to be partitioned and its elements pertaining to these domains should be defined im the following way: $\tau_{curr}(i, j, m, n) = 1$ only if $m = i$ and $n = j$. In terms of our earlier definition this amounts to the self-connectedness relation, i.e. the node (i, j) (where $(x_i, y_j) \in \mathcal{D}_r \ \forall \ r \neq s$) is connected to itself. This identity operation preserves the dopant distribution obtained so far by previous operations in the respective subdomains.

The present approach allows also a definition of interactions on a two- (and, in principle, also on a three-) dimensional lattice. Suppose the nodes (m, n) and (r, s) are initially not connected with one another, i.e. $\tau(m, n, r, s) = 0$ and $\tau(r, s, m, n) = 0$. The connection (i.e. interaction) can be activated at a later stage in the evolution, say at $t = t_o$ by turning the tensor elements to be, $\tau(m, n, r, s) = 1$ and/or $\tau(r, s, m, n) = 1$, depending on the kind of interaction considered, as soon as the values of the species density $g(i, j, t = t_o)$ under consideration at the nodes (m, n) and (r, s) are reaching specific values, for example. Construction of these interactions and their evaluation will be subject of future work.

3. Application of Drift-Diffusion

As an illustration of the extended formalism expounded in the preceding section, we consider drift diffusion in two structures, one consisting of simple conducting rectangular strip with x-axis along the drift direction Fig.1a, and a second structure, in which the latter rectangular domain has been extended in the transversal direction (y) by narrow rectangles, forming a cross structure, as shown in Fig.1b. In the initial stage of the drift-diffusion transport, the species, for example, electrons, are entering both structures at $x = 0$ and are allowed to spread in the structures by means of diffusion only. A typical diffusion profile for the cross structure is shown in Fig.2.

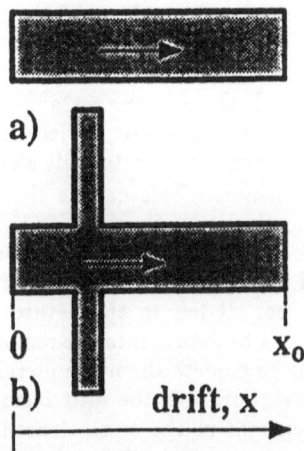

a)

b)

0   x_0

drift, x

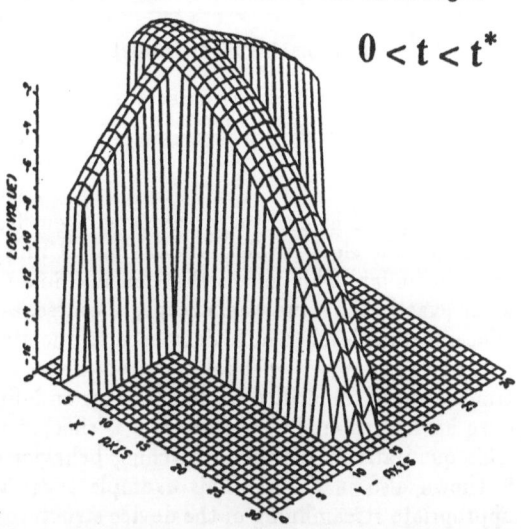

$0 < t < t^*$

Fig.1: Two structures used for drift diffusion studies: a) rectangular channel and b) rectangular channel with transversal arms (cross structure).

Fig.2: Diffusion profile of the species (electrons) in the cross structure prior to switching on of the drift field.

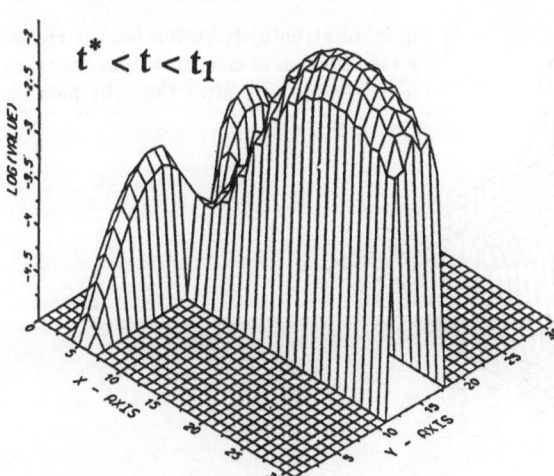

$t^* < t < t_1$

Fig.3: Electron distribution in the cross structure shortly after the drift field has been switched on.

It can be seen that the electrons spread also sideways into the lateral arms of the cross structure. At $t = t^*$ the drift (field) along the x-axis is switched on. After some time the diffusion profiles is shifted downwards under the influence of the drift field along the x-axis as shown in Fig.3. Let us now suppose that at a later time, $t = t_1 \geq t^*$, the supply of electrons at $x = 0$ has been cut-off. In the case of the simple rectangular structure, it is clear, that cutting-off the supply of electrons will let the current drop abruptly after a delay time $\Delta t = x_o/v$, where x_o is the length of the rectangular channel and v is the drift velocity. This means that Δt is

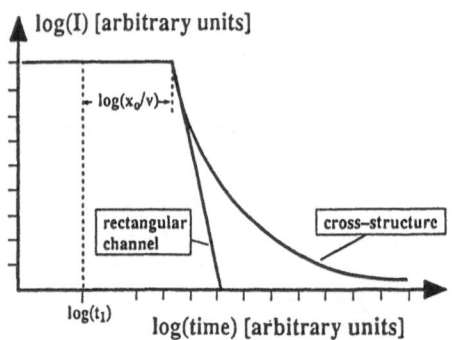

Fig.4: Current switching-off characteristics for rectangular channel and for the cross structure.

the time needed to deplete the channel completely. The current switch-off characteristics will exhibit some decaying tail, due to the fact that the cut-off electron packet has been broaden in the rectangular channel by diffusion. This is shown in Fig.4. The switching-off behavior for the cross structure is entirely different. Even at times considerably larger than $t = t_1 + \Delta t$, there will be still considerable level of current, flowing out of the channel, which is fed by the electrons outdiffusing from the lateral cross arms into the main rectangular channel. Once in the rectangular channel the electrons are subject not only to diffusion but also to drift along the x-axis, which, eventually, depletes them from the channel. This case is shown in Fig.5. The electron distributions for all times sufficiently larger than $t = t_1 + \Delta t$ are self-similar. This means that for any time $t \gg t_1 + \Delta t$ the electron distributions looks exactly the same way, except, of course, as should be expected, for the absolute concentration level, which decreases as the time goes on. It lies in the nature of diffusion (entropy) that it takes finite time for the electrons to diffuse into the regions transversal to the drift flow, but it takes an infinite time to deplete them completely, even having, as in the present case, a 'sink' at the intersection with the drift region. This qualitatively different switching behavior of the current in the cross structure is shown also in Fig.4. This example suggests, that at nanoscale dimensions an appropriate streamlining of the device structure will improve the switching properties of the device significantly.

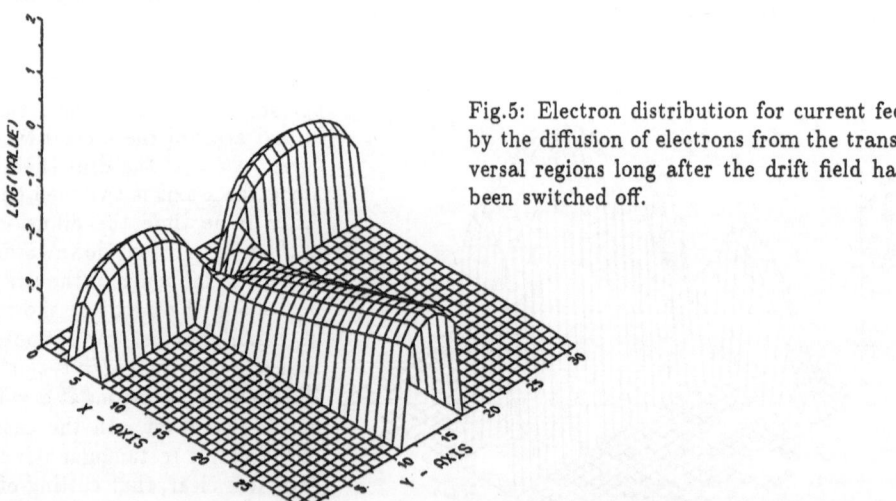

Fig.5: Electron distribution for current fed by the diffusion of electrons from the transversal regions long after the drift field has been switched off.

References

[1] Marius Orlowski, IEDM'92 Proceedings, p.161, San Francisco, 1992

SIMULATION OF SEMICONDUCTOR DEVICES AND PROCESSES Vol. 5   97
Edited by S. Selberherr, H. Stippel, E. Strasser – September 1993

Atomistic Evaluation of Diffusion Theories for the Diffusion of Dopants in Vacancy Gradients

S. List[‡], P. Pichler[‡], and H. Ryssel[‡†]

[‡]Fraunhofer-Institut für Integrierte Schaltungen
Artilleriestraße 12, D-91052 Erlangen, GERMANY
[†]Lehrstuhl für Elektronische Bauelemente, Universtität Erlangen-Nürnberg
Cauerstraße 6, D-91052 Erlangen, GERMANY

Abstract

A new approach is used for the the calculation of transport coefficients for dopants and vacancies from atomic jump frequencies in the presence of dopant or vacancy gradients. Results are shown for the diffusion under vacancy gradients with an attractive potential between the dopant and the defect. It is demonstrated that for a given vacancy gradient not only the absolute value but also the sign of the dopant flux depends on the range of the binding potential.

1. Introduction

Despite large efforts in recent years, the basic diffusion mechanisms are still under discussion. Basically, there are two common formulations of the diffusion equations for dopants including the effects of gradients of the point defect concentrations. According to pair diffusion theories [1], the flux of dopants is predicted in the same direction as the flux of the vacancies. On the other hand, atomistic theories neglecting pair diffusion [2] lead to an opposite direction of the two fluxes. If the dopant flux is written as a function of dopant and vacancy gradient as

$$ - \underline{F}_d = \frac{C_v}{C_{site}} \cdot D_d^o \cdot \underline{\nabla} C_d + \alpha \cdot D_d^o \cdot \frac{C_d}{C_{site}} \cdot \underline{\nabla} C_v \tag{1}$$

then the pair diffusion model [1] will lead to $\alpha = 1$ whereas atomistic theories [2] correspond to $\alpha = -1$. In (1), C_d and C_v denote the concentrations of dopants and vacancies, D_d^o is the normalized dopant diffusion coefficient (scaled with the probability of a site to be occupied by a vacancy). Both approaches have to be seen as idealizations and the underlying atomistic mechanisms remain unclear. Therefore the aim of the present work was to decide which of the two models can be derived from the vacancy hopping mechanism with an attractive potential between the dopant and the vacancy.

On the other hand, the point-defect gradient term has a non-negligible influence on the simulation of dopant diffusion when the vacancy gradients are large as, e.g., during transient diffusion after ion implantation.

2. Description of the Applied Method

The atomistic basis for the calculation of the dopant and vacancy fluxes is the vacancy hopping model. The only movement allowed is a vacancy hopping to a nearest neighbour site thus exchanging its place with either a silicon atom or the dopant atom. The probability of such a transition per time (the jump rate) may depend on the distance between the vacancy and the dopant atom thus taking into account some kind of force between the defect and the dopant. Special cases are, e.g., tracer diffusion where all jump frequencies are equal or the model of Hu [3] with 4 different jump frequencies which has been used to estimate the correlation factor for impurity diffusion via vacancies. For the calculation shown in this work, 7 different jump rates have been used taking into account interactions up to the 3rd coordination site. The method applied deals with a statistical ensemble of a limited, three-dimensional area of a silicon crystal which is represented by a time dependent two-particle probability distribution. $c(i,j,t)$ is the probability that the dopant atom is at site i and the vacancy is at site j. Since interactions of more than 2 particles are neglected, c contains all information about the statistical ensemble. Therefore the one-particle profiles as well as the dopant and vacancy fluxes can be calculated from the two-particle probability at each time. At the onset of diffusion, the correlated probability $c(i,j,0)$ is set to the product of the one-particle distributions $d(i,0)$ for the dopant and $v(j,0)$ for the vacancy. This corresponds to a statistical independence of dopant and vacancy distribution. Then the Master equation for the correlated probability $c(i,j,t)$ is solved to get the time evolution of $c(i,j,t)$. The boundary condition is particle conserving for the dopant, for the vacancy a constant flux at the surfaces can be simulated which prevents the vacancy gradient from flattening out too fast. The time evolution of $c(i,j,t)$ corresponds to a relaxation of the system towards a local equilibrium within a time scale which is small compared to the time necessary to get a major change in the 1-particle profiles $d(i,t)$ and $v(j,t)$. When the local equilibrium has been reached, transport coefficients can be calculated by dividing the fluxes through the corresponding driving forces (gradients).

The Master equation is solved numerically using an explicit timestep scheme with a stepsize of $0.1 \frac{1}{\tau}$ (τ is the jump frequency of a free vacancy).

3. Applications and Results

The method outlined in the previous section has been applied to the open question of diffusion theories adressed in the introduction. Calculations have been performed

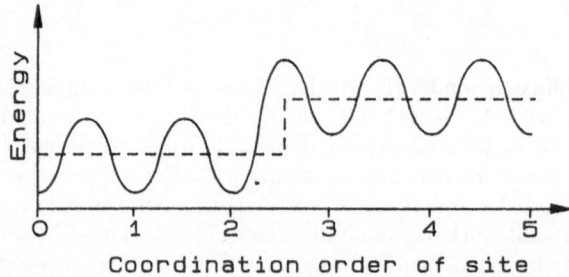

Figure 1: Schematic diagram of the vacancy potential as a function of distance from the impurity atom at the origin. The dashed line indicates the potential well caused by the impurity.

on a simulation area of 7 lattice constants in the x- and y-direction and 8 lattice constants in the z-direction (3136 sites) with different attractive potentials. For the sake of simplicity, rectangular potential wells are assumed. Fig. 1 shows the schematic drawing of a potential well reaching to the 2nd nearest neighbour site. Calculations have also been performed for a potential range up to the 1st and 3rd coordination site, and without a potential well (tracer diffusion).

Jump frequencies have been scaled to the jump frequency τ of a free vacancy, the lattice constant a ($5.431\mathring{A}$) was used as unit for lengths. This means that fluxes are given in units of τ/a^2, diffusion coefficients in $\tau \cdot a^2$, time in $\frac{1}{\tau}$, and densities in $1/a^3$. The initial conditions were a homogenous dopant profile of $2.5 \cdot 10^{-3} \cdot \frac{1}{a^3}$ and a constant vacancy gradient of $6.6 \cdot 10^{-4} \cdot \frac{1}{a^4}$ in the z-direction (the values result from a normalization condition). The constant surface vacancy flux was $6 \cdot 10^{-5} \cdot \frac{\tau}{a^2}$. This was chosen so that the vacancy profile changes as little as possible and the equilibrium establishes as fast as possible. All jump frequencies were set to τ except for the one which leads out of the potential well which was set to $\tau_{esc} = 0.02 \cdot \tau$. Thus, the

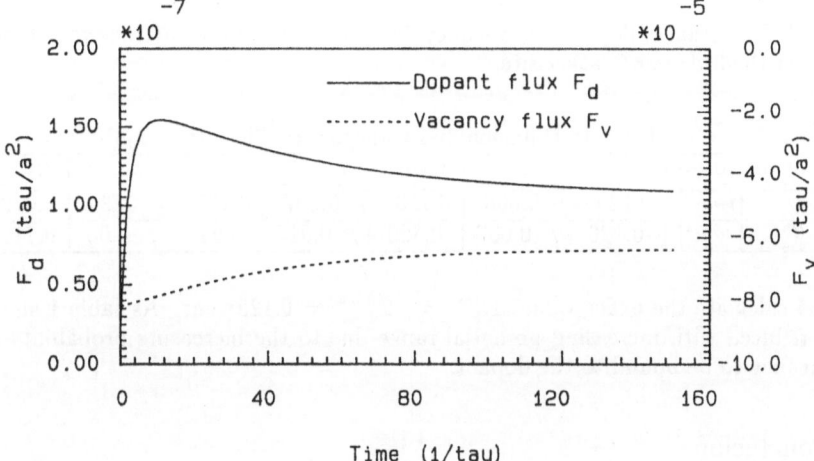

Figure 2: Evolution of dopant and vacancy fluxes with time for an attractive potential ending at the 2nd coordination site.

probability for the pair to dissociate was 50 times smaller than for staying together. Fluxes and diffusion coefficients are evaluated at a site in the center of the calculation area.

Fig. 2 shows the results for a potential range up to the 2nd coordination site. Local equilibrium is established after about 60 $\frac{1}{\tau}$. The absolute value of the dopant flux in the equilibrium is larger than at $t = 0$ due to the higher probability of a vacancy staying at neighbouring sites after the formation of pairs. The direction is opposite to the vacancy flux. In Fig. 3, the underlying potential ranges up to the 3rd coordination site. Due to the larger radius of the pairs, the time required for local equlibrium to establish is longer (about 120 $\frac{1}{\tau}$). At about $t = 20\frac{1}{\tau}$, the sign of the dopant flux changes, both fluxes are now in the same direction corresponding to the predictions of pair diffusion theories. The absolute value of F_d is also increased due to the larger vacancy density around the dopant.

The vacancy gradient stays constant throughout the calculation to within 10 percent. Table 1 lists the results for the vacancy diffusion coefficient D_v and the normalized dopant transport coefficient $T_d^o = \alpha \cdot D_d^o$. The errors are estimated from variations at sites around the centre. Notice that for tracer diffusion random walk theory can be

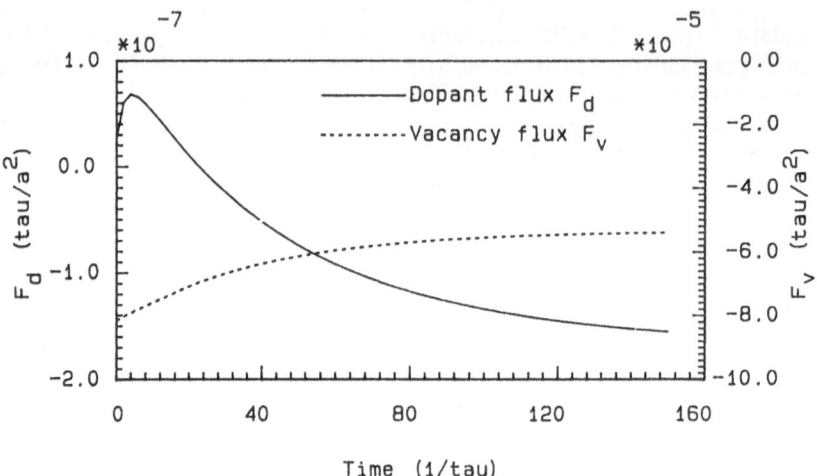

Figure 3: Evolution of dopant and vacancy fluxes with time for an attractive potential ending at the 3rd coordination site.

Table 1: Diffusion and transport coefficients

potential range		no potential	1st coo. site	2nd coo. site	3rd coo. site
D_v	$[\tau \cdot a^2]$	0.120 +/- 0.008	0.113 +/- 0.007	0.101 +/- 0.006	0.079 +/- 0.0
T_d^o	$[\tau \cdot a^2]$	-0.120 +/- 0.007	-0.300 +/- 0.015	-0.541 +/- 0.03	0.715 +/- 0.3

used to calculate the exact values $D_v^{trac} = -T_d^{o,trac} = 0.125\tau \cdot a^2$. As table 1 shows, D_v is reduced with increasing potential range due to the increasing probability for the vacancy to be bound to the dopant.

4. Conclusion

Dopant transport coefficients for diffusion under vacancy gradients have been calculated on the basis of the vacancy hopping mechanism with different binding potentials between dopant and vacancy. The results of this work show that both, pair diffusion models [1] and the theory of Maser [2], can be explained with the vacancy hopping mechanism, depending only on the type of interaction potential.

5. Acknowledgments

We wish to thank Dr. E. Geissler and H. Krausenberger from the RRZE (regional computation centre Erlangen) for their help with the adaption of the program used to a CRAY-YMP and a CDC4680. This work was part of **ADEQUAT** (JESSI project BT1B) and was funded as ESPRIT project 7236.

References

[1] M. Orlowsky, Appl. Phys. Lett. 53, 1323 (1988)

[2] K. Maser, Experimentelle Technik der Physik 39, 169 (1991)

[3] S. M. Hu, Phys. Rev. 180, 773 (1969)

SIMULATION OF SEMICONDUCTOR DEVICES AND PROCESSES Vol. 5
Edited by S. Selberherr, H. Stippel, E. Strasser – September 1993

Self Diffusion in Silicon Using the Ackland Potential

M. M. De Souza and G. A. J. Amaratunga

Department of Engineering, Cambridge University
Trumpington Street, Cambridge, CB2 1PZ, UNITED KINGDOM

Abstract

The study of self-diffusion using the Ackland potential yields an activation energy which lies within the experimental range of 4-5 eV. This study yields a configurational entropy for the interstitial defect of the order of 4.2K. The formation energy of both interstitials and vacancies has been found to be 3.9 ±0.1eV. The migration energy of interstitials is 0.7±0.1 eV and it is 0.2 eV for the vacancies.

1. Introduction

Silicon self-diffusion is mediated by point defects such as self-interstitials and vacancies. These point defects also influence the diffusion properties of dopants in silicon thereby causing anomalous oxidation enhanced diffusion, transient enhanced diffusion etc. Understanding the nature of these point defects is therefore important to VLSI technology.

In this paper, an attempt has been made to estimate the diffusion pathways for silicon point defects using a classical atomic potential approach. Most existing potentials for silicon are not well suited for diffusion studies[1, 2, 3]. For example, vacancy migration studies using these potentials reveal that the split vacancy configuration is more favourable compared to the isolated vacancy, indicating that none of these potentials correctly represent the ground state of the vacancy [4]. A recent study of self-diffusion using the Stillinger-Weber potential has yielded very high formation energies for hexagonal, split and bond-centered interstitials[5]. In their study, the proposed mechanism for interstitial migration is via an "extended" interstitial which moves from one low-energy configuration to another, via the tetrahedral site. However, this pathway by itself does not contribute to the activation energy for self-diffusion because it does not involve motion through a substitutional site [6].

In this paper, the study of diffusion has been carried out using the Ackland potential [7]. The Ackland potential is unique because of its bond-based nature and the fact that it allows for assymetric structured bonding, which is not easily performed using conventional LDA (Local Density Approximation) methods.

2. Computational Procedure

Simulations were performed at 300 K using a 2x2x2 unit cell with a total of 64 atoms. Periodic boundary conditions were employed in a constant NVT (Number of particles, Volume and Temperature) ensemble and energy minimisation was performed using the Metropolis Monte Carlo algorithm with multiple moves. The total energy in the Ackland potential is of the form:

$$E = \frac{1}{2} \sum_{i=1}^{N} \sum_{j=1}^{N} A e^{-\alpha r_{ij}} - \frac{1}{2} \sum_{i=1}^{N} \sum_{k=1}^{4} B r_{ik} e^{-\beta r_{ik}} + \sum_{i} \sum_{n=1}^{3} \sum_{m=n+1}^{4} C [\cos(\omega R_{k_m k_n}) + \frac{1}{3}]^2 \quad (1)$$

The last term in expression (1) is essential in order to stabilise reconstructed interstitial and vacancy formation energies. The parameter ω is adjusted so that the minimum of the function falls at the tetrahedral bond angle in the diamond structure for silicon.

During the calculation of the migration energies, the interstitial was constrained to move in a plane perpendicular to its direction of motion. At the saddle point, those bonds whose contribution to the total energy was very high were converted to "dangling" status. Such bonds do not contribute to the attractive as well as to the bond-bond repulsion terms in expression (1). In this plane, different bonding configurations were tried and the configuration which gave the least energy was selected to be the most probable one. Averages of the total energy after relaxation were taken, with an accuracy of ± 0.1 eV and plotted against the distance of the migrating atom from the center of mass of the simulation cell.

3. Results

A value of C=1.5 in expression (1) was found to give the most suitable values for the formation energies of defects. Both interstitials and vacancies were treated under identical conditions by using the same value of C. The formation energies of the T (Fig. 1a) and H (Fig. 1b) site interstitials were found to vary between 3.9 and 4.7 eV depending upon the bonding configuration. The formation energy of the split interstitial was found to be 4.6 eV. This configuration is three-fold coordinated, as shown in Fig. 1c and hence does not conform to the bonding rules for the ground state condition in the Ackland potential that every atom should be four-fold coordinated. In the <111> direction, a "buckled" configuration was found to be more favourable in comparison to the bond-centered configuration as shown in Fig. 2c . The formation energy of the vacancy was found to be 3.9 eV. The divacancy formation energy was found to be 8.5 eV for a typical configuration [4].

During the calculation of the migration energies, it was found that a migrating interstitial with one dangling bond has a higher energy (1.2-1.4 eV) as compared to an interstitial with two dangling bonds, except in the case of the split interstitial configuration, which has a comparable energy. The maximum possible saddle point energy along any pathway is hence the energy of this configuration with two dangling bonds at the interstitial. The migration energy was found to lie between 0.7-0.8 eV. The various pathways examined have been shown in Fig. (2), with the corresponding energy diagrams in Fig. 3. In addition the (T-B)' pathway (Fig. 2d) was also found to be favourable in contrast to [8]. The vacancy migration energy was found to have a low value of 0.2 eV. The split vacancy configuration was thus slightly higher in energy in comparison with the ground state vacancy.

The entropy of formation of a defect is the sum of the configurational entropy and the vibrational entropy. For the self-interstitial there are at least 68 different bonding configurations which yield the minimum energy of 3.9 eV. The configurational entropy for the self-interstitial component of self-diffusion is thus approximately 4.2K. This high value could explain the large entropy of 10K observed experimentally for silicon self-diffusion. This total entropy is a sum of the configurational, vibrational and migration entropies. No other calculation has ever yielded such a large value of the configurational entropy for the interstitial defect in silicon.

4. Conclusion

The Ackland potential yields an activation energy for self-diffusion that lies within the experimental range of 4-5 eV. The formation energy for interstitials and vacancies is equal to 3.9 eV. Both the types of defects could therefore contribute to self-diffusion. However, the migration energy of the silicon interstitial is 0.7 eV whereas that for the vacancy is 0.2 eV.

Acknowledgement

One of the authors (MMDS) would like to acknowledge the help and guidance of Dr. M. C. Payne of Physics Department, Cambridge University, U.K., and Dr. G.J. Ackland, Physics Department, University of Edinburgh, U. K. during the course of this study.

References

[1] J. Tersoff, *Phys. Rev. B*, **37**, 6991, 1988.

[2] R. Biswas and D. R. Hamman, *Phys. Rev. B*, **36**, 6434, 1987.

[3] F. H. Stillinger and T. A. Weber, *Phys. Rev. B*, **31**, 5262, 1985.

[4] S. V. Ghaisas, *Phys. Rev. B*, **43**, 1808, 1991.

[5] D. Maroudas and R. A. Brown, *Appl. Phys. Lett*, **62**, 172, 1993.

[6] R. Car, P. J, Kelly, A. Oshiyama and S. T. Pantelides *Phys. Rev. Lett*, **52**, 1814, 1984.

[7] G. J. Ackland, *Phys. Rev. B*, **44**, 3900, 1991.

[8] Y. Bar-Yam and J. D. Joannopoulos *Phys. Rev. B*, **30**, 2216, 1984.

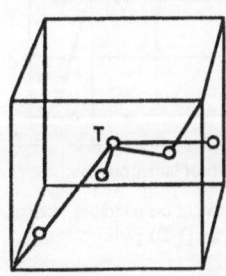 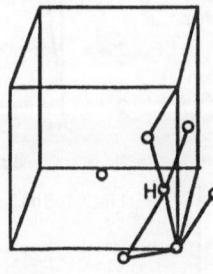 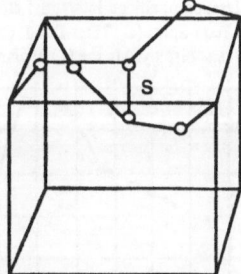

Fig. 1a: Tetrahedral site bonding configuration giving formation energy=3.9eV For a unit cell (0,0,0) to (1,1,1), T (0.5,0.5,0.5) is linked to (0.75,0.25,0.25), (1.0,0.5,0.5), (0.75,0.75,0.75) and (0.5,1.0,0.5).

Fig. 1b: Hexagonal site bonding configuration giving formation energy = 3.9 eV. For a unit cell (0,0,0) to (1,1,1) H (0.875,0.875 ,0.375) is bonded to (1.25, 0.75,0.25),(1,1,1), (0.75,0.75,0.75) and (0.5,1.0,0,0.5).

Fig. 1c: Split interstitial bonding configuration giving formation energy=4.6 eV. For a unit cell (0,0,0) to (1,1,1), S is bonded to (0.75,0.75,0.75), (0.25,0.25,0.75) and (0.5,0.5,1.0).

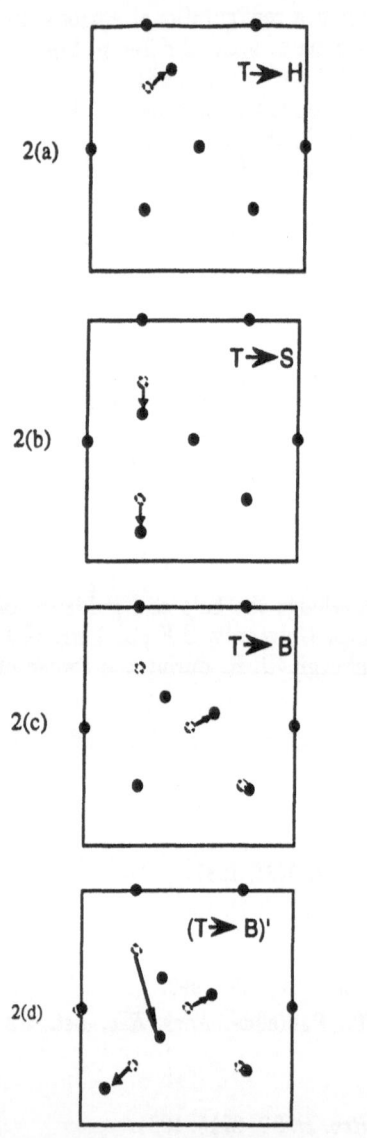

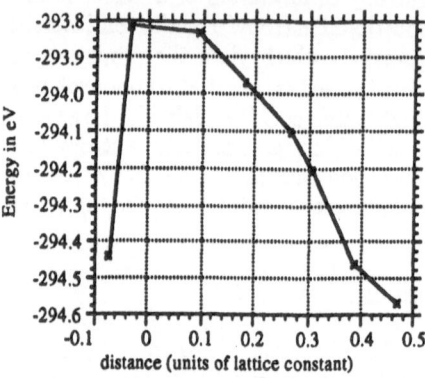

Fig. 3(b) Total energy curve of migrating interstitial
along the TS path.

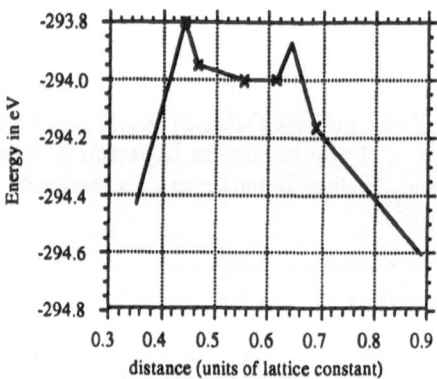

Fig. 3(c) Total energy curve of migrating interstitial
along the TB path.

Fig. 2. Migration of silicon interstitial along (a) Tetra-
Hex, (b) Tetra-Split (c) Tetra-Bond centered sites.
(d)Tetra-B' site displayed in the (110) plane.

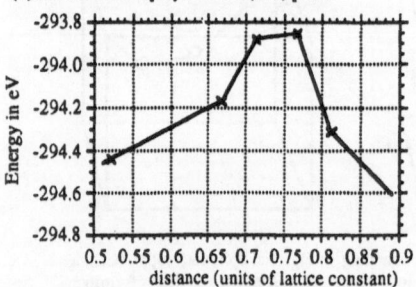

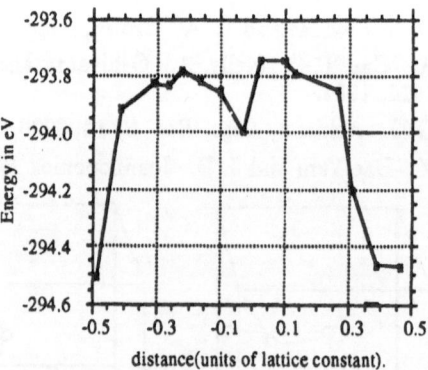

Fig. 3(d). Total energy curve of the migrating interstitial
along the (T-B)' path.

Fig. 3(a) Total energy curve of migrating interstitial
along the TH path.

SIMULATION OF SEMICONDUCTOR DEVICES AND PROCESSES Vol. 5 105
Edited by S. Selberherr, H. Stippel, E. Strasser – September 1993

Three-Dimensional Numerical Simulation for Low Dopant Diffusion in Silicon

C. S. Yun, O. K. Kwon[†], C. G. Hwang, and H. J. Hwang[†]

Advanced Technology Center, Memory Division, Semiconductor Business,
SamSung Electronics Co., Ltd.
KyungKi, KOREA

[†]Department of Electronic Engineering, Chung Ang University
Seoul, KOREA

Abstract

A numerical simulator for the calculation of redistribution of low dopant diffusion in silicon has been developed in three-dimensional(3D) geometry. The diffusion behavior of boron is investigated by using three mask structures and changing the contact hole sizes. The results of calculations show that 3D diffusion effects will be very important in the development of submicron process and small device.

1. Introduction

Process simulation has already established itself as an indispensable tool for process and device development of VLSIs. With shrinking device dimensions, high dimensional simulations are required to predict the acurate profiles and the 3D effects[1], because it is very difficult for 2D simulations to predict the 3D effects or understand the 3D behaviors. And it is usually restricted by analytical 3D impurity profiles rather than actual profiles. This comes partly from the inexistence of effective numerical 3D process simulators. With this point of view we have developed 3D numerical simulator (VLSIDIF) for low dopant diffusion in silicon. In this paper, 3D mask structure effects of boron diffusion were investigated and hole size effects were simulated and discussed in comparison with 3D behavior of oxide growth.

2. Numerical Solution Method

The equation normally used to describe dopant diffusion is

$$\frac{\partial C}{\partial t} = \frac{\partial}{\partial x}(D\frac{\partial C}{\partial x}) + \frac{\partial}{\partial y}(D\frac{\partial C}{\partial y}) + \frac{\partial}{\partial z}(D\frac{\partial C}{\partial z}) \qquad (1)$$

$$C(x,y,z) = C_o \quad : \text{ at the silicon surface} \qquad (2)$$

$$C(x,y,z) = 0 \quad : \text{ at the silicon bulk} \qquad (3)$$

$$\frac{\partial C}{\partial n} = 0 \quad : \text{ at the silicon sides and mask interfaces} \qquad (4)$$

where C is the dopant concentration, D is the diffusion coefficient, and t is time. Diffusion equation is scaled with dimensionless scaling method to prevent the underflow or overflow[2]. In the numerical computation, the weighted residual formula of the Finite Element Method(FEM) is adopted to discretize the space domain and the finite difference method is applied to discretize the time domain in equation (1) [3]. It is assumed that normal flux to the boundary and mask interfaces are taken to be zero and the atoms arriving at the silicon suface is constant C_o $Atoms/cm^3$ The final matrices formulation for diffusion equation (1) leads to

$$[\frac{1}{\Delta t_n} A_{ij} + K_{ij}] C_j^n = [\frac{1}{\Delta t_n} A_{ij}] C_j^{n-1} \tag{5}$$

where K_{ij} and A_{ij} are commonly known as the global stiffness and mass matrices and Δt is the time step. A fully implicit scheme is chosen in equation (5) as this can ensure an unconditionally stable and oscillation-free solution. A fast direct solver employing the frontal method is used to solve the global matrix and reduce the required core memory of computer[4]. This solver is particularly suited for solving the resultant matrix which is not only sparse but also banded. The FEM meshes based on automatic element subdivision of a few blocks were generated by 8 nodes rectangular parallelopiped elements.

3. Results and Discussion

Fig.1 shows 1D profile results of VLSIDIF compared with PROMIS1.5 and SUPREM-IV. The simulation was performed at 1000℃ and 1100℃ for 30 minutes and the surface concentration is maintained at 1e18 $Atoms/cm^3$ Fig.2 shows 2D simulated results of VLSIDIF compared with SUPREM-IV. In the case of VLSIDIF simulation, about 10,000 nodal points are used. The 1D & 2D simulated results of VLSIDIF show good agreement with those of other published 2D simulators. 3D diffusion behavior of low concentration dopant is investigated by using three mask structures which are called Hole , Island and Line Structure (HS, IS, LS). Fig.3 shows the three mask structure effects in the x-y planes along the silicon surface(z=0). In the LS, the 2D results of VLSIDIF are the same as normal 2D simulated results. But, In the IS, the dopants arround the mask region are concentrated at the corner edge of mask region. And in the HS, the dopants from mask window are spreaded into the mask corner region. From these results it has been found that the diffusion at the corner of mask edge is much enhanced in the IS because of multiple supply of dopants and retarded in the HS because of lateral spread of dopants from the mask window. These three mask structure effects was similar to the oxidation mechanism of 3D mask structures[5]. Fig.3(d) shows 3D profiles of HS which well define the depth and lateral profiles. Fig.4 shows the 3D boron distribution as a function of the hole (contact) size. Fig.5 shows hole size vs. simulated junction depth and oxide thickness[5] which were normalized with each maximum data. As the hole size is narrower than about 1.0 μm, the junction depth of dopant is decreased and the shape of doping profiles along the silicon surface is getting to be rounded at the mask corner region. These 3D hole size effects were explained by restricted dopants from narrow mask window. With its 3D VLSIDIF simulations, we found out Impurity Dilute Phenomena (IDP)[6] and Impurity Circular Phenomena (ICP) in narrow hole conctact structures. These results suggest that IDP and ICP of narrow hole area should be considered in the design of bit line or direct contact structures for submicron devices. In fig.6 elapsed time for simulation of the 3D diffusion process are demonstrated as a total grid number on SUN/IPC system. Elapsed time are drastically increased above the 5000 nodal points. To reduce the CPU time, more efficient matrix solver is required for 3D process simulation.

4. conclusion

The 3D diffusion simulator, VLSIDIF, has been developed to predict small –size process/device characteristics. The mask structure effect and the small contact hole size effect on the device fabrication have been found by varing simulation conditions. In addition, We know that the 3D diffusion phenomena was similar to the 3D effect on the oxide growth. ICP and IDP have been also found as a hole contact size and we know that they are very important in the small geometry processes.

Reference

[1] M.R. Pinto, J. Bude and C.S. Rafferty,"Simulation of ULSI Silicon MOSFET's," VPAD, pp22–25, May, 1993.
[2] J. Crank, The Mathematics of Diffusion, Oxford University Press, London, pp138–152, 1976.
[3] R. Ismail and G. Amaratunga," Adaptive Meshing Schemes for Simulating Dopant Diffusion," IEEE Trans. on CAD, Vol. 9, No. 3, pp276–289, March, 1990
[4] C. Taylor and T.G. Hughes,"Finite Element Programming of the Navier-Stokes Equations," Printed in Great Britain, pp120-153, 1981.
[5] H. Umimoto and S. Odanaka," Three-Dimensional Numerical Simulation of LOcal Oxidation of Silicon," IEEE Trans. on Elec. Devices, Vol. 38, No. 3, pp505-511, March, 1991.
[6] S. Onga and Taniguchi,"A Three-Dimensional Process Simulator and Its Application to Submicron VLSI's," in Digest Tech. papers Symposium on VLSI Technology, pp68–69, 1985.

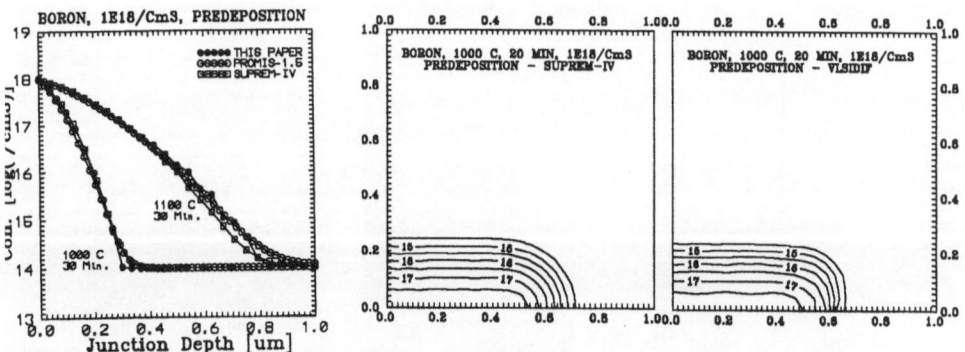

.1 1-dimensional simulated profiles VLSIDIF(●), PROMIS1.5(○) and PREM-IV(□).

Fig.2 Comparison of 2D simulated results of VLSIDIF with SPREM-IV after boron predeposition.

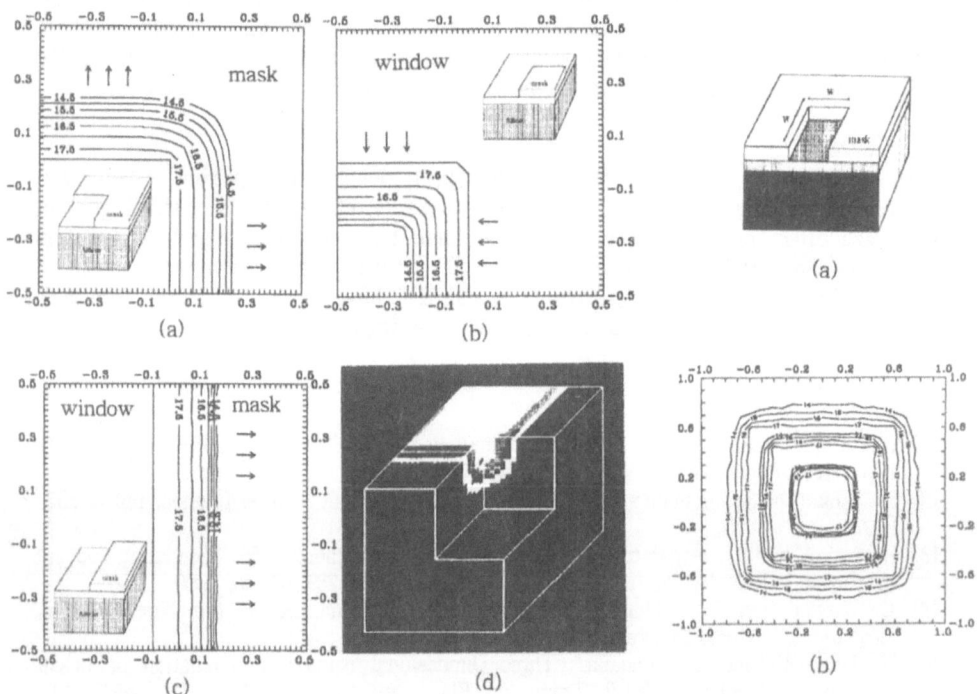

Fig.3 Simulation results of 3D distribution after boron predeposited 1000 ℃ for 30 minutes on the three different mask structures.

Fig.4 Simulated results of hole size effects on the silicon surface (hole size : 1.0x1.0 μm, 0.7x0.7 μm, 0.3x0.3 μm was overlaped).

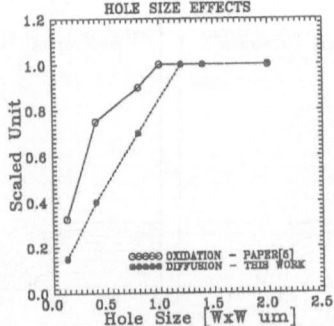

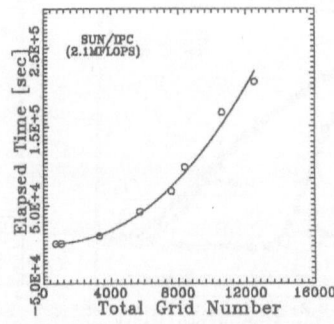

Fig.5 Hole contact size (WxW) vs. junction depth and oxide thickness [5].

Fig.6 Evaluated CPU time with varing the number of total nodal points.

SIMULATION OF SEMICONDUCTOR DEVICES AND PROCESSES Vol. 5
Edited by S. Selberherr, H. Stippel, E. Strasser – September 1993

3-D Diffusion Models for Chemically-Amplified Resists Using Massively Parallel Processors

E. Tomacruz, M. Zuniga, R. Guerrieri[†], A. Neureuther, and
A. Sangiovanni-Vicentelli

Department of Electrical Engineering & Computer Sciences, University of California
Berkeley, CA 94720, USA
[†]Dipartimento di Elettronica, Informatica e Sistemistica, Università di Bologna
Viale Risorgimento 2, I-40136 Bologna, ITALY

Abstract

Three-dimensional concentration dependent diffusions and simultaneous chemical reactions in chemically-amplified photoresists are simulated. Fickian, a linearly increasing diffusivity and an exponentially increasing diffusivity due to free volume increase with T-BOC are considered. Two different grid to processor mappings are proposed in implementing the simulator on massively parallel processors using the Scharfetter and Gummel discretization method. Experimental and simulation results which support an exponential diffusion coefficient are illustrated.

1. Introduction

The lithographic process of transferring an ideal layout to a less than ideal resist profile on a wafer is an increasingly important consideration in IC design. The ability of the nonlinearity of the resist to produce vertical line-edge profiles from low quality images in the presence of sharp vertical standing waves is an important technological improvement which has extended the working resolution of optical lithography. The improvement is especially significant for chemically-amplified resist systems which have both nonlinear chemical reaction kinetics and simultaneous concentration dependent diffusion. The level of complexity in modeling these reaction and diffusion effects is similar to that involved in modeling impurity concentration and point defect dependent diffusion in silicon. The use of high post-bake temperatures is preferred as resist sensitivity improves. However, it results in increased diffusion which must be examined in three dimensions. The nonlinearity of the reactions and the dependence of the diffusion on the local concentration can lead to improved performance. Understanding and balancing these mechanisms is the key goal in designing production worthy resists.

This research project models the 3D movement and reaction of species in the post-exposure bake of chemically-amplified resist systems. Both weak and strong dependencies of diffusion on species concentration are considered. Combinations of nonlinear reaction kinetics and concentration dependent diffusion scenarios are being considered for both acid-hardening (negative) resist and deprotection reaction (positive) resist. A bidirectional link to the SAMPLE-3D lithography simulator is established for facilitating studies of 3D effects.

2. Computational Models

A general model for two interacting and one diffusing species summarized by the equations below is used [1].

$$\frac{\partial C_1}{\partial t} = k_1(1 - C_1)C_2^{1.42} \ , \ \frac{\partial C_2}{\partial t} = \nabla \cdot (D_2 \nabla C_2) + k_2 C_2 \ , \ \frac{\partial}{\partial z}C_2(x, y, z) = 0|_{z = 0, T} \ ,$$

$$C_1(x, y, z) = 0|_{t = 0} \ , \ C_2(x, y, z) = C_{ss}|_{z = 0}$$

where C_1 is the concentration of activated sites, C_2 is the concentration of acid, k_1 is the reaction rate coefficient, and k_2 is the rate coefficient for the acid loss reaction. D_2 is the diffusion coefficient which may be dependent on C_1. The terms with k coefficients model a class of resist materials that rely on the acid catalyzed cross-linking of the resin matrix during post-exposure bake. The third equation, where T is resist thickness, specifies that no net flow of acid occurs across the boundaries. In a silylation process, silicon-containing compounds diffuse into the resist from an outside source. This is modeled by the fifth equation which forces the concentration of the species to be the constant C_{ss} at the resist surface.

The diffusion is not well understood and might be due to the nonlinear dependence of the diffusion rate on 1) the acid concentration itself (constant)[2], (2) the presence of deprotection sites which provide additional stepping stones (linear model), or (3) the increase in free volume with the deprotection reaction which creates a very rapid increase in diffusion pathways (exponential model)[3]. To explore the possibility of these various classes of acid concentration and material state dependent diffusion a general purpose 3-D reaction-diffusion simulator was written. For each of the three classes of increasingly higher nonlinearity an appropriate algorithm was chosen.

$$D_2 = \alpha \ , \ D_2 = \gamma + \beta C_1 \ , \ D_2 = \lambda exp(\omega C_1)$$

where α, γ, β, λ, and ω are constant parameters.

3. Discretization and Computational Steps

Since the exponential diffusion coefficient is a highly non-linear function of the variables, the standard difference discretization is not suitable for the task unless the grid spacings are made very small. To attain a more stable discretization, we adopt a technique proposed by Scharfetter and Gummel [4] which is now widely used for the discretization of the semiconductor device equations. The same approach is used for the discretization of the linear diffusion coefficient. Taking the limits of the linear diffusion coefficient discretization, the discretization for the constant diffusion coefficient is obtained.
Given that the flow of species C_2, is described as

$$J = \alpha \nabla C_2 \ , \ J = (\gamma + \beta C_1) \nabla C_2 \ , \ J = \lambda exp(\omega C_1) \nabla C_2$$

where C_1 is the other species interacting with C_2, two simplifying assumptions are made. First, C_1 is linearly discretized. Second, the flow of C_2 is constant between grid nodes in a one-dimensional grid. Using techniques used by Scharfetter and Gummel, the constant value of the flux can be extracted.

$$J = \alpha \frac{C_{2(i)} - C_{2(i-1)}}{x_i - x_{i-1}} \ , \ J = \frac{\beta(C_{1(i)} - C_{1(i-1)})}{\ln\left(\frac{\gamma + \beta C_{1(i)}}{\gamma + \beta C_{1(i-1)}}\right)} \frac{C_{2(i)} - C_{2(i-1)}}{x_i - x_{i-1}} \ ,$$

$$J = \lambda \frac{-\omega(C_{1(i)} - C_{1(i-1)})}{(exp(-\omega C_{1(i)}) - exp(-\omega C_{1(i-1)}))} \frac{C_{2(i)} - C_{2(i-1)}}{x_i - x_{i-1}}$$

Rectangular grids with nonuniformly spaced lines are used for the discretization. The nonlinear equations are solved using the Newton-Raphson method and the second order trapezoidal method is used for numerical integration. The Conjugate Gradient Squared (CGS) iterative algorithm is used to solve the unsymmetric sparse matrix. Incomplete LU decomposition with the partitioned natural ordering [5] is used as a preconditioner for the linear solver. At each time point, a nonlinear system of equations is solved using the Newton-Raphson method. For each Newton-Raphson iteration, a linear system of equations is solved. These computational steps are CPU time expensive for 3-D simulations but are expected to give convergent results that are accurate even for non-constant diffusion dependencies. Since mesh structures with over 100,000 nodes are not uncommon, super-computing machines are necessary. The Connection Machine 2 (CM-2) and the Connection Machine 5 (CM-5) computer architectures used for this study offer the computation power needed. Hence, massively parallel processors (MPPs) make 3-D simulations of reaction kinetics and diffusion in advanced resist processes practical.

4. Algorithmic Mapping

For the CM-2, each grid point in the mesh structure is mapped into a corresponding processor. Thus each processor stores the local values of C_1 and C_2, and the two rows of the matrix of the corresponding grid node. Using this allocation, the grid fits naturally on the organization of the machine and, at the same time, variables having a strong coupling due to spatial adjacency are tightly clustered. A reduction in the size of the matrix can be obtained by replacing C_1 with a function of C_2. With 256k bits of memory for each processor, a virtual processor ratio of 64 is obtained. Hence, an 8k CM-2 with 256k bits of memory for each processor would allow a user to simulate a 512k mesh structure.

The CM-5 mesh structure is divided into rectangular blocks each called a subdomain [6]. Each subdomain is mapped to a processor and the dimensions of the subdomains are equal in order to form cube partitions. This minimizes the total surface area which in turn minimizes the data length of communications between processors. A simple row ordering is used to map the subdomains to the CM-5 processors since the fat tree connections allow minimal penalty for communications between arbitrary processors [7].

5. Experimental and Simulation Results

Preliminary experimental data illustrated in Figure 1 show evidence of diffusion effects in I.B.M. APEX-E resist profiles. A Fisher F-test shows that a linewidth loss linear in time best describes Figure 1 since a linear model provides smaller sums of residuals compared to a quadratic model [8]. Also, the resist tends to have very vertical sidewalls regardless of exposure or bake time which would suggest a vertical front is propagating laterally through the resist. Figure 2 shows simulation results for the three proposed diffusion models. β is 1.0e-18 for *linear1*, β is 1.0e-17 for *linear2*, λ is 0.3e-19 for *exp1*, λ is 1.0e-19 for *exp2*, and λ is 1.0e-18 for *exp3*. α is 1.0e-18, γ is 1.0e-18, and ω is 5.0 for all simulations. Only the exponential model could show a linear trend and, as illustrated in Figure 3, it is the only model observed to have a vertical front propagating through the resist.

Current simulation data also show high average of acid concentration associated with the fraction of a standing wave cycle near the resist surface. This high average forms a foot and top-lip on the resist profile which are usually not seen in experimental results. Only the exponential diffusion coefficient has sufficient vertical acid movement to nearly eliminate foot and top-lip formations. More experimental results are presented in [8].

References

[1] R. Ferguson, Ph.D. Thesis, University of California, Berkeley, May 1991.
[2] T. Yoshimura et.al., *J. Vac. Sci. Technol. B*, Vol. 10, No. 6, 1992.
[3] W. Hinsberg et.al., *SPIE Proceedings*, Vol. 1925, 1993.
[4] D.L. Scharfetter, H.K. Gummel, *IEEE Trans. on Electron Devices*, Vol. ED-16, pp. 66-77, 1969.
[5] D. Webber et.al., *IEEE Trans. on CAD*, Vol. 10, pp. 1201-1209, 1991.
[6] E. Tomacruz et.al., *International Workshop on VLSI Process and Device Modeling*, pp. 20-21, 1993.
[7] Z. Bozkus et.al., *IEEE 4th Symp on the Frontiers of Massively Parallel Computation*, pp. 100-107, 1992.
[8] M. Zuniga et.al., Submitted to *Elec., Ion, and Photon Beams Conference*, San Diego, CA, USA, 1993.

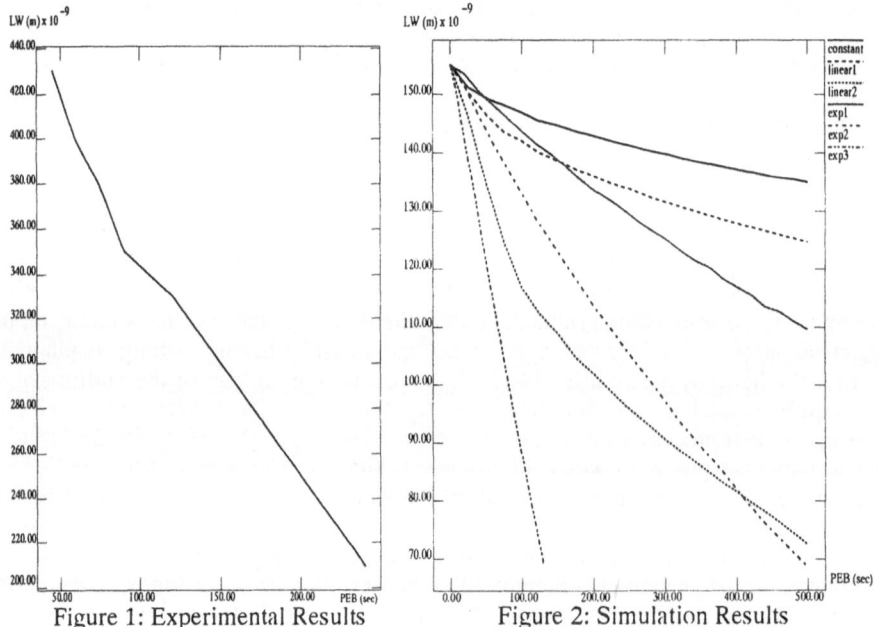

Figure 1: Experimental Results   Figure 2: Simulation Results

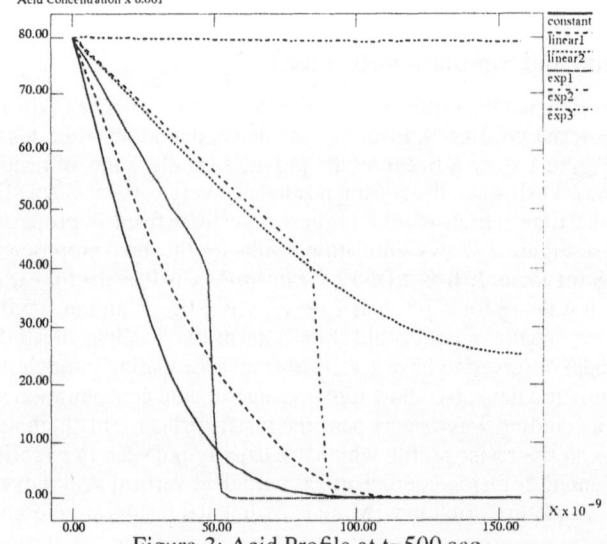

Figure 3: Acid Profile at t=500 sec

Improvement of Initial Solution Projection in Solving General Semiconductor Equations Including Engery Transport

L. L. So, D. Chen[†], Z. Yu[‡], and R. W. Dutton[‡]

Applied Theoretical Physics Division, Los Alamos National Laboratory
Los Alamos, NM 87545, USA

[†]Linear Process Development, National Semiconductor Corporation
2900 Semiconductor Drive, Santa Clara, CA 9502, USA

[‡]Integrated Circuits Laboratory, Stanford University
Stanford, CA 94305, USA

Abstract

An extension of Newton iteration method which provides an effective way to project an initial guess for a subsequent bias conditions – Newton projection Method (NPM) – is described. In particular, we applied the NPM to a system of semiconductor device equations including energy transport. The computational advantages of initial guess projection via NPM are illustrated through examples.

1. Introduction

It is well known that the closeness of the initial guess to the final solution is one of the most critical steps in solving nonlinear equations using the Newton-Raphson iterative method. Although there are several initial guess schemes such as "extrapolation" and "previous" currently adopted in popular semiconductor device simulators, each of these schemes has its limitation. Especially when the problem itself becomes more nonlinear, which is the case when the energy transport equations are added to the conventional drift-diffusion model, finding an effective way to project a "good" initial guess is very crucial.

We have endured a very difficult experience in achieving the convergence for solving the device characteristics once the carrier temperatures are included in addition to the electrostatic potential and carrier concentrations. In particular, when the device is in the operation region where the terminal currents are small, the bias steps have to be limited to about 50 mV for both diodes (e.g., ballistic $n^+ - n - n^+$) and BJTs. The underlying reason is that the carrier temperatures change very rapidly over a short distance, adding severe nonlinearity to the semiconductor problem.

In this work, we describe the extension of Newton Projection Method (NPM) [1] in solving semiconductor equations including energy transport equations[2, 3]. The central idea of the proposed algorithm is to treat biases, which enter into the system

---

[1]Presently with Applied Theoretical Physics Division, Los Alamos National Laboratory, Los Alamos, NM 87545.

[2]Presently with Linear Process Development, National Semiconductor Corp., Santa Clara, CA 95052.

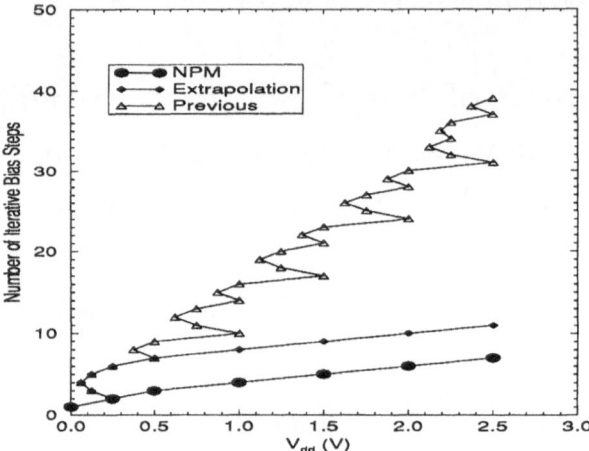

Figure 1: Comparison of the total number of required iterative bias steps of different initial guess schemes for a 1-μm silicon $n^+ - n - n^+$ diode.

is shown in Fig. 1. As expected, the case in which only the previous solution is used has the worst performance. A built-in mechanism which automatically cuts by half the magnitude of the bias step has to be frequently invoked whenever a convergence problem is encountered. Hence each back-tracking (converged or yet-to-be-convergenced) step is being counted into the total number of iterative steps required because of the actual computational cost involved. Although it is possible to use smaller bias steps initially to reduce the cost of back-tracking, it is rather difficult to achieve during a dynamic biasing cycle since it is nearly impossible to have a priori knowledge of what the optimal magnitude for the bias steps should be. While the "extrapolation" method renders some improvement especially in the high bias region, considerably smaller bias steps (about 50 mV) are required near the equilibrium, where the distribution of the carrier temperatures is very nonlinear with respect to the applied voltage. Fig. 2 illustrates the distribution of the majority carrier temperature along the device at small biases.

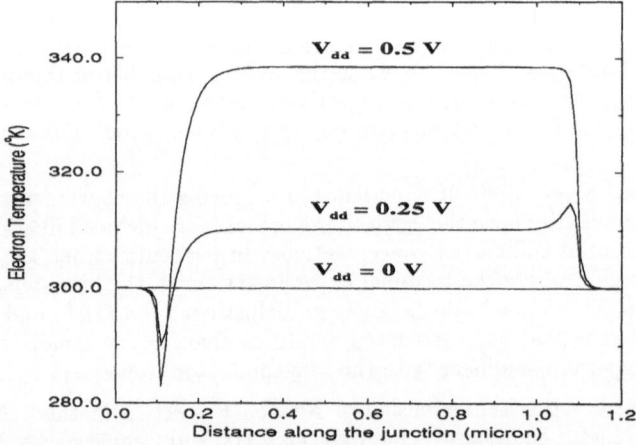

Figure 2: Electron temperature distribution along silicon $n^+ - n - n^+$ diode.

In a bipolar transistor, biases are required to change for both V_{BE} and V_{CE} in order to simulate the I-V characteristics. The number of iterative bias steps required for solution when employing different initial guess schemes as V_{CE} increases for an NPN transistor is shown in Fig. 3. Again, while the convergence behavior of the "extrapolation" method

of semiconductor equations through the boundary conditions, as variable parameters along with the basic equation variables, such as electrostatic potential, carrier concentrations and carrier temperatures. The variation of these quantities with respect to the change of applied bias at the current operation point is readily obtainable through a single Gaussian elimination step, given the Jacobian matrix. Thus by knowing the the functional dependence of the potential, carrier concentrations and temperature distribution on bias, a first order approximation of the solution for the next bias point can be constructed and can be used as an initial guess for the Newton iteration for a subsequent bias.

2. Proposed Algorithm

Consider a semiconductor system consisting of the basic quantities – potential ψ, the carrier concentrations n and p, the carrier temperatures T_n and T_p. The goal for a dc solution is to find a vector $\mathbf{x}$ for each fixed bias condition $\mathbf{V}_o$ for the following equation:

$$\mathbf{f}(\mathbf{x}, \mathbf{V}_o) = 0. \tag{1}$$

Here the vector $\mathbf{f}$ consists of the Poisson equation, carrier continuity equations and energy balance equations at each grid point. The solution vector has the form

$$\mathbf{x} = (\psi_1, n_1, p_1, Tn_1, Tp_1, \cdots, \psi_N, n_N, p_N, Tn_N, Tp_N)^T,$$

with the subscript N representing the number of grid points in the solution region. The bias $\mathbf{V}_o$ enters the equation as a boundary condition and is treated generally as a fixed parameters.

By the use of the Newton Projection Method, the first order approximation to the solution at a subsequent bias can be evaluated using the Taylor expansion of (1)

$$\mathbf{f}(\mathbf{x}_o + \Delta\mathbf{x}, \mathbf{V}_o + \Delta\mathbf{V}) = \mathbf{f}(\mathbf{x}_o, \mathbf{V}_o) + \left.\frac{\partial \mathbf{f}}{\partial \mathbf{x}}\right|_{(\mathbf{x}_o, \mathbf{V}_o)} \Delta\mathbf{x} + \left.\frac{\partial \mathbf{f}}{\partial \mathbf{V}}\right|_{(\mathbf{x}_o, \mathbf{V}_o)} \Delta\mathbf{V} = 0 \tag{2}$$

where $\mathbf{x}_o$ represents the dc solution. Because of the fact that $\mathbf{f}(\mathbf{x}_o, \mathbf{V}_o) = 0$, the above equation reduces to

$$\left.\frac{\partial \mathbf{f}}{\partial \mathbf{x}}\right|_{(\mathbf{x}_o, \mathbf{V}_o)} \Delta\mathbf{x} = - \left.\frac{\partial \mathbf{f}}{\partial \mathbf{V}}\right|_{(\mathbf{x}_o, \mathbf{V}_o)} \Delta\mathbf{V} \quad . \tag{3}$$

Here the Jacobian matrix $\partial\mathbf{f}/\partial\mathbf{x}|_{(\mathbf{x}_o, \mathbf{V}_o)}$ is already known from the last Newton iteration for the dc solution, and the vector on the right hand side $\partial\mathbf{f}/\partial\mathbf{V}|_{(\mathbf{x}_o, \mathbf{V}_o)}\Delta\mathbf{V}$ which relates the system $\mathbf{f}$ to the boundary condition $\mathbf{V}$ has a relatively simple form. Hence the solution $\Delta\mathbf{x}$ in Eq. (3) is straighforward since only one Guassion elimination step is involved. Since there is no iteration is required, the compution effort is minor.

In the case of a non-negligible magnitude of $\Delta\mathbf{V}$, such as bias step advances, $\mathbf{x} + \Delta\mathbf{x}$ represents a first-order approximation to the solution of system at the new bias of $\mathbf{V}_o + \Delta\mathbf{V}$. Thus it can be used as an initial guess for the final solution at the new bias.

3. Device Examples

We simulated a 1-μm silicon $n^+ - n - n^+$ diode using the ET model, starting from equilibrium condition up to V_{dd}=2.5 V using some fixed bias steps. The number of iterative bias steps required for solution when employing different initial guess schemes is comparable to NPM at large bias, the NPM is superior near the equilibrium. A consistent bias steps of 0.25 V can be used throughout the simulation region without incurring convergence problem.

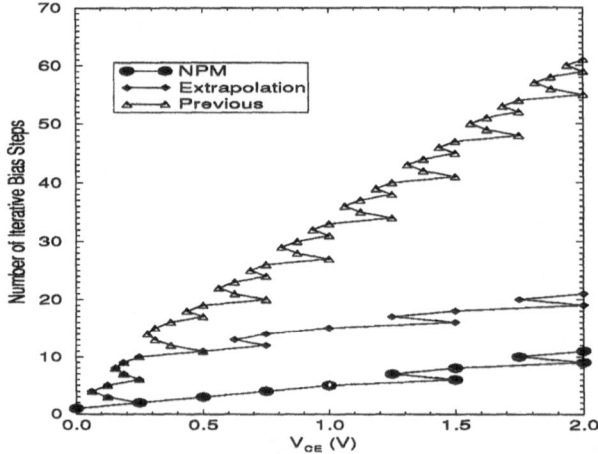

Figure 3: Comparison of the total number of required iterative bias steps of for an NPN bipolar transistor.

4. Conclusions

We have applied the NPM to semiconductor device modeling including energy transport equations. We have shown that the NPM can enhance the rate of convergency during bias advancing over conventional project methods by providing a better initial guess solution, especially when severe nonlinearity are present in the device equations. The improvement in the initial guess permits application of larger bias steps. Additionally, the advantages of the NPM over conventional projection method are two-fold. First, it is based only on the solution at the previous operation point, whereas the conventional extrapolation method needs at least two previous bias points. Hence the method can start even from the equilibrium solution. Second, since the change in bias is not limited to the scalar variable, the prediction can be made when the biases at the different contacts change with different magnitudes.

References

[1] Z. Yu, R. W. Dutton, and M. Vanzi, "An extension to Newton's Method in Device Simulators – On an efficient algorithm to evaluate small-signal parameters and to predict initial guess," *IEEE Trans. Computer-aided Design*, vol. CAD-6, no. 1, pp. 41-45, Jan. 1987

[2] D. Chen, E. C. Kan, U. Ravaioli, C. Shu and R. W. Dutton, "An improved energy transport model including non-parabolic and non-Maxwellian distribution effects," *IEEE Elec. Dev. Lett.*, vol. 13, no. 1, pp. 26-28, Jan. 1992.

[3] D. Chen, E. Sangiorgi, M. R. Pinto, E. C. Kan, U. Ravaioli and R. Dutton, "Analysis of spurious velocity overshoot in hydrodynamic simulations of ballistic diodes," *Proceeding of NUPAD V (Numerical Process and Device Modeling Workshop)*, Seattle, May 31 – June 1, 1992.

A Numerical Implementation of the Energy Balance Equations Based on Physical Considerations

W. Schoenmaker and R. Vankemmel

IMEC
75 Kapeldreef, B-3001 Leuven, BELGIUM

Abstract

We present an implementation of the energy balance equations in two dimensional device simulators which is based on a definite positive formulation of the ohmic energy sources for the carriers.

1. Introduction

The correct numerical treatment of the energy balance equations remains still a topic of active research. This is because naive discretizations of these equations lead to very unstable algorithms. Sometimes it may be very useful to return to the physical contents of the equations and derive a discretization prescription which is based on the underlying physics of the problem. A nice example which illustrates this statement is the Scharfetter-Gummel method for current discretization, which expresses charge conservation along a flux line segment. Starting from the physical meaning of the variables in the energy balance equations, we will derive a discretization prescription for them. Remarkably enough, we find a formulation of the energy balance equations which is well suited for considerations concerning the definiteness of the energy sources and sinks. The definite signs of these sources and sinks are very helpful in the construction of the Newton matrices, since the robustness improves considerably, if the linear solvers deal with matrices, which are semi-definite. [1] In this paper we will reformulate the energy balance equations in such a way that (1) the physical meaning of the variables is respected in the discretization procedure and (2) it becomes very straightforward to satisfy the condition that the Newton matrix is semi-definite.

2. The Hydrodynamic Model

The energy balance equations in steady state operation are (see Forghieri et al. [2])

$$\vec{\nabla}.\vec{S} - \vec{E}.\vec{J} + Uw + c\frac{w - w^*}{\tau_w} = 0 \tag{1}$$

where $c = p$ for holes and $c = n$ for electrons. The carrier energies are

$$w = \frac{1}{2}mv^2 + \frac{3}{2}kT \tag{2}$$

For the currents we use the thermodynamic relations

$$\vec{J} = q\mu c \vec{E} - sq\mu kT \vec{\nabla} c - sq\mu kc \vec{\nabla} T \tag{3}$$

The 'energy' flux vectors are

$$\vec{S} = -\kappa \vec{\nabla} T + \frac{s}{q}(w + kT)\vec{J} \tag{4}$$

The convective term in the energy flux vector has been neglected. This is motivated by the fact that we only take into account fluxes which are linear dependent of the driving mechanisms, $\vec{E}$, $\vec{\nabla} T$ and $\vec{\nabla} c$.

The vector $\vec{S}$ can not be considered as the correct energy flux vector. This is because the *physical* energy flux vector is given by

$$\vec{S}_{phys} = -\kappa \vec{\nabla} T + \frac{s}{q} w \vec{J} \tag{5}$$

Each term has a clear physical intepretation. The first term corresponds to heat conduction, i.e. energy transport without matter transport, whereas the second term represents the energy transport due to matter transport. As a consequence, the following expression,

$$\vec{\nabla}.(\vec{S} - \vec{S}_{phys}) = \frac{sk}{q}\vec{\nabla}.(T\vec{J}) \tag{6}$$

must be considered as a heat source in the energy balance equation. Indeed, since $\vec{J} = sc\vec{v}$, we have

$$\vec{\nabla}.(T\vec{J}) = \vec{J}.\vec{\nabla} T + sT\vec{v}.\vec{\nabla} c + sTc\vec{\nabla}.\vec{v} \tag{7}$$

and the energy balance equations become with using $\vec{F} = \frac{\vec{J}}{\mu c}$

$$\vec{\nabla}.\vec{S}_{phys} - \frac{\vec{J}}{q}.\vec{F} + wU + c\frac{w - w^*}{\tau_w} + \frac{kT}{q}c\vec{\nabla}.\vec{v} = 0 \tag{8}$$

The four energy sources/sinks have a straightforward physical intepretation. The first term,

$$\Sigma^{cur} = \frac{\vec{J}}{q}.\vec{F} = \frac{\mu c}{q}F^2 \tag{9}$$

corresponds to the work performed by the driving force. Note that this term is *positive definite*. Any numerical implementation should respect this property. In particular, in strong inversion layers, the numerical evaluation of F^2 is much less ambiguous than the evaluation of $\vec{E}.\vec{J}$. Furthermore, the quadratic appearance of this ohmic heat source term is very suitable for analysis of the signs of its contribution to the Newton-Raphson matrix. Indeed, it suffices to evaluate the signs of J_{ij} along the links.

The last term,

$$\Sigma^{com} = -\frac{kT}{q}c\vec{\nabla}.\vec{v} \tag{10}$$

corresponds to a change in energy density due to a compression of the carriers.

The reformulation with Σ^{cur} is also very instructive if we consider it together with Σ^{latt}. Since $\vec{v} = s\mu\vec{F}$, we obtain

$$\Sigma^{cur} + \Sigma^{latt} = (1 - \frac{1}{2}\frac{m\mu}{q\tau_w})\frac{\mu c}{q}F^2 - \frac{3}{2}kc\frac{T - T^*}{\tau_w} \tag{11}$$

There cannot be relaxed more kinetic energy to the lattice than the driving force can perform work, therefore we obtain the following consistency condition

$$1 - \frac{1}{2}\frac{m\mu}{q\tau_w} = 1 - \frac{1}{2}\frac{\tau_p}{\tau_w} \geq 0 \tag{12}$$

If this condition $\tau_p \leq 2\tau_w$, is not fulfilled, we will obtain carrier cooling. This phenomenon is not observed in Monte Carlo simulations. In a recent paper of Gardner [3], in which he compares the hydrodynamic model with Monte Carlo computations, this condition is indeed satisfied.

3. Scharfetter-Gummel fluxes.

The Scharfetter-Gummel method expresses the fluxes D_{ij}, J_{ij} and S_{ij} as functions of the endpoint variables: $\psi_i, c_i, T_i, \psi_j, c_j, T_j$. Since the fluxes are the linear responses to the driving mechanisms, we propose a decoupling which is based on this physical approximation, i.e. at each link of the mesh, the coefficients in front of the forces are taken constant. Then we find the following Scharfetter-Gummel fluxes for the links:

$$J_{ij}h = sq[D]\left(c_i B(X) - c_j B(-X)\right) \tag{13}$$

and

$$S_{ij}h = \frac{s}{q}[\epsilon_{kin}]J_{ij}h + [\kappa]\left(T_i B(Y) - T_j B(-Y)\right) \tag{14}$$

where B is the Bernoulli function. Furthermore we set $\bar{T} = (T_i + T_j)/2$ and $\Delta\psi = \psi_j - \psi_i$, etc. and

$$X = \frac{q}{k\bar{T}}s(\Delta\psi + \frac{k}{q}s\Delta T) \qquad Y = -\frac{3}{2}\frac{J_{ij}h}{q[\kappa]}s \tag{15}$$

4. Applications

In order to illustrate the proposed scheme for hydrodynamical modeling we have simulated a 0.5μ gate MOSFET. The structures is depicted in Fig. 1 and corresponds to a conventional LDD. Spacers are included. The following dimensions are used: oxide thickness: 10 nm, poly thickness: 350 nm, spacer width: 250 nm, contact hole: 0.7 μ, gate length: 0.5 μ, silicide thickness: 50 nm, contact-spacer gap: 0.04 μ, device height: 2 μ.

The silicides at the source and the drain are mimiced by added a very high dopant concentration to this region such that the resistivity becomes low. The polysilicon at the gate is represented by highly doped Silicon material.

In Fig. 2 the doping concentrations for the nmos is presented, whereas Fig. 3 shows the electron temperature distribution for $|V_{DS}| = 2.5V$ and $|V_G| = 2.0V$. In Fig.4 the IV characteristics are shown for the drift-diffusion model and the hydrodynamic model.

We observe that the hydrodynamic model gives rise to lower drain currents than the drift-diffusion model. The simulations confirm the experimental observation that there are also hotspots located at the source ofthe transistor.

Acknowledgements. Part of this work was funded by the ESPRIT project No. 7236 ADEQUAT. Discussions with Willy Schilders (PHILIPS) and Ludo Deferm (IMEC) and An Demesmaeker (IMEC) are gratefully acknowledged.

References

[1] W. Schilders see the proceedings SISDEP93

[2] A. Forghieri, R. Guerrieri, P. Ciampolini, A Gnudi, M. Rudan, G. Baccarani "A new discretization method of the semiconductor equations comprising momentum and energy balance." *IEEE Transactions on CAD Vol 7 231 (1988)*

[3] C.L. Gardner "Hydrodynamic and Monte Carlo Simulation of an Electron Shock Wave in a One Micron Semiconductor Device" *Proceedings of the International Workshop on Computational Electronics Beckman Institute. University of Illinois at Urbana-Champain, May 28-29, 1992*

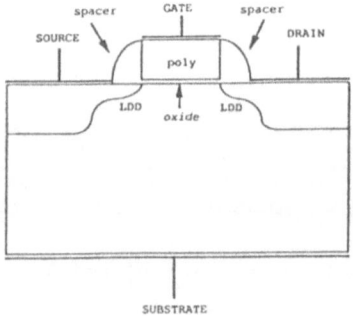

Fig.1: LDD MOS device lay-out.

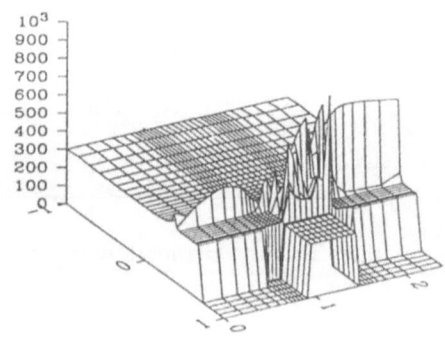

Fig.3: Electron temperature in nmos.

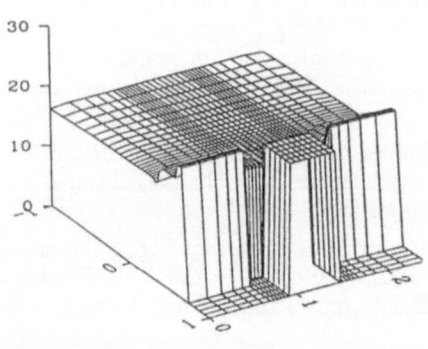

Fig.2: Log. of doping concentration.

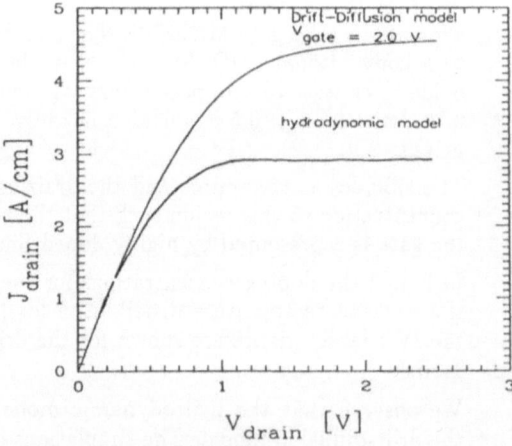

Fig.4: IV characteristic for nmos.

SIMULATION OF SEMICONDUCTOR DEVICES AND PROCESSES Vol. 5
Edited by S. Selberherr, H. Stippel, E. Strasser – September 1993

121

Construction of Stable Discretization Schemes for the Hydrodynamic Device Model

W. H. A. Schilders

Philips Research Laboratories
P.O.Box 80.000, NL-5600 JA Eindhoven, THE NETHERLANDS

Abstract

Several authors have reported numerical problems when simulating semiconductor devices using the hydrodynamic model. In this paper, the cause of these numerical problems is identified. Furthermore, a remedy consisting of adaptive quadrature rules is proposed. The resulting discrete schemes are stable, which is verified both theoretically and demonstrated through several examples.

1. Description of the model

The model considered in this paper has been described frequently in literature, and will therefore not be derived. It consists of the normal set of drift-diffusion equations, to which the following equations are added:

$$\nabla \cdot S_p = E \cdot J_p - R w_p - p \frac{w_p - w_0}{\tau_w^p}, \tag{1}$$

$$\nabla \cdot S_n = E \cdot J_n - R w_n - n \frac{w_n - w_0}{\tau_w^n}. \tag{2}$$

The energy flux densities are expressed in terms of the carrier temperatures, as follows:

$$S_p = -\kappa_p \nabla T_p + \frac{1}{q}(w_p + k T_p) J_p, \tag{3}$$

$$S_n = -\kappa_n \nabla T_n - \frac{1}{q}(w_n + k T_n) J_n. \tag{4}$$

In the above, w_p and w_n are the average energies of the carriers and the coefficients κ_p, κ_n are usually modelled using the Wiedemann-Franz law:

$$\kappa_p = C_{WF} k D_p p, \qquad \kappa_n = C_{WF} k D_n n.$$

In addition, the expressions for the current densities contain an extra term (due to temperature gradients) as compared to the classical drift-diffusion expressions:

$$J_p = -q D_p \nabla p + q \mu_p p E_p - q D_p^T p \nabla T_p, \tag{5}$$

$$J_n = q D_n \nabla n + q \mu_n n E_n + q D_n^T n \nabla T_n, \tag{6}$$

in which the thermal diffusion coefficients are often modelled by

$$D_p^T = \frac{D_p}{T_p} = \frac{k}{q} \mu_p, \qquad D_n^T = \frac{D_n}{T_n} = \frac{k}{q} \mu_n$$

2. The cause of numerical problems

The model described in the previous section is discretised using the finite volume method (or: box method). This means that, after a mesh (rectangular or triangular) has been constructed, each of the five partial differential equations is integrated over boxes around mesh points. In the resulting expressions, integrals of the form $\int \nabla \cdot Q$ (where Q is some vector quantity, for example J_n or S_n) are transformed into integrals over the boundary of the box. These boundary integrals can then be approximated by using low order quadrature rules, so that only normal components of the quantity Q at the midpoints between two nodes are required. These normal components are obtained by using Scharfetter-Gummel type expressions. This is very straightforward, and has been described in many publications.

The integrals of the right hand sides of the partial differential equations are usually approximated by applying the midpoint rule. Hence, the right hand sides only have to be evaluated at the mesh points. Both for Poisson's equation and the continuity equations, the midpoint rule can be shown to be the right choice in view of stability: the resulting discrete schemes yield non-oscillatory, physically relevant solutions, even for very coarse meshes. Because of the latter, adaptive meshing techniques starting from relatively coarse grids are possible. Clearly, this property would also be desirable for extended models, such as the hydrodynamic model. Unfortunately, this is not the case if the discretisation technique described is applied to equations (1)-(4). This is easily demonstrated using the following example, which was taken from [1]. For this problem (a pin-diode with a bias of 500 Volts), we display the discrete solution using 51 mesh points in Fig. 1. In addition to the oscillatory character of the discrete solution, we also found a rather slow convergence of the nonlinear solution process. These numerical problems were also observed for other devices; sometimes we even encountered severe cooling effects (carrier temperatures as low as 200 K).

Several authors have already expressed their opinion on the cause of these numerical problems. Some publications are devoted to a different discretisation of the normal components of S_p and S_n. This is, however, not the right angle of attack since it can be shown that the Scharfetter-Gummel type discretisation of these quantities leads to a system matrix with the right properties (M-matrix). Other authors have identified the discretisation of the term $E \cdot J_n$ as the cause of the problems, however, without giving a satisfactory explanation. Also, no adequate remedy has arisen from this observation.

After careful analysis of the system of equations, we found that the term $E \cdot J_n$ is indeed the cause of the numerical problems. More specifically, it is the additional term containing ∇T_n in the expression for J_n which causes instabilities. This can be verified mathematically by observing that, when expanding the term $E \cdot J_n$ into three terms, the use of the midpoint rule on the integrals over the boxes destroys the M-properties of the system matrix corresponding to the equations for the discrete carrier temperature T_n. A useful experiment to demonstrate this is the following: use a nonlinear solution strategy in which we first solve for the potential and the carrier concentrations (keeping the temperatures fixed), followed by the solution of the temperature equations. This iterative solution strategy may be viewed as a modified version of the wellknown Gummel's method. For the pin-diode used in Fig. 1, we find that the carrier temperatures are non-oscillatory after 1 iteration of this solution procedure. However, after 2 iterations we find 1 wiggle in the discrete solution and after 3 iterations there are 2 wiggles.

3. A stable discretisation scheme

The observation that the numerical problems are caused by the implicit dependence of the current density J_n on the carrier temperature T_n immediately points towards a possible remedy. Namely, the implicit dependence can be made explicit simply by expanding the term $E \cdot J_n$ into three terms and rewriting (2) in the form

$$\nabla \cdot S_n = qD_n E \cdot \nabla n + q\mu_n n \parallel E \parallel^2 + qD_n^T n E \cdot \nabla T_n - Rw_n - n\frac{w_n - w_0}{\tau_w^n} \qquad (7)$$

The third term in the right hand side of (7) being the only problematic term, we suggest to use the midpoint rule for all other terms, and a special quadrature rule for the integral of $qD_n^T n E.\nabla T_n$. We have developed an adaptive rule which takes into account the direction of the electric field E, full details of which will be given in a more extensive report (and at the conference). The effect of this rule on the coefficient matrices of the discretised system is such that contributions to the diagonal are non-negative, whereas contributions to the off-diagonal elements are non-positive. In this way, the M-matrix character [3] is conserved. In Fig. 2 we show the solution of the pin-diode obtained using this new quadrature rule, again with only 51 uniformly distributed mesh points.

4. Conclusion

In this paper, we have identified the true cause of the numerical problems encountered when using the hydrodynamic model. Having established, both theoretically and experimentally, that the implicit dependence of the current density on the gradient of the carrier temperature causes these problems, making this term explicit and using a different quadrature formula in the box integrals is shown to provide an adequate remedy. Although, in this paper, only a simple 1-d example has been given, experiments on 2-d examples have shown that the method performs well on more complicated devices (MOS-devices, bipolar transistors). The fact that the remedy has a sound mathematical basis is a guarantee that numerical problems are avoided irrespective of the grid size. In this respect, the proposed discretisation leads the way to adaptive meshing also for hydrodynamic simulations, without having to use extremely fine initial grids.

In addition to the non-oscillatory character of discrete solutions, we have found the nonlinear solution processes to converge much better. Indeed, vector extrapolation methods as used in [2] can be used to accellerate the modified Gummel method so as the obtain solutions even faster.

References

[1] M. Hamza, H. Morel, J.P. Chante, *Simulation of bipolar transport in semiconductor p-n junctions using the generalized hydrodynamic equations*, COMPEL, vol. 10, pp. 289-299 (1991)

[2] W.H.A. Schilders, P.A. Gough, K. Whight, *Extrapolation techniques for improved convergence in semiconductor device simulation*, Proc. NASECODE VIII Conference, Boole Press, Dublin, pp. 94-95 (1992)

[3] R.S. Varga, *Matrix iterative analysis*, Prentice Hall, London (1962)

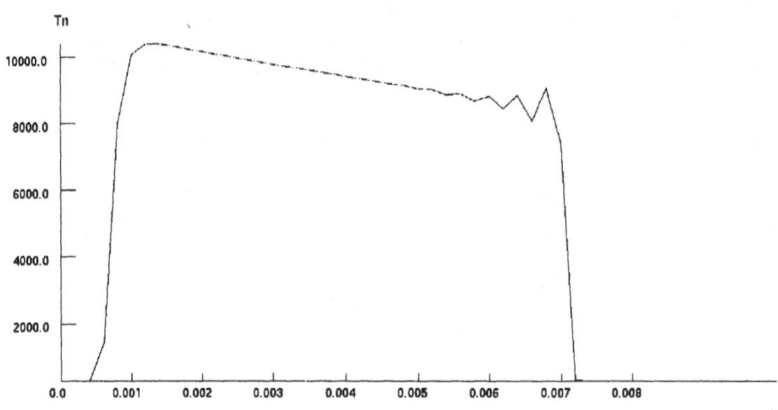

Figure 1: Electron temperature for pin-diode at -500 V using midpoint rule

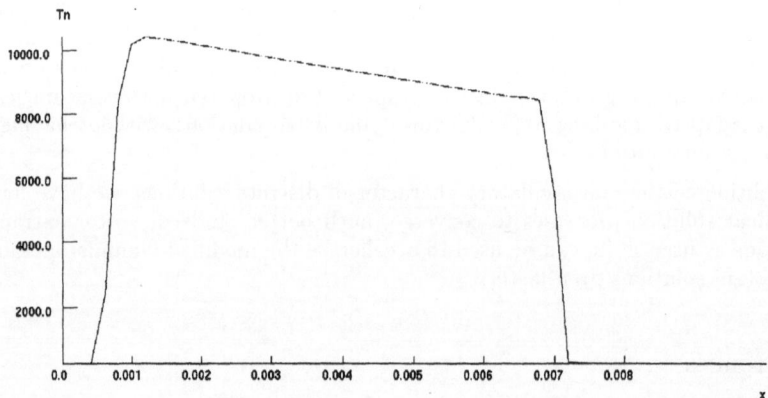

Figure 2: Electron temperature for pin-diode at -500 V using adaptive rule

SIMULATION OF SEMICONDUCTOR DEVICES AND PROCESSES Vol. 5
Edited by S. Selberherr, H. Stippel, E. Strasser – September 1993

Three-Dimensional Implementation of a Unified Transport Model

A. Pierantoni, A. Liuzzo, P. Ciampolini[†], and G. Baccarani

Dipartimento di Elettronica, Informatica e Sistemistica, Università di Bologna
Viale Risorgimento 2, I-40136 Bologna, ITALY
[†]Istituto di Elettronica, Università di Perugia
I-06131 Santa Lucia-Canetola, Perugia, ITALY

Abstract

This paper describes a unified transport model which self-consistently accounts for thermal effects and hot-carrier phenomena. Such result is achieved by including the energy balance equations for electrons, holes and lattice. The model has been incorporated into the three-dimensional device simulator HFIELDS-3D and its numerical efficiency is tested by simulating both unipolar and bipolar devices.

1. The physical model

The behavior of submicron devices is influenced by complex interactions among non-stationary phenomena, device self-heating, as well as fringing and proximity effects. In principle, therefore, such effects have to be self-consistently described within a unified framework: at present, however, only a few attempts to exploit such a model have been reported, mainly because of the computational burden associated with it. For this reason, Szeto and Reif [1], as well as Benvenuti *et al.* [2], limited themselves to one-dimensional analyses, whereas Katayama and Toyabe [3] developed a three-dimensional, finite-difference discretization scheme.

We present the inclusion of a generalized transport model into the more versatile environment provided by HFIELDS-3D, and compare the efficiency of some solution schemes. The model originates from the three BTEs which describe the dynamics of three interacting subsystems: namely electrons, holes and phonons. Energy-balance equations are obtained by taking the moments of order two of the BTEs:

$$\frac{\partial W_{n,p}}{\partial t} + \operatorname{div} \vec{S}_{n,p} = \vec{F} \cdot \vec{J}_{n,p} + \left(\frac{\partial W_{n,p}}{\partial t} \right)_{coll} \tag{1}$$

$$\frac{\partial W_L}{\partial t} + \operatorname{div} \vec{S}_L = \left(\frac{\partial W_L}{\partial t} \right)_{coll} \tag{2}$$

In the above equations, W represents the energy density and $\vec{S}$ the energy flow, defined as follows:

$$\vec{S}_{n,p} = -\kappa_{n,p} \operatorname{grad} T_{n,p} \mp \left(\frac{5}{2} - s \right) \frac{k_B T_{n,p}}{q} \vec{J}_{n,p} \tag{3}$$

$$\vec{S}_L = -\kappa_L \operatorname{grad} T_L \tag{4}$$

while the current densities $\vec{J}$ are given by:

$$\vec{J}_n = q\mu_n \left[\frac{k_B T_n}{q} \operatorname{grad} n + n \operatorname{grad} \left(\frac{k_B T_n}{q} - \varphi \right) \right] \tag{5}$$

$$\vec{J}_p = -q\mu_p \left[\frac{k_B T_p}{q} \operatorname{grad} p + p \operatorname{grad} \left(\frac{k_B T_p}{q} + \varphi \right) \right] \tag{6}$$

The collision terms, which account for the interactions among different subsystems, are described according to the relaxation-time approximation and fulfill the following relationship:

$$\left(\frac{\partial W_L}{\partial t} \right)_{coll} + \left(\frac{\partial W_n}{\partial t} \right)_{coll} + \left(\frac{\partial W_p}{\partial t} \right)_{coll} = E_G U \tag{7}$$

Finally, the model is completed by Poisson's and current-continuity equations. Spatial discretization is carried out by the Box Integration Method (in analogy with the scheme presented in [4]) applied to the hybrid mesh suggested by Conti et al. [5].

2. Simulation results

In this section, the simulation of two simple devices is discussed: namely, the MOS-FET shown in Fig. 1 and the BJT shown in Fig. 2 are considered. In both cases, comparisons among several transport models are carried out; more specifically, the following schemes are used: the standard drift-diffusion model (DD), the hydrodynamic model (HD) described in [4], the electrothermal (ET) model introduced in [6] and, finally, the "generalized" thermal-hydrodynamic (TH) model illustrated so far. Contour lines in Figs. 1 and 2 refer to the lattice temperature predicted by the TH model. Fig. 3 compares the MOSFET output characteristics predicted by the different models which exhibit some non-negligible discrepancies. In the low V_{DS} range, neither carrier nor lattice heating occurs, so that predictions given by the four models agree well; at intermediate V_{DS} values, carriers may attain energies well in excess of their equilibrium values, while no lattice heating occurs: HD and TH models consistently predict higher drain currents than DD and ET, due to the carrier velocity overshoot. Eventually the lattice heats up as well, thus resulting in increased lattice scattering probability; the $I_D^{(ET)}$ and $I_D^{(HT)}$ curves deviate downward, with respect to the corresponding isothermal models.

Fig. 4 illustrates the BJT turn-on characteristics: again a fairly large spreading of simulation results is found, depending on the model employed. Basically, carrier and lattice heating occur at the collector, reverse-biased junction, whereas a moderate cooling-down of the carriers can be observed at the emitter junction. In the simulated conditions, impact ionization phenomena play no significant role and are neglected. Thus, little influence of the carrier heating over the collector current is found, being the latter dominated by the cold carriers injected at the emitter. Consistently, the DD and HD predictions are fairly close: differences are mainly to be ascribed to the thermoelectric field appearing in Eqs. (5, 6). Lattice heating, instead, has a substantially larger impact on the characteristics, which is basically due to the increase of the minority carrier density in the base region.

3. Numerical strategies

A truly three-dimensional problem usually requires a large number of meshpoints, which makes it unpractical using a Newton-Raphson solution scheme. A block Gauss-Seidel scheme, analogous to the classical Gummel procedure, has thus necessarily to be used, which calls for a partition of the model equations and the unknown variables into separate sets. Table 1 illustrates our findings about different partitioning strategies: computational performances are compared, for the single-carrier case, assuming the plain DD scheme as a reference. In the second column, the abbreviations POI, ECC, LEB and EEB stand for the set of algebraic equations arising from the discretization of Poisson's equation, of the electron current-continuity equation, of the lattice energy balance and of the electron energy balance, respectively. The three-block TH scheme is simply obtained by supplementing the HD scheme with the decoupled solution of the LEB equation: however, due to the strong coupling among carrier and lattice temperatures, the convergence of such a simplistic algorithm turns out to be unacceptably slow. We found the two-block TH scheme to be much more effective: actually, in the unipolar case, by solving LEB and EEB in a coupled fashion, the computational overhead required by the model can be as low as 25 %, compared with the more traditional hydrodynamic solution. From a qualitative standpoint, similar results are exhibited by the bipolar transistor simulation, as shown in table 2^1: in this case, however, the performance ratio between HD and TH models, as well as between DD and ET models, is slightly worse, due to the larger sensitivity of the device behavior to the lattice temperature. Conversely, since the coupling among carrier energy-balance and continuity equations is weaker, the overall performance of the TH model, with respect to the DD scheme, improves.

The practicality of the TH model, as well as the completeness of the provided physical picture, is hence demonstrated.

References

[1] S. Szeto and R. Reif, "A unified electrothermal hot-carrier transport model for silicon bipolar transistor simulations," *Solid State Electronics*, vol. 32, pp. 307–315, 1989.

[2] A. Benvenuti, G. Ghione, M. Pinto, W. Coughran, and N. Schryer, "Coupled thermal-fully hydrodynamic simulation of InP-based HBTs," in *Proceedings of the IEDM 92 Conf.*, (San Francisco), pp. 737–740, 1992.

[3] K. Katayama and T. Toyabe, "A new hot carrier simulation method based on full 3D hydrodynamic equations," in *Proceedings of the IEDM 89 Conf.*, (Washington D.C), pp. 135–138, 1989.

[4] A. Pierantoni, P. Ciampolini, A. Gnudi, and G. Baccarani, "Three-dimensional evaluation of substrate current in recessed-oxide MOSFETs," *IEICE Trans. Electron.*, vol. E75-C, no. 2, pp. 181–188, 1992.

[5] P. Conti, N. Hitschfeld, and W. Fichtner, "OMEGA-an octree-based mixed element grid allocator for the simulation of complex 3D device structures," *IEEE Trans. on CAD of ICAS*, vol. CAD-10, pp. 1231–1241, 1991.

[6] A. Pierantoni, P. Ciampolini, and G. Baccarani, "Accurate modeling of electrothermal effects in silicon devices," in *Proceedings of the ESSDERC XXII Conf.*, (Lovanio), pp. 769–772, 1992.

---

[1]In table 2, abbreviations HCC and HEB indicate hole continuity equations and hole energy balance equations, respectively

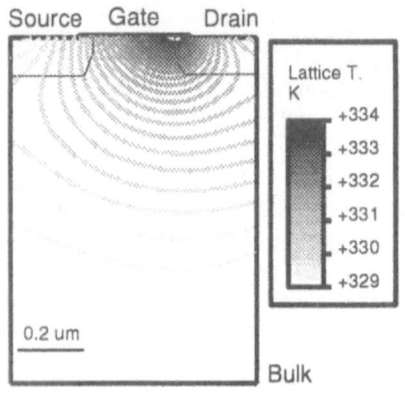

Fig. 1: Lattice temperature in the MOSFET ($V_{GS} = V_{DS} = 2$V).

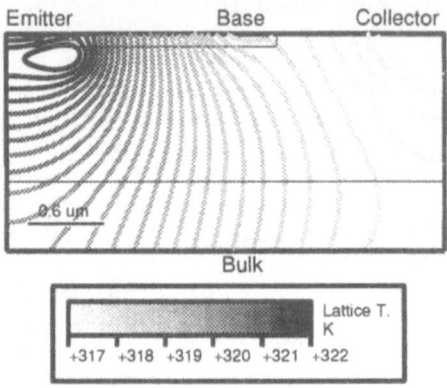

Fig. 2: Lattice temperature in the BJT ($V_C = 3$V, $V_{BE} = 0.95$V).

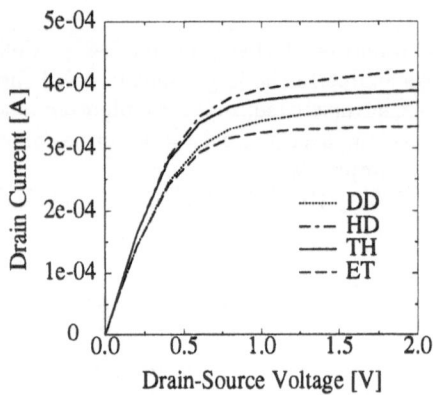

Fig. 3: MOSFET drain current comparison. ($V_{GS} = 2$V).

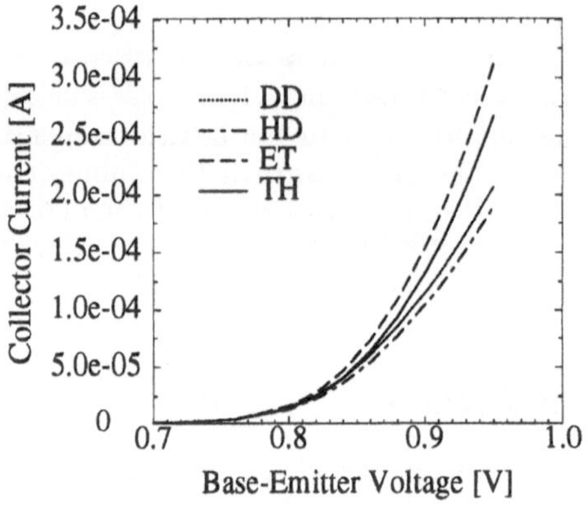

Fig. 4: BJT collector current comparison. ($V_C = 3$V).

Scheme	Block partition	Gummel Iterations	Normalized CPU time
DD	[POI,ECC]	-	1
ET	[POI,ECC] [LEB]	4	1.96
HD	[POI,ECC] [EEB]	18	4.63
TH three-block	[POI,ECC] [EEB] [LEB]	217	32.35
TH two-block	[POI,ECC] [EEB,LEB]	18	5.86

Table 1: MOS transistor

Scheme	Block partition	Gummel Iterations	Normalized CPU time
DD	[POI,ECC,HCC]	-	1
ET	[POI,ECC,HCC] [LEB]	6	2.25
HD	[POI,ECC,HCC] [EEB] [HEB]	17	2.35
TH	[POI,ECC,HCC] [EEB,HEB,LEB]	18	3.98

Table 2: bipolar transistor

SIMULATION OF SEMICONDUCTOR DEVICES AND PROCESSES Vol. 5
Edited by S. Selberherr, H. Stippel, E. Strasser – September 1993

129

Mixed-Mode Multi-Dimensional Device and Circuit Simulation

J. Litsios, S. Müller, and W. Fichtner

Integrated Systems Laboratory, ETH-Zürich
Gloriastraße 35, CH-8092 Zürich, SWITZERLAND

Abstract

A mixed-mode device and circuit modeling environment has been developed that allows the combined simulation 1D, 2D and 3D devices together with SPICE-like circuit elements. The work is based on the device simulator SIMUL [1] as well as the circuit simulator CAzM [2]. By extending SIMUL to handle simultaneously more than one device as well as giving it access to the circuit simulator's functionalities it has been possible to combine both simulators' features into a single program (1D, 2D, 3D devices, extensive physical models for power devices, thermal-electric effects and all basic SPICE models).

1. Introduction

Mixed-mode device and circuit simulators (MDC simulators) respond to the need of both device and system designers in the sense that device designers want to know how devices respond when surrounded by circuits and system designers want to design circuits in which some devices have no usable analytical model. To satisfy this large spectrum of users, MDC simulators need to offer as many features as the individual device and circuit simulators. This creates an ever increasing complexity on the part of the MDC simulators.

We describe here an MDC simulator that allows mixing 1D, 2D and 3D devices as well as circuit-level elements. The work is based on the device simulator SIMUL [1] as well as the circuit simulator CAzM [2]. While other mixed-mode device and circuit simulators have been presented (MEDUSA [3], GIGA [4], GENSIM [5], PISCES [6], CODECS [7]), the originality of this work lies specifically in the possibility to simulate together many devices of different dimensions as well as in its support of a full circuit simulator like CAzM.

Previous designs of MDC simulators have focused on connecting existing device and circuit simulators in a fairly rigid fashion. One of the main goals in the development of SIMUL has been flexibility. This has resulted in a very modular program that tries to separate concepts such as mesh dimension, physics, non-linear algorithms, linear solvers or sparse matrix formats.

The next section gives an overview of the organization of the simulator, followed by a description of the program's functionalities and input language. Finally some examples are presented.

2. Organization of SIMUL

A simplified structure of the simulator can be seen in Fig. 1. During a Newton-type solution the device and circuit equations are assembled together to form a linear system's right-hand side and Jacobian. SIMUL is written in C++ and uses an object oriented design. New equations are introduced as objects and are linked together through a data manager that works out their dependencies. This object oriented design of the code assures a large independence in the different parts of the code. Thus, for example, the non-linear solver does not need to know which linear solver it is using or the physical models do not need to know the dimension of the underlying mesh. It is this design philosophy which gives SIMUL its generality allowing, for example, the program to be linked simultaneously to many circuit simulators. Currently SIMUL is connected to the CAzM simulator and to a yet unnamed circuit simulator developed in our laboratory.

A key concept used in SIMUL is the possibility to merge variables from the different equations. This technique is inspired from the node merging used in the modified node analysis (MNA) method for circuit analysis. We define the merging of two variables x and y as replacing every occurrence of y by x and replacing the two equations associated with x and y by a single equation. Usually these equations represent the current, heat flow or charge mismatch, thus the merged equation is the sum of the individual equations. The use of variable merging allows the different device and circuit elements to be connected through the merging of connectivity nodes. Figure 2 shows how this is used to connect two devices with a resistive or non-resistive contact. Variable merging also facilitates implementing device structures that share common values such as floating gates.

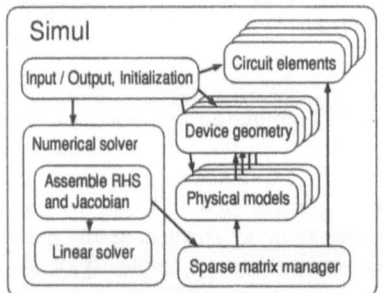

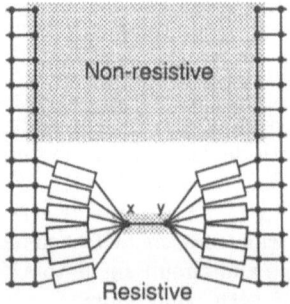

Figure 1: Structure of the mixed-mode multi-dimensional SIMUL

Figure 2: Variable merging is used for resistive and non-resistive device coupling

3. Functionality and Input language

The SIMUL input language offers all the necessary flexibility to perform complex simulations. Figure 3 shows an example of the SIMUL input language. The different device parameters such as the selection of the physical models or the desired precision can all be given independently for the different devices or circuits. A connectivity netlist is given as well as a series of numerical solve operations to be performed. The solver supports both coupled equations as well as Gummel iterations and a two level Newton solver (uses non-linear elimination). These can be performed on single devices, the full system or a specific set of devices and circuit elements. This flexibility is often necessary to construct a correct initial solution. Finally, SIMUL allows transient, quasi-stationary and small-signal analysis simulations using either a direct, iterative or domain decomposition linear solver. The domain decomposition solver splits the linear solve over the devices and the circuit and allows pivoting in the circuit part of the system.

```
System {                                    Device "mct" {
    Device "mct" "v1" (1="anode")               Electrode {
    Device "diode" "dio" (0="high" 1="anode")       { Number=0 Voltage=0 }
    CAZM ("anode"="anode" "high"="high"){           ...
                                                }
        ...                                     File {
        "l1 anode 0 -0.25e-5"                       Grid="v1"
        "v1 high 0 5000"
                                                    ...
        ...                                     }
    }                                           Physics {
}                                                   Recombination (
Solve {                                                 SRH(DopingDependence)
    Coupled { "v1".Poisson "v1".Electron "v1".Hole Circuit }   Auger Avalanche)
    Coupled { "dio".Poisson "dio".Electron "dio".Hole }    ...
    Coupled { Poisson Electron Hole Circuit }        }
    Transient ...                               ...
    }                                       }
```

Figure 3: Example of SIMUL's input language

4. Examples

Three examples of simulation results are presented. Figure 4 demonstrates the mixed-mode multi-dimensional features of SIMUL with a transient simulation of a recitifier in which each diode is modeled differently with a 1D, 2D or 3D device or with a SPICE model. Figure 5 shows the schematic used in simulating the influence of the gate resistance (Rg) on an IGBT's turn-off behavior when connected to a diode and snubber circuitry. Here the IGBT and the diode are simulated as 2D devices (700 and 300 vertices), the snubber diodes are 1D devices (228 vertices) and the additional circuitry use circuit models. Two simulation were performed with a gate resistance of 2Ω and 20Ω. They required 3972s (97 time steps) and 3329s (91 time steps), respectively, on a SPARCstation 10.

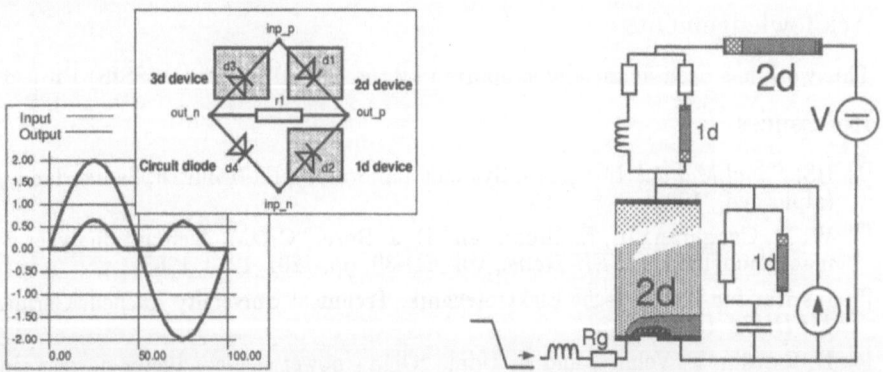

Figure 4: Simulation of a multi-dimensional mixed-mode rectifier

Figure 5: Schematic of the IGBT turn-off example

Figure 6a and 6b show the influence of the gate resistance on the IGBT's anode voltage and gate voltage when the gate is ramped from 15V to 0V in 500ns. Figure 7 shows the influence on the current through the IGBT.

Finally, Fig. 8 shows MOS C-V curves that were computed using SIMUL's small-signal capabilities. These curves were computed by connecting a MOS diode to a circuit voltage source. A quasi-static ramping of the voltage source is performed in which a small-signal analysis is computed.

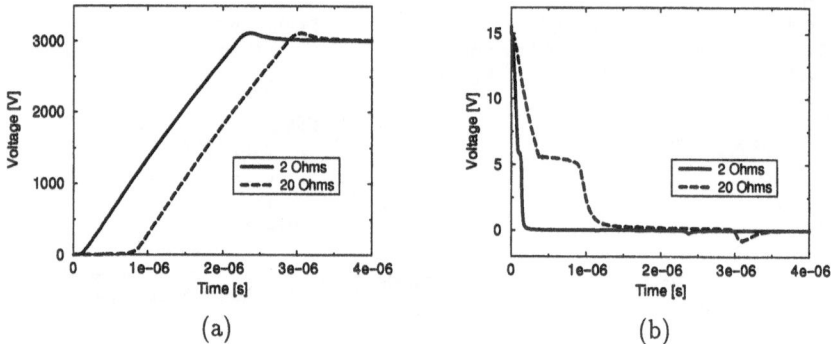

(a)   (b)

Figure 6: Influence of the gate resistance on the IGBT's switching behavior: (a) on the anode voltage, (b) on the gate voltage.

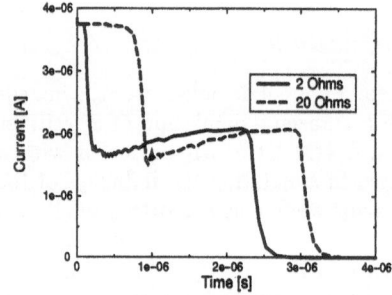

Figure 7: Influence of the gate resistance   Figure 8: MOS C-V curve simulation
on current through the IBGT.   with SIMUL.

Acknowledgements

This work has been financially supported by the ESPRIT-6075 (DESSIS) Project.

References

[1] IIS, *Simul Manual.* Integrated Systems Laboratory, ETH Zurich, Switzerland, 1.0 (alpha) ed., 1992.

[2] W. M. Coughran, Jr., E. Grosse, and D. J. Rose, "CAzM: A circuit analyzer with macromodelling," *IEEE Trans.*, vol. ED-30, pp. 1207–1213, 1983.

[3] Institut für Theoretische Elektrotechnik, Technical University Aachen, Germany, *MEDUSA User's Guide.*

[4] B. Freydin, E. Velmre, and A. Udal, "Giga - power semiconductor devices simulation software," *The Exterminator, Silvaco International*, 1990.

[5] J. Bürgler and W. Fichtner, *GENSIM User's Guide.* Integrated Systems Laboratory, ETH Zurich, Switzerland.

[6] M. R. Pinto, C. S. Rafferty, and R. W. Dutton, *PISCES II: Poisson and Continuity Equation Solver.* Stanford University, Stanford, CA 94305, 1984.

[7] K. Mayaram Tech. Rep. UCB/ERL M88/71, Electronic Research Laboratory, University of California at Berkeley, 1988.

SIMULATION OF SEMICONDUCTOR DEVICES AND PROCESSES Vol. 5   133
Edited by S. Selberherr, H. Stippel, E. Strasser – September 1993

Modeling High Concentration Boron Diffusion with Dynamic Clustering: Influence of the Initial Conditions

B. Baccus and E. Vandenbossche

IEMN-ISEN
41 Boulevard Vauban, F-59046 Lille Cédex, FRANCE

Abstract

Boron diffusion and activation at high concentrations are key problems in the formation of shallow P^+ junctions. Therefore, it is needed to understand and to predict accuratly the dopant behaviour under these conditions. In this paper, the modeling of boron is discussed, by the use of a non-equilibrium point-defect model, including amorphization and a dynamic clustering component. The initial conditions are of major importance not only for the transient enhanced diffusion, but also for the amount of active dopant. As a result, it is possible to obtain activation levels greater than the solubility limit, as observed experimentally.

1. The point-defect diffusion model

A non-equilibrium point defect-model has been used throughout the study. It solves rigorously the interactions between dopants and point-defects (including charged species) and takes into account the ion implantation damage. Five coupled equations are needed to solve the diffusion of boron : substitutional boron Bs, interstitial boron Bi, interstitial I, vacancy V, and Poisson equation [1]. In this frame, the simplest modeling of a boron cluster is given by : Bs + Bi ↔ BsBi , where the cluster is considered as immobile and in non-equilibrium. Such cluster has been already proposed in a static form [2] or a dynamic one [3] as chosen here. This adds a sixth equation to the system. It is possible to include clusters containing more atoms, but it does not give a significant improvement.

2. Initial conditions

In the case of high-dose boron ion implantation, it is difficult to determine exactly the initial conditions. There is no amorphization, however it is known that extended defects can appear as well as activation levels greater than the solubility limit (Csol) [4]. The modeling used in the present study is a generalisation of the amorphization model proposed in [5]. In the following, the 'disordered' zone is considered as a region close to amorphization in which, (1) the boron is completely active and (2) some supersaturation of point-defects still exists (in particular interstitials). This is then an intermediate case between a relatively low-dose implant (high concentrations of I and V) and the situation after solid phase epitaxy of an amorphized region. A typical example is given in fig. 1 for a 5.10^{15} at/cm^2 20 keV boron implant.

3. Comparison with experiments

The calculations have been compared to several experiments including the ones from Solmi [4] and Cowern [6]. Fig. 2 shows the different species after 10 s of diffusion at 850^0C (initial condition in fig. 1). The interstitial level is severely decreased in the 'disordered' zone due to the formation of the Bi and BsBi species. A high level of activation is still maintained. A good agreement between simulation and SIMS is found for different diffusion times (fig. 3) and also concerning the activation level (fig. 4). The activation is strongly dependent on the initial interstitial supersaturation (fig. 5). If the supersaturation is extremely high, all the substitutional boron will be converted rapidly to the interstitial component and the activation will be below Csol, reaching this value for long times. On the contrary, if the zone would be completely amorphized (interstitial level close to equilibrium), a very high level of activation could be obtained even for long times. As expected, the cases under study fall between these two limits, enabling activations slightly higher than Csol.

4. Conclusion

It has been shown that the level of activation resulting from high-concentration boron diffusion after post-implant annealing, can be controlled by the initial amount of point-defects in the 'disordered' zone. Such modeling allows to reproduce experiments for amorphizing and non-amorphizing ambients.

References

[1] B. Baccus et al., IEEE Trans. Electron Devices, ED-39, 648 (1992).
[2] D. Mathiot and J.C. Pfister, J. Appl. Phys., 55 (10), 3518 (1984).
[3] M. Hane and H. Matsumoto, IEDM Techn. Digest, 701 (1991).
[4] S. Solmi, F. Baruffaldi and R. Canteri, J. Appl. Phys., 69 (4), 2135 (1991).
[5] H. Kinoshita and D.L. Kwong, IEDM Techn. Digest, 165 (1992).
[6] N.E. B. Cowern, K.T.F Janssen and H.F.F. Jos, J. Appl. Phys., 68 (12), 6191 (1991).

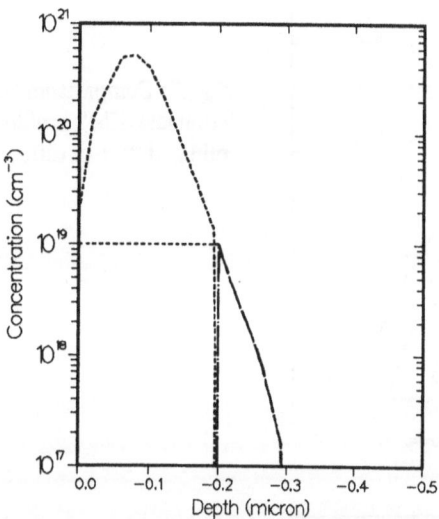

Fig. 1 : Initial conditions for a 5.10^{15} at/cm^2,
20 keV boron implant.

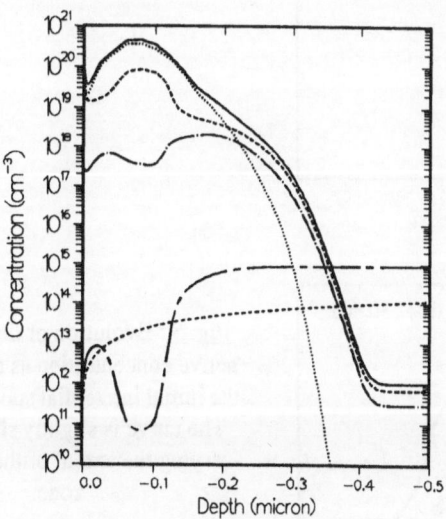

Fig. 2 : Species after 10 s diffusion at 850^{0}C.

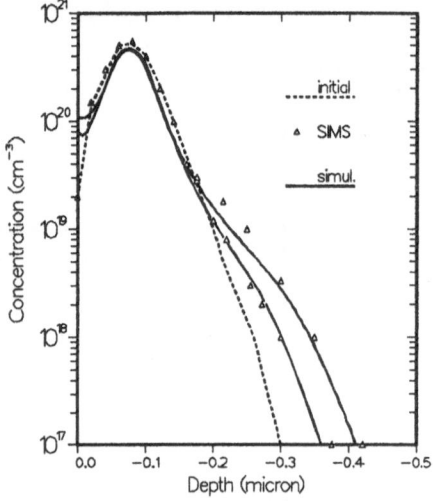

Fig. 3 : Comparisons between simulation and SIMS profiles [4], after 30 min and 2hrs of diffusion at 850^{0}C.

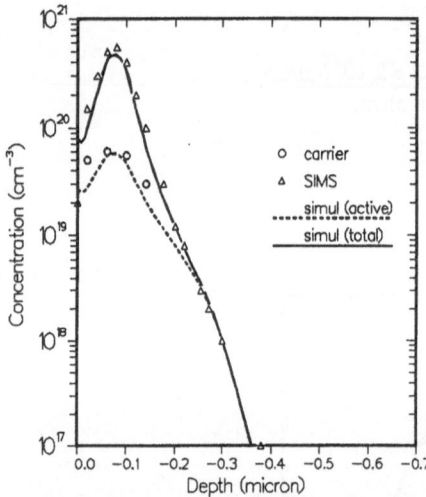

Fig. 4 : Comparison between simulation and experiments (SIMS and carrier profiles). After 30 min of diffusion at 850^{0}C.

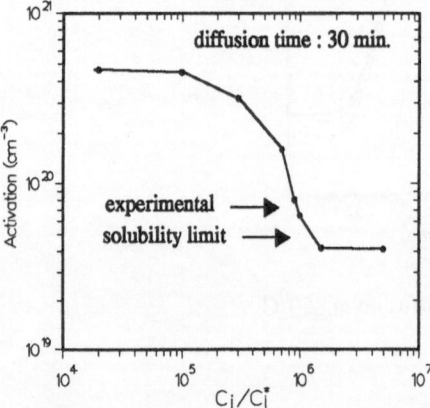

Fig. 5 : Evolution of the maximum active concentration as a function of the initial interstitial supersaturation. The curve is slightly shifted when varying the extent of the disordered zone.

Simulation of High-Dose Ion Implantation-Induced Transient Diffusion and of Electrical Activation of Boron in Crystalline Silicon

H. U. Jäger[‡†]

[‡]Research Center Rossendorf Inc.,
Institute for Ion Beam Physics and Materials Research
Postfach 510119, D-01314 Dresden, GERMANY
[†]Guest Scientist in the Fraunhofer-Institute of Microelectronic Circuits and Systems
Grenzstraße 28, D-01109 Dresden, GERMANY

Abstract

Coupled diffusion-reaction equations for boron and for point defects and rather simple initial conditions are used to model the implantation-induced transiently enhanced diffusion and the electrical activation of high-dose boron distributions during annealing.

1. Introduction

This investigation is dealing with the formation of shallow p[+]-type regions in crystalline silicon by low-energy high-dose boron ion implantation and subsequent furnace annealing at low temperature (800°C). Figs. 1–3 show typical experimental data [1] as well as the respective results of our simulations. At 800°C, the normal boron diffusion is known to be negligible. In case of post-implantation annealing, however, both low-dose and high-dose boron profiles become modified in their tail regions up to a critical concentration c_{enh} of about 4×10^{18} cm^{-3}, which is far below the boron solid solubility limit $c_{sol}(800°C) = 3.2\times10^{19}$ cm^{-3}. This implantation-induced tail broadening relaxes within a time period of about 30 min. The non-diffusing boron in the profile peak region above c_{enh}, but below c_{sol}, is found to be partially electrically inactive. It becomes electrically active only after annealing periods of many hours.

The models applied to simulate the transient diffusion and electrical activation of boron at low temperatures range from pure phenomenological ones [1] to formulations [2, 3, 4, 5] which include explicitly reactions between dopant species and silicon point defects under nonequilibrium conditions. Our approach [5] which provided good results for low boron doses $D \lesssim 5\times10^{14}$ cm^{-2} has to be modified if the boron peak concentration in the sample significantly exceeds the solid solubility limit c_{sol}. The present conference contribution is aimed at a brief explanation of these model extensions. Some new simulation results for high boron doses are also presented (figs. 1–3), but a more complete discussion must be given elsewhere [6].

---

*Guest scientist in the Fraunhofer-Institute of Microelectronic Circuits and Systems, IMS2, Grenzstr. 28, 01109 Dresden, Germany

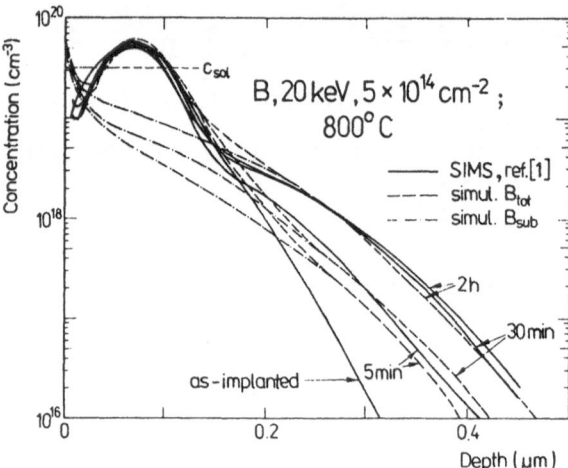

Figure 1: Time evolution of boron atomic (B_{tot}) and electrical (B_{sub}) profiles during 800°C furnace annealing after 20 keV 5×10^{14} cm^{-2} B ion implantation. The simulations are compared with profiles which have been measured by Solmi et al. [1] using secondary ion mass spectroscopy (SIMS).

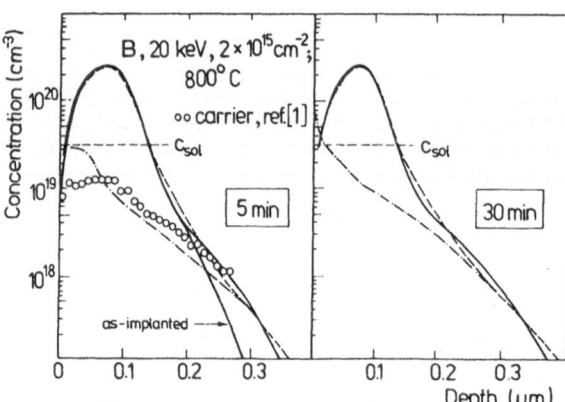

Figure 2: Model predictions and experimental profiles for a boron dose of 2×10^{15} cm^{-2}. The carrier profile has been measured [1] using anodic stripping followed by incremental sheet resistance and Hall effect measurements; for the other details, see fig. 1.

2. Basic model

The boron diffusion and activation have been modelled by solving a system of coupled diffusion-reaction equations. The species considered are boron atoms in solution on substitutional and interstitial sites (B_{sub}^-, B_{int}) as well as vacancies V^z and silicon self-interstitials I^z in various charge states z. The substitutional boron atoms and the point defects are assumed to form pairs ($B_{sub}^- I^z$), ($B_{sub}^- V^z$) in accordance with the mass action law. These pairs are the only diffusing vehicles (point defect impurity pair diffusion, [7]). The reactions, which are taken into account to simulate the change in activation during diffusion, are the kick-out reaction $B_{int} \Longleftrightarrow B_{sub} + I$, the Frank-Turnbull mechanism $B_{int} + V \Longleftrightarrow B_{sub}$, and the interstitial-vacancy annihilation and generation $I + V \Longleftrightarrow 0$.

The complicated details of defects evolution during ion implantation and during heating up to the annealing temperature are behind the scope of this study. Therefore a key problem of our approach is the finding of adequate initial conditions for the point defects and for the electrically active boron concentration B_{sub}. Let us postulate that a very early stage of annealing exists where diffusion is still unimportant,

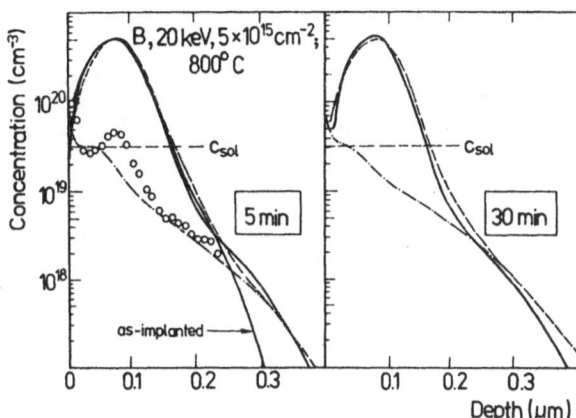

Figure 3: Model predictions and experimental profiles for a boron dose of 5×10^{15} cm^{-2}. For details, see figs. 1 and 2.

but local equilibrium between the three reactions mentioned above is realized just at this moment. For a given total boron concentration, this local equilibrium is uniquely characterized by the concentration difference $Q = c_I - c_V - c_{B,sub}$, which represents the characteristic quantity of the given reactive species ensemble.

For low boron doses, an assumption of $Q = 0$ proves to be reasonable for all depth intervals in defining the initial conditions . A value $Q = 0$ can be thought to be originally realized by $c_I = c_V$, $c_{B,sub} = 0$ or by by $c_I = c_{B,sub}$, $c_V = 0$. Computing the local equilibrium for $Q = 0$ [5], one obtains initial conditions of the type $c_{B,sub} \approx c_I$, $c_V \ll c_I$.

3. High boron doses

For higher boron doses, of course clustered boron atoms B_{clus} must be taken into account. In our code, the boron concentrations exceeding c_{sol} are assumed to be completely clustered in the beginning

$$B_{tot} = B_{sub} + B_{int} + B_{clus} , \quad \text{with } B_{clus} = B_{tot} - c_{sol} \text{ for } B_{tot} > c_{sol} ;$$

precipitation models [1] have not yet been introduced.

The main problem of a straightforward point defect diffusion model seems to be the explanation of the experimental result that the boron profile broadening in the tail becomes nearly independent on the actual value of the high boron dose (compare, e.g., figs. 2 and 3). If our values [5] for the diffusion coefficients, the reaction rates and the point defect equilibrium concentrations c_I^*, c_V^* are not essentially changed, the initial conditions for implantation-induced diffusion in high-dose boron profiles can't be defined by $Q = 0$. The interstitial oversaturation affecting boron diffusion must be limited. In the present simulations we have used interstitial distributions whose peaks have been shifted in dependence on boron dose, so that the interstitial and boron profile tails will nearly coincide. A constant interstitial area density of 3.3×10^{14} cm^{-2} has been used. Furthermore, a complete activation of the solved boron atoms and equilibrium vacancy concentrations (c_V^*) have been assumed in the beginning. These details are illustrated by fig. 4a. The effect of the subsequent local equilibration is shown in fig. 4b. The model modifications explained so far are not yet sufficient to reproduce a dose-independent boron profile broadening. The outdiffusion of the interstitial oversaturations is retarded for increasing boron doses if increasing depths of the initial interstitial distributions are used. In order to compensate this effect,

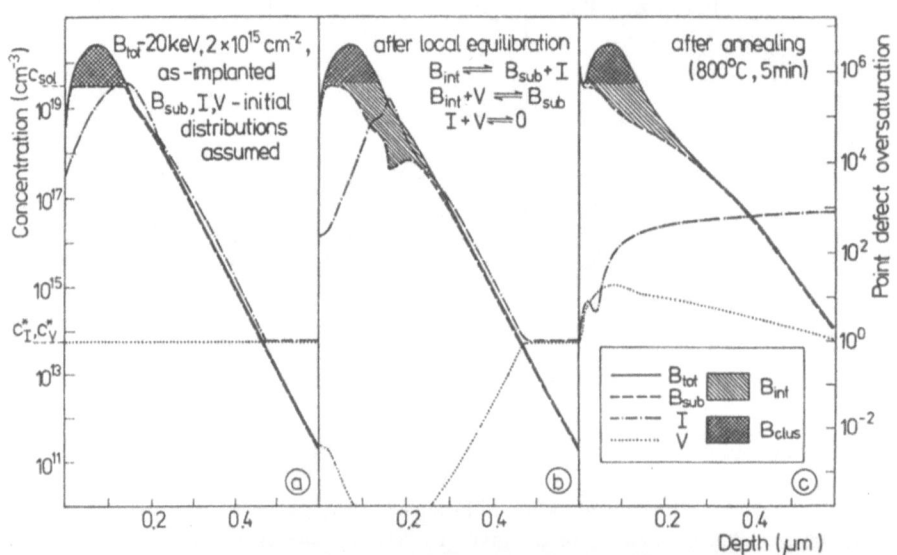

Figure 4: Illustration of the model components and of the predicted relations between boron profiles (left-hand scale) and point defect profiles (left- and right-hand scales) for an 800°C 5 min annealing after 20 keV 2×10^{15} cm^{-2} B ion implantation.

one can take into account that transitions of boron atoms from clusters into solution should affect directly the silicon point defect distributions. Such a correlation exists also in the dynamic clustering model of Cowern et al. [2]. We have assumed that the transition of one boron atom from a cluster onto an interstitial site (B_{int}) and the inverse process are accompanied by the annihilation or generation, respectively, of one silicon point defect. This process, which is described by the reactions $B_{clus} + I \Longleftrightarrow B_{int}$, $B_{clus} \Longleftrightarrow B_{int} + V$, seems to be necessary to obtain in the framework of our approach reasonable diffusion profiles for high boron doses (figs. 2, 3).

References

[1] S. Solmi, F. Baruffaldi and R. Canteri, J. Appl. Phys. 69(1991)2135

[2] N. E. B. Cowern, H. F. F. Jos, K. T. F. Janssen and A. J. H. Wachters, Mat. Res. Soc. Symp. Proc. 163(1990)605

[3] M. Hane and H. Matsumoto, IEDM Techn. Dig., p. 701(1991)

[4] B. Baccus, T. Wada, N. Shigyo, M. Norishima, H. Nakajima, K. Inou, T. Iinuma and H. Iwai, IEEE Trans. Electr. Dev. 39(1992)648

[5] H. U. Jäger, 8th International Conference on Ion Beam Modification of Materials, Heidelberg, Germany, Sept. 7 - 11, 1992, contr. paper P232; to be published in Nucl. Instr. & Meth. Phys. Res. B

[6] H. U. Jäger, to be published

[7] B. J. Mulvaney and W. B. Richardson, Appl. Phys. Lett. 51(1987)1439

SIMULATION OF SEMICONDUCTOR DEVICES AND PROCESSES Vol. 5   141
Edited by S. Selberherr, H. Stippel, E. Strasser – September 1993

Physical Modeling of the Enhanced Diffusion of Boron Due to Ion Implantation in Thin Base npn Bipolar Transistors

M. Mouis, H. J. Gregory[†], S. Denorme, D. Mathiot, P. Ashburn[†], D. J. Robbins[†],

and J.L. Glasper[‡]

CNET-CNS, France Telecom
Chemin du Vieux Chêne, BP 98, F-38243 Meylan Cédex, FRANCE
[†]Deptartment of Electronics and Computer Science, University of Southampton
Southampton, S09 5NH, UNITED KINGDOM
[‡]Defence Research Agency
St. Andrews Road, Malvern, WR14 3PS, UNITED KINGDOM

Abstract

Using the most advanced physical models of diffusion, we have simulated boron diffusion in the context of a low thermal budget technology for thin-base integrated bipolar transistors. We demonstrated that simulation was able to account for the base broadening due to arsenic implantation in a monocrystalline emitter. Moreover, even in polysilicon emitter bipolar transistors, where the effect of the emitter implantation is suppressed, we found that the extrinsic base implantations could still induce a non negligible base broadening.

The trend in bipolar technology is towards highly doped thin base devices, where both base transit time and base resistance are low, resulting in higher operation frequencies. This is achievable only if the base fabrication and the whole technology process allow the fabrication of very steep doping profiles. Low thermal budgets are of course essential. But it is well known that some technological steps induce an enhanced diffusion of boron and a detrimental broadening of the base.[1,2,3] There is a general agreement to attribute this base broadening to the acceleration of boron diffusion in the presence of implantation defects.[4] Although there has been a lot of progress in the analysis of the physical mechanisms involved in defect and dopant diffusion, there has been to our knowledge no attempt to apply the corresponding physical models to the enhanced diffusion of boron in devices. Yet, it is essential to predict such an effect, especially in bipolar transistors where the base width is a critical parameter. In this paper, we evaluate base broadening by using the most advanced physical models of diffusion, which have recently been implemented in the process and device simulation software TITAN developed at CNET.[5] First, we studied the boron diffusion induced by arsenic implantation in a monocrystalline emitter. We simulated a bipolar transistor based on the layer structure of Table 1. Simulation results show very good agreement with the SIMS profiles of non implanted annealed structures (curves a-b, Fig. 1). For implanted structures, a standard simulation gives the same amount of diffusion whereas the SIMS profile shows a very strong enhancement of boron diffusion, consistent with a defect assisted diffusion mechanism (curve c, Fig.1).

Si emitter	n = 10^{18} cm^{-3}	300 nm
Si cap		20 nm
Si spacer		50 or 100 nm
Si base	p = 7x10^{18} or 2x10^{19} cm^{-3}	25 nm
Si spacer		50 or 100 nm
Si buffer		
Substrate		

Table 1 :
Layer structure of the bipolar structures. As (30 keV) was implanted in the monocrystalline emitter (except for some reference wafers). A rapid thermal anneal (RTA) was then performed at 800, 850 or 900 °C.

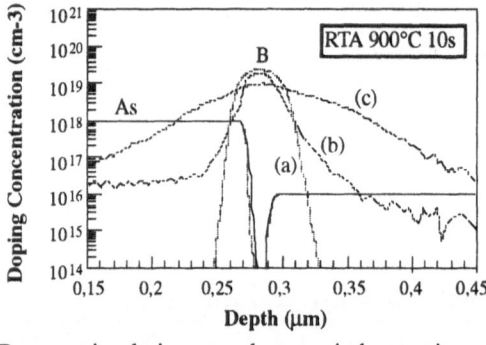

Figure 1 :
Dopant profiles measured by SIMS in (b) the non-implanted and (c) the implanted wafers. The result of a standard simulation neglecting coupling with defects is shown for comparison (a). The nominal base doping is 2x10^{19} cm^{-3}

Process simulation was then carried out using advanced diffusion models.[6] We assumed perfect recrystallisation of the amorphised layer at the beginning of the RTA and defects were therefore introduced by implanting free interstitials beyond the amorphisation limit. The defects were first assumed to be free at t=0. Simulation results are shown in Fig. 2. Although significant, the calculated base broadening could not reach the measured values, unless non realistic (i. e. much too high) initial interstitial supersaturations were used. However, implantation related point defects are in reality aggregated into small clusters and during RTA, free defects are only progressively released by cluster dissolution.

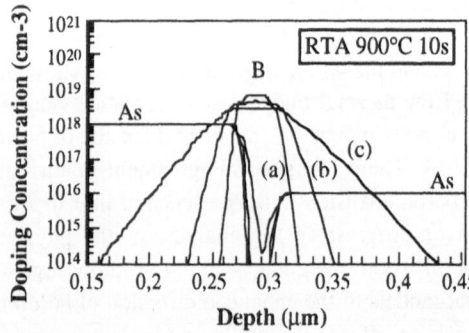

Figure 2 :
Calculated dopant profiles when coupling with implantation defects is ignored (a) or is taken into account (b, c : maximum supersaturation for the initial defect distribution of about 4x10^5 and 4x10^6; all defects assumed free at t=0). The nominal base doping is 7x10^{18} cm^{-3}.

As a first attempt at modelling this situation, we simulated the cluster dissolution kinetics by a time discretization technique. Fig. 3 compares the final dopant profiles obtained after the implantation of the same initial supersaturation, either at the beginning of the RTA or at several successive stages of the RTA. It shows clearly that the kinetics of cluster dissolution is not a negligible phenomenon : it strongly enhances diffusion by maintaining the interstitial supersaturation in the base during a longer time. Therefore, significant diffusion is obtained without the need for non realistically high supersaturations. Accounting for cluster dissolution will be necessary for precise modelling.

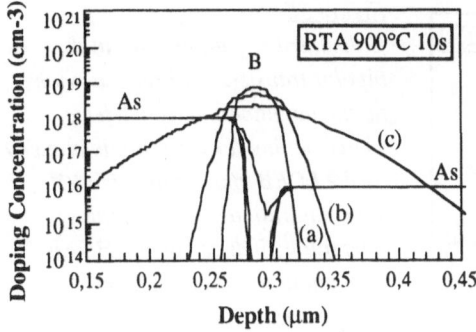

Figure 3 :
Calculated dopant profiles with an interstitial supersaturation of $4x10^5$ applied once at $t=0$ (curve b) or 40 times during the RTA (curve c). The diffusion without coupling with defects is given as (a).

Finally, in most integrated bipolar and BiCMOS circuits, the contacts are implanted and their distance from the active base is usually short compared to the defect diffusion length (about 30-40 μm). Typically, the collector contact is around 2 μm from the intrinsic base, while the extrinsic base may even be self-aligned.[7]

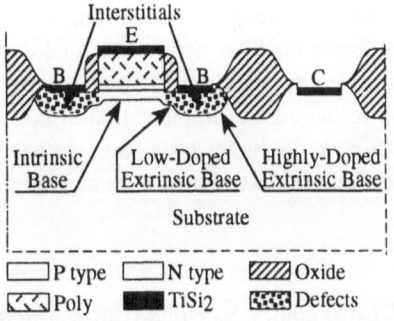

Figure 4 :
Schematic view of the simulated polysilicon-emitter bipolar transistor : the defects induced by the lightly-doped extrinsic base (LDEB) and of the highly-doped extrinsic base (HDEB) self-aligned implantations enhance B diffusion in the intrinsic base.

Therefore, it is useful to evaluate the influence of the contact implantations. We studied a self-aligned polysilicon-emitter bipolar transistor with a thin epitaxial base (Fig. 4). To simulate the whole structure, we used the latest version of TITAN, with a precise modelling of the grain structure. Coupled diffusion of dopants and defects was of course accounted for. The base broadening induced by collector implantation was found to be negligible due to the small solid angle offered to defects coming from the collector to the intrinsic base. In contrast, the base implantation generates defects close to the intrinsic base. It has been shown that the defects induced by boron implantation were mainly associated to the activated boron atoms themselves and were of the interstitial type.[8] In addition, we found that, when accounted for, vacancies relaxed in the first 0.5 s by diffusing towards the silicon surface, and did not affect significantly boron diffusion. Therefore, the only defects which have to be accounted for are interstitials. Fig 5 shows the base profile calculated with no defects and with two different intestitial supersaturations, respectively generated by the lightly-doped extrinsic base and by the highly-doped extrinsic base implantations. With a typical 10^{17} cm^{-3} doping concentration in the collector, the calculated base width was 60% larger when accounting for the defects generated by the extrinsic base implantation.

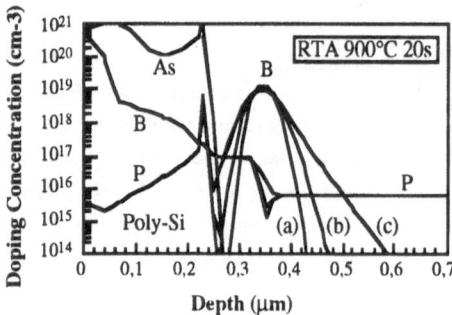

Figure 5 :
Calculated dopant profiles in the bipolar transistor of Fig. 4, a) with no implantation induced defects, b) accounting for interstitials induced by the LDEB implantation ($5x10^4$ supersaturation), c) accounting for interstitials induced by the HDEB implantation ($3.5x10^6$ super-saturation).

To conclude, we have simulated boron diffusion in the context of a low thermal budget technology for thin base integrated bipolar transistors. Using the most advanced physical models of diffusion presently available, we demonstrated that simulation was able to account for the diffusion enhancement arising from coupling with crystal defects. Information about the global distribution of defects can be found (e.g. from Monte-Carlo simulations of implantation) but it was found that accounting for their arrangements into clusters and for the cluster dissolution kinetics during the RTA will be essential to obtain quantitative agreement with measurement. Valuable information can however be obtained through qualitative comparisons. In addition to the emitter implantation effect which leads to a large enhancement of the boron diffusivity in the intrinsic base, we have also investigated the possible broadening induced by extrinsic base and collector implantations.

Acknowledgments :
The authors are thankful to S. Martin (CNET/CNS) and P. Scheiblin (CISI), who are in charge of TITAN software development at CNET/CNS, for their efficient cooperation. This work has been carried out in a CNET/CNS department associated with CNRS (Centre National de la Recherche Scientifique, France) and the Laboratoire de Physique de la Matière at INSA (Institut National des Sciences Appliquées, France).

References :
1 P. R. Pukite, S. S. Iyer, G. J. Scilla, Appl. Phys. Lett. **54** (10) (1989) 916-918
2 D. J. Gravesteijn et al., MRS Symposium Proceedings **vol. 220** (1991) 3-14
3 A. Pruijboom et al., Microelectronics Engineering **19** (1992) 427-434
4 A. E. Michel, 2nd Int. Symp. on Modelling in Semiconductor Technol., Proceedings **vol. 91-4** (1991) 242-253
5 A. Gerodolle et al., Proceedings of NASECODE VI, 56-57 (1989, Boole Press)
6 D. Mathiot, S. Martin, J. Appl. Phys. **70** (6) (1991) 3071-3080
7 A. Marty et al., Microelectronics Engineering **19** (1992) 547-550
8 F. Marou et al., Nuclear Instr. and Methods in Phys. Research **B55** (1991) 655-660

SIMULATION OF SEMICONDUCTOR DEVICES AND PROCESSES Vol. 5
Edited by S. Selberherr, H. Stippel, E. Strasser – September 1993

Simulation of Denuded Zone Formation in CZ Silicon

J. Esfandyari, G. Hobler, S. Senkader, H. Pötzl, and B. Murphy[†]

Institut für Allgemeine Elektrotechnik und Elektronik, TU Vienna
Gußhausstraße 27-29, A-1040 Wien, AUSTRIA
[†]Wacker-Chemitronic GmbH
D-84489 Burghausen, GERMANY

Abstract

In ULSI device processing technology, internal gettering (IG) of metallic contaminants is an important issue. The structure of IG wafers consists of the
bulk microdefect region and a defect-free subsurface region, termed as "denuded zone" (DZ). In our work, we present a detailed analysis of the coupled
diffusion-oxygen precipitation problem encountered in the simulation of the
denuded zone formation.

1. Introduction

In ULSI device processing technology, internal gettering (IG) of metallic contaminants
is considered to be an important issue. A well-controlled IG process can lead to a
significant improvement in yield and device performance. In general, the structure of
IG wafers consists of the bulk microdefect region and a defect-free subsurface region,
termed as "denuded zone" (DZ).

In a controlled IG process, two important factors need to be kept in mind. (i) Only
a certain amount of bulk microdefects should be nucleated for gettering of metallic
contaminants because excessive precipitation would lead to a reduction in dissolved
oxygen concentration and loss in resistance against wafer warpage and (ii) a well-
defined DZ must be created to minimize the growth of residual microdefects near the
wafer surface. A typical controlled IG cycle consists of three annealing steps. The
first step, termed as "denudation", is to deplete the oxygen from the subsurface region
via out-diffusion. The second and third steps are necessary to nucleate and grow the
microdefects to the size that is thermodynamically stable.

Defects engineering, including the modeling of the IG process, becomes increasingly
important in the development of ULSI device fabrication processes. The experimental
determination of DZ, oxygen precipitate sizes and microdefect densities is toilsome.
The theoretical prediction and the use of computer models can help to interpret the
results from precipitation experiments. It also saves total process development time
and cost.

In several models for oxygen precipitation and DZ-formation proposed [1], [2], it is
assumed that oxygen precipitation in silicon occurs through homogenous nucleation

despite the fact that more and more evidence points towards a heterogenous mechanism [3]. In homogenous models, steady-state equations [2] have been commonly used to describe the precipitation behaviour. On the other hand, these models cannot explain the dissolution of preexisting precipitates and do not take into account the size distribution of precipitates.

In our work based on the model of Schrems et al.[4] for oxygen precipitation we present a detailed analysis of the coupled diffusion-precipitation problem encountered in the simulation of the denuded zone formation. Oxygen precipitates are described by a size distribution function $f(n)$, i.e. by the number of precipitates per cm^3 containing n oxygen atoms as a function of n. Growth and dissolution is described by chemical rate equations for the smallest precipitates ($n \leq 20$) in combination with a single Fokker-Planck equation for all larger precipitate sizes. These equations are coupled with the diffusion equation for interstitial oxygen. In this work the model is applied for studying oxygen precipitation not only in the bulk, but also as a function of the depth perpendicular to the wafer surface.

2. Model

A detailed description of the model for bulk oxygen precipitation can be found elsewhere[4]. The model describes the growth and dissolution of oxygen precipitates statistically. It combines chemical rate equations (RE) for an accurate description of the smallest clusters with an approximating Fokker-Planck equation (FPE) for describing all larger precipitate sizes. The coefficients in the resulting system of differential equations are related to the growth and dissolution rates of the oxygen precipitates containing the Gibbs free energy of an individual oxygen precipitate. Assuming spherical precipitates composed of SiO_2, the Gibbs free energy $G(n,t)$ of a precipitate with n oxygen atoms is modelled as the sum of volume energy G_O and interfacial energy G^{if},

$$G_O = -nkT\ln\left(\frac{C_O}{C_O^{eq}}\right) \qquad G^{if} = 4\pi r^2 \alpha \left(1 + \left(\frac{\zeta_1}{n}\right)^{1/3} + \left(\frac{\zeta_2}{n}\right)^{2/3}\right) \qquad (1)$$

C_O and C_O^{eq} denote the concentration of interstitial oxygen atoms and their solubility. G^{if} is proportional to the square of the precipitate radius r and to the unit surface energy $\alpha = 0.31 Jm^{-2}$ [3] corrected by a factor containing free parameters ζ_1 and ζ_2 [5]. The values $\zeta_1 = 0.22$ and $\zeta_2 = 0.33$ have been used.

The model for bulk oxygen precipitation has been extended for DZ simulation by taking oxygen diffusion into account and by treating the precipitation process as a function of the distance from the wafer surface. The differential equation for interstitial oxygen (C_O) contains a diffusion and a precipitation ($C_{O,OP}$) term,

$$\frac{\partial}{\partial t}C_O = D_O \frac{\partial^2}{\partial x^2}C_O - \frac{\partial}{\partial t}C_{O,OP} \qquad C_{O,OP} = \int n \cdot f(n)\,dn. \qquad (2)$$

D_O denotes the effective diffusivity. This equation is solved together with the RE/FPE for the oxygen precipitates by PROMIS [6]. As initial value $C_O(t=0) = 9.5\cdot10^{17}cm^{-3}$, as boundary condition $C_O(x=0) = C_O^{eq}$ is assumed. In order to avoid troubles due to the large range of n ($1\ldots10^{11}$) a transformation $\rho(n)$ has been applied. ρ equals n for the smallest precipitates, and $\rho(n)$ is a cubic function for larger n, carefully selected to avoid both large nonlinearities and an excessive range of n.

3. Results and Discussion

Since the small nuclei have an ultimate importance for the following precipitation the numerical simulations take into account the thermal history of samples including cooling down in crystal growth approximately described by an exponential decrease of temperature from $1400°C$ to $450°C$ in 1h. Figures 1–3 show the size distribution function after each step of a typical HI-LO-HI anneal used for denuded zone formation $(1100°C/16h + 650°C/16h + 1000°C/16h)$. The logarithm of the size distribution is shown as a function of $\log(n)$ and x. In the first step (Fig. 1), due to the high temperature mainly outdiffusion occurs while only small precipitates emerge. In the second step (Fig. 2), due to the low temperature almost no diffusion occurs while important precipitation (nucleation) takes place. It can be seen that less precipitation occurs near the surface because of the lower interstital oxygen concentration than in the bulk. The precipitates in the bulk may grow in the third step (Fig. 3) because the radius of the larger ones is larger than the critical radius at the higher temperature of this step. In contrast, the radius of all precipitates at the surface is smaller than the critical radius. Therefore they tend to dissolve.

Figures 4 and 5 show the interstitial oxygen concentration depicted as a function of depth after each step of two different IG annealing cycles. The shape of the interstitial oxygen profile can be explained by out-diffusion of the oxygen from the subsurface region during the first HI anneal at $1100°C$ and by precipitation of oxygen in the bulk at successive anneals. The final interstitial oxygen distribution is in good agreement with the experimental data of Isomae et al. [8]. The experimental data could only be explained by assuming the value by Takano and Maki [9] for long range diffusion coefficient D_O and the value by Craven[7] for oxygen solubility C_O^{eq}.

4. Conclusion

The effect of the IG anneal cycles on the oxygen precipitation and DZ formation in CZ-silicon was studied. Extensions of the recently developed model of Schrems et al. were discussed. The results imply that long range out-diffusion has to be assumed and crystal growth condition and thermal history of the wafer play a significant role with respect to DZ formation.

References

[1] B.Rogers et al.: "VLSI Science and Tech.1984", The Electr. Soc. Proc. 74,1984.

[2] N.Inoue et al.: Semicond. Silicon 1981, 4890, 1981.

[3] H.Bender et al.: in Handbook of Semicond., 2nd ed., Vol.3, in press.

[4] M.Schrems et al.: Proc. SISDEP91, 113, 1991.

[5] K.Nishioka: Phys. Rev. A, 16, 2143, 1977

[6] P.Pichler: IEEE Trans. CAD, 4, 384, 1985.

[7] W.A.Craven: Semicond. Silicon 1981, 254, 1981.

[8] S.Isomae et al.: J. Appl. Phys., 55, 817, 1984.

[9] Y.Takano, M.Maki: Semicond. Silicon 1973, 469, 1973.

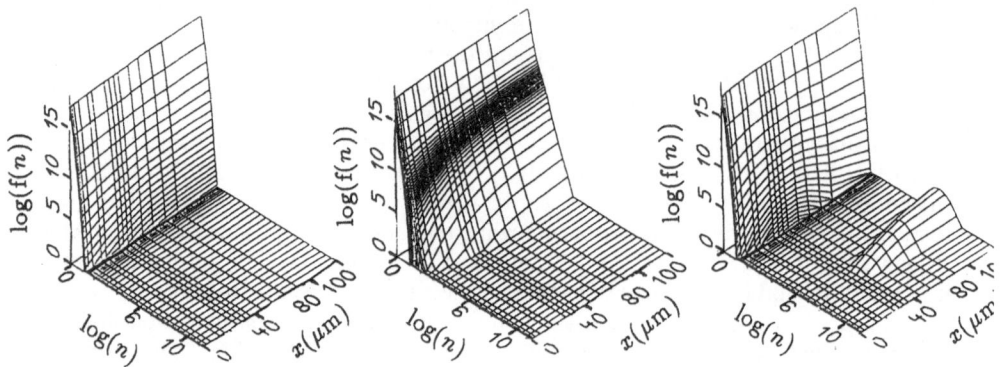

Figure 1: Size distribution function as a function of log(n) and x after 1100°C/16h anneal.

Figure 2: Size distribution function as a function of log(n) and x after 1100°C/16h+ 650°C/16h anneal.

Figure 3: Size distribution function as a function of log(n) and x after 1100°C/16h+ 650°C/16h+1000°C/16h anneal.

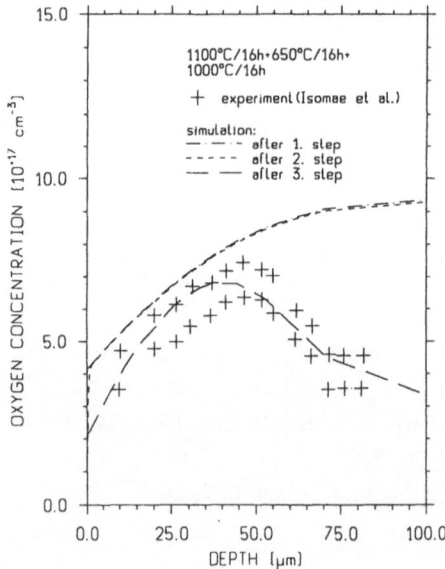

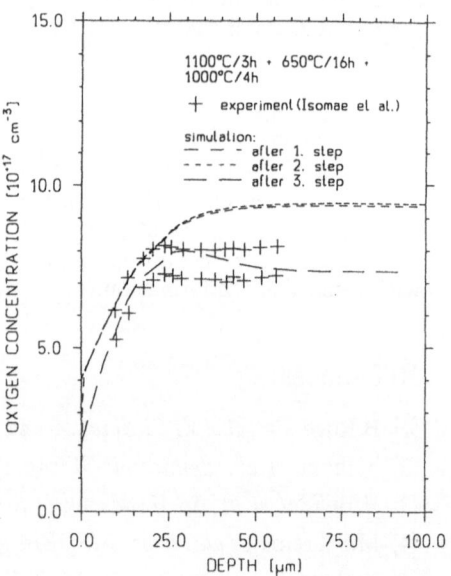

Figure 4: Depth profile of interstitial oxygen concentration in the wafer subjected to three-step annealing (1100°C/16h+650°C/16h+1000°C/16h)

Figure 5: Depth profile of interstitial oxygen concentration in the wafer subjected to three-step annealing (1100°C/3h+650°C/16h+ 1000°C/4h)

A Closed Hydrodynamic Model for Hot-Carrier Transport in Submicron Semiconductor Devices

V. Gružinskis, E. Starikov, P. Shiktorov, L. Reggiani[†], M. Saraniti[†], and L. Varani[†]

Semiconductor Physics Institute
A. Goštauto 11, LT-2600 Vilnius, LITHUANIA
[†]Dipartimento di Fisica ed Istituto Nazionale di Fisica della Materia,
Università di Modena
Via Campi 213/A, I-41100 Modena, ITALY

Abstract

We develop a closed hydrodynamic model within the framework of the single-electron gas approach. Calculations are presented for the small-signal response, diffusion coefficient, spectral density of velocity fluctuations of bulk semiconductors as well as for the concentration, velocity, energy and electric-field profiles of n^+nn^+ structures. The validation of the present model is confirmed by a favourable comparison with analogous Monte Carlo simulations.

1. Introduction

In recent years significant efforts have been devoted to develop hydrodynamic (HD) approaches to be used for the modeling of submicron semiconductor devices [1,2]. These approaches are based on a set of conservation equations obtained under the so called single-electron gas approximation. To describe the dynamics of carrier heating and to account for the nonparabolicity and multivalley effects, the velocity and energy relaxation rates as well as the energy dependent effective mass are introduced in more or less similar ways. However, the diffusion effects are often included by assuming a Maxwellian shape of the hot-carrier distribution function thus introducing the concept of an effective temperature. This is far from being an acceptable assumption, especially for multi-valley semiconductors [3]. The aim of this work is to go beyond the use of an effective temperature. To this purpose, we propose a total energy scheme which consistently accounts for diffusion and convective contributions within the single-electron gas model.

2. Theory and results

The procedure is a generalization of the moment method. It consists in multiplying by an arbitrary function $A(\mathbf{p})$ the Boltzmann equation and then integrating all over the momentum space $\mathbf{p}$ thus providing a conservation equation for the macroscopic quantity $< A(z,t) >$ in the form:

$$\frac{\partial}{\partial t}(n < A >) + \frac{\partial}{\partial z}(n < vA >) - eE\, n\left\langle \frac{\partial A}{\partial p_z} \right\rangle = -\nu_A(< A > -A_{th})n \qquad (1)$$

where the brackets $< ... >$ mean ensemble-averaging over the distribution function $f(\mathbf{p}, z, t)$, $n(z, t)$ is the carrier concentration, $v(\mathbf{p})$ the carrier velocity along the field, e the electron charge, $E(z, t)$ the electric field taken along the z direction, ν_A the relaxation rate of $< A >$ to its equilibrium value A_{th}. It is evident that the term under the spatial derivative always involves moments of higher order. Therefore, to close the system of equtions we assume that $< vA >$ consists of a dynamic and chaotic contributions as:

$$< vA >=< v(z,t) >< A(z,t) > +Q_A(< \epsilon(z,t) >) \qquad (2)$$

where $< v(z,t) >$ is the average drift-velocity, $< \epsilon(z,t) >$ the mean energy, and $Q_A =< \delta v \delta A >_0$ the covariance of the instantaneous fluctuations of $v(\mathbf{p})$ and $A(\mathbf{p})$ averaged over the stationary distribution function of the homogeneous case at the same mean energy. By substituting 1, $v(\mathbf{p})$ and $\epsilon(\mathbf{p})$ in place of $A(\mathbf{p})$ into Eqs. (1) and (2), one obtains, respectively, the concentration, velocity and energy conservation equations:

$$\frac{\partial n}{\partial t} = -\frac{\partial}{\partial z}(n < v >) \qquad (3)$$

$$\frac{\partial < v >}{\partial t} = eE < m^{-1} > - < v > \nu_v - < v > \frac{\partial < v >}{\partial z} - \frac{1}{n}\frac{\partial}{\partial z}(nQ_v) \qquad (4)$$

$$\frac{\partial < \epsilon >}{\partial t} = eE < v > -(< \epsilon > -\epsilon_{th})\nu_\epsilon - < v > \frac{\partial < \epsilon >}{\partial z} - \frac{1}{n}\frac{\partial}{\partial z}(nQ_\epsilon) \qquad (5)$$

The model contains five parameters, namely: the average of the reciprocal effective-mass $< m^{-1} >$, the velocity and energy relaxation rates, ν_v and ν_ϵ, the instantaneous velocity-velocity and velocity-energy fluctuations, $Q_v =< \delta v^2 >_0$ and $Q_\epsilon = < \delta v \delta \epsilon >_0$. All the parameters are assumed to depend on the local instantaneous energy only, and as such they are obtained from a stationary Monte Carlo (MC) simulation of the bulk semiconductor [3].

The small signal analysis is performed by linearizing the homogeneous and stationary conservation equations with respect to small perturbaions of field, velocity and energy [3]. As an application, Fig. 1 shows the frequency dependence of the real part of the differential mobility calculated with the HD and MC approaches for the case of $n - InP$. It should be noticed that the model gives good quantitative agreement for the threshold of negative differential mobility that is of great importance for the Gunn device simulations.

The diffusion and noise analysis is based on the knowledge of the correlation function of velocity fluctuations. These functions can be constructed from the general solution of the linearized equation used for the small-signal analysis. Under hot-carrier conditions, both $< \delta v^2 >_0$ and $< \delta v \delta \epsilon >_0$ give contributions to velocity fluctuations. Figure 2 shows the field dependence of the longitudinal diffusion coefficient calculated by the present HD and MC approaches. Figure 3 presents the comparison between the spectral density of velocity fluctuations calculated in the framework of the HD and the MC approaches. The overall good agreement so found between the HD and MC results validates the present model and supports its application to nonhomogeneous situations. Moreover, one can use the $S_v(f)$ provided by this model as a local noise sources for a self-consistent calculations of the voltage spectral density using the impedance field method [4]. As application of present results to the modelling of submicron devices we have considered the case of n^+nn^+ structures. As an example, Fig. 4 shows the concentration, velocity, energy and electric field distributions along an InP diode calculated by the HD and MC methods ($n^+ = 10^{18}$, $n = 3 \times 10^{16}$ cm^{-3}, $l_1^+ = 0.1$, $l_n = 0.5$, $l_2^+ = 0.3$ μm). Apart from some discrepancies which appear as a rule near the anode contact, good agreement is found between the results obtained with the two methods.

3. Conclusions

A closed hydrodynamic model which accounts for the nonparabolicity of the band and the non-Maxwellian shape of the hot carrier distribution function is developed. The small-signal and diffusion-noise characteristics of bulk n-InP as well as a n^+nn^+ diode modelling have been evaluated and compared with a full Monte Carlo approach. The excellent agreement found fully supports the physical reliability of the proposed model which has the advantage of requiring reasonably cheap computanional facilities (e.g a 486 pc) and short running times (e.g. few minutes for characterizing a n^+nn^+ structure as reported above).

This work has been partially supported by the CEC CIPA 3510PL921499 contract.

References

[1] G. Baccarani, M. Rudan, R. Guerrieri and P. Ciampolini, in "Process and Device Modeling", ed. by W. L. Engl (Elsevier/North-Holland, New York, 1986).

[2] S. Selberherr, "Analysis and Simulation of Semiconductor Devices" (Springer, Wien-New York, 1984).

[3] V. Gružinskis, E. Starikov, P. Shiktorov, L. Reggiani, M. Saraniti and L. Varani, Semicond. Science Technol., in press (July, 1993).

[4] V. Gružinskis, E. Starikov, P. Shiktorov, L. Reggiani, M. Saraniti and L. Varani, to be presented at the 12th Noise Conference (St Louis Missouri, 1993).

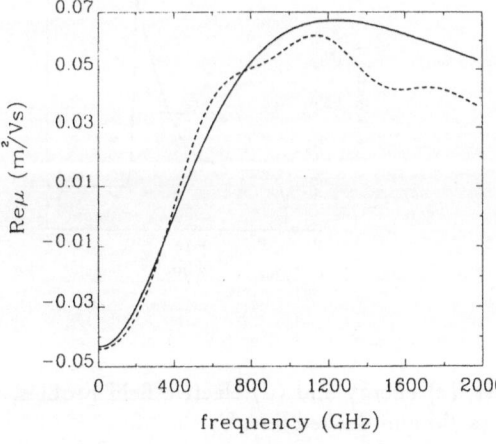

Figure 1: Real part of the differential mobility $Re\mu(f)$ calculated by the HD and MC approaches (solid and dashed lines, respectively). n-InP, $n = 10^{16}$ cm^{-3}, 300 K, $E = 25$ kV/cm.

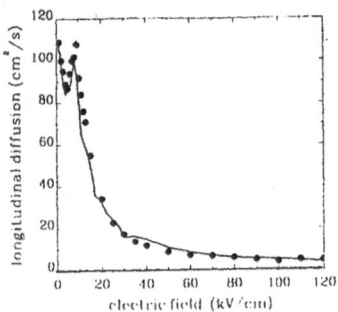

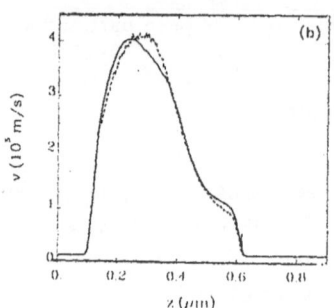

Figure 2: Longitudinal diffusivity obtained with HD and MC approaches (solid lines and dots).

Figure 3: Spectra of velocity fluctuations obtained with HD and MC approaches (solid and dashed lines).

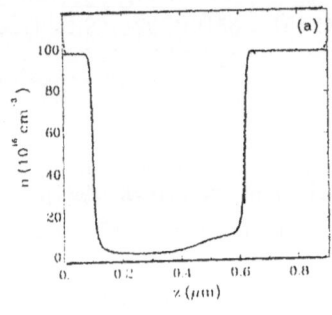

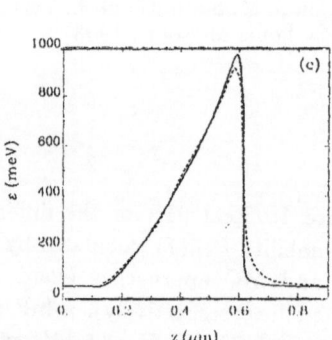

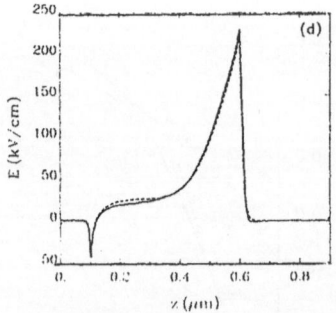

Figure 4: (a) concentration, (b) velocity, (c) energy and (d) electric field profiles, obtained with HD and MC approaches (solid and dashed lines).

SIMULATION OF SEMICONDUCTOR DEVICES AND PROCESSES Vol. 5
Edited by S. Selberherr, H. Stippel, E. Strasser – September 1993

Critical Assessment of Different Hydrodynamical Models for Avalanche Multiplication Calculation in Silicon Bipolar Transistors

A. D. Sadovnikov and D. J. Roulston

Electrical and Computer Engineering Department, University of Waterloo
Waterloo, Ontario, N2L 3G1, CANADA

Abstract

Different hydrodynamical numerical and analytical models for avalance multiplication coefficient calculation in silicon BJTs are considered. A comparison with experimental data is made.

1. Introduction

Non-uniform electron energy distribution has a significant effect on impact ionization phenomena in silicon bipolar transistors. A number of different analytical and numerical hydrodynamic (HD) models were suggested and used by different authors during the last few years [1-4] to calculate ionization currents in silicon devices. The purpose of this report is to investigate the applicability of these models for avalanche multiplication coefficient M_e calculation in silicon n-p-n BJT.

2. Model descriptions

For the sake of simplicity we will consider a one-dimensional case and use the simplest HD model [3] which will be referred to as HDM1:

$$\frac{dJ}{dx} = q(R - G) \tag{1}$$

$$J = q\mu n F + \mu \frac{dnk_B T}{dx}, \tag{2}$$

$$\frac{dSn}{dx} = JF - \frac{3}{2}nk_B \frac{T - T_0}{\tau_w}, \tag{3}$$

$$Sn = -\frac{k_B T}{q}[C_1 J + C_2 \mu n \frac{dk_B T}{dx}], \tag{4}$$

where all notations have their usual meaning. In general, coefficients τ_w, C_1, C_2 are functions of electron energy W, but they are taken as constants in [3] and we will use the values $\tau_w = 0.3$ ps, $C_1 = 1.7$, $C_2 = 1.2$ which were obtained by fitting simulation results with experimental and Monte-Carlo data [5]. We ignore the impact ionization rate term in (3) and therefore restrict ourselves to small M_e values.

To close this set of equations, we add Poisson's equation for electric potential, the continuity equation for hole concentration (for which we shall assume constant hole temperature distribution) and corresponding boundary conditions.

To calculate M_e values one should know the $W(x)$ distribution in the BJT because ionization coefficients depend on W but not on local field. The HD model calculates this distribution but takes much more time than the drift-diffusion (DD) model. That is why there were several attempts to find a simple formula for the $W(x)$ calculation using $F(x)$ obtained with the DD model [1,4]. Let us briefly trace the method of derivation of these forlmulas.

Neglecting the temperature diffusion term in (4) and assuming that $div(J) = 0$, equations (4) and (3) reduce to the first order differential equation

$$C_1 v \frac{k_B T}{dx} = -qvF - \frac{3}{2} k_B \frac{T - T_0}{\tau_w} \qquad (5)$$

where $v = -J/qn$ is the electron velocity. We found out that the first assumption significantly influences the $v(x)$ distribution but much less the $T(x)$ distribution. The second assumption is correct because in the base-collector space-charge region (SCR) we have nearly one-dimensional electron flow without recombination. To proceed, one should make an approximation for $v(x)$. Using $2/3 C_1 v(x)\tau_w = \lambda_w = const$, it is possible to integrate (5) analytically and obtain [1]:

$$T(x) - T_0 = -\frac{q}{k_B C_1} \int_0^x F(z) exp(\frac{x - z}{\lambda_w}) dz \qquad (6)$$

This is obviously a poor approximation because v is position dependent and C_1 and τ_w in general depend on electron energy.

One can also use (2) to obtain v and, after substituting into (5), obtain the non-linear differential equation

$$\frac{dT}{dx} = -\frac{1}{2}[\frac{q}{k_B}F(1 + \frac{1}{C_1}) + T\frac{dln(n)}{dx}] - \sqrt{[\frac{q}{k_B}F(1 - \frac{1}{C_1}) + T\frac{dln(n)}{dx}]^2/4 + \frac{1.5q}{k_B} \frac{T - T_0}{C_1 \tau_w \mu}} \qquad (7)$$

which should be solved by numerical integration. For $W = 3/2k_B T$, $C_1 = 5/2$ and $dln(n)/dx = 0$ (7) reduces to the equation, used in [4] for MOSFETs. However we cannot neglect the electron concentration gradient in the collector-base SCR and hence using the $n(x)$ distribution calculated by the DD model, will obviously be a poor approximation.

3. Results and discussion

A special numerical code has been designed for 1D BJT simulation using the DD model, the HDM1 and the more complicated HD model from [2] (which will be referred to as HDM2). For the DD calculations we used (6) or (7) to calculate the $T(x)$ distribution and in all cases for *effective field* calculation we use an equilibrium relationship

$$F_{eff} = \sqrt{V_T 1.5(T/T_0 - 1)/\mu/\tau_w} \qquad (8)$$

which follows from (3) for $dSn/dx = 0$. The $F_{eff}(x)$ dependence was used for the ionization integral and, finally, for M_e calculation. Calculations using (6) were done with $\lambda_w = 34$ nm which corresponds to the assumption $v(x) = v_s = 10^7$ cm/s. Shown on Fig.1 are calculated $W(x)$ distributions for the BJT from [1] with $N_C = 6 \times 10^{17}$ cm^{-3}. A few remarks are appropriate:

1. Calculations using (6) and (7) give higher maximum W value and significantly broader $W(x)$ distribution near maximum value than do HD calculations. This is the consequence of a poor $v(x)$ distribution approximation in both methods. Accounting for the $n(x)$ gradient, calculated by the DD model, in (7) helps only a little.

2. The difference between $W(x)$ distributions calculated by HD and DD models depends on V_{CB} (compare Fig. 1a and 1b). That is why even treating C_1 and λ_w in (6) as fitting parameters, it is impossible to achieve good agreement for different V_{CB} values even for the same BJT.

3. Both HD models agree well in spite of the difference in their coefficients.

Finally, we calculated $M_e(V_{CB})$ dependencies using these models and $\alpha(F)$ dependence from [6] (see Fig. 2). We found out that (6) with $C_1 = 2.5$ and (7) with $C_1 = 2.4$ give maximum $W(x)$ values which are very close to those calculated by the HD models (but the shape of the $W(x)$ distributions remain different). Roughly speaking this corresponds to a λ_w value increased to 50 nm and partially explains why in previous work [1,5,7] significantly larger λ_w values were used in order to achieve agreement with experimental data. The other reason for the λ_w difference is that the real C_1 and τ_w values for the high-energy regime can be larger than those we used.

We see that agreement between all calculated and experimental curves is quite good and can be further improved by using other $\alpha(F)$ dependencies, available in the literature. Formula (7) seems to give a slightly better result than (6) because it uses a slightly better approximation for the $v(x)$ distribution.

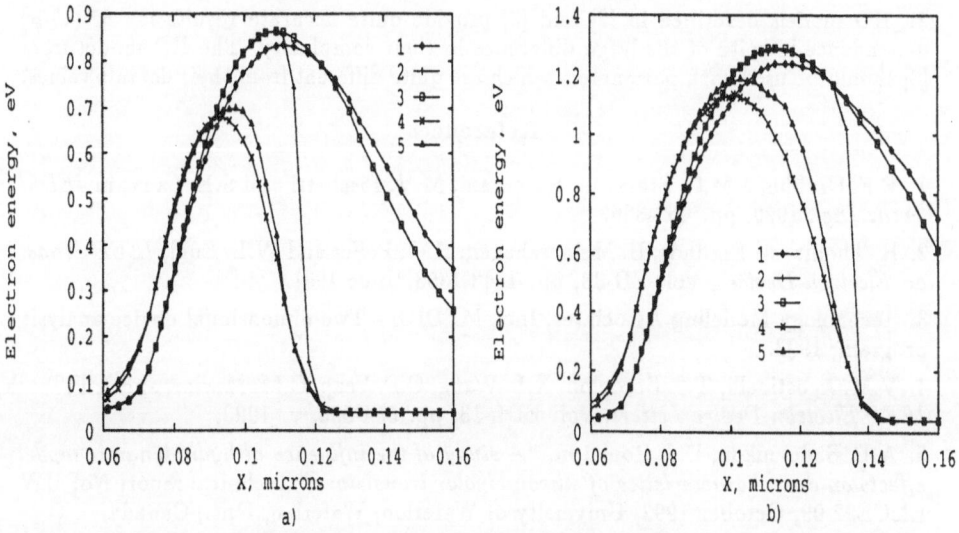

Fig.1. Calculated dependencies of electron energy for a) $V_{CB} = 1$V, b) $V_{CB} = 2.5$V, using: 1 - (6); 2 - (7); 3 - (7) with $dln(n)/dx = 0$; 4 - HDM1[3]; 5 - HDM2 [2].

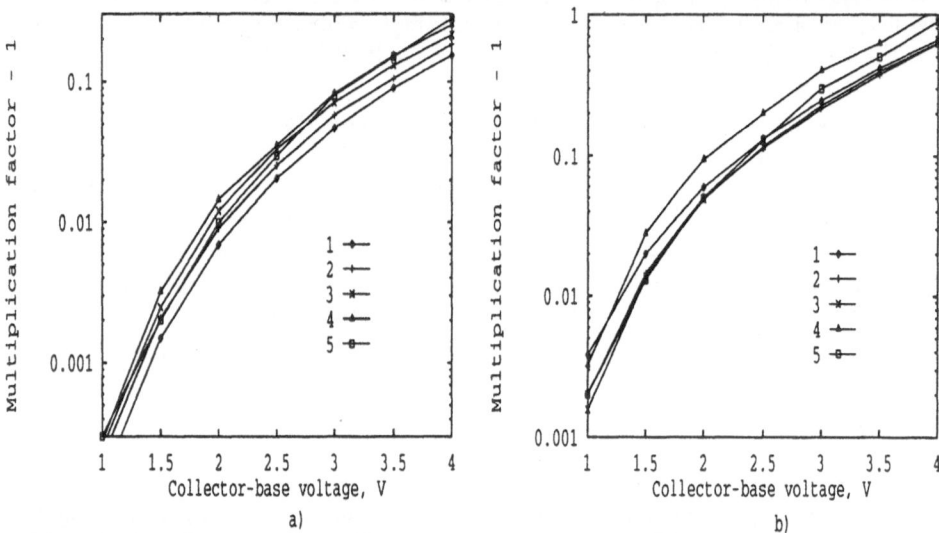

Fig. 2. Dependence of the avalanche multiplication factor M_e on the collector-base voltage V_{CB} for two BJTs with constant collector doping: a) $2 \cdot 10^{17}$ and b) $6 \cdot 10^{17} cm^{-3}$ calculated using: 1- (6); 2 - (7); 3 - HDM1 [3]; 4 - HDM2 [2]; 5 - experiment [1].

4. Conclusions

1. Simple analytical formulas (6) and (7) do not describe the $W(x)$ electron energy dependence correctly. However with the proper choice of coefficients they seems to be a good engineering tool for M_e value calculations in silicon BJTs.

2. HD models described in [2] and [3] provide quite accurate results for $M_e(V_{CB})$ dependence in spite of the large difference in their complexity. The HD model from [3] should be used with parameters which are quite different from their default values.

References

1. E.F. Crabbe, J.M.C. Stork, G. Baccarani, M.V. Fischetti and S.E. Laux, in *IEDM Tech. Dig.*, 1990, pp. 463-366.

2. R. Thoma, A. Emunds, B. Meinerzhagen, H.-J. Peifer and W.L. Engl, *IEEE Trans. on Electron Devices*, Vol. ED-38, pp. 1343-1353, June 1991.

3. Technology Modeling Associates, Inc., MEDICI - Two-dimensional device analysis program, 1992.

4. V.M. Agostinelli, T.J. Bordelon, X.L. Wang, C.F. Yeap, C.M. Maziar, and A.F.Tasch, *IEEE Electron Device Letters*, Vol. EDL-13, pp. 554-556, v. 1992.

5. A.D. Sadovnikov, D.J. Roulston, "*A study of the influence of hydrodynamic model effects on d.c. characteristics of silicon bipolar transistors,*" Technical report No. UW E&CE92-09, October 1992, University of Waterloo, Waterloo, Ont., Canada.

6. C.R. Crowell and S.M. Sze, Appl. Phys. Lett., vol. 9, pp. 242-244, Sept. 1966.

7. J.W. Slotboom, G. Streutker, M.J. v. Dort, P.H. Woerlee, A. Pruijmboom, D.J. Gravesteijn, in IEDM Tech. Dig., 1991, pp. 127-130.

Dual Energy Transport Model with Coupled Lattice and Carrier Temperatures

D. Chen, Z. Yu, K.-C. Wu, R. Goossens, and R. W. Dutton

Integrated Circuits Laboratory, Stanford University
Stanford, CA 94305, USA

Abstract

A Dual Energy Transport (Dual ET) model was developed, that includes Poisson's equation, carrier continuity equations and the energy balance and thermal diffusion equation. Six variables (electric potential, electron and hole concentrations, electron and hole temperatures, and lattice temperature) can be obtained, describing all electro-thermal effects in the electrons, holes and lattice subsytems. Results for diode breakdown are shown.

1. Introduction

With progress in technology, progressively more electrical and thermal effects play a role. These effects also interact. To describe them accurately, an electro-thermal model including carrier and lattice energy transport is needed. Here, a dual energy transport (DUET) model is developed. The model considers the particle and energy transport within and between the three subsystems (lattice, electrons and holes) by solving six equations: Poisson's equation, the continuity equations for electrons and holes, the energy balance equations for electrons and holes, and the thermal diffusion equation for lattice. The electric potential ψ, the electron and hole concentrations n and p, the electron and hole temperatures T_n and T_p, and the lattice temperature T_L can be obtained. This model has been implemented in PISCES. Simulation results for a $n - p$ diode are presented, to show the advantages of the new model.

2. Equations

The following six equation are used in this model:

$$\nabla^2 \psi = \frac{q}{\epsilon}(n - p + N_A - N_D) \tag{1}$$

$$q\frac{\partial n}{\partial t} - \nabla \cdot \mathbf{J}_n = -qU \; ; \tag{2}$$

$$q\frac{\partial n}{\partial t} + \nabla \cdot \mathbf{J}_p = -qU \; ; \tag{3}$$

$$\frac{\partial(n < E_n >)}{\partial t} + \nabla \cdot \mathbf{S}_n = \mathbf{F} \cdot \mathbf{J}_n - W_n \tag{4}$$

$$\frac{\partial(p < E_p >)}{\partial t} + \nabla \cdot \mathbf{S}_p = \mathbf{F} \cdot \mathbf{J}_p - W_p \tag{5}$$

$$C_L \cdot \frac{\partial T_L}{\partial t} - \nabla \cdot (\kappa \nabla T_L) = H \tag{6}$$

whith ψ, $\mathbf{F}$ the electrical potential and field; n, p the electron and hole concentrations; $\mathbf{J}_n$, $\mathbf{J}_p$ the electron and hole currents; E_n, E_p the electron and hole energies; $\mathbf{S}_n$, $\mathbf{S}_p$ the electron and hole energy flows; T_L the lattice temperature; C_L, κ the heat capacity and thermal conductivity for the lattice; U the net carrier recombination rate; W_n, W_p the net energy loss rate for electrons and holes; H the net heat source for lattice.

The currents and energy flows can be approximated by [1, 2]

$$\mathbf{J_n} = -q\mu_n(n\nabla\psi - V_n\nabla n - n(1 + \gamma_n)\nabla V_n) \tag{7}$$

$$\mathbf{J_p} = -q\mu_p(p\nabla\psi + V_p\nabla p + p(1 + \gamma_p)\nabla V_p) \tag{8}$$

$$\mathbf{S_n} = qC_{en}\mu_n V_n(n\nabla\psi - V_n\nabla n - n(2 + \gamma_n)\nabla V_n) \tag{9}$$

$$\mathbf{S_p} = qC_{ep}\mu_p V_p(p\nabla\psi + V_p\nabla p + p(2 + \gamma_p)\nabla V_p) \tag{10}$$

where μ_n and μ_p are the electron and hole mobilities, $V_n = k_B T_n/q$ and $V_p = k_B T_p/q$ with T_n and T_p being the electron and hole temperatures, and $C_{en} = 5/2 + \gamma_n$ and $C_{ep} = 5/2 + \gamma_p$ with $\gamma_n(T_n, T_L)$ and $\gamma_p(T_p, T_L)$ being defined by $\gamma_n = (T_n/\mu_n)\partial\mu_n/\partial T_n$ and $\gamma_p = (T_p/\mu_p)\partial\mu_p/\partial T_p$, respectively.

3. Exchanging Terms

The terms U, W_n, W_p and H need to be carefully evaluated, regarding all the important carrier and energy exchanging mechanisms. The net recombination rate can be written as

$$U = R_{srh} + R_n^{Aug} + R_p^{Aug} - G_n^{ii} - G_p^{ii} \tag{11}$$

where R_{srh} is the SRH recombination rate, R_n^{Aug} and R_p^{Aug} are the Auger recombination rate related to electrons and holes, and G_n^{ii} and G_p^{ii} are the I.I. rates due to electrons and holes, respectively. The conventional expressions for these terms can be found in [3]. The energy loss for electrons, W_n, includes the following terms:

$$W_{lat} \simeq n(< E_n > -E_0)/\tau_{wn}$$

$$\simeq \frac{3}{2}nk_B(T_n - T_L)/\tau_{wn}$$

$$W_{shr} \simeq \frac{3}{2}k_B T_n R_{srh} \tag{12}$$

$$W_{imp} \simeq (E_{gap} + \delta_n K_B T_n)G_n^{ii} - \delta_p K_B T_p G_p^{ii} \tag{13}$$

It can be approximated by

$$W_{imp} \simeq E_{gap}G_n^{ii} \tag{14}$$

Auger recombination can be considered the reverse procedure of the impact ionizations:

$$W_{aug} \simeq E_{gap} R_n^{Aug} \tag{15}$$

The total energy loss for electrons and holes becomes

$$
\begin{aligned}
W_n &= W_{lat} + W_{shr} + W_{imp} - W_{aug} \\
&\simeq \frac{3}{2} n \frac{k_B (T_n - T_L)}{\tau_{wn}} + \frac{3}{2} k_B T_n R_{srh} + E_{gap} (G_n^{ii} - R_n^{Aug})
\end{aligned}
\tag{16}
$$

$$
\begin{aligned}
W_p &\simeq \frac{3}{2} p \frac{k_B (T_p - T_L)}{\tau_{wp}} + \frac{3}{2} k_B T_p R_{srh} \\
&\quad + E_{gap} (G_p^{ii} - R_p^{Aug})
\end{aligned}
\tag{17}
$$

Finally, we find the expression for the net heat source, H,

$$
\begin{aligned}
H &= W_n + W_p + E_{gap} U \\
&\simeq \frac{3n}{2} \frac{k_B (T_n - T_L)}{\tau_{wn}} + \frac{3p}{2} \frac{k_B (T_p - T_L)}{\tau_{wp}} + (\frac{3}{2} k_B (T_n + T_p) + E_{gap}) R_{srh}
\end{aligned}
\tag{18}
$$

4. Simulations and Discussions

The DUET model can be approximated by simpler models for special applications. Assuming that T_L is constant, the DUET model defaults to the energy transport (ET) model [1, 2] (a hydrodynamic [4] -like model), which includes only the hot-carrier effects. Assuming that $T_n = T_p = T_L$, the thermodynamic (TD) model [5] can be derived from the DUET model; this choice would be appropriate for cases involving lattice heating and thermal diffusions. The Dual ET model has been implemented in PISCES. Excellent convergency behavior was observed in all cases.

Simulation results for a $n - p$ diode are shown below. Its doping profile is plotted in Fig. 1. In Figure 2 and 3, from DUET simulations, the distributions of T_n, T_p

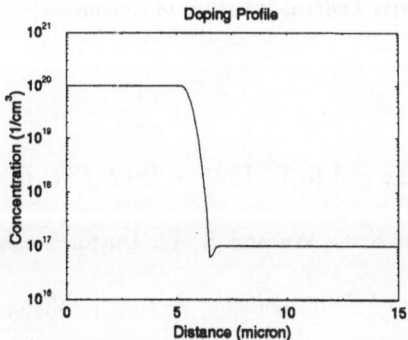

Fig. 1: Doping profile for a $n - p$ diode

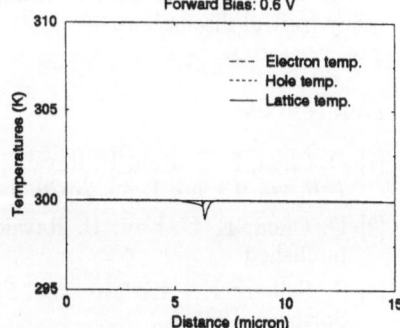

Fig. 2: Temperature distributions at forward bias: $V = 0.6V$

and T_L, for the forward biases $0.6V$ and $1V$, respectively, are plotted. As shown in these figures, $T_n \approx T_p \approx T_L$ under the forward biases, where the hot carrier effect

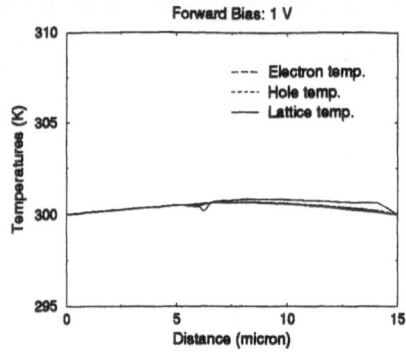
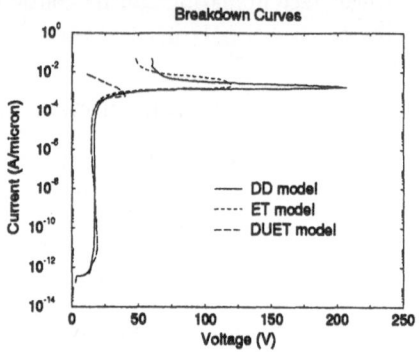

Fig. 3: Temperature distributions at forward bias: $V = 1V$

Fig. 4: I-V breakdown curves from DD, ET and Dual ET simulations.

is negligible, so that both TD and DUET models are valid. The reversed biased breakdown curves for this diode, using Drift-Diffusion (DD) model, ET model and Dual ET model, respectively, are included in Fig. 4, where the field-dependent impact ionization model was used for DD, and the carrier temperature-dependent I.I. model used for ET and DUET.

For these models breakdowns due to impact ionization happen at reverse biases around $18 - 20V$. A sharp snap-back (burn-out) occurs at high current levels for all the models. For different models, the burn-out voltage, V_{bo} shows a big difference. The DD model predicts V_{bo} at $208V$. The ET model, predicts V_{bo} at $148V$. For the Duel ET model, taking energy transport for both the carriers and lattice into account, V_{bo} is only about $40V$.

5. Acknowledgements

This work is supported by the Semiconductor Research Corporation(SRC) under contract no. 90-DJ116A20, by the Army Research Office under Grant DAAL03-91-G-0123 and by the National Science Foundation(NSF) through the National Center for Computational Electronics(NCCE) grant no. NSFECS-9200560. The authors would like to thank Prof. Gerhard Wachutka of Swiss Federal Institute of Technology for many fruitful discussions.

References

[1] D. Chen, E. C. Kan, U. Ravaioli, C.-W. Shu and R. W. Dutton, *IEEE Elec. Dev. Lett*, vol. 13, no. 1, pp. 26-28, Jan. 1992.

[2] D. Chen, E. C. Kan, U. Ravaioli, Z. Yu, K.-C. Wu and R. W. Dutton, to be published.

[3] S. Selberherr, *Analysis and Simulation of Semiconductor Devices*, Vienna: Springer, 1984.

[4] K. Bløtekjær, *IEEE Trans. Elec. Dev.*, vol. 17, p. 38, 1970.

[5] G. K. Wachutka, *IEEE Trans. CAD*, vol. 9, no. 11, pp. 1141-1149, Nov. 1990.

Inclusion of Electron-Electron Scattering in the Spherical Harmonics Expansion Treatment of the Boltzmann Transport Equation

D. Ventura, A. Gnudi, and G. Baccarani

Dipartimento di Elettronica, Informatica e Sistemistica, Università di Bologna
Viale Risorgimento 2, I-40136 Bologna, ITALY

Abstract

A new methodology is proposed to include the short range electron-electron interaction in the Spherical Harmonics Expansion approach to the Boltzmann Transport Equation, in the frame of spherical and non-parabolic bands. The electron energy distribution is computed in the uniform field case, and large corrections in the high-energy tail are observed.

1. Introduction

Short-range electron-electron (e-e) scattering is currently thought to have a major influence on the high-energy part of the electron population in semiconductor devices. Its implementation is therefore almost mandatory in every model of carrier transport aimed at the investigation of typical hot-electron effects such as current multiplication factors due to impact ionization and gate oxide injection. It is well known that this mechanism has been found difficult to treat even in Monte Carlo (MC) simulators, due to the non-linearity of the effect and to the cumbersome integrals required to find the scattering rates.

The Spherical Harmonics Expansion (SHE) scheme [1, 2, 3] has been developed recently as an alternative to both moment-based solutions of the Boltzmann Transport Equation (BTE) and Monte Carlo simulators since it allows a deterministic approximate solution of the BTE by reducing the dimension of the argument of the unknown carrier density $f(\mathbf{r}, \mathbf{k})$.

The purpose of this work is that of incorporating the e-e scattering in the program described in [1], which deals with the homogeneous case and spherical symmetric bands, without introducing any simplifying assumptions on the scattering integral calculations other than those intrinsically involved in the SHE method.

2. The mathematical procedure

Consider the expression of the e-e scattering operator

$$- f(\mathbf{k}) \int S(\mathbf{k}, \mathbf{k_o}, \mathbf{k'}, \mathbf{k'_o}) \, f(\mathbf{k_o}) \, d\mathbf{k_o} \, d\mathbf{k'} \, d\mathbf{k'_o} + \int S(\mathbf{k'}, \mathbf{k'_o}, \mathbf{k}, \mathbf{k_o}) \, f(\mathbf{k'}) \, f(\mathbf{k'_o}) \, d\mathbf{k_o} \, d\mathbf{k'} \, d\mathbf{k'_o}$$

$$(1)$$

with the scattering matrix (Born approximation)

$$S(\mathbf{k}, \mathbf{k_o}, \mathbf{k'}, \mathbf{k_o'}) = c_{ee}\,\delta(\mathbf{k} + \mathbf{k_o} - \mathbf{k'} - \mathbf{k_o'})\,\delta(E + E_o - E' - E_o')\,\frac{1}{(\beta^2 + q^2)^2} \qquad (2)$$

where $\mathbf{k}$ and $\mathbf{k_o}$ are the initial states of the two scattering electrons, the corresponding primed vectors are the final states, $\mathbf{q}$ is defined as $\mathbf{q} = \mathbf{k} - \mathbf{k'}$, and β is the inverse screening length. The following procedure is adopted:

1) as requested by the SHE scheme $f(\mathbf{a})$ is expanded as $f_o(a) + f_1(a)\cos(\theta)$, where θ is the angle between $\mathbf{a}$ and the reference axis, and the expansion is truncated at the second term;

2) the integration over $d\mathbf{k_o'}$ in the first integral and over $d\mathbf{k_o}$ in the second integral is performed by eliminating the momentum δ function;

3) the $d\mathbf{k_o}$ integral in the first term of (1) and the $d\mathbf{k_o'}$ integral in the second term are computed analytically in their angular parts using the energy δ function;

4) in both terms in (1) the angular part of the integral in $d\mathbf{k'}$ is performed analytically;

5) the remaining double integrals in $dk_o\,dk'$ for the first term, and $dk_o'\,dk'$ for the second term, are numerically computed.

The most peculiar part of this scheme is point 3) above. Actually, the energy δ function has usually been eliminated by an integration over the modulus of one of the $\mathbf{k}$ vectors involved in (1). This choice brings about overwhelming mathematical complications when applied to a band shape other than parabolic, thus forcing a choice between using an oversimplified band shape which fails at high energies, and a fully numeric integration scheme. We have taken a different stance by integrating the energy δ function on the cosine of one of the angles involved in (1). For instance, in the first integral in (1) we can write

$$\delta\left(E + E_o - E' - E_o'\right) =$$

$$= \frac{\gamma'(E + E_o - E')}{2\sqrt{q^2\gamma(E_o)}}\,\delta\left(\cos(\widehat{k_o q}) - \frac{\gamma(E + E_o - E') - \gamma(E_o) - q^2}{2\sqrt{q^2\gamma(E_o)}}\right) \qquad (3)$$

where

$$\gamma(E(\mathbf{k})) \equiv k^2 \qquad (4)$$

and γ' stands for the derivative of γ with respect to the energy. We proceed in an analogous way with the second integral in (1).

The non-linearities of quadratic type in the unknown density function are treated via an iterative method, where the density functions inside the integrals are frozen from the previous iteration step.

We stress the fact that no assumption on the band-shape other than spherical symmetry is required. In fact a set of non-parabolic bands is used in our simulations.

The procedure described above can partially be applied in the frame of the MC method. Since MC does not extract the angular dependence of f, we are left with double integrals in $dk_o\,dk'$ to be performed numerically in the computation of the total scattering rate.

3. Results

Figs. 1 and 2 show the results of the homogeneous SHE simulations for two different electric fields and various electron densities; β is set equal to the inverse Debye length. Even for low densities the significant effect of e-e scattering on the high-energy tail of the electron distribution is clearly seen. The effect is much less pronounced at electric fields higher than 300 kV/cm and lower than 30 kV/cm, since in this case the distribution is closer to a maxwellian shape.

Consider the expression (1) divided by $f(\mathbf{k})$. The zero order term of the SHE of such term describes the electron balance due to e-e scattering at a given energy, and represents an e-e effective scattering rate (ESR). Notice that such ESR is positive if the electron in-rate is larger than the electron out-rate. In Fig. 3 the positive part of the ESR is plotted for the same electric field and electron densities as for Fig. 1. The ESR for energies lower than 0.3 eV, not represented in the figure, is negative, reflecting the fact that cold electrons tend to be scattered to more energetic states. At larger energies one can note a very rapid growth of the ESR, which in the tail becomes comparable with the optical phonon scattering rate. This accounts for the major changes exhibited by the tail of the electron distribution in Figs. 1 and 2. It should be noticed that the increase of the ESR is related to the deviation of the distribution function from the maxwellian shape: a maxwellian distribution would result in a zero ESR.

The accuracy of the SHE method strictly depends on the fact that the coefficients of the expansion of the unknown distribution become negligible after a few terms. A comparison with MC data, where the e-e scattering was neglected, showed good agreement when two terms of the expansion were considered, and an excellent agreement with three terms [1]. Fig. 4 shows the ratio of the first two terms f_0 and f_1 for a 200 kV/cm field. It is apparent that f_1 looses much of its weight in the high-energy region, thus suggesting that the e-e scattering leads to a more spherical distribution. This makes us confident about the appropriateness of the truncation at the second term of the SHE.

4. Acknoledgments

This work has been supported by IBM, General-Technology Division, Essex Junction VT, 05452-USA.

References

[1] D. Ventura et al. *An Efficient Method for Evaluating the Energy Distribution of Electrons in Semiconductors Based on Spherical Harmonics Expansion*, IEICE Trans. Electron., vol. E75-C, no. 2, p. 194, 1992

[2] A. Gnudi et al. *One-Dimensional Simulation of a Bipolar Transistor by Means of Spherical Harmonics Expansion of the Boltzmann Transport Equation*, Proc. of the SISDEP '91 Conf., p. 205, September 1991, Zurich

[3] D. Ventura et al. *Multidimensional Spherical Harmonics Expansion of Boltzmann Equation for Transport in Semiconductors*, Appl. Math. Lett., vol. 5, no. 3, p. 85, 1992

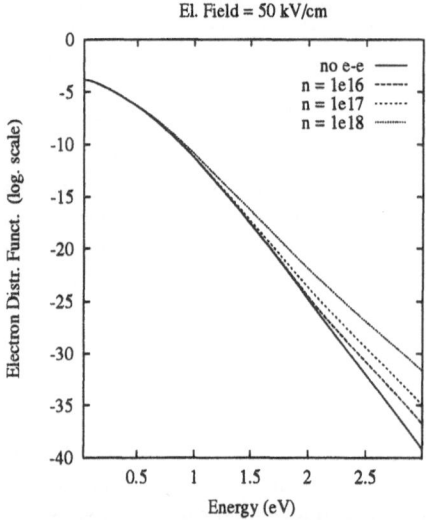

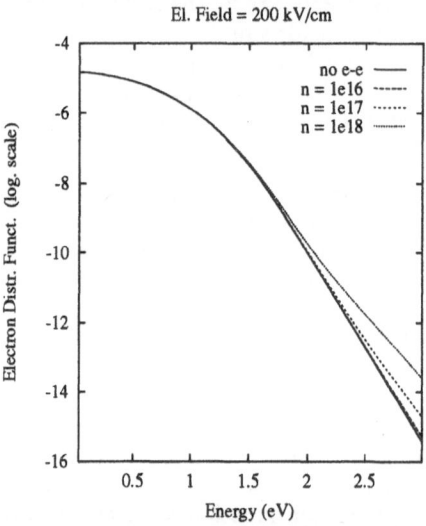

Figure 1: Electron distribution function for a 50 kV/cm electric field and various electron concentrations.

Figure 2: Electron distribution function for a 200 kV/cm electric field and various electron concentrations.

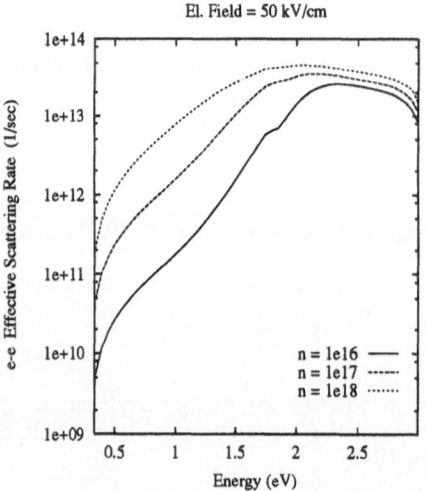

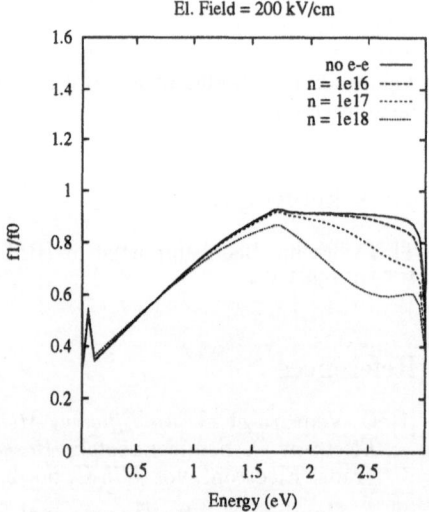

Figure 3: Short-range e-e effective scattering rate for various electron concentrations and a 50 kV/cm electric field.

Figure 4: The ratio between the first two SHE components for various electron concentrations and for a 200 kV/cm electric field compared with the ratio obtained without e-e scattering.

SIMULATION OF SEMICONDUCTOR DEVICES AND PROCESSES Vol. 5   165
Edited by S. Selberherr, H. Stippel, E. Strasser – September 1993

Quantitative 2D Stress Dependent Oxidation with Viscoelastic Model

V. Senez, D. Collard, and B. Baccus

IEMN-ISEN
41 Boulevard Vauban, F-59046 Lille Cédex, FRANCE

Abstract

For the proper modelling of the stress in two-dimensional local oxidation, oxide and nitride have to be considered as viscoelastic materials. This paper presents an original calibration of the viscoelastic model. It is based on the effect of stresses on the grown oxide thickness.

Viscoelastic treatment of thermal oxidation has the valuable advantage to take into account the mechanical properties of IC manufacturing materials in a large temperature range. Various models [1,2] have been introduced in the past, but their calibration just includes the oxide properties and neglects nitride modelling, or the oxide is assumed purely viscous [3]. In this paper, a viscoelastic oxidation model is compared to Kao's results and field oxide thickness reduction experiments. A self-consistent method is exposed which allows the coupled adjustment of oxide and nitride properties. The entire model is then applied to practical cases.

1. Numerical aspect

The 2D-oxidation model is implemented in the multilayer process simulator IMPACT 4 [4]. The stress dependent oxide motion is solved with a Gauss method, combined to an algorithm for profile and wavefront reduction of the mechanical matrix [5]. The benefits of this algorithm is obvious, compared to CG [6], at high temperature kinetics where oxide behaves like an incompressible viscous material giving a near undefined expression to the strain-stress relation. Oxidation simulations at 800 $^\circ$C and 1100 $^\circ$C with respective oxide viscosity equal to 10^{16} and 10^{12} poises were treated by both solvers : the CPU time is reduced by a factor 3 at 800 $^\circ$C and by a factor 10 at 1100 $^\circ$C.

2. Calibration

The stress dependence of oxide diffusivity and reaction rate constant is introduced according to Kao and Sutardja [7,8]. The corresponding activation volumes are V_d and V_k. Non-Newtonian behavior are assumed for oxide and nitride. The Eyring's plasticity formula [8] is used and requires the determination of the low stress viscosity and plasticity activation volume, respectively V_o, V_{po} for oxide and V_n, V_{pn} for nitride.

The calibration of Kao's experiments starts at low temperature with arbitrary value of V_d. Concave structure with large radii provides the adjustment of V_o while V_{po} is obtained for small radii. V_k is defined by comparison with the convexe structure results. The influence of diffusivity lowering on the oxide growth, V_d, is deduced through LOCOS thinning with large nitride mask aperture. V_d is modified and the previous procedure is repeated. Table 1 presents the best oxide parameters. The simulated concave structure is given in figure 1. Figure 2 shows the agreement with Kao's data at 900°C and 1000°C.

Since specific direct experiment for nitride does not exist, the modelling is achieved by simulations of the local oxide thinning phenomenon in structures with narrow nitride mask opening. They reveal that V_{pn} has rather no influence in oxide thickness reduction for thick nitride, V_n is the dominant parameter. On the contrary, V_{pn} can be deduced from thin nitride experiments where its effect is pronounced (see figure 3). Numerous iterations are necessary to define the nitride properties.

3. Application to practical cases:

Two real examples were chosen in literature to demonstrate the validity of the modelling. A LOCOS structure [9] was simulated, in steam ambient, at 1000°C to grow 0.47 µm thick field oxide. The oxide thickness variation with decreasing nitride mask opening is presented in figure 4 and is compared to experiment. Figure 5 corresponds to the growth of a ROI structure [10] in the case of a very small nitride length. For these examples, the set of parameters is : V_d =75 A^3, V_k=15 A^3, V_o=2.10^{14}poises, V_{po}=425 A^3, V_n=5.10^{15} poises, V_{pn} = 170 A^3 and good agreement is obtained that validates the global approach.

4. Conclusions

A new method for the calibration of the viscoelastic oxidation model has been developped. Its procedure couples together the adjustment of Kao's experiments with the fitting of the oxide thinning phenomenon in the LOCOS structure. Consequently, the respective influence of oxide and nitride in the generation of stresses are more accurately modeled. The extension of the method to other IC materials is desirable to study advanced isolation structures.

Acknowledgements

The authors would like to thank M. Brault and J. Lebailly from PHILIPS COMPONENTS, Caen for providing experimental results.

References

[1] J.P. Peng et al., COMPEL, vol. 10, p. 341, 1991.
[2] A. Poncet, IEEE Trans. on Computer-Aided Design, vol. 4, p. 41, 1985.
[3] C.S. Rafferty, IEDM Tech. Digest, p. 741, 1990.
[4] B. Baccus et al., Solid-State Electronics, vol. 32, p. 1013, 1989.
[5] S.W. Sloan, Int. J. Numeric. Methods Eng., vol. 28, p. 2651, 1989.
[6] D. Collard et al., in NUPAD IV Proceedings, p. 21, 1992.
[7] D.B. Kao et al., IEEE Trans. on Electron Devices, vol. 35, p. 25, 1988.
[8] P. Sutardja et al., IEEE Trans. on Electron Devices, vol. 36, p. 2415, 1989.
[9] B. Coulman et al., Proceedings of the 2nd Int. Symp. on ULSI Sci. and Techno., p. 759, 1989.
[10] M. Brault and J. Lebailly, private communication, 1992.

Temperature (°C)	VD (Å^3)	VK (Å^3)	V_0 (poises)	V_{pp} (Å^3)
800			9.10^{15}	300
900	75	15	6.10^{15}	390
1000			2.10^{14}	425
1100			4.10^{13}	1000

Table 1: Best parameters for KAO's data.

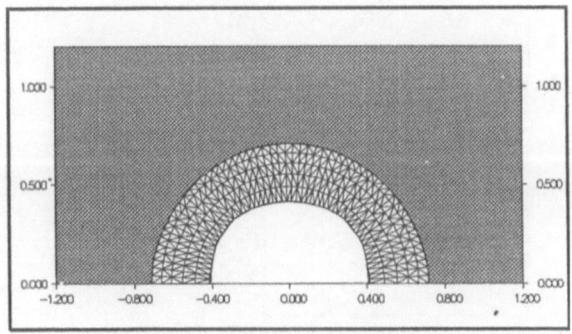

Figure 1 : Simulated concave structure.

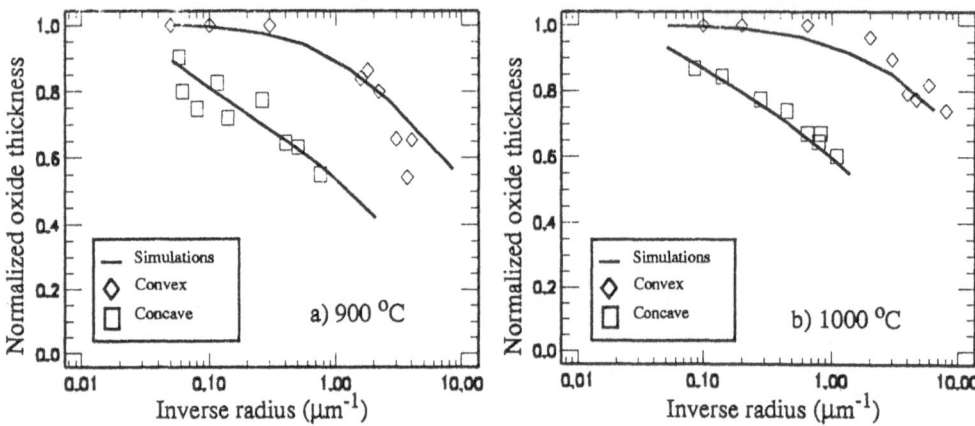

Figure 2 : Adjustment to KAO's data at a) 900 °C and b) 1000 °C.

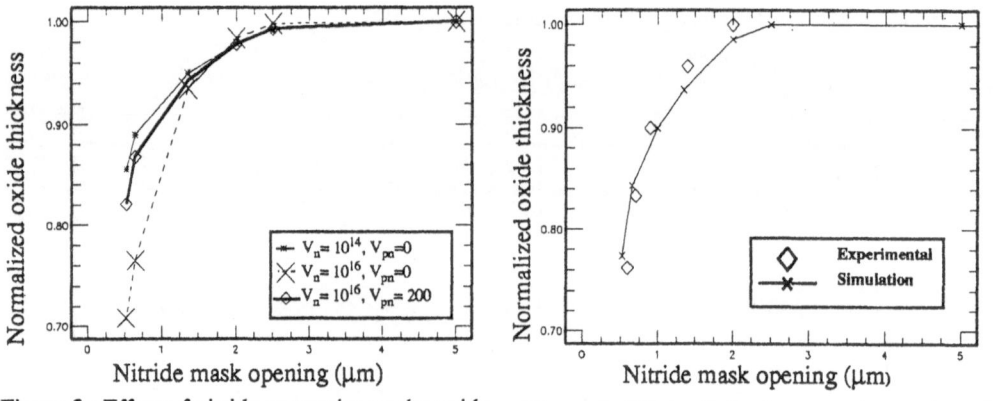

Figure 3 : Effect of nitride properties on the oxide thickness reduction in case of thin film (0.09 μm)

Figure 4 : Effect of nitride properties on the oxide thickness reduction in case of thin film (0.09 μm)

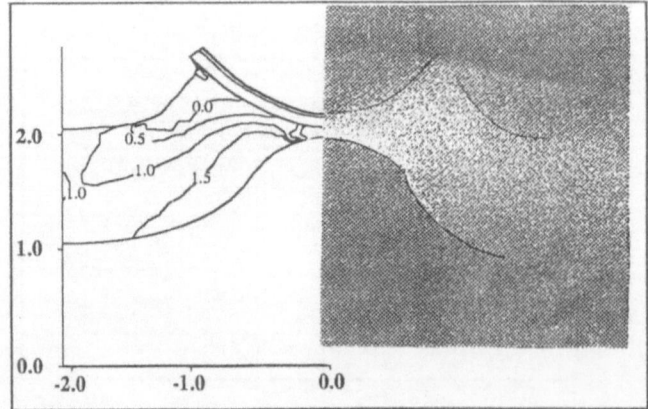

Figure 5 : Recessed oxide structure at 1000 °C with pad-SiO$_2$ (20 nm)/ Si$_3$N$_4$ (120 nm). The final field oxide is 1. μm. Simulated pressure distribution is given. Stress units are 100 MPa.

SIMULATION OF SEMICONDUCTOR DEVICES AND PROCESSES Vol. 5
Edited by S. Selberherr, H. Stippel, E. Strasser – September 1993

Oxidation Simulation and Growth Kinetics of Thin SiO_2 in Pure N_2O

S. C. Sun and H. Y. Chang

National Nano Device Laboratory, Chiao Tung University
Hsinchu, Taiwan, TAIWAN

Abstract

This paper reports the oxide growth behavior of silicon in an N_2O ambient using conventional furnace method. With all three different wafer orientations and wider temperature and time ranges used in our experiments, it has been found that the oxide growth can be simulated by the linear-parabolic model. This is in direct contrast with previous reports that the oxide growth is self-limited, or it was described only by the linear rate constant in oxidation.

1. Introduction

NH_3-Nitrided SiO_2 has been studied in the past ten years to improve the electrical properties of silicon dioxide layer. A re-oxidation is required immediately after nitridation step to minimize the high electron trapping efficiency and high electron trap generation rate due to a large amount of hydrogen atoms (H) in the oxide. Though reoxidation has been proved to be effective in reducing hydrogen concentration., the optimization of this process is complicated. Recently there has been tremendous interest in N_2O oxidation using either conventional furnace [1] or rapid thermal processor [2] to form high quality gate dielectrics. The growth of SiO_2 was reported to be self-limiting [3]. However, we have found the non-saturation growth behavior over a wide range of temperatures and different crystal orientations in N_2O oxidation [4].

In this work we present an oxidation model and discuss some simulation results.

2. Experimental

N_2O oxidation was performed in a conventional resistive-heated furnace in high purity gas (moisture less than 0.5 ppm) (>99.9995%) for (100), (111) and (110) oriented Si substrates. Oxide thicknesses were measured using ellipsometer with an assumed refractive index of 1.462. They were also confirmed by a high-frequency C-V measurement with less than 5% discrepancy. Fig. 1 shows oxide thickness versus oxidation time for (111) Si over the temperature from 900 C to 1100 C. (100) and (110) oriented substrates exhibit similar behavior but with different thicknesses.

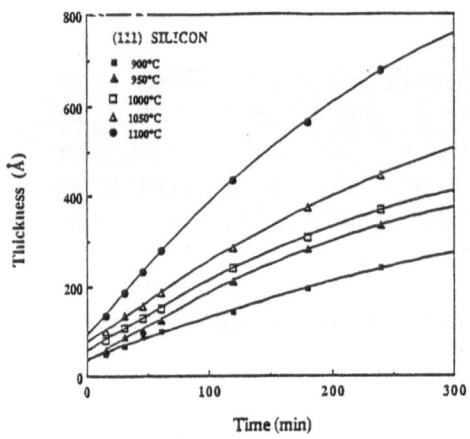

Fig. 1 Oxide thickness vs. oxidation time for (111)-oriented silicon

3. Results and Discussion

Several important features can be observed from experimental results. First, the growth of SiO_2 was found to be not self-limited. Second, the N_2O oxidation rates are much slower than those in dry O_2. Third, the (110) and (111) oxidation exhibits the "crossover" behavior in which the (110)-oriented silicon has a higher oxide growth rate than (111)-oriented wafer in the initial oxide growth. After oxide thickness exceeds 175 A, the (111)-oriented silicon has highest growth rate. This phenomenon was previously also observed in the dry oxygen oxidation [5].

Since oxygen from N_2O decomposition at high temperatures ($N_2O \rightarrow N_2 + 1/2O_2$) is primarily responsible for the oxidation and NO is responsible for nitogen incorporation in the Si/SiO_2 interface [6] which may retard the oxide growth, the rather complex oxide growth behavior in N_2O can be modeled by using oxidation of silicon in diluted dry oxygen. Empirically it was found that the growth rate in pure N_2O is essentially identical to that in 10% dry O_2 in argon over the temperature range from 900 C to 1000 C. The experimental data of 10%O_2/Ar [5] are compared with N_2O data as shown in Fig. 2.

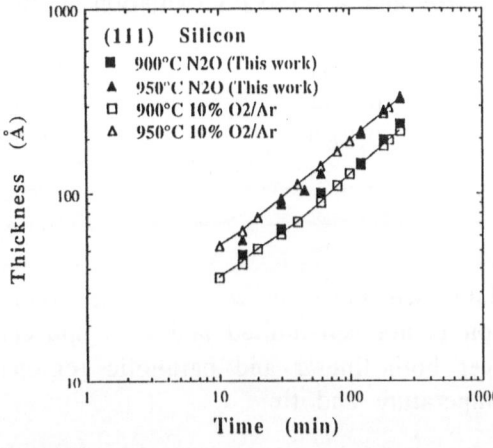

Fig. 2 Comparison of N_2O oxidation and 10% dry O_2-argon mixture oxidation for (111)-oriented silicon

Both N_2O and 10% O_2 in argon follow the same linear-parabolic model. Based on the experimental data, except for the very short oxidation time, the linear rate constant (B/A) and parabolic rate constant (B) can be determined and are plotted as a function of temerature as shown in Fig. 3 and Fig. 4 respectively.

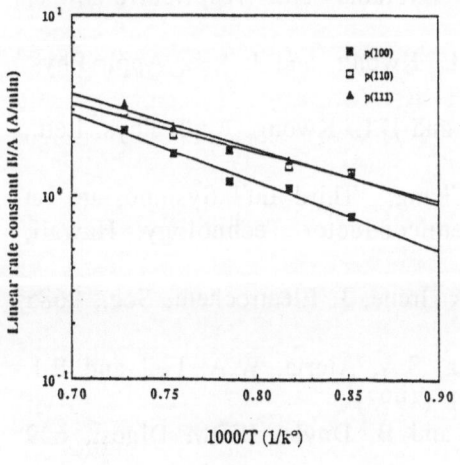

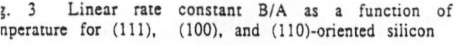

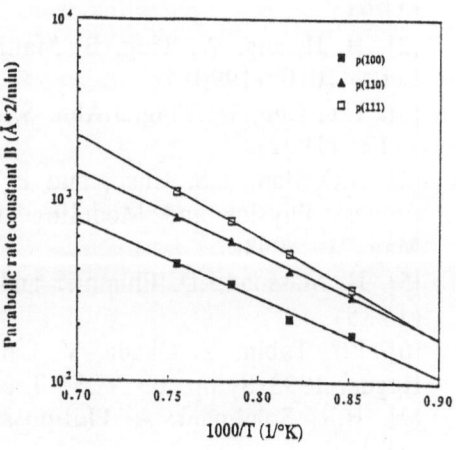

з. 3   Linear rate constant B/A as a function of nperature for (111), (100), and (110)-oriented silicon

Fig. 4   Parabolic rate constant B as a function of temperature for three different orientations

Previous study of N_2O oxidation covers only· a limited temperature range, time and oxide thickness [7]. Therefore the growth behavior in that case was characterized only by the linear rate constant (B/A). However it is evident that as oxidation time gets longer or oxidation temperature is greater than 1000 C, the single linear rate constant representation is inadequate. In order to fit a much broader temperature range and oxidation time, parabolic growth behavior must be considered. The activation energy E_a of B/A and B for all three orientations is shown in Table 1.

Table 1.   Activation energy E_a of B/A and B for N_2O oxidation in the Linear-Parabolic Model

	(100)	(110)	(111)
B/A	1.005	0.734	0.860
B	0.800	0.880	1.100

4. Summary

Using the experimental data, we have shown that the oxide growth of silicon in N_2O oxidation is not self-limited and an empirical model has been proposed to cover both linerar and parabolic regions of oxide growth over a wide temperature and time

Acknowledgements: This research was supported by NSC 82-0404-E-009-234 of R.O.C.

References
[1]. W. Ting, G.Q. Lo, J. Ahn, and D.L. Kwong, Proc., 1991 Int'l Symposium on VLSI Technology, Systems and Applications, 47 (1991)
[2]. H. Hwang, W. Ting, B. Maiti, D.L. Kwong, and J. Lee, Appl. Phys. Lett., 1010 (1990)
[3]. T.Y. Chu, W. Ting, J.Ahn, S. Lin, and D.L. Kwong, Appl. Phys. Lett., 1412 (1991)
[4]. S.C. Sun, T.S. Chao, and H.Y. Chang, Third Int'l Sysmposium on Process Physics and Modeling in Semiconductor Technology, Hawaii, May 21, 1993.
[5]. H. Massoud,J.D. Plummer and E.A. Irene, J. Electrochem. Soc., 2685 (1985)
[6]. P.J. Tobin, Y. Okada, V. Lahkotia, S.A. Ajuria, W.A. Feil and R.I. Hegde, 1993 Symp. on VLSI Tech. 51 (1993)
[7]. H.R. Soleimani, A. Philipossian, and B. Doyle, IEDM Digest, 629 (1992)

Accurate Simulation of Mechanical Stresses in Silicon During Thermal Oxidation

A. Poncet

CNET-CNS, France Telecom
Chemin du Vieux Chêne, BP 98, F-38243 Meylan Cédex, FRANCE

Abstract

The aim of this paper is to present viscoelastic models to accurately simulate mechanical stresses which result from volume expansion during thermal oxidation or temperature ramps in silicon technology. Comparisons are made with wafer curvature measurements and it is shown that mechanical stresses can explain the "anomalously" fast initial regime during dry oxidation, without involving any additional chemical mechanism.

1. Introduction

Numerical models for local thermal oxidation have been developed for a decade [3,4] starting from the well known Deal&Grove 1D model [1]. Even at a very early stage of development, these 2D models were successfully applied to advanced isolation techniques like Sealed Isolation LOCOS (SILO) process [6]. This process is nowadays a candidate to compete with Poly-buffered LOCOS (PBL) for quarter micron CMOS technology [12]. Simulations of PBL have been performed more recently, due to the complexity resulting from the polysilicon layer which bends while being oxidized [10,11]. The effects of mechanical stresses have been introduced first in [6] to explain short bird's beaks, and the expression of stress induced reductions on diffusivity, reaction rates and viscosity are nowadays classically taken from the works of Kao, Suturdja et al. [8,9]. However, some progress was still expected for the accurate modelling of the complete mechanical stresses in Silicon, i.e. not only the additional stresses which result from local 2D effects and which are classically used in the present stress-reduced oxidation models; it includes stresses which result, even in 1D, from the expansion of silicon to form new oxide.

The aim of this paper is to present viscoelastic models to simulate more accurately this volume expansion. This approach includes the modelling of thermal stresses to account for temperature ramps. Comparisons are made with experiments on wafer curvature [2,7], and it is finally confirmed that, as anticipated by Fargeix and Guibaudo [5], mechanical stresses which are present in the thermally grown dioxide layers even in planar structures can explain the "anomalously" fast initial regime during dry oxidation.

2. A Comprehensive Model for Moving Oxide/Silicon Interfaces

The standard algorithm for LOCOS simulation consists in solving first the diffusion-reaction equation at each time step in order to compute oxidant concentration, next growing a new oxide layer without any material deformation, and finally solving some equations (elasticity, Stokes, or viscoelasticity) in the upper layers in order to express the mechanical reaction of the rigid substrate by means of imposed velocities on the Si/SiO2 interface.

Such a procedure fails when the silicon itself *bends* during oxidation, like in PBL for instance . Moreover, mechanical stresses computed by this way are *uncomplete* insofar as they

represent only 2D effects in the overlayers. For instance, expressing sources of point defects versus stresses in silicon requires a complete stress computation. For these reasons, a new algorithm is proposed to deal with this paradox: to handle materials which can expand while they must keep uncompressible at given temperature and composition.
- the Si/SiO2 interface is moved to express silicon consumption without any dilatation;
- next, viscoelasticity equations are provided with suitable extra terms to account for the expansion of the newly grown oxide;
- finally, the large *tensile* stresses which come out from a straighforward evaluation from velocities in the newly grown dioxide layer are updated, depending on what is needed:
* to get back usual profiles (extra 2D stress), stresses must return to zero in 1D areas;
* to get the *exact* stress tensor in all layers, and to update hydrostatic pressure (which is used in Uzawa algorithm to express uncompressibility), stresses which would result from unconstrained dilatation must be substracted; so doing, the lateral stresses in dioxide layers return to compressive as expected. This method has been validated first on thermal dilatation to check that, on one hand tangent stresses at the interface between two layers with different dilatation coeffficients are well balanced, and on the other hand that the volume expansion does not depend on boundary conditions: it must be for instance the same in a 1D structure with boundary conditions which impose displacements in only one direction, and a complete 2D structure; it is not the case with the "elastic" model which allows some compressibility of materials.

3. Alternative models for mechanics

Two models have been investigated: an incremental linear elastic one with maxwelian stress relaxation, and a complete elasto-visco-plasticone; stress-reduced viscosity from Suturdja & Oldham [9] is available in both models. For sake of simplicity, the first model will be called "elastic", and the second one "viscous". Comparisons are shown on fig. 1.

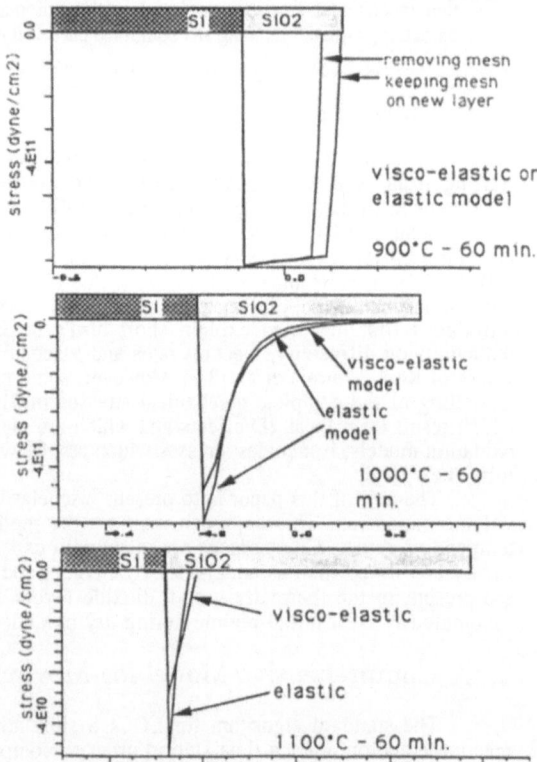

With the "elastic" model, it is obviously impossible to decrease the magnitude of stresses in the newly formed oxide at each time step, because stress relaxation is applied only at the end of the time step; therefore, the stress dependent viscosity just leads to a stiffer stress reduction inside the oxide film, but this does not affect significantly stress in silicon.

On the other hand, as a linearisation loop is used in the "viscous" model, at least to express uncompressibility, the effect of reduced viscosity is observed all through the film thickness. In that case, mechanical stresses in silicon dioxide along the silicon/oxide interface are reduced accordingly.

However, in both cases the stress reduction is far from what is needed to decrease the wafer curvature significantly (fig. 2); this is in contradiction with a recent paper [14]. In the present model, the only way to get realistic curvatures is to put an anisotropy factor in the dilatation term: with 100% anisotropy, mechanical stresses reduce to the extra 2D stresses, as defined above.

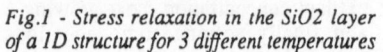

Fig.1 - Stress relaxation in the SiO2 layer of a 1D structure for 3 different temperatures

Unfortunately, the anisotropy factor, which is always greater than 90%, exhibits a very strong temperature dependence which is still to be fitted with a wider range of experiments.

4. Stress computation in full wafer oxidation

The most usefull results published up to now in the literature to validate stress computations during thermal oxidation remain data on wafer curvature from Eernisse [2] because stress relaxation and stress dependent viscosity lead to stiff variation of mechanical stresses in the oxide film, and then to large uncertaincies when simulated stresses are compared with mean stress values found in some publications. On the other hand, experiments on micro-Raman analysis [11,13] still suffer from a lack of accuracy to investigate local effects below 0.1 micron.

Curvature radii reported by Eernisse are large compared with the size of the area to simulate which, in turn, is large compared with the oxide film thickness (typical orders of magnitude are respectively 10^6, 10^2 and 10^{-1} micrometers). While the behaviour of the "elastic" model is numerically stable, the "viscous" model exhibits large round off errors (fig. 2). In order to overcome this difficulty, a scaling has been designed: the curvature radius and the tangent component of stress in silicon along the silicon / silicon dioxide interface have been plotted versus wafer thickness. Two fair linear relationships versus the substrate thickness have been observed for the square root of the curvature radius and for the inverse of the tangent stress in silicon along the oxide intreface, provided that the dimension of the simulated area is large compared with this thickness (at least a factor of 5), and that the mesh is fine enough (fig. 3). This allows the dimensions of the simulated zone to be drastically reduced, and then curvature radii to be larger and less sensitive to round-off errors. As a consequence, mechanical stresses deduced from such simulations are more reliable as well.

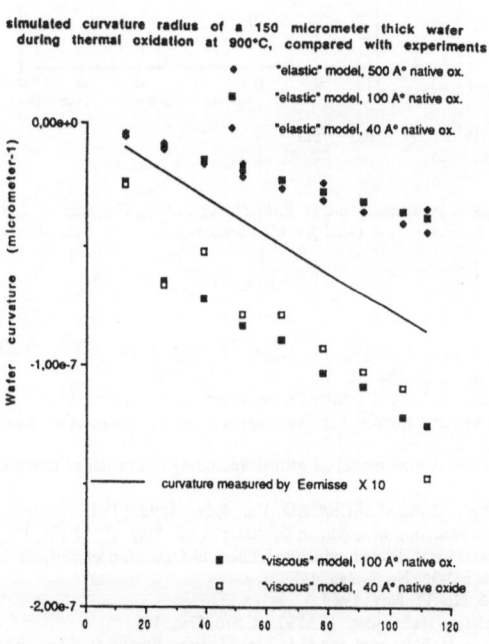

Fig. 2 - *comparison between curvature from Eernisse and simulated curvatures using two alternative models*

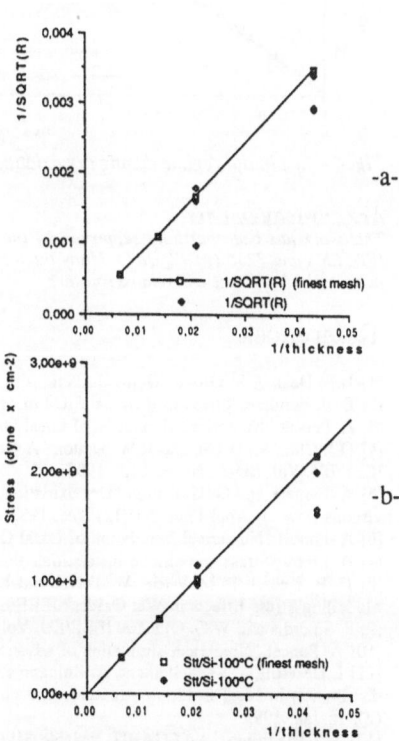

Fig. 3 - *curvature*$^{-1/2}$ *(a) and tangent stress in Si along SiO2 interface (b) versus inverse substrate thickness.*

5. Application to the early regime in dry oxidation

As recommended by several authors, the best way to exhibit the fast initial regime during dry oxidation is to plot the inverse of oxide growth velocity versus oxide thickness; comparison between figs. 4-a and 4-b plots illustrate this recommendation. It is shown on fig. 4-b how this fast initial regime can be explained by using only stress reduced parameters without adding new physical or chemical mechanisms (like diffusion of two oxidizing species).

As mentionned above, total stresses are overestimated by more than one order of magnitude in the "elastic" model with isotropic dilation forces; let notice that this model is valid at temperature used in the simulations reported on fig. 2 (900°C); therefore the activation volume to use in order to account for stress reduced diffusivity must be reduced by the same factor; for this reason, numerical results depicted on fig. 4 were obtained with an activation volume of 0.5Å^3 More accurate stress profiles at Si/SiO2 interface are necessary to obtain quantitatively good results with standard values of activation volumes (around 50Å^3).

-a-   -b-

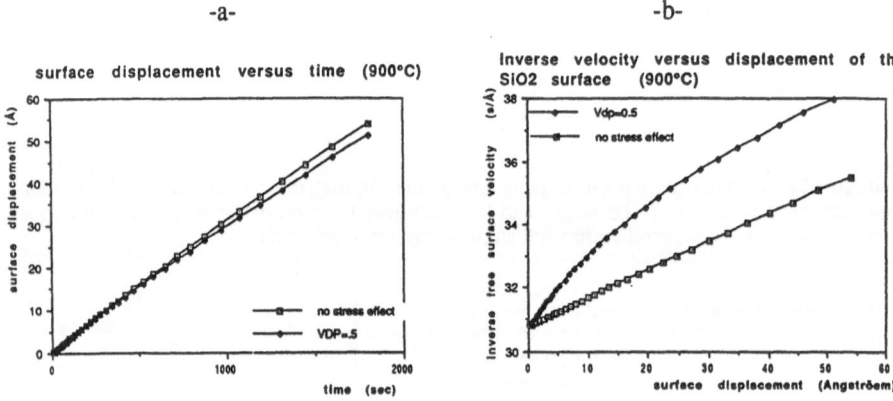

Fig. 4 - fast initial regime during dry oxidation explained by stress effect

Acknowledgements
This work has been partially supported by the European community under ESPRIT regulation, Projects n° 2197 (STORM) and 7236 (ADEQUAT). Many thanks to S. K. Jones from GEC for his participation to the validation of models presented here and related software

References

[1] B.E. Deal, A.S. Grove "Genaral relationship for the thermal oxidation of silicon" J. Appl. Phys., 36 (13), 1965
[2] E. P. Eernisse "Stress in theremal SiO2 dring growth" Appl. Phys. Letter 35 (1) 1 July 1979
[3] A. Poncet "Numerical Simulation of Local Oxidation of Silicon" Summer Course IMEC Leuven , June 1983
[4] D.J. Chin, S.-Y. Oh, and R.W. Dutton "A General Solution Method for Two-Dimensional Non Planar Oxidation" IEEE/ED, Vol. ED-30 No. 9, Sept. 1983
[5] A. Fargeix and G. Guibaudo "Dry oxidation of silicon: A new model of growth including relaxation of stress by viscous flow" J. Appl Phys. 54 (12) Dec. 1983
[6] A. Poncet "Numerical Simulation of Local Oxidation of Silicon" IEEE/CAD, Vol. 4, No. 1, Jan. 1985
[7] B. Leroy "Stress and silicon insterstitials during the oxidation of a silicon substrate" Phil. Mag. B 55 (2) 1987
[8] D.-B. Kao, J. P. McVittie, W. D.Nix and K.C. Saraswat " Two-dimensional Thermal Oxidation of Silicon -II. Modelling Stress Effects in Wet Oxides" IEEE/ED Vol. ED-35, No. 1, Jan. 1988
[9] P. Suturdja and W.G. Oldham. IEEE/ED, Vol ED-36, No. 11, Nov. 1989
[10] A. Poncet "Numerical simulation of advanced isolation techniques" ICM'91 , Cairo, Dec. 1991
[11] I. De Wolf, J. Vanhellement, A. Romano-Rodriguez, H. Norström and H.E. Maes "Micro-Raman study of stress distribution in local isolation structures and correlatation with transmission electron microscopy, J. ppl. Phys. 71 (2), 15 Jan. 1992
[12] S. Deleonibus "A GIGABIT scalable SILO field isolation using Rapid Thermal Nitridation (RTN) of silicon" ESSDERC'92 Conf. Leuven, Sept. 1992
[13] S. K. Jones, P.J. Pearson, C. Hill and A.V. Hetherington - STORM Internal report August 1992
[14] T. Uchida, N. Kotani and N. Tsubouchi "Verification of the Viscoelastic Oxidation Model Using Simple Test Structures" VPAD Conf. Proc. May 1993

Mechanical Stress Simulation During Gate Formation of MOS Devices Considering Crystallization-Induced Stress of p-Doped Silicon Thin Films

H. Miura, N. Saito, and N. Okamoto

Mechanical Engineering Research Laboratory, Hitachi, Ltd.
502 Kandatsu, Tsuchiura, Ibaraki 300, JAPAN

Abstract

Mechanical stress in silicon substrates caused by thin-film deposition of gate material of MOS transistors is analyzed by the finite element method. The results reveal that to predict precise stress distribution, it is very important to take into account the intrinsic stress of the thin films used as gate material as well as thermal stress .

1. Introduction

With the trend towards high integration of LSIs, it has become increasingly important to minimize or control the mechanical stresses that occur in LSI device structures during manufacturing processes [1]. This is because the device structure has become very complicated, with a lot of sharp corners, where mechanical stress easily concentrates. In addition, newly employed gate materials of MOS transistors, such as P-doped poly-Si films and WSix films, are found to hold very high intrinsic stress of about 1000 MPa [2]. Thus, the internal stress of the new device has been increasing drastically, causing not a few mechanical failures such as film cracking, delamination, and dislocation generation in silicon substrates. Since these failures deteriorate device characteristics, they have to be eliminated during manufacturing. These mechanical failures have been eliminated by many trial and error efforts, but it usually takes more than several months to find an appropriate countermeasure. To minimize device development cost and time, a mechanical stress design system has been developed by applying finite element analysis and stress measurement [3]. The system consists of three subsystems: an experimental database construction system, a stress simulation system and a stress evaluation system. This system is effective for optimizing both device structure and manufacturing process without mechanical failures. In this paper, stress optimization of MOS (metal oxide semiconductor) structures using P-doped Si thin films as gate material by applying this system is discussed .

2. Stress measurement of gate materials

The internal stress of thin films used as gate material was measured by detecting changes in substrate curvature after film deposition. A scanning laser microscope was used for the curvature measurements. P-doped silicon thin films and WSix (2.4<x<2.7) thin films were deposited on 10-nm thermally oxidized silicon wafers by chemical vapor deposition

Specimens were cut from the wafers as strips, 30-mm long, 3-mm wide and 0.55-mm thick. The resolution of the stress measurement was about 15 MPa.

An example of data obtained from the phosphorous of $4E20/cm^3$-doped silicon thin film 150 nm thick is shown in Fig. 1. The abscissa is temperature and the ordinate is the internal stress of the film. In this case, the film was deposited in an amorphous phase and held compressive stress of about 200 MPa. During crystallization of the film, the internal stress changed drastically to tensile stress of about 600 MPa at about 600°C. This stress change is due to volume shrinkage of the film. The developed stress decreased with annealing at temperatures above 700°C because of the viscoelastisity of the film. On the other hand, the internal stress of the conventional POCl3 treated polysilicon thin films did not change from the initial tensile stress of about 300 MPa during annealing. Therefore, it is important to choose the annealing temperature so as to control the internal stress of the P-doped silicon thin films.

Figure 2 shows example internal stress changes of 150-nm thick WSix thin films. Though the initial stress depends on the film deposition temperature, the final stress after 900°C annealing reached the almost constant value of 1500 MPa. The stress change of the film deposited at 400°C occurred during silicidation due to volume shrinkage of the film.

3. Stress simulation

Mechanical stress was simulatedusing the finite element method [4]. To take into account the internal stress of thin films, Maxwell's viscoelastic model was modified as shown in Fig. 3. In this model, the stress-strain relationship can be expressed as

$$\Delta\sigma = (D + \Delta D)(\Delta\varepsilon - \Delta\varepsilon_\theta - \beta\Delta\varepsilon_v) + \Delta D \cdot D^{-1}(\sigma - \sigma_i),$$

where, σ is stress, σ_i is internal stress of the film, ε is the total strain, ε_θ is thermal strain and ε_v is viscous strain. D is the material moduli matrix and D^{-1} is its inverse matrix. β is the elastic ratio of Young's moduli, E1 and E2. The internal stress is treated as initial strain, which is assumed to be a parallel term of the model (Fig. 3). In this model, the internal stress is a function of temperature to take into account the internal stress change of thin films.

The model MOS structure is shown in Fig. 4. In this model, P-doped silicon and WSix films are integrated as the gate material. The space between two gates is fixed at 0.7 μm. The gate width effect on the substrate stress is analyzed. An example of the predicted stress is shown in Fig. 5. The abscissa is the gate width and the ordinate is the averaged normal stress along the gate width direction. The predicted stress is averaged in the area of the gate space region, 0.7 μm in width and 0.5 μm in depth, because of the spatial resolution of stress measurement using microscopic Raman spectroscopy [5]. The predicted stress without considering the internal stress of films, i.e., considering only thermal stress, differs completely from the measured result. The predicted thermal stress shows little gate width dependency on the substrate stress, but the predicted result considering both thermal stress and the internal stress of both films agrees very well with the measured result. This example clearly indicates that the residual stress in the substrate after gate formation using P-doped silicon and WSix films is mainly determined by the internal stress of the films. It is important, therefore, to take into account the internal stress of thin films in order to obtain precise simulation results.

4. Optimization of gate formation process

The leakage current level of a MOS structure between N^+ drain and p-type substrate using the P-doped silicon film as gate material was evaluated (Fig. 6). When the annealing temperature after the deposition of the gate was lower than 800°C, abnormal leakage current was measured, whereas the leakage current level of the MOS structure using the conventional POCl3 treated polysilicon film was low and stable. The abnormal leakage current was caused because the dislocations that were generated and grew at the gate edge after the annealing.

To discuss the dislocation generation mechanism, stress simulation of the MOS structure was performed by considering the internal stress change of the P-doped silicon films as shown in Fig. 1. The predicted substrate stress at the gate edge is summarized in Fig. 7, where the abscissa is the annealing temperature and the ordinate is the normalized resolved shear stress along [111] slip plane. In this case, the stress level predicted from the conventional POCl3 treated polysilicon gate is used for the normalization. As shown in this figure, the substrate stress level at the gate edge increases with lowering annealing temperature. Dislocations were observed when the stress level exceeded 2.0, i.e., the annealing temperature was lower than 800°C. To eliminate dislocations, the annealing temperature was set at 950°C. Applying the revised process, the leakage current level became low enough to assure device reliability, as shown in Fig. 8. No dislocations were observed at the gate edge after 950°C annealing. It is concluded that the mechanical stress design is very important in improving both the device and the process reliability.

5. Summary

Mechanical stress simulation was performed with the aim of eliminating dislocations at the gate edge of MOS structures using P-doped Si and WSix thin films. It is very important to take into account the internal stress of the films so as to obtain precise simulation results.

Acknowledgements

The authors would like to thank Dr. Toru Kaga, Dr. Hiroo Masuda, Mr. Shuji Ikeda and Ms. Chiemi Hashimoto for their helpful discussions and the sample preparation.

References

[1] S. M. Hu, *J. Appl. Phys.* vol. 70, (1991), R53.
[2] H. Miura, et al., *Appl. Phys. Lett.*, vol. 60, No. 22, (1992), 2746.
[3] H. Miura, et al., *Proc. Int. Workshop on VLSI Process and Device Modeling*, Nara, Japan, (1993), 60.
[4] N. Saito, et al., *Proc. of the Int. Conf. on Computational Engineering Science*, Melbourne, Australia, (1991), 880.
[5] H. Sakata, et al., *Proc. of 9th Int. Conf. on Experimental Mechanics*, vol. 3, (1990), 1307.

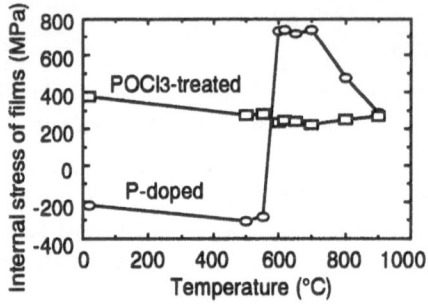

Fig. 1 Internal stress change of Silicon thin films

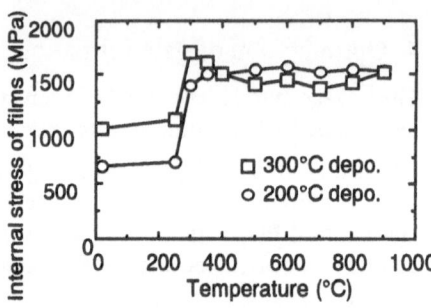

Fig. 2. Internal stress change of WSix thin films (2.4<x<2.7)

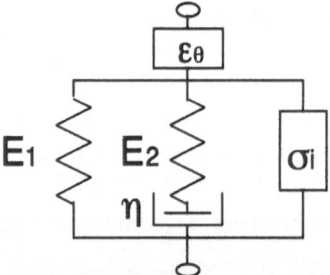

Fig. 3. Maxwell's viscoelastic model

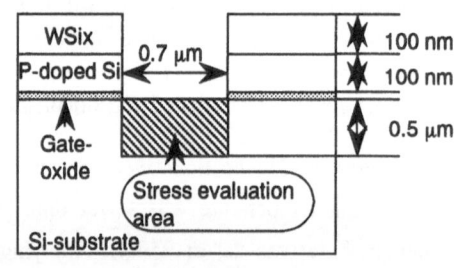

Fig. 4 Model MOS structure

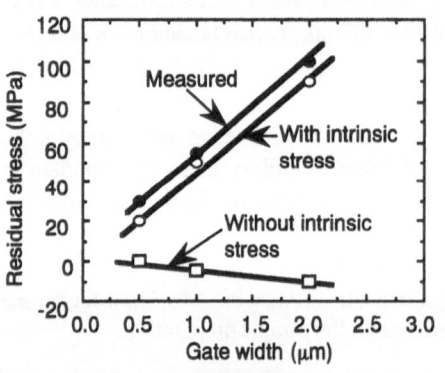

Fig. 5 Preicted gate width effect on substrate stress

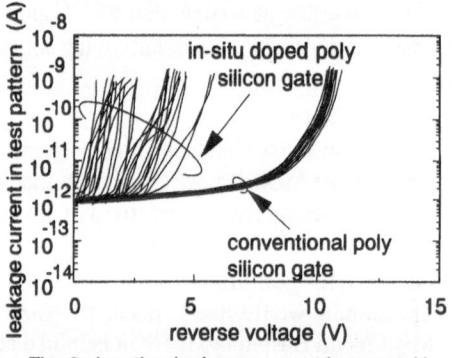

Fig. 6. Junction leakage current between N+ drain and P-type substrate of MOS structure after 800°C annealing

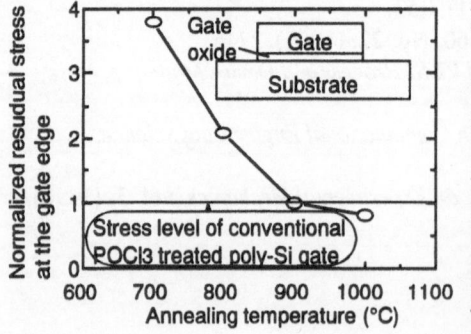

Fig. 7 Predicted annealing temperature dependence of the substrate stress at the gate edge.

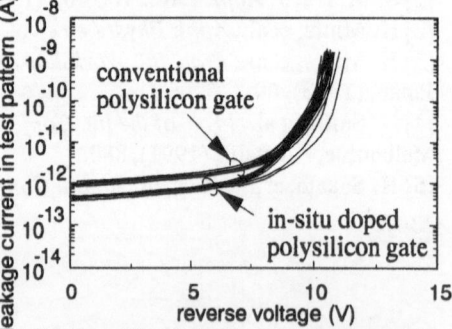

Fig. 8 Junction leakage current between N+ drain and P-type substrate of MOS structure after 950°C annealing

Monte Carlo Simulation of Carrier-Carrier Interaction for Silicon Devices

A. Abramo, R. Brunetti[†], C. Jacoboni[†], F. Venturi[‡], and E. Sangiorgi

Dipartimento di Elettronica, Informatica e Sistemistica, Università di Bologna
Viale Risorgimento 2, I-40136 Bologna, ITALY
[†]Dipartimento di Fisica, Università di Modena
Via Campi 213/A, I-41100 Modena, ITALY
[‡]Dipartimento di Ingegneria dell'Informazione, Università di Parma
Viale delle Scienze, I-43100 Parma, ITALY

Abstract

This paper presents a Monte Carlo approach to electron-electron $(e - e)$ scattering in silicon which is suitable for applications to device analysis. A generalization of the theoretical approach of Bohm and Pines [1] to the case of an arbitrary isotropic multiband model is used. Results for the effect of the Coulomb interaction on the electron distribution function and on the energy-loss properties of the electron gas are presented.

1. Introduction

Hot-carrier effects still represent the major threat for MOSFET reliability. In order to effectively design modern submicron technologies such effects must be correctly simulated since the knowledge of the carrier distribution function is necessary to quantitatively describe physical phenomena, such as carrier injection and trapping into the SiO_2, responsible for device degradation. In this frame, the Monte Carlo method can be considered the most settled, but particular care must be taken to the microscopic transport physics in order to have quantitative confidence on the modeling of the distribution function. Inter-carrier Coulomb interaction is generally neglected in Monte Carlo simulations since it is considered sufficiently low inside most of the channel region to make the interaction irrelevant. However, since in a MOSFET the point of maximum carrier heating falls just inside the drain junction, where the concentration of cold carriers is high, the Coulomb scattering can be very significant. In addition, due to the high carrier concentrations in the channel of modern MOSFETs, it is also questionable whether this interaction can be neglected in this region. E-e interaction has been considered by several researchers so far, with as few approximations as possible [2, 3, 4], but usually the amount of CPU time needed imposes very severe limitations to its practical applicability.

2. Theoretical approach and the Monte Carlo simulator

In the present paper, following the approach of Bohm and Pines [1], the inter-carrier Coulomb interaction has been analyzed in terms of short and long-range components. The short-range contribution can be described by two-particle collisions through the introduction of a screened Coulomb potential. The long-range contribution can be formally transformed into the coupling of a single carrier with the *sea* of the other electrons, whose coherent behavior gives rise to plasma oscillations. We have studied, in particular, the generalization of this approach to the case of an arbitrary isotropic multiband model [5, 6, 7], which allowed us to consistently include the Coulomb interaction in the Monte Carlo simulator for MOSFET devices described in [8, 9].

The scattering probability for the short-range two-particle interaction is given by

$$P_{e-e}(\mathbf{k}_{1b}) = \frac{2\pi}{\hbar}\left(\frac{e^2}{V_\epsilon}\right)^2 \sum_{\mathbf{k}_{2b},n_{2b}} \sum_{\mathbf{k}_{1a},n_{1a}} \sum_{\mathbf{k}_{2a},n_{2a}} f(\mathbf{k}_{2b}) \frac{1}{||\mathbf{k}_{2a}-\mathbf{k}_{2b}|^2 + q_D^2|^2} \tag{1}$$
$$\delta(\mathbf{k}_{1b}+\mathbf{k}_{2b}; \mathbf{k}_{1a}+\mathbf{k}_{2a})\delta(\varepsilon_{k_{1a}} - \varepsilon_{k_{1b}} + \varepsilon_{k_{2a}} - \varepsilon_{k_{2b}})$$

where q_D is the inverse of the Debye screening length, 1 and 2 are labels for the two particles involved in the scattering process (supposed to be distinguishable), b and a refer to states "before" and "after scattering", respectively, and n are the band indices.

Since the distribution function $f(\mathbf{k})$ of the counterparts in any given two-carrier scattering is not known a priori, an overestimation-rejection of the short-range scattering probability has been used in order to compute the carrier free-flight [10], thus allowing to treat the short-range term without any analytical assumption on the shape of $f(\mathbf{k})$. This numerical procedure, that will be described in details in a longer paper, has the effect of reasonably contain the CPU time necessary for the Monte Carlo simulation (about 2-3 CPU hours on an IBM RISC/6000).

The carrier-plasmon interaction has been explicitly included among the considered scattering mechanisms by using the following scattering probability:

$$P_{el-pl}(\mathbf{k}) = \frac{2\pi}{\hbar} \frac{e^2\hbar^3}{8m^{*2}V\epsilon\omega_p} \sum_{q<q_c} \frac{1}{q^2}(2\mathbf{q}\mathbf{k}+q^2)^2(1-n_{k+q})n_k \tag{2}$$
$$\{n_q\delta[\varepsilon(k+q)-\varepsilon(k)-\hbar\omega_p] + (n_{-q}+1)\delta[\varepsilon(k+q)-\varepsilon(k)+\hbar\omega_p]\}$$

where ω_p is the plasma frequency.

The long-range Coulomb interaction has been accounted for as an explicit scattering mechanism because, as described in [8, 9], Poisson equation is solved for carrier densities averaged during intervals much longer than the plasma frequency. This avoids the double estimation of the long-range coupling. The plasma frequency must be evaluated, in this case, for each space cell in the simulator, provided that the cell dimensions are much larger than the carrier screening length.

3. Numerical results

Numerical results have been obtained from Monte Carlo simulations of electrons in homogeneous silicon. Fig.1 reports the electron distribution function for homogeneous silicon at low temperature. The results obtained including the short range $e - e$ interaction are compared with those obtained by considering only optical phonon interaction: the $e - e$ mechanism clearly washes out the oscillations introduced by the periodicity of the optical phonons [11]. For this example a simple parabolic band was used.

Next the model of the Coulomb interaction has been applied to homogeneous silicon simulations using the multiband model described in [5]. Fig.2 reports, for the first conduction band, the short range $e - e$ scattering rate as obtained *a posteriori* from the simulations in the two cases of electron density $n = 10^{17} cm^{-3}$ and $n = 10^{20} cm^{-3}$. The data show similar behavior for the scattering rates, but the actual effect of the mechanism on the distribution function is markedly different for the two densities. To show this point, Fig.3 reports the number of $e - e$ interactions as a function of the energy exchanged in the interaction between the colliding carriers. While in the case of $n = 10^{17}$ the number of the events with high energy exchange is very limited, the opposite happens for the high density case. As a consequence, the effect of the $e - e$ interaction is noticeable in the shape of the distribution function only in the case of $n = 10^{20} cm^{-3}$, as shown in Fig.4 for different electric field strengths.

Fig.5 shows the effect of the interaction on the average quantities of the electron gas, namely mean energy and drift velocity, for the high density case. The mean energy is marginally affected by the interaction while a noticeable reduction is observed for the drift velocity, probably due to the enhanced symmetrical shape of the k distribution function.

Finally, the model has been applied to the inhomogeneous case of the silicon $n - i - n$ diode. Fig.6 shows the effect on the distribution function of the introduction of electron-plasmon interaction together with the short-range $e - e$ scattering. The distribution function refers to the space point located $10 nm$ inside the heavily doped $(10^{20} cm^{-3})$ drain junction: the hot electron gas interacts with the "sea" of cold electrons in the drain region and the long-range mechanism dominates the electron cooling effect.

References

[1] D. Pines and D. Bohm, *Phys. Rev.* **85**(2), 338, 1952.

[2] M. Fischetti, S. Laux, and W. Lee, *Sol. State Electron.,* **32**, 1723, 1989.

[3] M.V. Fischetti, *Phys. Rev.* B,**44**, 5527, 19891.

[4] C. Jacoboni and P. Lugli, *The Monte Carlo method for Semiconductor Device Simulation*, Springer Verlag, New York. 1989

[5] R. Brunetti, C. Jacoboni, F. Venturi,E. Sangiorgi, and B. Riccò, *Solid State Electron.*, **32**, 1663, 1989.

[6] A. Abramo, C. Fiegna, F. Venturi, R. Brunetti, E. Sangiorgi, C. Bergonzoni, and B. Riccò, *SISDEP Tech. Dig.*, Hartung-Gorre Verlag (Zurich), p. 257, 1991.

[7] A. Abramo, F. Venturi, E. Sangiorgi, J.M. Higman, C. Fiegna, and B. Riccò, *4th Workshop on Numerical Modeling of Processes and Devices for Integrated Circuits - NUPAD IV*, p.85, 1992.

[8] F. Venturi, R.K. Smith, E. Sangiorgi, M.R. Pinto, and B. Riccò, *IEEE Trans. Computer-Aided Design*, vol. CAD-8, p.360, 1989.

[9] F. Venturi, E. Sangiorgi, R. Brunetti, W. Quade, C. Jacoboni, and B. Riccò, *IEEE Trans. Computer-Aided Design*, vol. CAD-10, p.1276, 1991.

[10] R. Brunetti, C. Jacoboni, A. Matulionis, and V. Dienys, *Physica B*, **134**, 369, 1985.

[11] A.A. Grinberg, S. Luryi, N.L. Schryer, R.K. Smith, C. Lee, U. Ravaioli, E. Sangiorgi, *Phys. Rev.* B,**44**, 10536, 1991.

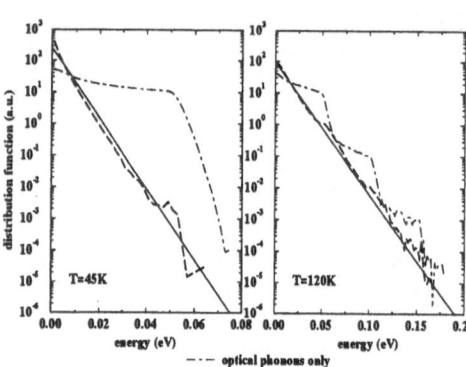

Fig.1: electron distribution function in bulk silicon at low temperature.

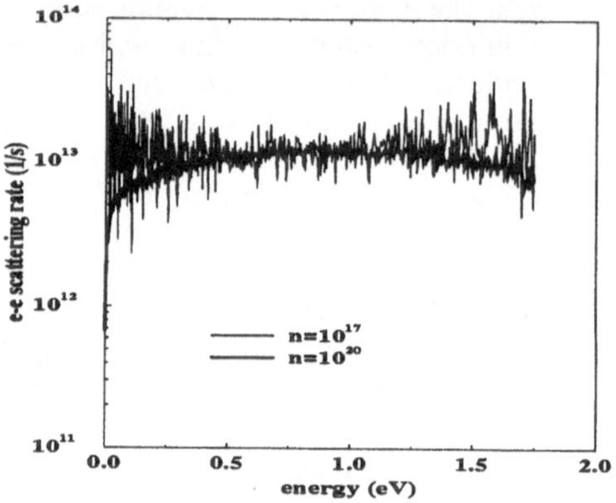

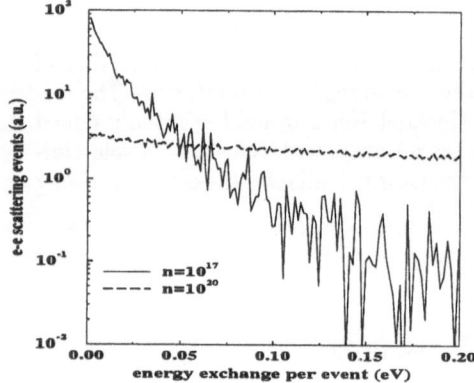

Fig.3: number of short range $e - e$ interactions as a function of the energy exchanged between the colliding carriers.

Fig.2: short range $e - e$ scattering rate for the first conduction band of the model [5].

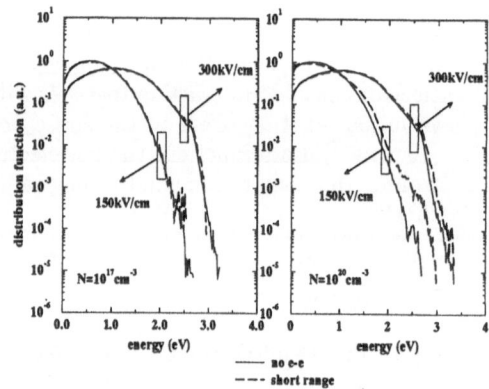

Fig.4: electron distribution function at different electric fields and different electron densities.

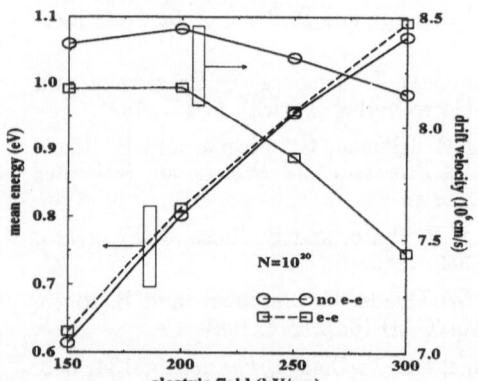

Fig.5: Mean energy and drift velocity at different electric field strengths.

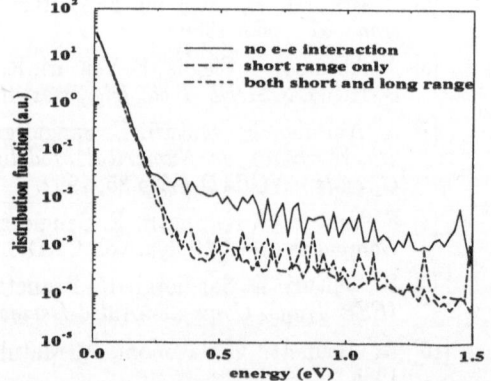

Fig.6: electron distribution function $10 nm$ inside the heavily doped drain junction of an $n - i - n$ diode.

Monte Carlo MOSFET Simulator Including Inversion Layer Quantization

J. Voves

Department of Microelectronics, Faculty of Electrical Engineering,
The Czech Technical University of Prague
Technickà 2, CS-16627 Praha 6, CZECH REPUBLIC

Abstract

A two-dimensional ensemble Monte Carlo Si-n-MOSFET simulator, considering two-dimensional electron gas behaviour in inversion layer, is presented. Description of inversion layer quantization is solved in the new complex way. Standard bulk Monte Carlo transport model is switched to the two-dimensional electron gas model always when electron reaches the quantized region. Electron energy profiles along the MOSFET channel are studied. Comparison with experimental results, measured on the short channel MOSFET test structures is performed.

1. Introduction

In the latest most advanced Monte Carlo MOSFET simulators [1,2] inversion layer quantization effects are not treated. Until now the problem of the electron transport in the MOSFET inversion layer has been solved only by the separate quasi-two-dimensional Monte Carlo simulation with uniform inversion layer thickness [3].

2. Present model

The present simulator solves this problem in a new complex way. Initial conditions are calculated by "drift -diffused" two-dimensional simulator PISCES-2B. Appropriate

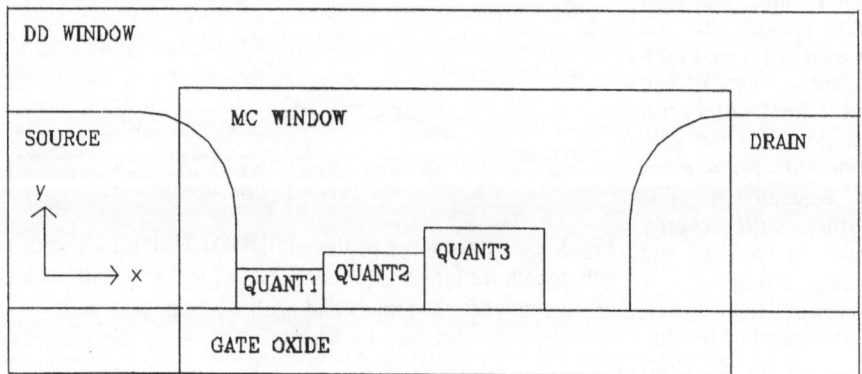

Fig.1 Windows for "drift-diffused" (DD) and Monte Carlo (MC) simulation and quantum regions (QUANT1-3).

window is chosen for Monte Carlo simulation. Quantized region near to Si-SiO$_2$ interface is divided into several parts as shown in Fig.1. Subband energy levels are calculated using actual values of inversion and depletion layer charges. These levels are schematically drawn in Fig.2. Standard bulk Monte Carlo transport model is switched to two-dimensional electron gas model always when electron reaches the quantized region. Surface roughness, interface oxide charge, intravalley phonon and intervalley phonon scattering rates are calculated using following assumptions:

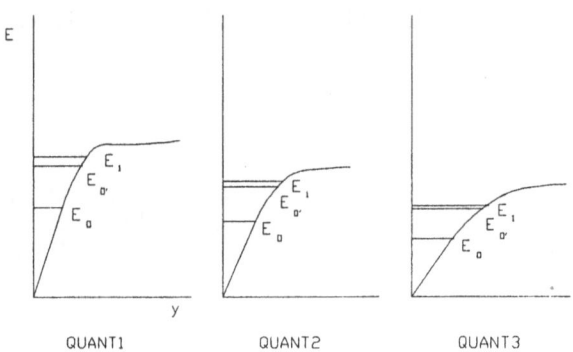

Fig.2 Schematic silicon conduction band profiles and three lowest subband energy levels E_0, E_1 and $E_{0'}$ in different quantized inversion layer regions QUANT1-3.

1/ Only three lowest subbands E_0, $E_{0'}$ and E_1 are considered in the Hartree approximation. 2/ Energies E_0 and $E_{0'}$ subbands are assumed to be equal. 3/ No dependence of quantization on the lateral electric field is supposed [4].

3. Simulation results

Higher electron mobility in the MOSFET inversion layer due to 2-D electron gas transport is observed especially for smaller drain bias. Comparison of corresponding electron energy profiles is depicted in Fig.3. 2D electron energy profiles in the MOSFET channel region are analyzed with respect to the drain junction avalanche breakdown and hot electron emission into the gate oxide [5]. Fig.4 gives an example of this profile with energy peaks at source and drain junctions. Comparison with experimental results, measured on the short

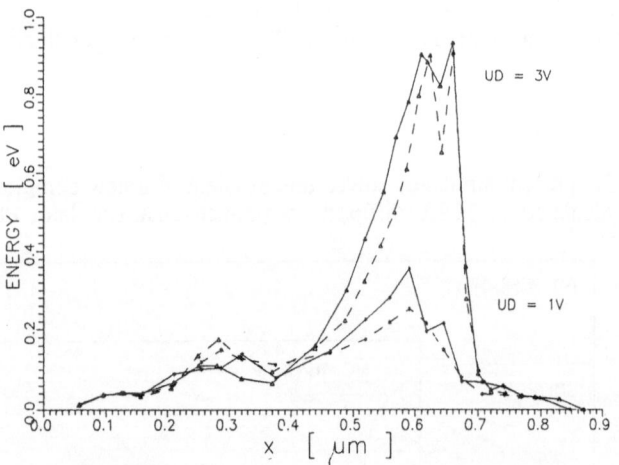

Fig.3 Electron energy profiles in DDD1 NMOSFET with 1um gate length for bias $V_D=1V$ and 3V, $V_G=1.46V$ with inclusion of quantum effects (solid) and without them (hatched).

channel MOSFET test structures is performed. Better agreement of calculated and

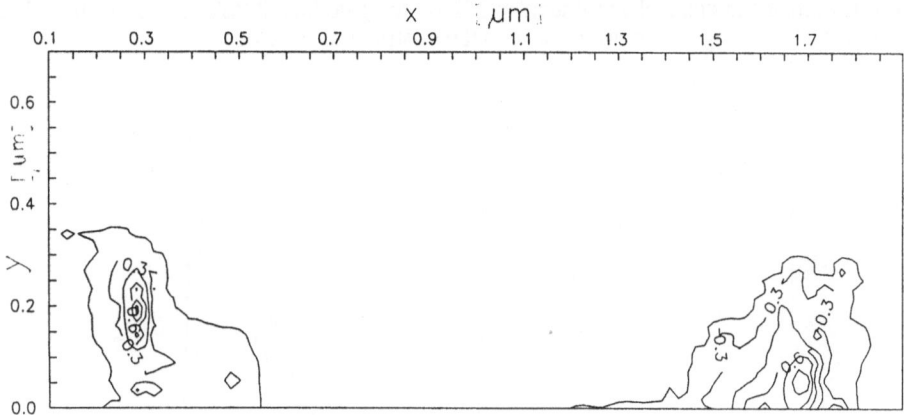

Fig.4 Equienergetical lines of electrons in DDD1 NMOSFET for bias $V_D=3V$, $V_G=2.26V$ and gate length 2 um.

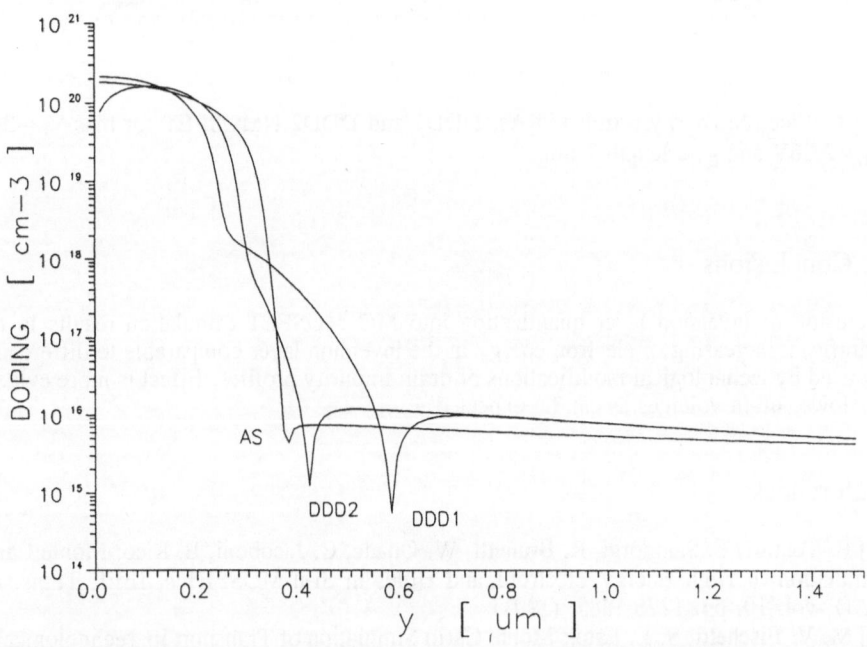

Fig.5 Doping profiles of S/D regions for different NMOSFET modifications (As drain, DDD1, DDD2 "double diffused drain").

measured data by inclusion of inversion layer quantization is observed.

The MC Simulator is used in MOS IC process evaluation. Transistors fabricated by standard As-drain NMOS technology are compared with two modifications of DDD ("double diffused drain") MOSFET. Source/drain regions concentration profiles in the transversal 1-D section are shown in Fig.5. Corresponding average electron energy profiles along the channel are drawn in Fig.6 for gate bias 2.26V and drain bias 3V. Noticeable lowering of energy peak for DDD1 structure is evident.

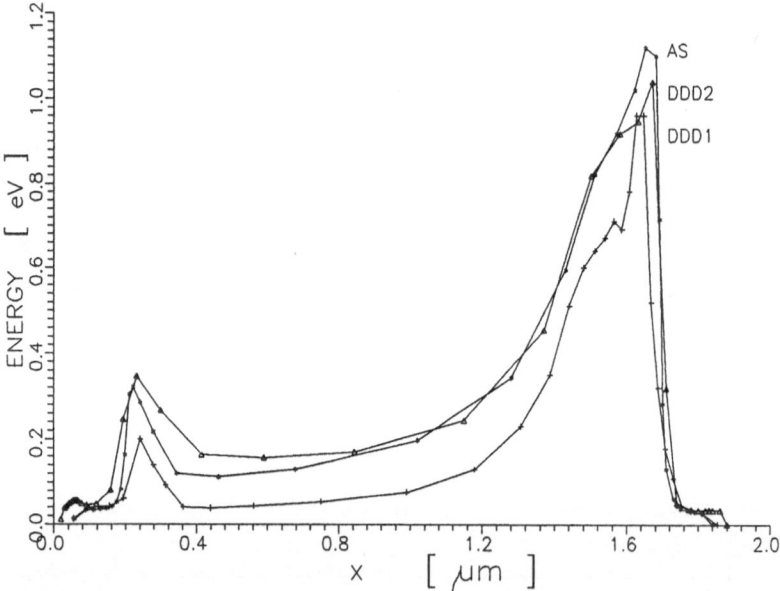

Fig.6 Electron energy profiles in As, DDD1 and DDD2 NMOSFET for bias V_D=3V, V_G=2.26V and gate length 2 um.

4. Conclusions

Inclusion of inversion layer quantization into MC MOSFET Simulation results in the significant increasing of electron energy in the inversion layer comparable to differences induced by technological modifications of drain impurity profiles. Effect is more evident for lower drain voltages as can be expected.

References:

[1] F. Venturi, E. Sangiorgi, R. Brunetti, W. Quade, C. Jacoboni, B. Rico: Monte Carlo Simulation of High Energy Electrons and Holes in Si-n-MOSFET's, IEEE Trans. on CAD, vol. 10, pp. 1276-1285, (1991)
[2] M. V. Fischetti, S. E. Laux: Monte Carlo Simulation of Transport in Technologically Significant Semiconductors of the Diamond and Zinc-Blende Structures-Part II: Submicrometer MOSFET's, IEEE Trans on El. Dev., vol. 38, pp. 650-660, (1991)
[3] Ch. Hao, J. Zimmermann, M. Charef, R. Fauquembergue, E. Constant: Monte Carlo Study of Two-Dimensional Electron Gas Transport in Si-MOS Devices, Solid-St. El., vol. 28, pp. 733-740, (1985)
[4] J. Voves: Inversion Layer Quantization in Monte Carlo MOSFET Simulation, 13th Gen.Conf.Cond.Matter Div.EPS, Regensburg, (1993)
[5] J. Voves, J. Veselý: MOSFET Gate Current Modelling Using Monte Carlo Method, ESSDERC 88, pp.791-794, Montpellier, (1988)

Three-Dimensional Monte Carlo Simulation of Submicronic Devices

C. Brisset, P. Dollfus, N. Chemarin, R. Castagné, and P. Hesto

Institut d'Electronique Fondamentale, CNRS URA022, Université Paris-Sud
Bât 220, F-91405 Orsay Cédex, FRANCE

Abstract

Our particle Monte-Carlo program for device modelling (MONACO) has been extended to the third dimension in geometric space. We describe the main features of this model. It has been used to simulate a 0.1 μm-gate-width N-channel MOSFET. Important geometric edge effects occur and induce a reduction of the effective gate width.

1. Introduction

The scaling down in present ULSI technologies' constrains the device physicists to take into account at the same time the non-stationary transport [1-3] and the three-dimensional effects in the geometric space [4,5]. The particle ensemble Monte-Carlo technique is actually the most accurate device modeling method that can meet these requirements. So, we have extended our particle Monte-Carlo model to three-dimensional (3D) simulations. Some important points of this new 3D model are presented. We described also some precautions that must be taken to obtain accurate simulation results. Finally, we present a steady-state analysis of a N-channel MOSFET having a short gate width (W=0.1 μm) and a small ratio W/L (W/L=0.1). This analysis shows off important edge effects that influence the device performances.

2. The model

The carrier motion, that was previously described [2,6], is unchanged. The potential distribution in the device is calculated from Poisson's equation using a finite-element formulation in a non-uniform rectangular meshing. Poisson's equation is then solved using a L.U. method. The electric field components are calculated by derivation of the distribution potential. A rectangular meshing facilitates the detection of the cell changes during the motion of each carrier, which is required to up-date the local electric field acting on the carrier. The boundary conditions in the solution of Poisson's equation consist in imposing the potential at every contact surface (Dirichlet condition). Elsewhere on the boundary the normal component of the electric field is taken equal to zero (Neumann condition). Consistently with Neumann condition a carrier reaching such a boundary is reflected. A carrier reaching an ohmic contact is free to leave the device through this contact. In the cells adjacent to an ohmic contact the equilibrium carrier concentration is assumed to be recovered ($n=N_D$ or $p=N_A$). So, before each resolution of Poisson's equation, a lack of carriers appearing in such a cell is compensated by injection of carriers. Its initial energy and momentum are specified by a Maxwellian distribution. This last boundary condition is the only condition of injection of carriers in the device.

As in all dynamic model based on the solution of Poisson's equation the time and spatial grids must meet requirements related to physical constants of time and spatial relaxation. In one hand, the time step Δt between two up-dates of the electric field distribution must not be greater than the dielectric relaxation time in the heavily doped regions ($\Delta t \approx 5$ fs for a doping level of 10^{18} cm^{-3} in Si). This ensures that the carrier population can relax after every local perturbation without error due to the "frozen" local field. In the other hand, in regions where the potential is likely to vary spatially, the mesh size must be less than the length of variation, i.e., the Debye length (L_D) that can be as small as 4 nm for a carrier concentration of 10^{18} cm^{-3}.

Furthermore the choice of the volume of the cells adjacent to ohmic contacts appears (especially in 3D modelling). It is related to the particle nature of the model and to the boundary conditions applied at ohmic contacts. As above mentioned, a lack of carrier in a cell adjacent to an ohmic contact is quasi-immediately compensated by injection of carrier. Such an algorithm leads to an average excess of about one carrier compared to the equilibrium number n_0 (that corresponds to a number of particle n equal to the number of impurities N_D). The system tends to relax the relative excess of carrier equal to $1/n_0$ by re-adjustment of the electric field. This effect is of course all the more important that n_0 is low.

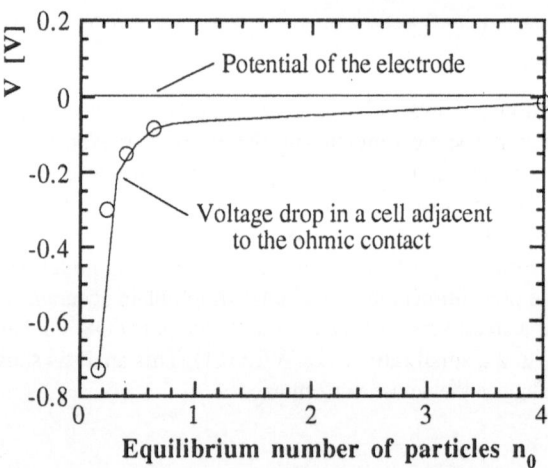

Figure 1: Voltage drop in a cell adjacent to an ohmic contact ($N_D=10^{18}$ cm^{-3}) as a function of the equilibrium number of particles n_0.

In Fig.1 the voltage drop in a cell adjacent to an ohmic contact ($N_D=10^{18}$ cm^{-3}) is plotted as a function of the equilibrium number of particles n_0. The variation of n_0 was obtained by varying the mesh spacing. For $n_0 \leq 1$ the voltage drop increases rapidly with decreasing n_0, which can lead to strong perturbations in the overall potential and carrier distribution in the device. A value of n_0 greater than one seems to be required to maintain low levels of potential fluctuations in ohmic contacts.

The choice of the cell sizes L is thus subject to two conditions: (i): $L \leq L_D$ in regions where the potential is likely to vary. (ii) the cell volume must be large enough near ohmic contacts so that $n_0 > 1$.

3. 3D modeling of a N-MOSFET

To point out the three-dimensional effects, we have compared the behavior of two N-channel MOSFETs using the 3D algorithm. The gate length and the oxide thickness are respectively 1 μm and 20 nm. The doping level of acceptor impurities in the channel is $N_A = 4 \times 10^{16}$ cm^{-3}. The N$^+$ regions of source and drain contacts are doped to $N_D = 10^{18}$ cm^{-3}. The gate width of the first MOSFET is assumed to be boundless. In fact we have simulated only a 0.3 μm wide slice with imposing Neumann boundary conditions at the edge of the device perpendicular to the gate length. The gate width of the second MOSFET is limited to 0.1 μm (W/L=0.1) all other things being equal. The drain characteristics in both cases well saturate. The transfer characteristics of both transistors are plotted in Fig.2. For comparison, the currents are plotted for a normalized gate width of 1mm. One can notice that the drain currents are smaller for the 0.1 μm-gate-width MOSFET (at least 30% smaller according to V_{GS}). The lower currents are due to a reduction of the effective gate width, which is shown in Fig.3. It is a plot of the potential and electron concentration in the inversion layer along the width of the device ($V_{GS} = V_{DS} = 5$ V). The voltage drop between the gated and the ungated region is partially applied on the gate edges, which originates the reduction of the effective gate width. This effect induces also a shift of threshold voltage. The values of V_T can be derived from the square of the drain current against V_{GS} curves. The 0.1 μm-gate-width device exhibits a value of V_T greater than the boundless gate device of about 0.3 V (1.3 V instead of 1 V).

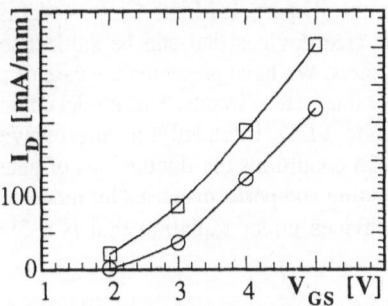

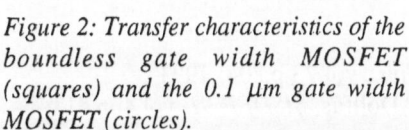

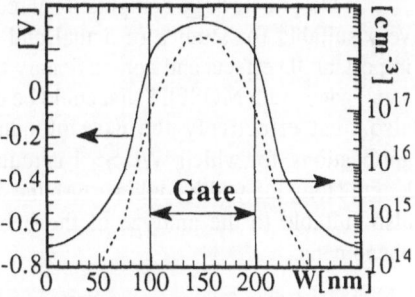

Figure 2: Transfer characteristics of the boundless gate width MOSFET (squares) and the 0.1 μm gate width MOSFET (circles).

Figure 3: Potential (solid line) and electron concentration (dashed line) in the inversion layer along the width in the middle of the device.

Another 3D effect occurs beyond the pinch-off point at the end of the channel near the drain of the 0.1μm gate width MOSFET; the current partially flows down into the bulk of the device (Fig.4a), as in large gate width MOSFET, but also spreads outside the gate (Fig.4b).

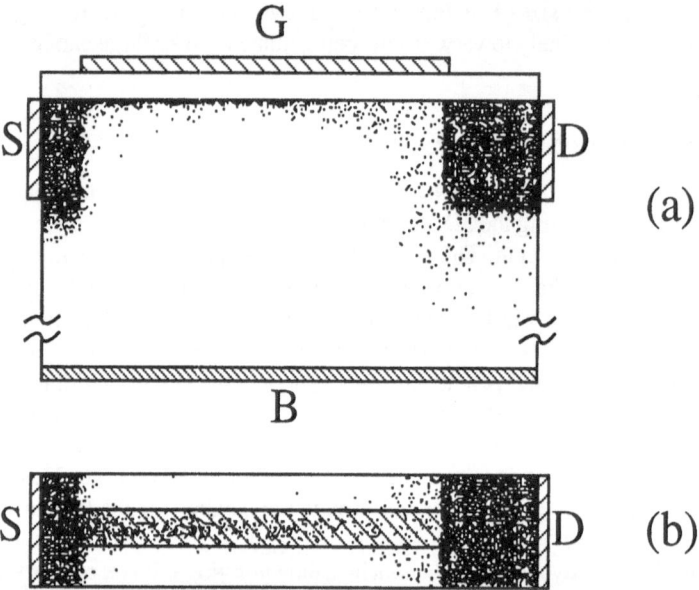

Figure 4: The electron distribution in the 0.1 μm gate width MOSFET for $V_{GS}=5$ V and $V_{DS}=5$ V.

4. Conclusion and perspectives

Despite its large computational requirements, the 3D particle Monte-Carlo model is very suitable for studies of actual and future low size devices that can be subject to important 3D effects and non-stationary transport effects. We have presented the case of a small gate width MOSFET that could be designed for integrated circuits. This model could also treat effectively the case of nanometric gate FETs intended for microwave applications, for which W/L >> 1 is required. In such conditions the fluctuations of gate length along the width could perturb the device operating and performances. Our model is also suitable to the analyze of the behavior of devices under radiation that is a 3D phenomena.

References

[1]   P.Dollfus, C.Bru, and P.Hesto, *J.Appl.Phys.*, vol.73,p.804, 1993

[2]   L.Rajaonarison, P.Hesto, J.F.Pône, and P.Dollfus, *SISDEP 91*, vol.4, p.513, 1991

[3]   H.Sheng, R.Guerrieri, and A.Sangiovanni-Vincentelli *SISDEP 91*, vol.4, p.285, 1991

[4]   H.C.Chan, and T.J.Shieh, *IEEE Trans. Electron Devices*, vol.38,p.2427, 1993

[5]   H.Matsuo, J.Tanaka, A.Mishima, K.Tago, and T.Toyabe *SISDEP 91*, vol.4, p.165, 1991

[6]   P.Hesto, J.F.Pône, M.Mouis, J.L.Pelouard, and R.Castagné, *NASECODE IV, (Boole Press, Dublin, 1985)*, p.315, 1985

SIMULATION OF SEMICONDUCTOR DEVICES AND PROCESSES *Vol. 5*
Edited by S. Selberherr, H. Stippel, E. Strasser – September 1993

193

Monte Carlo Analysis of Voltage Fluctuations in Two-Terminal Semiconductor Devices

T. Gonzàles, D. Pardo, L. Varani[†], and L. Reggiani[†]

Departamento de Fisica Aplicada, Facultad de Ciencias, Universidad de Salamanca
Plaza de la Merced s/n, E-37008 Salamanca, SPAIN
[†]Dipartimento di Fisica ed Istituto Nazionale di Fisica della Materia,
Universitá di Modena
Via Campi 213/A, I-41100 Modena, ITALY

Abstract

Electronic noise in two-terminal semiconductor devices is investigated by an original Monte Carlo procedure which provides a spatial map of voltage fluctuations in the structures. Voltage-noise operation is employed. The results obtained for submicron n^+nn^+ structures and Schottky-barrier diodes show that the high-resistivity regions are responsible for the low-frequency noise, while the low-resistivity regions mostly contribute to the noise at the highest frequencies.

1. Introduction

The analyses of velocity fluctuations in electronic devices are of great importance when trying to optimize their performances. In fact, the magnitudes employed for the characterization of fluctuations provide specific information about the transport processes which limit the efficiency of the devices. In this paper we present an original Monte Carlo method for the analysis of voltage fluctuations in one-dimensional semiconductor devices. More traditional methods, such as the impedance-field [1] or the transfer-impedance [2], introduce approximations related to the statistical properties of the microscopic noise sources. On the contrary, by employing an ensemble Monte Carlo simulation coupled with a one-dimensional Poisson solver, fluctuations in carrier velocity and electric field are self-consistently accounted for, thus avoiding any approximation. Moreover, from the simulation we provide a spatial map of voltage fluctuations in the devices. To this end, the spectral density of voltage fluctuations of the open circuit, $S_V(x,f)$, is determined as a function of different positions x inside the device as measured from one of the terminals.

2. Theoretical analysis

The theory underlying the present method is based on the following. In a one-dimensional structure of length L, the total current, $I(t)$, is given by [3, 4]:

$$I(t) = I_c(t) - \frac{\varepsilon_0 \varepsilon_r A}{L} \frac{d}{dt} \Delta V(L,t) \tag{1}$$

where ε_0 is the free space permittivity, ε_r the relative static dielectric constant of the material, A the cross-sectional area, $\Delta V(L,t)$ the instantaneous voltage drop between the terminals, and $I_c(t)$ the conduction current defined by:

$$I_c(t) = -\frac{e}{L}\sum_{i=1}^{N_T(t)} v_i(t) \tag{2}$$

with e the absolute value of the electron charge, $N_T(t)$ the total number of carriers inside the device, and $v_i(t)$ the instantaneous velocity along the field direction of the ith particle.

By imposing that the total current is constant in time, $I(t)=I_0$, from Eq. 1 we obtain:

$$\frac{d}{dt}\Delta V^I(L,t) = \frac{L}{\varepsilon_0\varepsilon_r A}\left[I_c(t) - I_0\right] \tag{3}$$

where the superscript I is to recall the use of voltage-noise operation. From this expression the instantaneous voltage drop between the terminals can be calculated with the following procedure:

(i) Starting from the stationary operation point in the device corresponding to I_0, $[\Delta V^I(L,0), I_0]$, one solves the Poisson equation, simulates one time step Δt and gets the conduction current $I_c(\Delta t)$.

(ii) Once $I_c(\Delta t)$ is evaluated, Eq. 3 is integrated by employing a finite-differences scheme. The new instantaneous voltage drop between the terminals, $\Delta V^I(L,\Delta t)$, is thus calculated as:

$$\Delta V^I(L,\Delta t) = \Delta V^I(L,0) + \frac{L}{\varepsilon_0\varepsilon_r A}\left[I_c(\Delta t) - I_0\right]\Delta t \tag{4}$$

(iii) With the new value $\Delta V^I(L,\Delta t)$ one solves the Poisson equation and obtains the value of the voltage drop at each position x of the structure as measured from the first terminal $\Delta V^I(x,\Delta t)=V^I(x,\Delta t)-V^I(0,\Delta t)$.

(iv) A successive time step is then simulated to obtain the new value of I_c, and the process is iterated by repeating from point (ii). The number of simulated time steps must be enough to get a sufficient resolution in the calculation of the autocorrelation function of voltage fluctuations, $C_V(x,t)=\overline{\delta\Delta V^I(x,0)\delta\Delta V^I(x,t)}$, where the bar indicates time average and $\delta\Delta V^I(x,t)=\Delta V^I(x,t)-\overline{\Delta V^I(x)}$. By Fourier transformation, the corresponding spectral density as a function of frequency, $S_V(x,f)$, is obtained.

3. Results

This method has been applied to two types of devices: Si and GaAs n^+nn^+ structures, and a GaAs Schottky-barrier diode. The calculations are performed by coupling self-consistently a Monte Carlo simulator with a one-dimensional Poisson solver. The microscopic models for Si and GaAs are the same of [5] and [6], respectively. The devices are divided into equal cells of 100 Å each. The value of the time step is 10 fs for the case of Si and 2.5 fs for GaAs. Periodic boundary conditions have been employed for the n^+nn^+ structures [5], so that $N_T(t)=N_T$ is constant with time. In the case of the

Schottky diode one of the terminals is an ohmic contact which updates $N_T(t)$ in each time step, and the other terminal is the Schottky contact, which acts as a perfect absorbing boundary. The cross-sectional area adopted in the simulations is 10^{-13} m^2 for the n^+nn^+ structures and $2\cdot10^{-13}$ m^2 for the Schottky-barrier diode, which means an average number of simulated carriers between 4200 and 8000 depending on the device.

The results for the spectral density of voltage fluctuations in Si and GaAs n^+nn^+ structures are shown in Fig. 1. The dopings are $n^+=10^{17}$ cm^{-3} and $n=10^{16}$ cm^{-3}, and T=300 K. The fluctuations are around an average voltage $\Delta V^l(L)=0.4$ V. The effects observed in the voltage fluctuations are similar for both materials. When the operation point is near equilibrium most of the low-frequency noise originates from n region of the device, due to its lager resistance. At increasing voltages (like in the case of Fig. 1) the onset of hot-carriers conditions leads to the penetration of the noise sources into the drain region, and an important part of the low-frequency noise comes from the beginning of the second n^+ region. This effect is more pronounced in the case of GaAs [Fig. 1(b)], due to the presence of carriers in the upper satellite valleys, with higher effective mass, which involves a deeper penetration of hot carriers into the drain before they can thermalize, making this region highly resistive and thus an important source of noise. When going to higher frequencies the contribution of the n region in the structures decreases, while that of the n^+ regions increases, reaching its maximum value for the associated plasma frequency (around 1300 GHz for Si, and 3000 GHz for GaAs) [3].

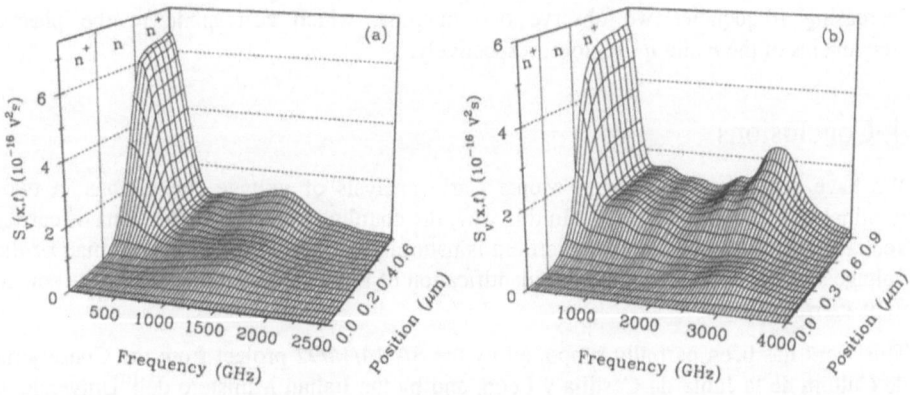

Fig. 1. Spectral density of voltage fluctuations as a function of frequency and position around an average voltage of 0.4 V in (a) Si and (b) GaAs n^+nn^+ structures at T=300 K, with $n^+=10^{17}$ cm^{-3} and $n=10^{16}$ cm^{-3}. The length of each of the regions is (a) 0.20-0.25-0.25 μm and (b) 0.15-0.25-0.50 μm, respectively.

The GaAs Schottky-barrier diode that is simulated consists of a first n^+ region (10^{17} cm^{-3}) of 0.35 μm, and a second n region (10^{16} cm^{-3}) of 0.35 μm at which end is the Schottky barrier. The barrier height is 0.735 V, which leads to an effective built-in voltage at equilibrium of 0.640 V. Fig. 2(a) shows the diode current-voltage characteristic. Only the forward-bias range can be simulated with the present model. Two different regions can be clearly observed: a first exponential region where the current is determined by the thermoionic emission of carriers over the barrier, and a second one where the current is determined by the series resistance and tends to assume a linear behavior due to the disappearance of the barrier.

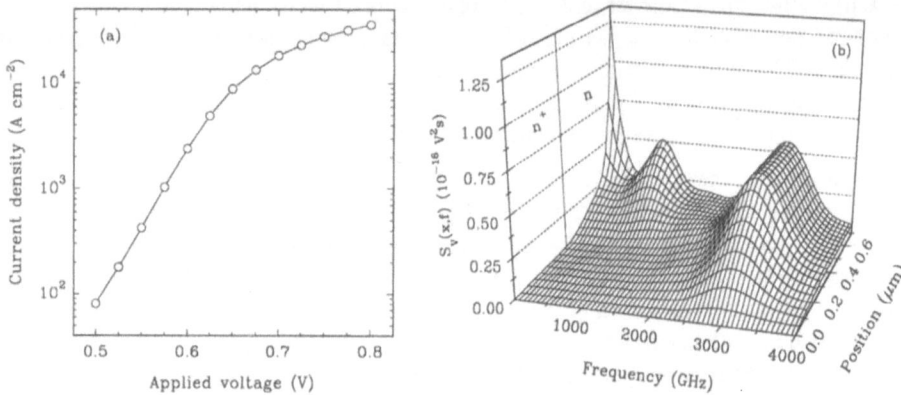

Fig. 2. (a) Current-voltage characteristic under forward-bias conditions, and (b) spectral density of voltage fluctuations as a function of frequency and position for an average voltage of 0.575 V, at T=300 K, in the GaAs Schottky-barrier diode described in the text.

Fig. 2(b) shows the spectral density of voltage fluctuations in the Schottky-barrier diode for an average voltage of 0.575 V, corresponding to the exponential region of the I-V characteristic. At low frequency the main contribution to S_V comes from the n region, more specifically from the section of the device close to the barrier. This is due to its high differential resistance, originated by the space-charge region near the barrier. At increasing frequencies we observe two maxima, which correspond to the plasma frequencies of the n and n^+ regions, respectively.

4. Conclusions

We have presented an original Monte Carlo analysis of voltage fluctuations in two-terminal semiconductor devices. In this way, the coupling between fluctuations of carrier velocity and self-consistent electric field is naturally accounted for. A spatial map of the voltage spectral density enables the identification of the local strength of the noise source to be carried out.

This work has been partially supported by the SA-14/14/92 project from the Conserjería de Cultura de la Junta de Castilla y León, and by the Italian Ministero dell' Università e della Ricerca Scientifica e Tecnologica (MURST).

References

[1] W. Shockley, J. A. Copeland and P. James in *Quantum Theory of Atoms, Molecules and the Solid State*, Ed. P. O. Löwdin (Academic, New York, 1966), p. 537.
[2] K. M. van Vliet, A. Friedmann, R. J. J. Zijlstra, A. Gisolf and A. van der Ziel, J. Appl. Phys. **46**, 1804 (1975).
[3] J. Zimmermann and E. Constant, Solid-State Electron. **23**, 915 (1980).
[4] L. Reggiani, T. Khun and L. Varani, Appl. Phys. **A54**, 411 (1992).
[5] L. Varani, T. Kuhn, L. Reggiani and Y. Perlès, Solid-State Electron. **36**, 251 (1993).
[6] T. González, J. E. Velázquez, P. M. Gutiérrez and D. Pardo, Appl. Phys. Lett. **60**, 613 (1992).

Simulation of Sputter Deposition Process by DUPSIM

M. Seifert, F. Richter[†], and R. G. Spallek

Institut für Grundlagen der Elektrotechnik/Elektronik, Fakultät Elektrotechnik,
Technische Universität Dresden
Mommsenstraße 13, D-01069 Dresden, GERMANY
[†]Fraunhofer-Institut für Mikroelektronische Schaltungen und Systeme 2
Grenzstraße 28, D-01109 Dresden, GERMANY

Abstract

A new deposition simulator is presented that allows modeling of sputter deposition processes. Various materials as well as several types of ring shaped magnetron sources can be considered. Some applications are presented to demonstrate the possibilities of the new program. It is part of the two-dimensional process simulator DUPSIM, which is able to simulate many fabrication processes of semiconductor devices in integrated circuits. Thereby a great flexibility and ease of use is reached.

1. Introduction

During the last several years, especially two-dimensional process simulators were developed which achieved a great propagation in simulating VLSI-structures in a realistic manner [1] [2]. The modeling of deposition and etching steps, by that time probably the least developed area of IC topography simulation, became an essential part of a complete process simulation.

Because of the trend in decreasing device sizes there is an increasing need to replace simple analytic models of deposition process simulation with more far reaching solutions to investigate the typically metallization effects (see below) of submicrometre designed structures [3]. The results obtained with appropriated simulators can also be used to solve e.g. current and temperature distributions in the deposited layer [4].

This paper includes a possibility of simulating the sputter deposition process of magnetron diodes. There are no restrictions in geometry of the substrate, so every process during IC fabrication can be considered. The second part shows some applications to demonstrate the possibilities of the program by comparing calculated deposition profiles with experimentally obtained results.[8].

2. Modeling of Deposition Source

The distribution function of the oncoming vapor flux is calculated using a method described in [5] which we especially adjusted to magnetron sputtering planar diodes. Following assumptions for this model are necessary and meaningful:

1. The mean free path of atoms is large compared to the distance between target and substrate [5].

2. The sticking coefficient is assumed to be 1, i.e. there is no resputtering rate [5].

3. The direction distribution of the emitted atoms on target level is given by

$$f(\alpha) = 1/\pi \cos \alpha \qquad (1)$$

4. The density of deposited layer is equal to the target material density.

5. Deposition source has a ring shaped target located parallel to the substrate (other target shapes in preparation).

6. Substrate structure is infinitely extended into the third Dimension, i.e. two-dimensional simulation.

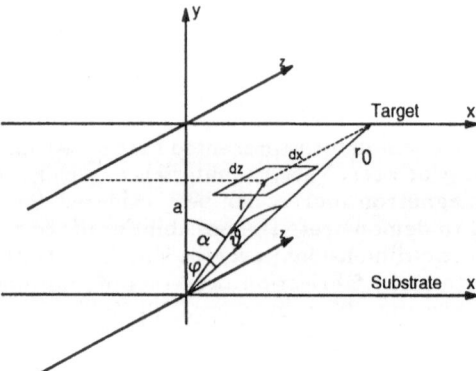

Figure 1: Main Sputter Geometry

A possibility to describe the atomic flux in the three-dimensional main geometric schema shown in (Fig. 1) is:

$$\vec{di} = dxdz \; f(\varphi,\vartheta) \; h(x,z) \; \cos\vartheta \; d\varphi d\vartheta \qquad (2)$$

Here φ and ϑ represent direction α in (1) and $h(x,z)$ is given as a function modeling the target surface influenced by sputter eroding (see below).

The atomic flux density which is required to calculate the growth at a substrate point is defined as the atomic flow per an area unit (assumption 1)

$$dA = r_0 d\varphi = r^2 \cos\vartheta \; d\varphi d\vartheta \qquad (3)$$

(see Fig. 1) as follows

$$\vec{dj} = \frac{dxdz \; f(\varphi,\vartheta) \; h(x,z)}{r^2} \qquad (4)$$

At this point a simplification is appropriated to limit computational expenditure. Instead of flux density $\vec{dj}$ (4) the component lying in the x-y plane

$$dj_{xy} = \int \cos\vartheta \; \vec{dj} = dx \int_{-\infty}^{+\infty} \frac{f(\varphi,\vartheta) \; h(x,z) \; \cos\vartheta}{r^2} \; dz = dx \; V(\varphi) \qquad (5)$$

is calculated. That means simulation is exact only on structures infinitely extended into z-direction (assumption 6). With some geometric relations (see Fig. 1) and (1) the distribution function $V(\varphi)$ finally is presentable as

$$V(\varphi) = \frac{\cos \varphi}{\pi r_0^2} \int\limits_{-\infty}^{+\infty} \frac{h(a \tan \varphi, z)}{1 + z^2/r_0^2} \, dz \qquad (6)$$

After numerical solution of this expression it is possible to calculate the layer growth as described in the following section. The process parameters at this step are the distance between target and substrate, the position of substrate towards deposition source and the measures of the used up target. If needed, a special routine hereby simulates the moving substrate on circular multi-wafer plants.

3. Numerical Model of Deposition

DUPSIM uses the wide-spread string algorithm to describe the surface of layers and their movement in case of two-dimensional simulation . The total layer thickness D is defined as follows

$$D = r \int\limits_{x_1}^{x_2} (\vec{e}, \vec{dj}_{xz}) \qquad (7)$$

where $\vec{e}$ is a unit vector located perpendicular to the wafer surface. To obtain the layer growth it is necessary to solve the expression (7) numerically for each node in both directions. In the algorithm there is involved a calculation of the limits x_1 and x_2 which are determined by the effect of shadowing. The quantified deposition rate r is determined either by the user or (as an estimation) by a special tool which includes the parameters discharge current and sputtering yield (specified by measurement).

In order to balance numerical problems such as loops, needles etc. which occurred during deposition algorithm a special interface to the layer system of DUPSIM was provided in which node management and layer handling algorithms are implemented.

4. Software Aspects

DUPSIM is a modular software system which was developed at the TU Dresden in cooperation with industrial partners. It consists of an input interpreter, several processing units [6], a graphic output package [7] and an interface to device simulation programs. The system is running on DEC-VAX computers, SUN sparc-stations and HP-Apollo workstations.

The sputter deposition program is integrated in a collection of numerical deposition models within the simulator. It is written in FORTRAN 77 in view of great portability to several computers.

5. Applications

To examine the practical possibilities of the simulator we compared simulation results with a series of deposition experiments performed on several machines. Fig. 2 shows an in z-direction infinitely extended structure with the appropriated simulation results. According to assumption 6 there is nearly exact agreement between simulation and reality to be found.

Figure 2: Infinitely Extended Structure

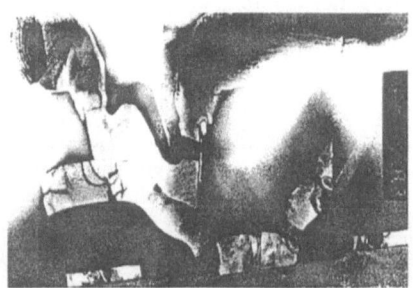 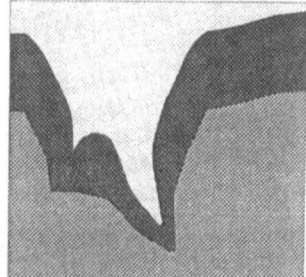

Figure 3: Via Contact

The above mentioned cracks e.g. appear in complicated structures like via-contacts as shown in Fig. 3.

Because of the reached agreement between calculation and experiment the effect of crack development approximately can be considered as a consequence of shadowing and directional dependence of the oncoming vapor stream in case of magnetron sputtering metallization.

References

[1] W. G. Oldham et al., "A general Simulator for VLSI Lithography and Etching process: Part II", *IEEE Trans. on ED*, vol 27, August 1980, pp. 1455-1459

[2] J. Lorenz, J. Pelka, H. Ryssel, A. Sachs and M. Svoboda, "COMPOSITE – A complete modeling program of silicon technology",*IEEE Trans. on CAD*, vol. 4, October 1985, pp. 421-430

[3] H. Joswig, A. Kohlhase and P. Küchler, "Advanced Metallization of Very-Large-Scale Integration Devices", *Thin Solid Films*, 175 (1989) pp. 17-22

[4] T. Smy et al., "Simulation of the Effect of Thin Film Microstructure on Current and Temperature Distributions in VLSI Structures", *J. Vac. Sci. Technol.*, B10(5), Sept/Oct 1992, pp 2267-2276

[5] I. A. Blech, "Evaporated Film Profiles over Steps in Substrates", *Thin Solid Films*, 6 (1970) pp. 113-118

[6] Dirk Bothmann, "Beiträge zur Technologiesimulation mit dem 2D-Simulator DUPSIM", Dissertation, Technische Universität Dresden, Dresden, 1990

[7] P. Kaden, "DEGRAF: Grafisches Auswertesystem für mehrdimensionale Prozeß- und Bauelementesimulation", Diplomarbeit, Technische Universität Dresden, 1989

[8] SEM-Photographs: Zentrum Mikroelektronik Dresden GmbH, ZMD/CV-55/90, Abt. HSV

SIMULATION OF SEMICONDUCTOR DEVICES AND PROCESSES Vol. 5
Edited by S. Selberherr, H. Stippel, E. Strasser – September 1993

Process Simulation for Nonplanar Structures with the Multigrid Solver LiSS

R. Strasser, G. Nanz, and W. Joppich[†]

Campus Based Engineering Center Vienna, Digital Equipment Corporation
Favoritenstraße 7, A-1040 Wien, AUSTRIA
[†]Gesellschaft für Mathematik und Datenverarbeitung (GMD)
Schloß Birlinghoven, Postfach 1240, D-53757 St. Augustin 1, GERMANY

Abstract

This paper presents the application of LiSS, the 2D multigrid solver for parabolic and elliptic differential equations developed by the GMD, to the simulation of diffusion processes in complex nonplanar structures. A domain-splitting method is presented. The behaviour of different multigrid cycles applied to rectangular and curvilinear grids of a trench structure is investigated.

1. Introduction

Multigrid methods have been applied successfully to planar diffusion [1] and oxidation problems [2]. In our approach we used the general purpose multigrid solver LiSS [3], implemented a discretization scheme for the diffusion equations and applied it to several complex structures.

The underlying equations for the N_{EQ} diffusing species are

$$\frac{\partial C_i}{\partial t} + \text{div}\left(\sum_{j=1}^{N_{EQ}} a_{i,j} \cdot \text{grad} C_j + b_i \cdot C_i \text{grad} \psi\right) + \gamma_i = 0 \qquad i = 1, \ldots, N_{EQ} \quad (1)$$

with assumed local charge neutrality. The applied diffusion models are fully compatible with those of PROMIS [4]. The discretization of the equations has been done using a finite volume approach on a nine point stencil, the resulting set of nonlinear equations is solved by a Newton iteration scheme.

In this approach the whole domain is divided into blocks. For each block a boundary-fitted logically rectangular grid is generated with a biharmonic generator also using a fast multigrid algorithm. This technique provides a high geometric flexibility. Thus it enables the solution of a wide range of problems and reduces numerical problems due to distortions.

2. Multigrid Strategy

Multigrid algorithms have been developed to overcome two disadvantages of iterative solvers such as SOR, first the so called h-dependent convergence behaviour which results in slower convergence for finer grids and second the initially good reduction of both residual and error norms which becomes worse with an increasing number of iterations.

A multigrid solver mainly consists of four parts: the smoothing, the restriction, the injection method, and finally the cycle type. The *smoother* makes the error a smooth function over the whole domain. Relaxation methods are used for this purpose, because they damp high-frequent residual–"modes" very efficiently. By means of the *restriction* method the residuals are transferred from one level (or grid) to the next coarser level. The *prolongation* method defines how to transfer the residuals from one level to the next finer level. Finally the *cycle type* defines how many smoothing

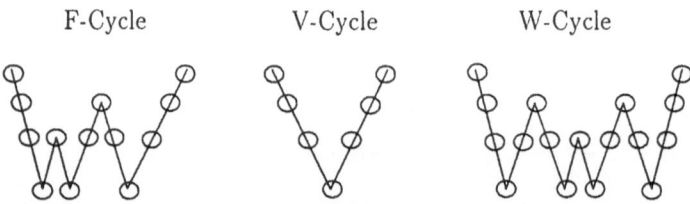

Figure 1: Multigrid Cycle-Types

steps should be performed on each level and the sequence of the different levels as shown in Fig. 1. For example a $V(\nu_{down}, \nu_{up})$ cycle performs ν_{down} smoothing steps before restricting to the next coarser level and ν_{up} smoothing steps after prolongation from the coarser level. The proper choice of these components has great impact to the efficiency of the algorithm.

3. A Trench Problem

A practical application is a "simple" trench as shown in Fig. 2 (boron background doping, arsenic source/drain implant with a 7° tilt angle). In this case the domain is

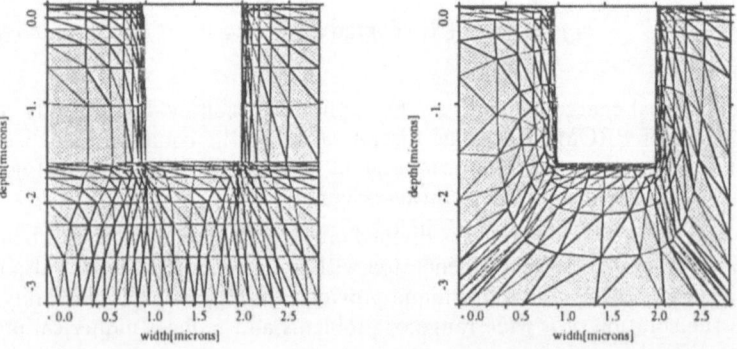

Figure 2: The trench geometry with a block structured grid (a) and a conventional grid (b) (The triangles are an effect of the visualization of non-orthoproduct grids)

split into three nearly rectangular blocks where a grid is generated separately for each of them. In regions near to the boundaries, where high dopant gradients occur, the grids have been refined. From Fig. 2a the advantage of block–splitting, mentioned in section 1 (distortion), becomes obvious. This becomes even more important when treating more complicated structures such as undercuts.

The multigrid method used for this example consists of the following components. As *smoother* a Gauss–Seidel relaxation has been applied. *Restriction* was done by "full weighting", which uses the local average of the residuals on the finer level. The *prolongation* is a simple linear interpolation. The cycle is of V-type with different parameters ν_{down} and ν_{up}, respectively. The table of Fig. 3 shows the residual history, the runtimes and the convergence rates of various combinations of cycle parameters. Note that for higher values of $\nu_{down} + \nu_{up}$ the convergence is faster, but it should be noted that the computational effort rises, too. As these results show, the multigrid-cylce takes about half the computation time of a single-grid cycle. The experiments showed that a $V(2,1)$-Cycle provides excellent results for a wide range of problems. For a small number of cycles the convergence rate of the single-grid cycle is in the same

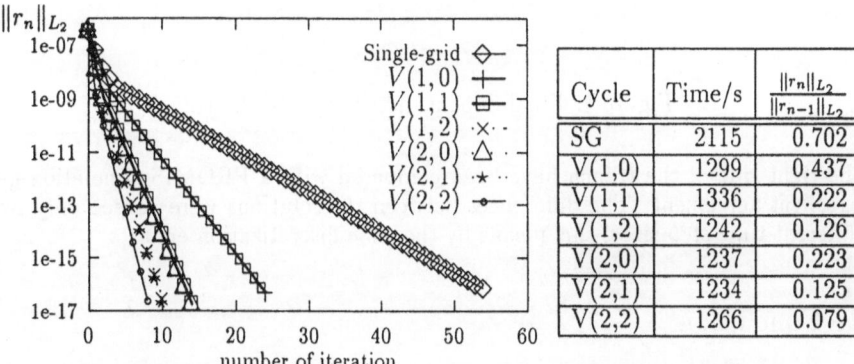

Cycle	Time/s	$\dfrac{\|r_n\|_{L_2}}{\|r_{n-1}\|_{L_2}}$
SG	2115	0.702
$V(1,0)$	1299	0.437
$V(1,1)$	1336	0.222
$V(1,2)$	1242	0.126
$V(2,0)$	1237	0.223
$V(2,1)$	1234	0.125
$V(2,2)$	1266	0.079

Figure 3: Residual reduction of a single-grid solver, multigrid solver with various cycles, their runtimes and their convergence rates

range as the convergence rate of a multigrid cycle (see Fig. 3 for the first 3 iterations). After a few iterations the high frequent residual–modes have been smoothed and the remaining residual is dominated by low–frequent modes. For these low-frequent modes single-grid methods are very inefficient as Fig. 3 shows. In contrast to the single-grid method the multigrid-method does not show this effect. Although the computational effort of a multigrid cylce is a multiple (almost twice as much) of the effort a single-grid method (depending on the cycle-parameters ν_{down}, ν_{up}), the multigrid method is significantly faster because of the constant and fast convergence for all cycles.

These tests have been performed with both types of grids shown in Fig. 2. The run-times for the same number of unknowns are approximately the same for both grid types. Against the expectations the curvilinear grid from Fig. 2b did not lead to convergence or accuracy problems. However, the advantage (with respect to compu-tation time) of the blockstructured grid is obvious since the blocks can be handled in parallel only exchanging information at their common boundaries.

The physical results of the computation are presented in Fig. 4. The diffusion was performed at 1000°C for 20 minutes. The results for the one-dimensional part on

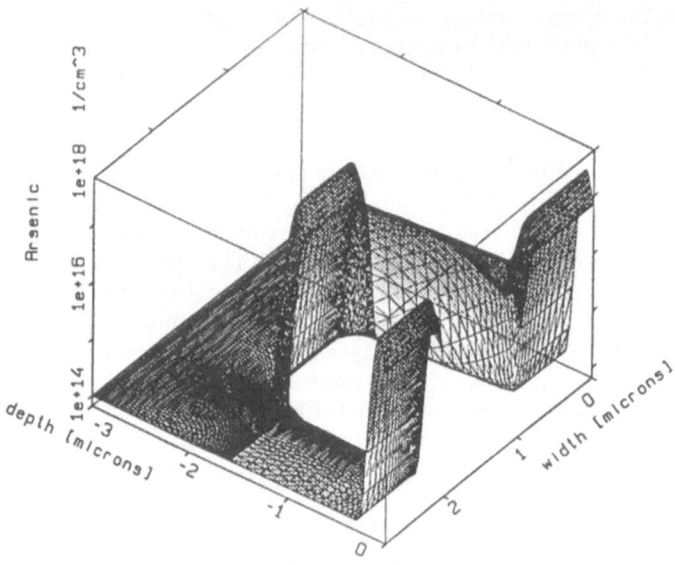

Figure 4: Final dopant distribution in the trench

the right side of the trench have been compared with a PROMIS simulation giving excellent agreement. The differences between the solutions were caused only by the different time step sizes, this means by the time discretization error.

4. Conclusion

This work has been an investigation about the suitability of the general purpose solver LiSS for the simulation of diffusion problems. The results obtained fit well with theorectical predictions. Convergence properties of the multigrid methods are very good, assuming that suitable multigrid components are in use.

Acknowledgement

The authors would like to thank Siegfried Selberherr, Claus Fischer, Stefan Halama and Walter Tuppa for the support within this project.

References

[1] Mijalković, S.; *An adaptive multigrid algorithm for simulation of diffusion processes in semiconductor device fabrication,* Electrosoft, Vol. 1, No. 4, pp. 277-290 (1990)

[2] Joppich, W.; *Mehrgitterverfahren für Diffusionsprobleme der Prozeßsimulation,* PhD-Thesis Universität Bonn (in german)

[3] Lonsdale, G.; Stüben K.; *The LiSS Package,* Gesellschaft für Mathematik und Datenverarbeitung m.b.II

[4] Hobler, G.; Pichler, P.; Wimmer, K.; *Promis 1.5,* Technical University Vienna, Institute for Microelectronics

SIMULATION OF SEMICONDUCTOR DEVICES AND PROCESSES Vol. 5
Edited by S. Selberherr, H. Stippel, E. Strasser – September 1993

Multizone Adaptive Grid Generation Technique for Multilayer Multistep Process Simulation

M. K. Moallemi and H. Zhang

Department of Mechanical Engineering, Polytechnic University
6 Metrotech Center, Brooklyn, NY 11201, USA

Abstract

A multizone adaptive grid generation technique is developed and used with a curvilinear finite-volume approach to simulate multistep IC processes on non-planar multilayer structures with moving boundaries. The capabilities of the numerical scheme is demonstrated by simulating silicon oxidation and impurity diffusion in trench structures with corner angles equal to, and greater than 90°.

1. Introduction

An IC process simulator, as a design and development tool for today's VLSI technology, must be able to model multistep processes on multilayer nonplanar structures which may also involve moving boundaries. Several two- and three-dimensional [1-3] process simulators have been reported in the literature, each with its own merits and demerits. The principal shortcomings of these simulators are *a)* their inability to handle generalized geometries (*e.g.*, structures with corner angles $\geq 90°$), and *b)* their inaccurate or inefficient treatments of moving boundaries or irregular domains. These deficiencies are mostly due to the simplistic approach to the discretization of the problem domain and/or the model equations [1-4]. Recent trend in the evolution of process simulators has been towards the use of numerical grid generating techniques with desirable characteristics such as boundary-fitting capability [3-4], and adaptivity to the solution development [4-5]. These grid generating routines provide geometric flexibility, but require special consideration in the discretization of the governing equations. Simple analytical transformation of the equations to the computational domain is known to fail in simulation of processes in structures with corner angle $\geq 90°$.

In this paper, a numerical methodology is presented that utilizes a *multizone adaptive grid generation technique* [5] for the discretization of the physical domain, and a *curvilinear finite-volume approach* [5] for the discretization of the governing equations (in the physical space) and development of finite difference equations. The numerical scheme is validated by simulating silicon oxidation process (with the inclusion of stress effects) on a 50° trench and comparing the results with the published SEM photographs [6]. The capabilities of this numerical method is demonstrated by simulating oxidation of silicon on a 90° trench and an undercut structure.

2. Physical Model

Simultaneous and sequential silicon oxidation and impurity diffusion in nonplanar structures are modeled. The oxidation model is based on a steady state oxidant diffusion, and a slow incompressible viscous flow of oxide [7]. The effects of oxidation-induced stress on the oxide growth which are significant in the nonplanar structures are accounted for by the use of stress-dependent physical parameters [6]. Impurity diffusion in the silicon and oxide is assumed to be transient, and nonlinear effects are considered by taking the effective diffusion coefficient to be function of concentration [8]. Under oxidizing conditions, the impurity segregation, moving boundary flux, and oxidation-enhanced diffusion effects are also included in the model. The simulations presented here were performed with the input physical parameters used in [6] and [8] to permit comparison. The details of the physical model and the values of the input parameters are provided in [5].

3. Numerical Method

The major building block of the numerical method is a multizone adaptive grid generating technique. The original technique was developed by Brackbill and Saltzman [9] who used the variational method to minimizing a linear combination of integrals which are measures of different grid characteristics, including smoothness, orthogonality and weighted cell area of the grids. To add the multizoning feature, this technique was modified by using constrained variational minimization on the grid line that separates two zones, thus, permitting the grid nods to move along this interface line only. To improve the orthogonality of the grid system near the moving and free boundaries, as well as its adaptivity to the movement of the oxide interface, geometric and local weighting functions were also included [5].

Over this nonorthogonal grid system, the governing equations are discretized using a curvilinear finite-volume approach. This approach is based on flux discretization in the physical domain, therefore, circumvents the limitations of direct (analytical) transformation (*i.e.*, large artificial source terms). More detailed information about the numerical methodology and its implementation are provided in [5].

4. Results and Discussions

The accuracy of the numerical scheme was validated by performing simulation of oxidation on a 50° trench (1 μm deep and 3 μm wide). As shown in Fig. 1, the predictions of this simulation are in excellent agreement with the SEM data [6]. A second simulation was performed on a 90° trench, with initial grid system and boron distribution of Fig. 2. The grid distribution, the trench shape, and boron redistribution after 40.0 minutes of oxidation in wet O_2 at 1000 °C are presented in Fig. 3. The figure clearly indicates the effects of stress, surface curvature and orientation on oxide growth. A comparison between Figs. 2 and 3 reveals the evolution (adaptivity) of the grid system with time and growth of the oxide layer. Figure 3 shows a big drop in the boron concentration in the silicon in the vicinity of oxide interface that is attributed to the larger diffusion coefficient for boron in the oxide compared to the silicon substrate. The third simulation was performed on an asymmetrical trench structure with initial width of 1.8 μm and corner angles 112° and 90°. The cross section of the structure and the corresponding grid system after 7.2 minutes of oxidation in a wet ambient at 1000 °C is shown in Fig. 4. The figure indicates nonuniform oxide growth due to stress and orientation dependence of oxidation rate parameters.

5. Conclusion

A numerical scheme was developed utilizing a multizone adaptive grid generation technique, and a curvilinear finite-volume approach. The capabilities of the method was demonstrated by simulating silicon oxidation and impurity diffusion in trench structures with different corner angles. The work proved that this methodology has potential for the development of a robust and versatile tool for IC process modeling.

References

[1] B. Baccus, D. Collard, E. Dubois, D. Morel, "IMPACT4 -- A general two-dimensional multilayer simulator," in Proc. Inter. Conf. on Semiconductor Devices and Processes, G. Baccarani, and M. Rudan (Eds.), Vol. 3, pp. 255-266, 1988.

[2] M. E. Law, C. S. Rafferty, and R. W. Dutton, SUPREM-IV. Technical report integrated circuits laboratories, Stanford University, Stanford, CA, 1988.

[3] S. Odanaka, H. Umimoto, M. Wakabayashi, and H. Esaki, "SMART-P, Rigorous three-dimensional process simulator on a supercomputer," *IEEE Trans. Computer Aided Design*, Vol. 7, No. 6, pp. 675-683, 1988.

[4] K. Wimmer, R. Bauer, S. Halama, G. Hobler, and S. Selberherr, "Transformation methods for nonplanar process simulation," *Simulation of Semiconductor Devices and Processes*, Vol. 4, pp. 131-138, 1991.

[5] M. K. Moallemi, and H. Zhang, "Application of multizone adaptive grid generation technique for simulation of multistep IC processes in nonplanar structures", *IEEE Trans. Computer Aided Design* (submitted).

[6] H. Umimoto, S. Odanaka, I. Nakao, and H. Esaki, "Numerical modeling of non-planar oxidation coupled with stress effects," *IEEE Trans. Computer Aided Design*, Vol. 38, No. 3, pp. 505-511, 1991.

[7] D. Chin, S. Y. Oh, S. M. Hu, R. W. Dutton, and J. L. Moll, "Two-dimensional oxidation," *IEEE Trans. Electron Devices*, Vol. ED-30, No. 7, pp. 744-749, 1983.

[8] D. Anderson, and K. O. Jeppson, "Nonlinear two-step diffusion in semiconductors," *J. Electrochem. Soc.*, Vol. 131, No. 11, pp. 2675-2679, 1984.

[9] J. U. Brackbill, and J. S. Saltzman, "Adaptive zoning for singular problems in two dimensions," *J. Comp. Phys.*, Vol. 46, pp. 342-368, 1982.

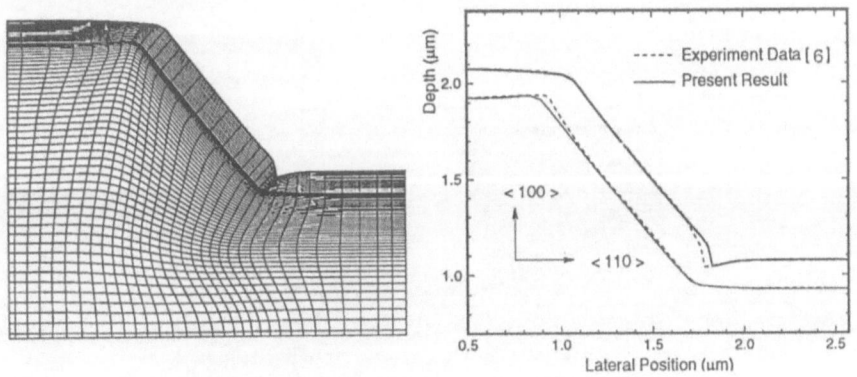

Fig. 1 Grid distribution and comparison of oxide profile in a trench after 20 minutes oxidation in a wet ambient at 1000 °C.

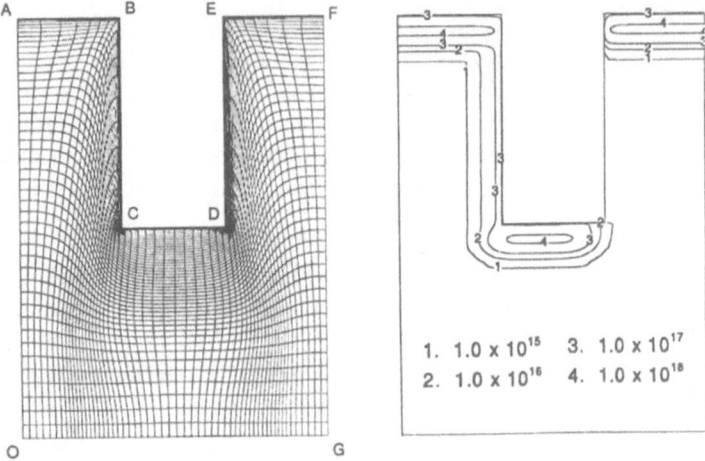

Fig. 2 Initial grid system and boron distribution (OA = 2.0 μm, OG = 1.5 μm).

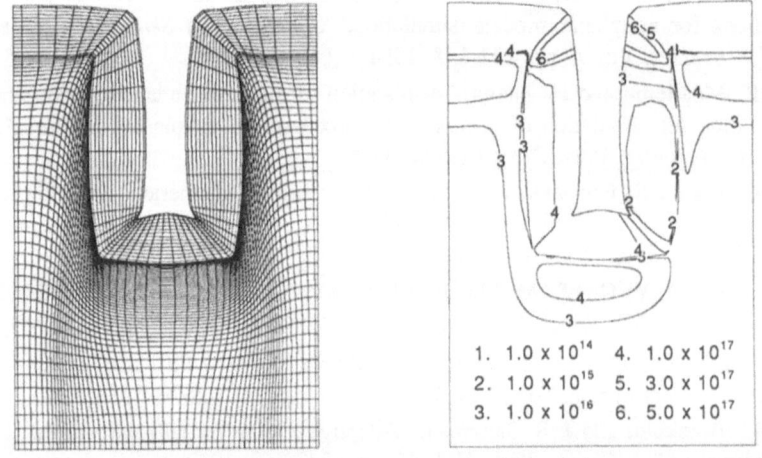

Fig. 3 Grid distribution, oxide profile and boron distribution after 40 minutes.

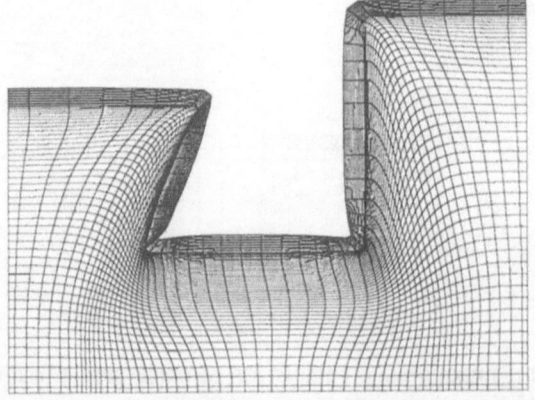

Fig. 4 Grid system and oxide profile after 7.2 minutes oxidation.

Process Optimization in a Production Environment Using Simulation and Taguchi Methods

M. McGee and A. J. Walton[†]

Motorola Ltd.
Colvilles Road Kelvin Industrial Estate, East Kilbride, Glasgow, G75 0TG,
UNITED KINGDOM
[†]Edinburgh Microfabrication Facility, Department of Electrical Engineering,
University of Edinburgh
Kings Buildings, Edinburgh, EH9 3JL, UNITED KINGDOM

Abstract

This paper discusses the use of Taguchi Method in conjunction with process simulation for optimising IC processes in a high volume production environment. The use of orthogonal arrays is presented and two applications detailed.

1. Introduction

In the past, process and device simulation has been routinely used in R&D laboratories, but this has not been the case in the production environment. However, structured experimentation is widely used in manufacturing plants with Taguchi methods being very popular. This paper describes the application of simulation in conjunction with Taguchi methods to two problems experienced in a high volume production facility. While both of the examples are relatively simple they do demonstrate that simulation, when used with Taguchi methods, has the potential to save a significant amount of money as well as speed up the time taken to identify the causes of problems.

2. The Taguchi Method

The Taguchi method has been the subject of much interest over recent years [1]. While statisticians have argued about its merits there is a plethora of papers that report successful applications [2, 3]. The method uses orthogonal arrays [4] to reduce the number of experimental runs required to characterise a process. These arrays have special properties that allow the effect of control factor upon responses to be calculated. An example of an orthogonal array which can be used to examine the effect of control factors A to G is given in table 1.

Each of the control factors has two level settings and the effect these have upon the output responses is easily determined because the orthogonal array has been designed to effectively balance out the influence of all the other factor settings. For example the effect of level setting 1 of control factor A is simply obtained by calculating the average of experiments 1, 2, 3 and 4. Similarly, the effect of level setting 2 for factor A is determined by taking the average of experiments 5, 6, 7 and 8. The impact of all the other factor settings can be calculated in a similar manner and the results plotted as illustrated later.

Expt	Control Factor Setting						
No	A	B	C	D	E	F	G
1	1	1	1	1	1	1	1
2	1	1	1	2	2	2	2
3	1	2	2	1	1	2	2
4	1	2	2	2	2	1	1
5	2	1	2	1	2	1	2
6	2	1	2	2	1	2	1
7	2	2	1	1	2	2	1
8	2	2	1	2	1	1	2

Table 1: Table 1. A $L_8(2_7)$ orthogonal array.

3. Implant Monitor

In order to confirm the correct functioning of the implanters, test wafers are processed at specified intervals. The process consists of growing an oxide, performing the implant followed by an anneal, after which, the sheet resistance is measured. To speed up the monitoring cycle time the anneal process was changed from a conventional furnace to an RTA system. At the same time the variation in sheet resistance was observed to increase and this was initially attributed to oxide thickness variation. This was of obvious concern because it reduced the effectiveness of the monitor for identifying potential implant problems.

SUPREM-3 was used to simulate the process described above and an $L_8(2_7)$ orthogonal array used to examine the effect of varying what were thought to be the key control factors. The effect of all these factors on the final sheet resistance is shown in figure 1.

As expected these results show that the oxide thickness does influence the final sheet resistivity. However, somewhat surprisingly, the same change in oxide thickness has a proportionately larger effect when the RTA system is used. All the other control factors have a roughly similar effect upon the sheet resistance except for the dose where the RTA wafers appear to be less sensitive to dose variation.

These results suggest that the perceived problem of oxide thickness variation is not in fact the case. The move to an RTA anneal has made the monitor wafers more sensitive to any changes in oxide thickness, whereas previously, variations of this magnitude did not have any significant effect. Of equal concern is that the RTA wafers were less sensitive to dose variations and hence the monitor, when RTA processed, is less effective. The complete process of designing the "experiment", performing the simulations and processing the results was performed in a single afternoon. This compared with the cost of processing up to 24 wafers with its typical one week cycle time.

4. Depletion Threshold Variation

As part of the continual improvement programme the variation of depletion threshold voltage for the 4μm nMOS products was targeted. This work commenced before

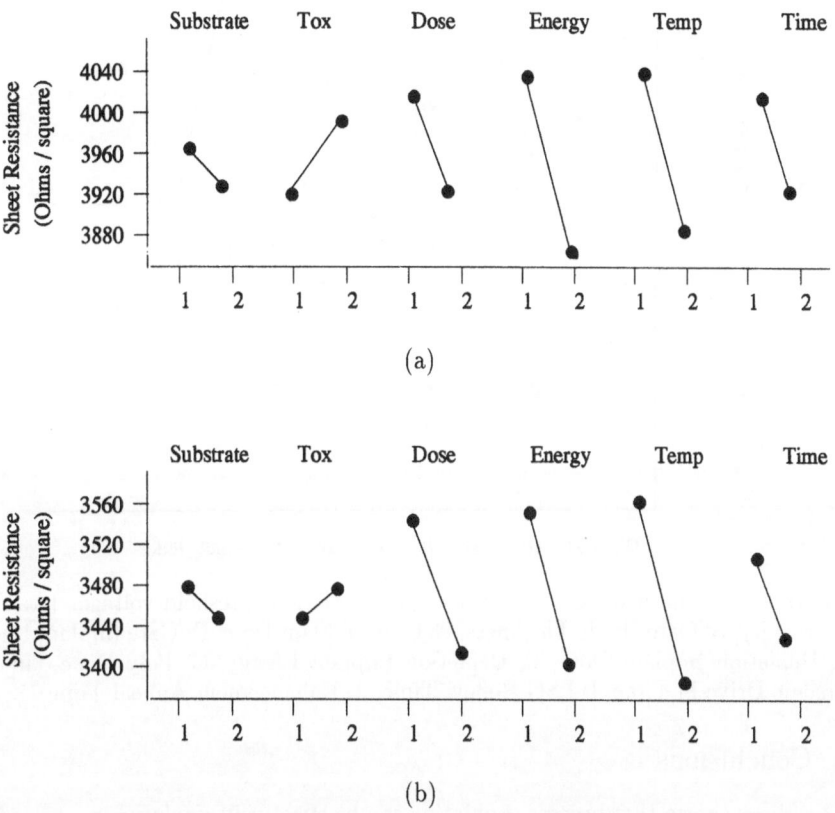

Figure 1: The effect of control factors upon sheet resistivity. (a) RTA, (b) Furnace.

simulation was available on site and so an experiment was run using Taguchi techniques to identify the main controlling factors. A $L_{16}(2_{15})$ experiment, consisting of 16 runs, was performed using 48 wafers. The time taken to run this experiment from the initial design through processing and final analysis was approximately 4 months.

At about the time that final measurements were being performed simulation facilities became available and the experiment was repeated using SUPREM-3. Two days was spent building the data file for the complete nMOS process including de-bugging and characterisation. The various runs as defined by the $L_{16}(2_{15})$ Taguchi array were subsequently modelled on a workstation within 3 hours and the results are displayed in figure 2.

The results obtained from this simulation identified the same controlling factors (and their relative significance) as those identified by the previous silicon experiment. As expected, depletion dose and gate oxide thickness were the most significant factors. However, the exercise also highlighted that the normal variations in wafer substrate resistivity significantly influenced the threshold voltage and this had not been previously considered an issue.

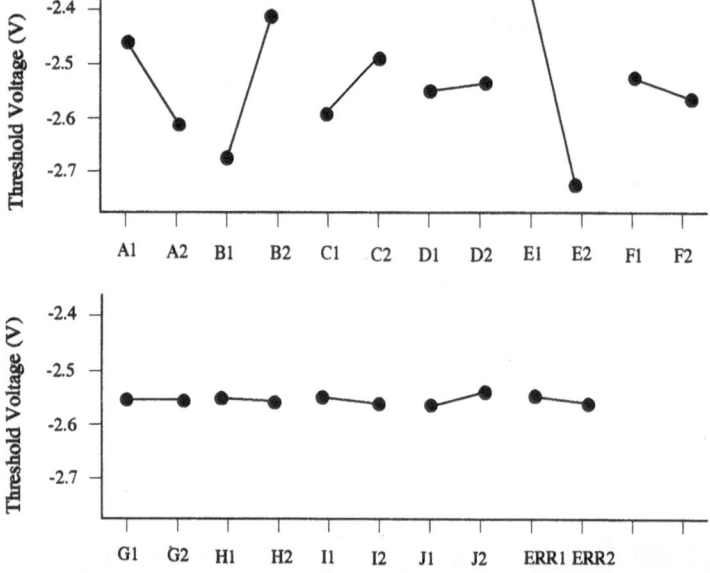

Figure 2: The effect of control factors upon depletion threshold voltage. (A: Wafer Resistivity, B: Gate Oxide Thickness, C: Gate Implant Dose, D: Gate Implant Energy, E: Depletion Implant Dose, F: Depletion Implant Energy, G: Poly Dope Time, H: Arsenic Drive-in Time, I: PSG Reflow Time, J: Enhancement Anneal Time)

5. Conclusions

It has been shown that process simulation can be effectively employed in a large MOS production facility to solve problems. In addition it has been demonstrated that the combination of experimental design and simulation is extremely powerful and can significantly reduce the number of simulation runs required. The two examples detailed above have had significant cost benefits as well as reducing the cycle time involved in the problem solving process.

6. Acknowledgements

The authors would like to acknowledge the support of TMA who supplied the software for this work.

References

[1] G. Taguchi,em System of Experimental Design: Engineering Methods to Optimize Quality and Minimize Costs, UNIPUB White Plains/ ASI Dearborn, 1987.

[2] M.S. Phadke, *Quality Engineering using Robust Design*, Prentice Hall, London, 1989.

[3] *Taguchi Methods: Selected Papers on Methodology and Applications*, ASI, Dearborn, 1988

[4] G. Taguchi, S. Konishi, *Taguchi Methods: Orthogonal Arrays and Linear Graphs*, ASI, Dearborn, 1987"

Optimization of DMOS Transistors for Smart Power Technologies by Simulation and Response Surface Methods

W. Kanert, N. Krischke, and K. Wiesinger[†]

Semiconductor Group, Siemens AG
Otto-Hahn-Ring 6, D-81739 München, GERMANY
[†]Semiconductor Group, Siemens AG
Siemensstraße 2, A-9500 Villach, AUSTRIA

Abstract

DMOS transistors for smart power technologies were investigated by extensive use of process and device simulation. For the task of simultaneously optimizing a multitude of parameters, experimental designs and response surface methods were used.

1. Introduction

The necessity to reduce development time and costs in the semiconductor industry enforces an increase in simulation efforts. CMOS technologies constitute the main driving force for the development and implementation of advanced physical models. Beside these advanced technologies, smart power technologies have gained much interest in recent years [1,2,3]. Although from the point of view of physical modeling not as demanding as e.g. far submicron CMOS, these technologies pose severe difficulties due to their complexities, the variety of devices and the combination of high voltage/power with logic and analog circuit capabilities.

The DMOS transistor is of central importance in smart power technologies. Fig. 1 shows a DMOS cell together with the periphery of the cell field. Targets for optimization include threshold voltage, saturation current, on-resistance, and breakdown voltage. These targets are influenced by process parameters, for instance epi thickness and concentration, body implant and body and source diffusion, and geometrical parameters such as cell dimension and poly field plate length on field oxide. The task of simultaneously optimizing a multitude of parameters leads to departing from the traditional "one-factor-at-a-time" method and to using statistical methods instead [4].

Extensive simulation was used to optimize such a device. Simulations were carried out with SUPREM-3 [5], TSUPREM-4 [6], and MEDICI [7]. Parameters were calibrated using experimental data. Fig. 2 compares simulated and measured values of threshold voltage. A systematic difference between both data sets can be seen, which may be

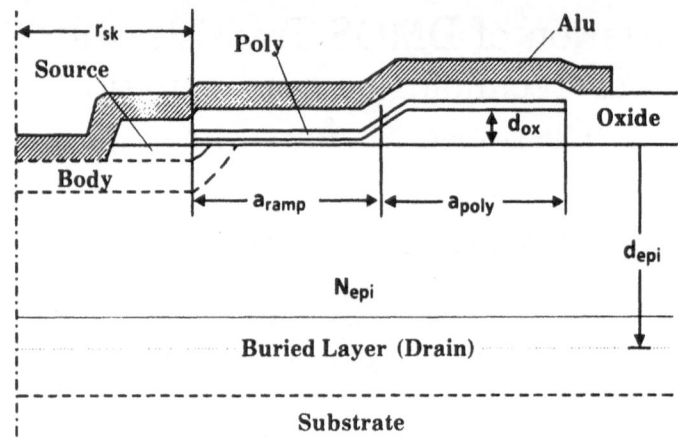

Figure 1: DMOS cross section. Half of the inner cell and periphery are shown. Process and geometrical parameters as used in the optimization of breakdown voltage are indicated (see text).

explained by oxide charge and other effects that were not taken into account in the simulation.

The factors with the major influence vary according to the targets chosen. Therefore, different experimental designs were used for intrinsic device performance data (inner DMOS cell) and breakdown optimization (peripheral structure).

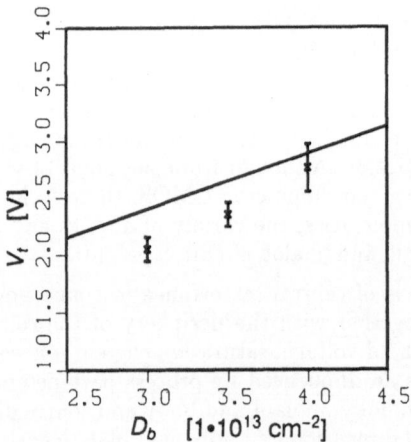

Figure 2: DMOS threshold voltage as a function of body implant dose. Comparison of simulated and experimental data.

Four factors were taken into account for threshold voltage: body implant dose (D_b) and energy (E_b), body (t_b) and source (t_e) diffusion time. Fig. 3 shows contour lines for constant threshold voltage as a function of body implant dose and diffusion time. The other factors (E_b, t_e) are held constant.

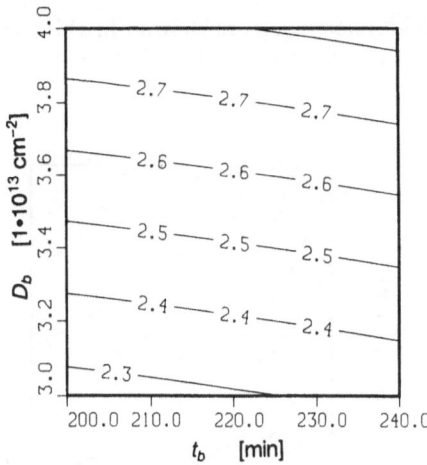

Figure 3: DMOS threshold voltage as a function of body implant dose and body diffusion time. Contour lines correspond to constant threshold voltage. Body implant energy and source diffusion time are kept constant.

The breakdown voltage is not only influenced by epi thickness (d_{epi}) and concentration (N_{epi}), but also by geometrical parameters. The gate polysilicon acts as a field plate to increase the breakdown voltage source/body to drain. A Box-Behnken design with the six factors as defined in Fig. 1 was used for the optimization. Fig. 4 shows contour lines for breakdown voltage as a function of d_{epi} and N_{epi}. For an assumed 10 percent variation of the process parameters the fitted polynomials obtained with the response surface method immediately give information on the corresponding variation of the target values, which is indicated in Fig. 4 by a box drawn around the central point.

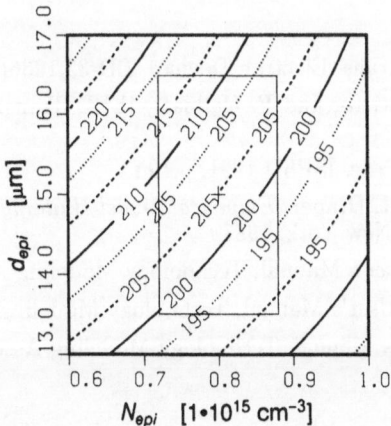

Figure 4: DMOS breakdown voltage. Contour lines of constant breakdown voltage as a function of epi thickness and concentration. Other parameters are held constant. Different line types correspond to different sets of parameters.

Fig. 5 gives an example of the influence of the geometrical parameters on the breakdown voltage. Contour lines as a function of cell contact dimension r_{sk} and distance of the gate edge to field oxide a_{ramp} are shown. In this case, too, the contour lines correspond to a specific set of values for the other parameters.

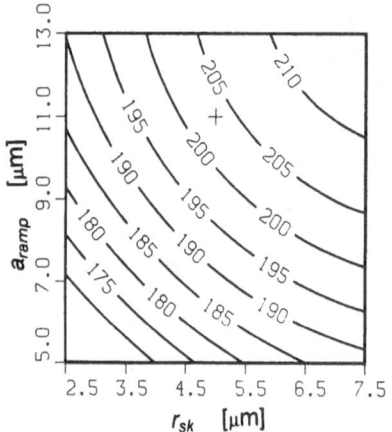

Figure 5: DMOS breakdown voltage. Contour lines of constant breakdown voltage as a function of cell contact length and distance of gate polysilicon edge to field oxide. Other parameters are held constant.

In conclusion we have used process and device simulation to optimize a DMOS transistor for a smart power technology. Classical simulation models are sufficient for this purpose, but the structures, nevertheless, constitute a challenge due to their complexities. Response surface methods were applied to efficiently treat several parameters simultaneously and thereby reduce computing resources.

References

[1] B. J. Baliga, IEEE Trans. Electron Devices, ED-33,1936(1986)

[2] A. Andreini, C. Contiero, and P. Galbiati, IEEE Trans. Electron Devices, ED-33,2025(1986)

[3] A. Preußger et. al., Proc. ISPSD 1991, p.195

[4] G. E. P. Box and N.R. Draper, *Empirical Model-Building and Response Surfaces*, John Wiley & Sons, New York, 1987

[5] TMA SUPREM-3 User's Manual, Technology Modeling Associates, 1990

[6] TMA TSUPREM-4 User's Manual, Technology Modeling Associates, 1991

[7] TMA MEDICI User's Manual, Technology Modeling Associates, 1992

SIMULATION OF SEMICONDUCTOR DEVICES AND PROCESSES Vol. 5   217
Edited by S. Selberherr, H. Stippel, E. Strasser – September 1993

Improved Technology Understanding through Using Process Simulation and Measurements

T. Feudel, W. Fichtner, N. Strecker, R. Zingg, G. Dallmann[†], and E. Döring[‡]

Integrated Systems Laboratory, ETH-Zürich
Gloriastraße 35, CH-8092 Zürich, SWITZERLAND
[†]Institut Fresenius, Angewandte Festkörperelektronik GmbH
Königsbrücker Landstraße 159, D-01109 Dresden, GERMANY
[‡]EM Microelectronic Marin SA
Rue des Sors 3, CH-2074 Marin, SWITZERLAND

Abstract

This paper will provide an example for improved technology understanding through a combination of simulation and measurements. Sample misprepara-tions and design errors could be detected very fast with process simulation, an extremly helpful tool to correlate physical and electrical measurements.

1. Introduction

In the past few years, process and device simulation has become indispensable for the development of semiconductor technologies. In contrast to device simulation, how-ever, process simulators sometimes lack in accuracy and robustness. The reason is that many effects cannot be measured directly, especially in multi-dimensional (2D and 3D) situations. In order to be useful a modern process simulation environment must fulfill at least two criteria: (1) the physical models have to be accurate be-cause of the shrinking device dimensions, and (2) to realistically describe influences between neighboring structures, large areas have to be handled. In this contribution, we present our approach in successfully applying a simulation environment together with sophisticated characterization techniques to the development of a complex VLSI BiCMOS process.

2. Device Technology

In a joint effort between the Integrated Systems Laboratory and EM, a modular BiCMOS technology is being developed, suitable for both low voltage (1.5 V) and high voltage (up to 120 V) analog-digital applications. Apart from conventional (5 V) n- and p-channel MOS devices, a variety of different bipolar and FET devices have been designed (e.g. vertical npn, lateral and vertical pnp transistors, JFET, EMOS, DMOS, diodes, resistors).

3. Simulation and Experiments

Compared with a conventional VLSI CMOS process, this BiCMOS technology is considerably more complex. As examples, we might mention the formation of dual (n^+/p^+) buried layers, the epitaxial substrate, the base/emitter optimization for npn and pnp, and, last but not least, the highly intricate isolation schemes. For a first round of experiments, a complete set of design rules for the active device regions was "extracted" from a large number of process and device simulations. The fabrications resulted in good MOS devices and in poor, but working bipolar transistors. In the extreme cases of low n-well and low base implants, electrical measurements on npn transistors gave an Early Voltage of 2 V, a current gain of 1000, a collector-emitter breakdown voltage of 4.5 V and a relatively high leakage current (nA) from base to substrate.

Therefore a second round of simulations was necessary. For this, we have used a system which automatically performs 1D and 2D simulations based on design information (CIF files, Fig. 2) and the process description [1]. Our in-house simulators TESIM (1D) and DIOS (2D) provide a consistent model base, which has been verified for this process by a series of SIMS, SEM and EBIC measurements. Figure 1 shows a comparison of SIMS and TESIM results for the p-base with and without the influence of the n^+-emitter. In this case, no emitter dip effect could be found.

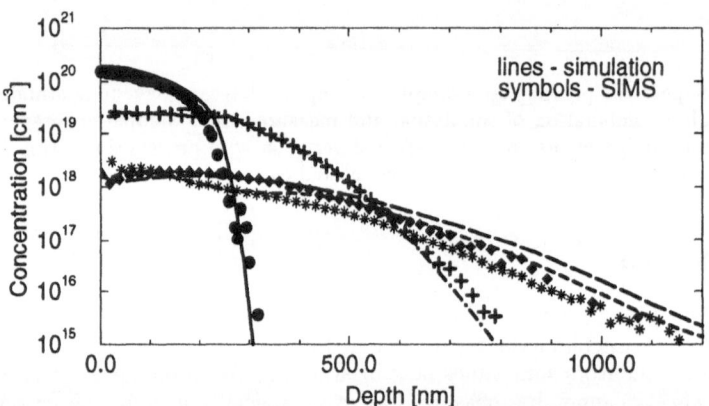

Figure 1: Cut through emitter (As: filled circles and solid lines, P: crosses and dash dotted, B: stars and short dashed) and base region (B: diamonds and long dashed)

Using the advanced automatic grid adaption strategies of DIOS and a modified pragmatic approach for local oxidation [2] of arbitrary shaped structures, rather complex "complete" BiCMOS devices including isolation areas could be simulated in a single run. Fig. 2 shows the cross-sectional view of an n-p-n transistor (65k grid points, 25 CPU-h on a SUN-Sparcstation 10).

Both the simulation and the EBIC images show the reason for the high leakage currents in the too low doped n-well (upper right corner of Fig. 2 and 4). Also shown are n^+-regions on the lateral borders of the p-base regions because of a too small polysilicon width. The reason for the low breakdown voltage has been found in the strong influence of the n^+-buried layer on the active areas. In this case, the EBIC image showed a junction depth for the buried layer of 12 μm and a plateau in the linescan profile at 7.5 μm detecting an internal electric field maxima (Fig. 3). This

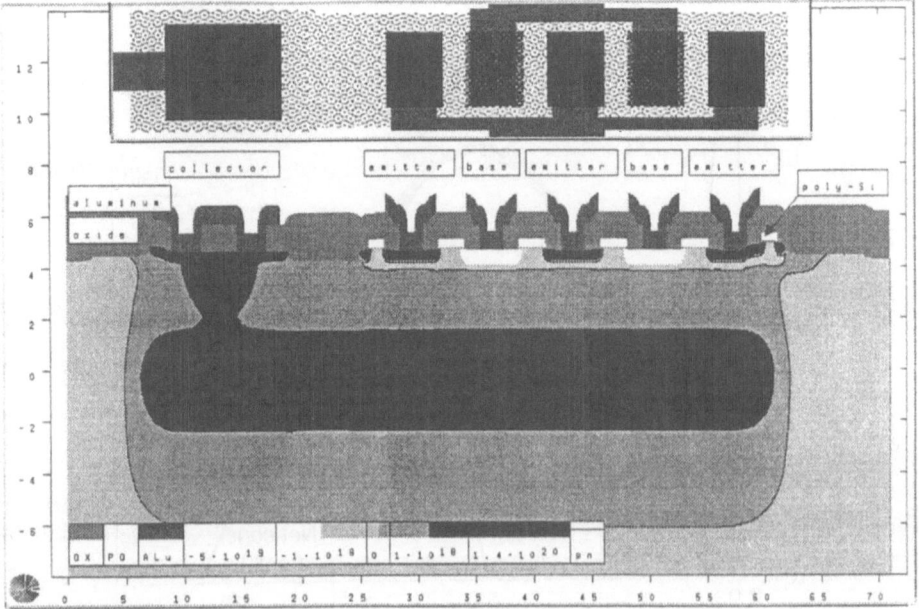

Figure 2: Layout and cross section of a simulated n-p-n transistor

was in bad agreement with first simulation results. The reason was found through the SIMS profiles showing an additional phosphorus implant below the arsenic buried layer. This implant could be traced to an error in the process flow.

4. Conclusions

The simulation system has been shown to be robust enough for automatic simulation of large BiCMOS structures. It was necessary to improve the grid adaption strategies and to refine the 2D oxidation models to handle complex device structures. A comparison with experimental data showed the bottlenecks of the used technology. Especially EBIC was found to be a valuable technique for the characterization of the 2D shape of pn-junctions.

Acknowledgement

The project was funded by the Swiss LESIT program. The authors wish to thank the EM processing team for sample preparation and the Institut Fresenius Dresden for performing SIMS and EBIC measurements.

References

[1] W. Fichtner et al., *An Approach to Three-Dimensional VLSI Process Simulation*. Electrochem. Soc. Spring Meeting, Honolulu, Hawaii, May 1993, Extended Abstracts, Volume 93-1, p. 1062.

[2] N.Guillemot et al., IEEE Trans. Electron Dev., ED-34,1033(1987)

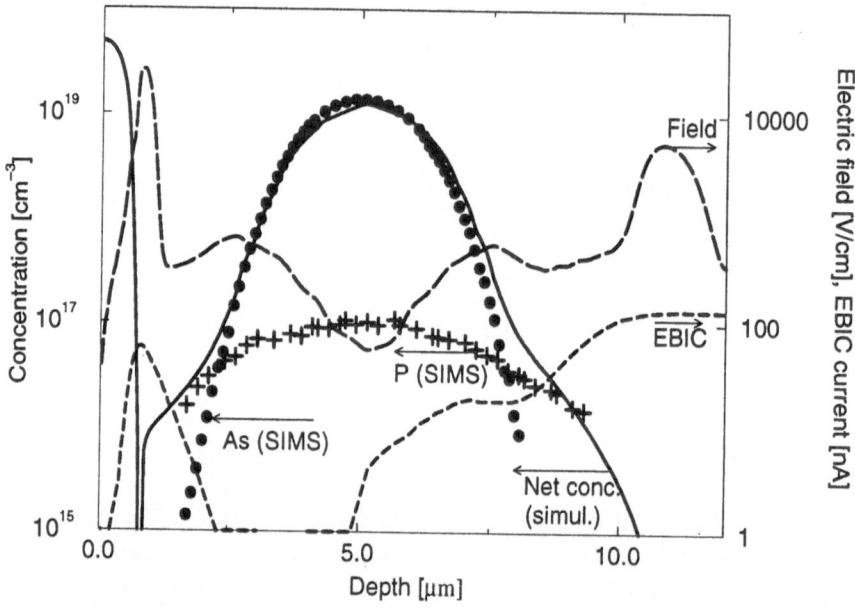

Figure 3: Measured (SIMS, EBIC) and simulated (net doping, electric field) profiles in the base contact region

Figure 4: SEM and EBIC images of the bevelled n-p-n transistor (beam voltage 3 kV)

SIMULATION OF SEMICONDUCTOR DEVICES AND PROCESSES Vol. 5
Edited by S. Selberherr, H. Stippel, E. Strasser – September 1993

Hot Carrier Suppression for an Optimized 10V CMOS Process

A. Bauer, H. Noll, and H. Pimingstorfer[†]

Reasearch and Development, AMS Austria Mikro Systeme International AG
Schloß Premstätten, A-8141 Unterpremstätten, AUSTRIA
[†]Institute for Microelectronics, TU Vienna
Gußhausstraße 27-29, A-1040 Wien, AUSTRIA

Abstract

An advanced 2μm twin-well CMOS process for mixed analog/digital designs with operating voltages up to 10 Volts is presented. The process uses an optimized LDD-structure with high-energy n- implant for the n-channel transistors which drastically reduces the substrate current by a factor of approximately 50 compared to a standard 2μm/5V process technology. A number of relevant process parameters have been optimized to find the ideal balance between driving capability/speed and long-time reliability.

1. Introduction

The maximum supply voltage of most CMOS VLSI processes is limited to 5V or less. However, certain applications require higher operating voltages, e.g. telecommunication and data-conversion products with the need of an extended signal range or improved signal-to-noise ratio. Whilst a 5V process with 2μm channel length does not require advanced transistor design, a higher operating voltage can lead to similar problems as known in submicron process technology.

In order to achieve 10V supply voltage the substrate current and hot-carrier generation has to be reduced drastically. This will increase the bipolar breakdown voltage and ensures long-term reliability. In general, a smooth doping profile at the drain/channel junction gives a wider spread of the drain potential and reduces the electric field and therefore the hot-carrier degradation.

Further improvements can be obtained, if the main part of the drain current bypasses the high-field region, as both, hot-carrier degradation and substrate current generation are proportional to the carrier density [1]. The degradation is also a function of the distance of the hot-carrier generation center from the surface and the gate edge, so that a low substrate current value does not necessarily imply a long lifetime [2]. Therefore, besides the reduction of substrate current, it is absolutely essential, to develop a hot-carrier resistant structure as reported in [3, 4].

2. Process Development

The original 2μm/5V process uses a DDD-structure (doubly doped drain) for the n-channel transistors. At higher drain voltage, the substrate current increases rapidly and causes bipolar breakdown at about 8.5V. As a simple adjustment of the DDD-implant parameters did not provide any improvement, an LDD structure was introduced.

By using the PROMIS [5] and MINIMOS [6] simulation tools and statistical design of experiments a sensitivity analysis of the substrate current was carried out (Fig. 1, 2, 3). Based on this analysis and subsequent experiments a high-energy / low-dose LDD-implant (170keV, 5E12 to 1E13/cm^2) with 0.5μm wide spacer was chosen. For this process condition, an exceptionally low substrate current was measured. It was reduced by a factor of 50-100 (1-2nA/μm measured at 5V) compared to the standard process (100-150nA/μm) with snapback voltages of more than 14 Volts.

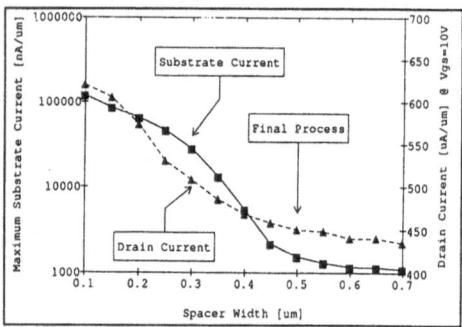

Fig. 1: Simulated substrate and drain current vs. spacer width (Vds=10V, LDD-implant parameters: 8E12/cm^2/170keV).

The high implant energy is fairly uncommon in LDD technology but it seems to be a key parameter for this process. Unfortunately, without further attention, the implant may channel through the gate and generate low-threshold spots in the channel region resulting in increased leakage current or even shorted transistors. Measurements on dedicated test structures showed that this behavior occurs 10-50 times for 10^{6}μm of transistor width. Oxidizing the polysilicon (about 50nm) before doing the LDD-implant reduced the probability to 0.1-1 defects per 10^6 μm gate length - but this was still unacceptable.

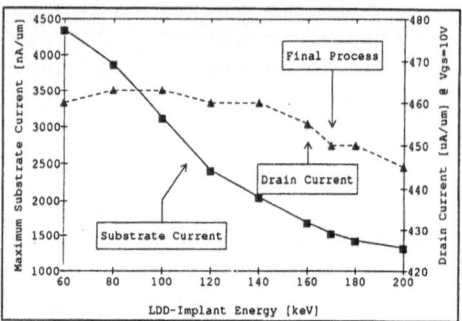

Fig. 2: Simulated substrate and drain current vs. LDD-implant energy (Vds=10V, implant dose=8E12/cm^2, spacer width=0.5μm).

The problem was avoided by using the remaining photoresist which is left on top of the polysilicon after gate-etch, as a self-aligned implant mask. As the n-channel LDD-implant also goes into the p-channel source/drain regions, a p-LDD implant (1E13/cm^2 dose) has been added for compensation.

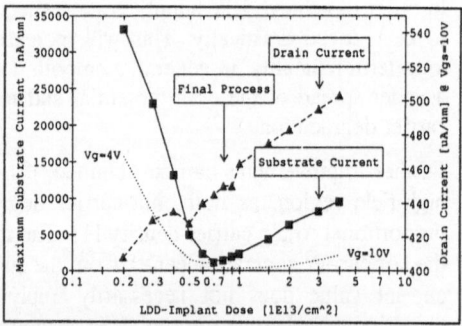

Fig. 3: Simulated substrate and drain current vs. LDD-implant dose. (Vds=10V, implant energy=170keV, spacer width=0.5μm).

Beside the p-LDD implant, process modifications for the p-channel transistors involve a shallow, high-dose (6E15/cm^2,

55keV) BF_2 source/drain implant. The implant is driven by the full source/drain diffusion (1000°C, 50min) under the spacer to form a smooth doping profile at the drain junction which effectively suppresses PMOS hot-carrier generation.

The final process parameters were optimized for maximum driving capability at minimum 10 years extrapolated DC lifetime (10% gm degradation) under all bias conditions [7].

3. Process and Device Simulation

The effective suppression of substrate current is due to the fact that the drain current is not coincident with the location of the electric field maximum. The peak lateral field is located deep in the substrate (Fig. 4a) and remains nearly unchanged as the gate voltage is swept from average to high voltage.

At low gate voltage the drain current bends down after leaving the channel and bypasses the high-field region (Fig. 4b). As a result the carrier-generation rate is low and also sufficiently far away from the surface (Fig. 4c). As the gate voltage is increased the drain current increases but shifts towards the surface. Overall, the integral carrier-generation rate is reduced.

The balance of the reduction of the electric field and the increase of current-density depends on the LDD-implant dose. For a lower implant dose the substrate current curve bends up again at high gate voltages (Fig. 5). As at high gate voltages the center of hot-carriers also comes closer to the surface, it is obvious, that the device is more sensitive for this bias-condition. Actually, a bad lifetime has been measured for the $5E12/cm^2$ group at high gate voltages. This behaviour is reproduced by simulation quite well (Fig. 6).

It has been reported, that the upward-bending of the substrate-current curve is related to an electric-field peak on the source side resulting primarily from a low n- implant dose [8]. For our process, we did not observe a significant electric-field peak nor carrier-generation at the source. Instead, simulation shows that the carrier density becomes comparable to the n- doping concentration at high gate voltage. Therefore, the potential drop is shifted from the LDD region to the n-/n+ section which results in an increase of the electric field.

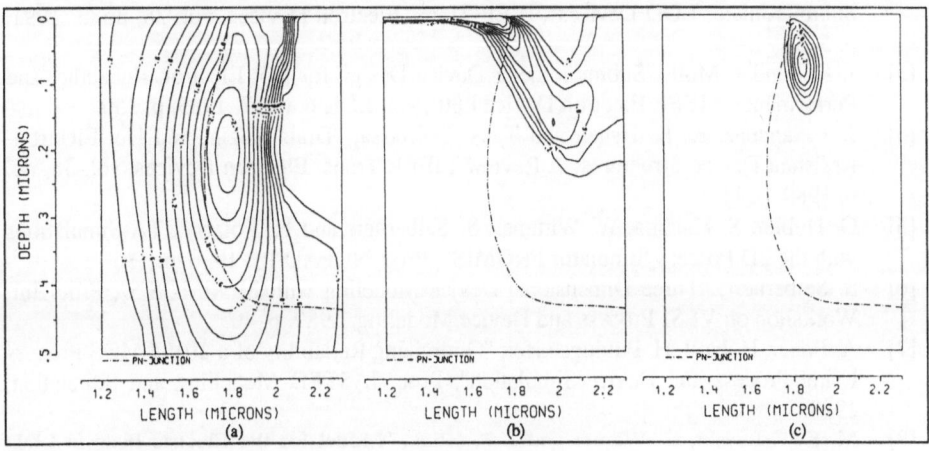

Fig. 4: Simulated electric field (a), electron current (b) and carrier generation (c) for LDD-implant parameters $8E12/cm^2$, 170keV (Vds=10V, Vgs=4V).

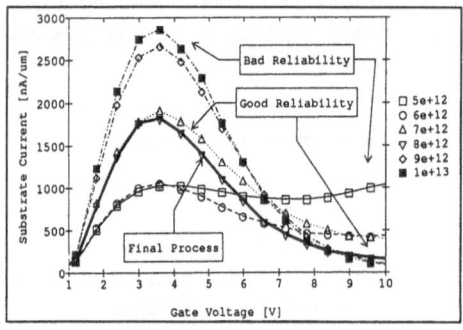
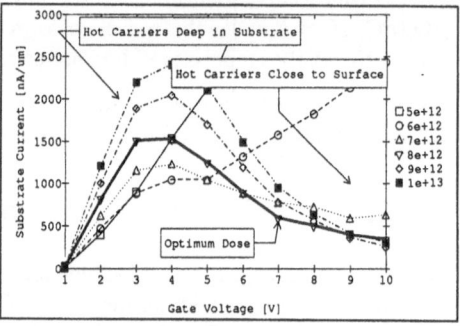

Fig. 5: Measured substrate current vs. gate voltage var. LDD-implant dose (Vds=10V).

Fig. 6: Simulated substrate current vs. gate voltage var. LDD-implant dose (Vds=10V).

4. Conclusions

The development of a new 10V CMOS process has been presented. The LDD-structure itself and the integration in the standard process was optimized by means of process and device simulation and statistical design of experiments.

Except for the improved breakdown behavior and the reduced substrate current all electrical parameters are very close to the original 2μm/5V process. The process does not require any special design rules. Therefore, almost all standard-cell libraries can be used without any modification.

References

[1]   M.-L. Chen, C.-W. Leung, W. T. Cochran, W. Jüngling, C. Dziuba and T. Yang, "Suppression of Hot-Carrier Effects in Submicrometer CMOS Technology", IEEE Trans. Electron Devices, vol. 35, no. 12, 1988, p. 2210.

[2]   W. Hänsch, C. Mazure, A. Lill and M. Orlowski, "Hot Carrier Hardness Analysis of Submicrometer LDD Devices", IEEE Trans. Electron Devices, vol. 38, no. 3, 1991, p. 512.

[3]   J. Hui and J. Moll, "Submicrometer Device Design for Hot-Electron Reliability and Performance", IEEE Electron Device Lett., vol. EDL-6, no. 7, 1985, p. 350.

[4]   J. J. Sanchez, K. K. Hsueh and T. A. DeMassa, "Drain-Engineered Hot-Electron-Resistant Device Structures: A Review", IEEE Trans. Electron Devices, vol. 36, no. 6, 1989, p. 1125.

[5]   G. Hobler, S. Halama, W. Wimmer, S. Selberherr and H. Pötzl, "RTA-Simulations with the 2D Process Simulator PROMIS", Proc. NUPAD III, 1990, p. 13.

[6]   S. Selberherr, "Three Dimensional Device Modeling with MINIMOS 5", Proc. Int. Workshop on VLSI Process and Device Modeling, 1989, p. 40.

[7]   A. Bauer, H. Noll, H. Pimingstorfer, "Optimizing Reliability of a 10V CMOS Process Using Process and Device Simulation", Proc. IASTED Modelling and Simulation, 1993.

[8]   M. K. Orlowski, C. Werner and J. P. Klink, "Model for the Electric Field in LDD MOSFET´s - Part I: Field Peaks on the Source Side", IEEE Trans. Electron Devices, vol. 36, no. 2, 1989, p. 375.

SIMULATION OF SEMICONDUCTOR DEVICES AND PROCESSES Vol. 5
Edited by S. Selberherr, H. Stippel, E. Strasser – September 1993

Electrical Parameter Sensitivity of Deep Submicron and Micron MOSFET Devices with Variation in Processing Conditions

R. Sitte, S. Dimitrijev, and H. B. Harrison

School of Microelectronic Engineering, Faculty of Science and Technology,
Griffith University
Nathan, Queensland, 4111, AUSTRALIA

Abstract

A sensitivity analysis, to examine the influence of processing conditions on device parameters, has been performed and the results are presented here. These results show that effects on micron size devices do not necessarily hold for deep submicron size devices. It is therefore necessary to review to what extent the heuristics which applied for micron size devices are applicable for deep submicron devices.

1. Introduction

Integrated circuit manufacturing is complex because it involves a variety of processing materials and equipment throughout a myriad of processing steps. Each of these steps is a potential source of fluctuations in device characteristics, caused primarily by variations in process conditions, brought about by changes aimed to adjustment, wear, ageing, etc., servicing of equipment, or for example from the purity and concentrations of chemical reactants. Collectively these fluctuations cause variations in the product itself, as for example, fluctuations in the thickness of oxide film grown on the wafer shifting the threshold voltage of any one single transistor. The processing specifications of a wafer are based on parameters developed for a single device. It is therefore important to know to which extent fluctuations of individual process parameters influence the device parameters of the end product. This notion is necessary, not only to decide which processing steps will benefit from tighter control, but also to apply forward correction effectively [1]. While the effects of processing parameters are in the main understood for devices currently produced, little is known about deep submicron devices because little manufacturing experience has yet been accumulated. To gain insight into the effects of process parameters on deep submicron devices a comparative study, between 0.1 μm and 1.0 μm size devices has been carried out. The work is based on simulation using MINIMOS [2]. The structure of the simulation experiment and the results obtained are explained in subsequent sections of this paper.

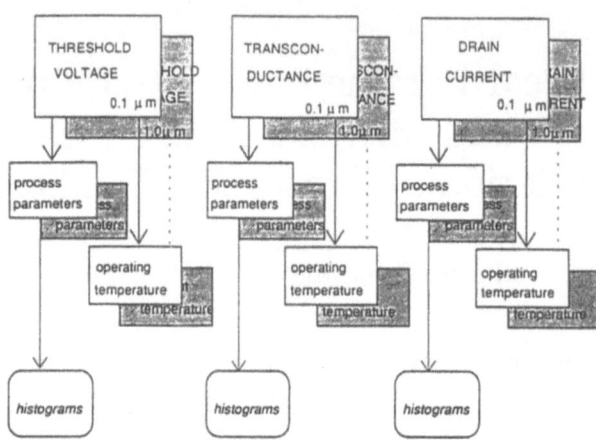

Figure 1: The three main simulation modules of the sensitivity analysis for the considered device parameters, and the operating temperature modules: all simulations were done twice, once for micron (clear front boxes) and once for submicron size devices (shown as shaded background boxes). For each module, comparative histograms were produced with the normalized results

2. Structure of the Simulation Experiment

The function of MOSFET devices is controlled by the joint effects of many parameters and conditions. In our study we have chosen the three main device parameters and operating temperature to be focused on. The device parameters to be simulated were the threshold voltage, the transconductance, and the drain current when the transistor is off. For each of these parameters a simulation module has been implemented to study the variation of any of these device parameter, when the given processing parameter fluctuates in a range typically found in manufacturing. The processing parameters used were gate oxide thickness, gate length, gate oxide charge, bulk doping, and each of the implant parameters for threshold voltage adjustment and source/drain implant: dose, energy, and annealing time and temperature.

The operating temperature is also important to the device because integrated circuits are usually part of other instruments, and thus may have to operate at other temperatures than the optima for which they were. To study this effect another set of simulation modules has been performed.

All the simulations were done twice: for a 0.1 μ deep submicron MOSFETs designed at IBM [3] and for a typical 1.0 μm micron size MOSFET.

Figure 1 summarizes the modular structure of the simulations, showing the submicron device set of simulations modules as clear front boxes, and the micron size device simulation modules as shaded background boxes.

3. Results and Discussion

The purpose of this study is to *compare* the device parameter sensitivity of deep submicron and micron MOSFET devices on processing parameters. In order to establish comparisons a common metric to express processing fluctuations is necessary, thus for each simulation run of a device parameter vs. processing parameter pair the obtained data were plotted, normalized and fitted. A linear fit was possible in almost all the cases for the plotted data. This is because simulations were done for the relatively small range of fluctuations, typically found in manufacturing, for example $\sigma = 10$ %, that is 20 % or more to either side. The slope of all sets of fitted data was then collected to produce histograms, which make a comparison of the impact of processing parameter fluctuations possible.

Figure 2 shows one such histogram for the threshold voltage. The axis of the histogram measures the sensitivity rate (% vs %) on a processing parameter. Negative sensitivity rates towards the left side, and positive rate towards the right.

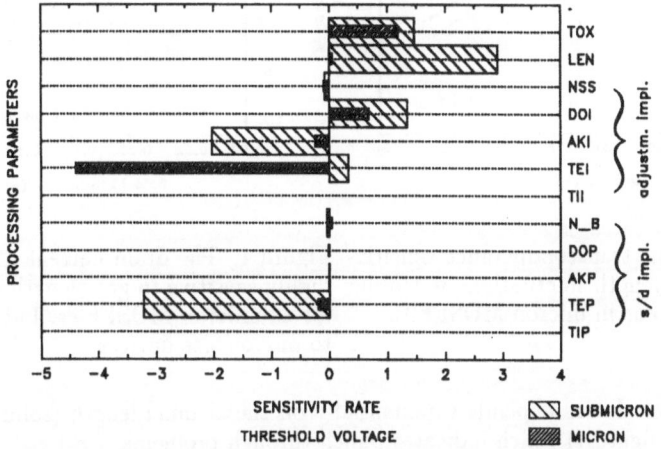

Figure 2: Example of a fluctuation sensitivity histogram

In this study it has been found that a parameter's influencing effect can vary widely: from a strongly influencing parameter in say, micron size devices - to negligible influence in deep submicron devices, or vice versa. For example the implant energy in threshold voltage adjustment is a highly sensitive parameter for the deep sub- micron device, but of little sensitivity for the micron size device. It has also been found that for some processing parameters, a fluctuation - say an increase - can have one effect on a device parameter for one size MOSFETs and the opposite effect for the other size considered. For example a fluctuation of increased oxide thickness causes the off current to increase in the micron size device, it however decreases it in the deep submicron device.

In the case of gate length linear approximation was good for the transconductance, but not for the off current. As can be seen from the figure 3 the threshold voltage and the transconductance sensitivity on the gate length sensitivity is similar for the submicron and micron size device: a 10 % increase in the gate length, as it could happen from slight under etching, causes a decrease in transconductance of approximately 5 % for the deep submicron device, and 7 % decrease for the micron size device.

The off current (drain current, with zero gate voltage) has a completely different sensitivity on the gate length as can be seen from fig 4. For the micron size device it exhibits very little sensitivity to fluctuations (dotted line, left axis scale). For the submicron size device however, it exhibits extremely high sensitivity, but it does not respond linearly (dashed line) to the gate length fluctuation considered. Moreover, we found that this dependence could not be linearized even when the drain current was plotted vs. the reciprocal value of the channel length (1/L). The off current of

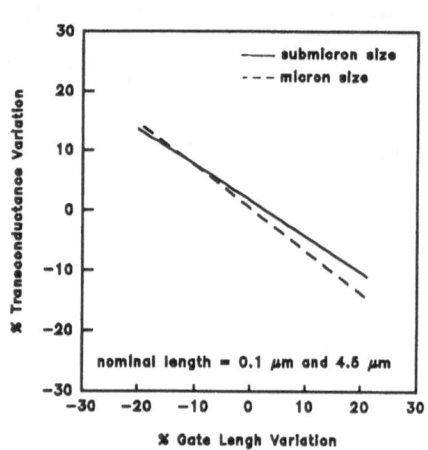 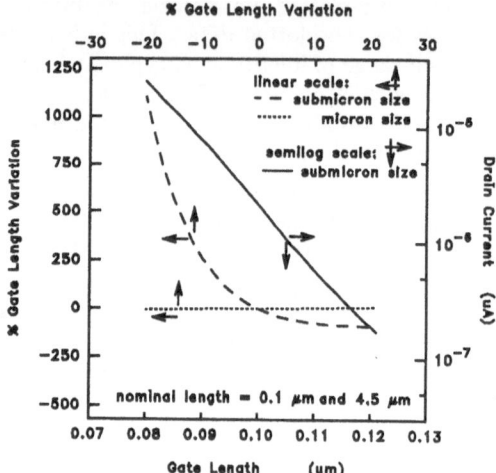

Figure 3: The transconductance sensitivity on gate length fluctuations is similar in submicron as in micron MOSFETs.

Figure 4: The drain current at $V_g = 0$ is highly sensitive to gate length fluctuation for submicron MOSFETs, but very little to micron size devices.

the submicron device depends exponentially on the channel length (solid line, right axis scale in figure 4) which indicates punch through problems.

4. Conclusion

In this paper a sensitivity study aimed at comparing parameter sensitivity of deep submicron and micron MOSFET devices on processing parameters has been presented. This study shows that parameter sensitivity information is necessary for the advancing submicron technology because the response of device parameters to fluctuations of processing parameters is not necessarily the same as in micron size devices. It is thus important to review to what extent the heuristics which applied for micron size devices, will apply for submicron devices.

References

[1] R. Sitte, H.B. Harrison ,"Techniques to Enhance Yield in VLSI Fabrication", *Proceedings 9th Microelectronics Conference* (The Institution of Radio and Electronics Engineers Australia) 1990, pp 141 - 146

[2] S.Selberherr, A.Schütz, H.Pötzl, "Investigation of Parameter Sensitivity of Short Channel MOSFETs", *Solid State Electronics*, Vol 25, No.2, pp 85 - 90, 1982

[3] G.A. Sai-Halasz, M.R. Wordeman, D.P. Kern, E. Ganin, S. Rishton, D.S. Zichermann, H. Schmid, M.R. Polcari, H.Y. Ng, P.J. Restle, T.H.P. Chang, R.B. Dennard, "Design and Experimental Technology for 0.1-μm Gate-Length Low Temperature Operation MOSFETs", *IEEE Electron Device Letters* vol 8, pp 463 - 466, 1987

Predicting Manufacturing Variabilities for Deep Submicron Technologies: Integration of Process, Device, and Statistical Simulations

Z. Krivokapic and W. D. Heavlin

Advanced Micro Device
P.O. Box 3453, Sunnyvale, CA 94088, USA

Abstract

Process and device simulators are used to project the long term manufacturing distributions of 0.35 µm, planarized, concave transistors. Particular care is taken to calibrate the simulators, and to quantify the contribution from each manufacturing source of variation.

1. Introduction

To introduce a new integrated circuit technology efficiently, the performance characteristics and manufacturing specifications must be determined as early as possible. This allows circuit design to proceed in parallel with technology development, and manufacturing hardware to be chosen appropriately. A major limitation to technology development is the complexity of the technologies themselves, and their correspondingly slow manufacturing cycle times. Process and device simulators, well calibrated, promise some relief, by allowing us to test simple theories more easily, to consider more alternatives than silicon, and to determine optimum targets.

In this paper, we extend the application of such simulators to project the long term manufacturing distributions of transistor parameters[1,2]; our example is a 0.35 µm planarized and concave device, for which we are able to make predictions before achieving first silicon. Our approach features minimum silicon, TSuprem4 and Medici process and device simulations, an objective calibration method, and some adapted statistical methods. We estimate not only the long term total manufacturing variation, but also decompose it into components associated with different manufacturing steps. By knowing the total expected manufacturing distribution, we can design circuits to be more robust. By quantifying the components of this distribution, we can determine process performance objectives and select equipment more appropriately.

2. Technology

The device is a 0.35 µm self-aligned channel transistor(Fig. 1). Note that it is fully planarized and concave. LDD diffusion comes from a doped polysilicon layer. Since the gate area is etched through the oxide layer, the damaged underlying silicon has to be removed; once removed, a trench in the silicon results.

By selectively limiting threshold adjust and punchthrough stop implants to the channel region only, we are able to self-align the channel to the source and drain. This decreases the source/drain resistance, and improves hot carrier immunity.

3. Method

Our method consists of five distinct steps: (a) Using an earlier generation technology, we run simulations over a broad range of possible factors, in order to identify those parameters with the most influence on device performance. (b) In silicon, we design and execute a factorial experiment varying these parameters.

(c) We complement the splits of the silicon experiment with simulations. Wherever feasible, two-dimensional device simulations are used. In the case of the p-channel device, sidewall leakage current along the trench isolation requires that we use the three-dimensional simulator DaVinci[3].

By correlating silicon and simulation results, we define a calibration relationship for subsequent simulations, broadly valid over a range of parameter values. This calibration approach is to be distinguished from the usual industry calibration practice, which fits limited experimental data - often just one device - by changing selected parameters in the process and device simulators. In our case, our experimental data is from a factorial lot with a deliberately broad range of process variations, not directly amenable to the usual approach. Further, uncontrollable, unmeasurable variations from contamination, plasma damage, and mechanical stress can effect device characteristics dramatically, yet with present-day tools we are not able to simulate them. Finally, critical dimension control is not fully mature for sub-half-micron devices, and we need to compare devices with the same effective channel lengths.

For all these complications, our calibration approach is straightforward. All process and device simulations use only default parameter values. Simulations of devices at nominal dimensions are compared to the corresponding structures in silicon - without taking into account effective channel lengths. These are compared in scatterplots, such as in Fig. 2. For each scatterplot, we calculate the linear relationship by least squares. This allows us to transform all simulated results into predictions of what is most likely to occur in silicon. In the simulation experiments subsequently described, we use these transformations throughout, always report calibrated values, and make no further reference to calibration.

(d) Following a Latin hypercube design[4], we characterize the non-linearities of the parameter space with additional simulations. The "noise factors" are those parameters most difficult to control in manufacturing, and most likely to induce substantial variation in device performance. For our device, four of the noise factors (ranges) are gate oxide thickness (8.5 and 9.5 nm), gate length, gate trench depth (50 and 100 nm), and spacer width (30 and 50 nm). A fifth noise factor, the source and drain resistance, is a combination of doping, critical dimension, and mask registration error variation.

(e) From these simulations, we predict the distributions of the device's current and voltage characteristics. Using a multivariate interpolator[5] we estimate the improvement in distribution that results from perfect control of each noise factor singly, that is, the amount of manufacturing variation each factor induces.

4. Results

Our example assesses a technology in an early stage. We conclude from Table 1 that n-channel V_t's are most affected by gate oxide thickness variation, linear transconductance by gate critical dimensions, drive current and large signal transconductance by gate trench depth variations, yet can be controlled sufficiently.

In the case of p-channel devices, the blanket punchthrough stop implant, not selective to the channel region only, causes problems. As shown in Table 2, the control of threshold voltage, linear transconductance, and drain-source current are of the order of 10 percent one-sigma. Further, any variation in gate trench depth drastically affects device characteristics. We conclude that to ensure manufacturability one must use either a self-aligned punchthrough stop implant, or drastically improve trench depth control.

5. Summary

Any methodology evaluating manufacturability must be objective and reproducible. Our approach fulfills is so in several ways: The simulations are grounded in silicon over a manufacturing window, robust to any single observation, yet based on default parameters. As a result, our calibration approach is easily automated, reproduced, and commercialized. In the analysis, the percent of variation each factor contributes to the total is quantified, complementing the projected distributions.

6. References

(1) J.H. Kibarian, and A.J. Strojwas, "Using spatial information to analyze correlations between test structure data," *IEEE Trans. Semiconductor Manufacturing,* vol. 4, August 1991, pp. 219-225.
(2) I.C. Kizilyalli, et al., "Predictive worst case statistical modeling of 0.8-μm BICMOS bipolar transistors: a methodology based on process and mixed device/circuit level simulators", *IEEE Trans. Electron Devices,* vol. 40, pp. 966-973, May 1993.
(3) Technology Modeling Associates, *TMA DaVinci Manual,* Nov. 1991.
(4) M.D. McKay, et al., "A comparison of three methods for selecting values of input variables in the analysis of output from a computer code," *Technometrics,* vol. 21, August 1979, pp. 239-245.
(5) G. Matheron, *The Theory of Regionalized Variables,* Centre de Morphologie Mathématique de Fontainbleau, Paris, 1971.

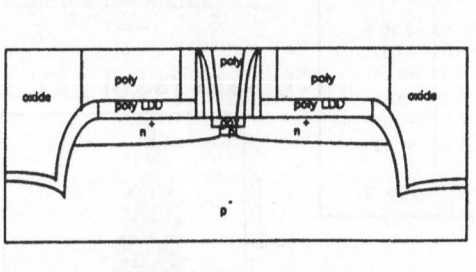

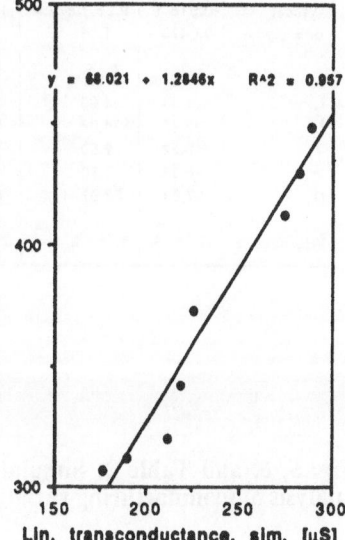

Fig. 1: Cross-sectional view of an n-channel concave device with recessed gate oxide and selectively implanted channel region.

Fig. 2 (on the right): Calibration curve for linear transconductance for 0.5 μm concave devices.

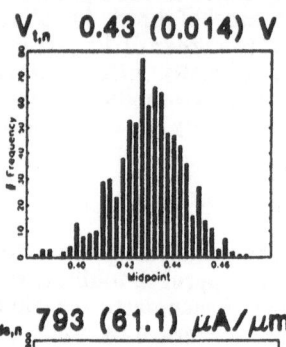

N-Channel Analysis of Manufacturing Variation

	V$_t$	g$_m$	g$_{ds}$	I$_{ds}$
typical	0.43 V	31.1 μS/μm	377 μS/μm	793 μA/μm
one sigma	0.014 V	2.17	20.2	61.1
L$_s$	12.66 %	7.16 %	19.46 %	21.48 %
L$_D$	0.95	13.43	11.18	14.39
L	25.26	44.93	17.59	24.07
T$_{ox}$	57.15	11.93	1.99	3.13
W$_{ox}$	11.67	1.97	0.03	5.12
D$_c$	-2.11	22.34	44.29	34.27
Interactions	5.56 %	1.76 %	-5.47 %	2.46 %

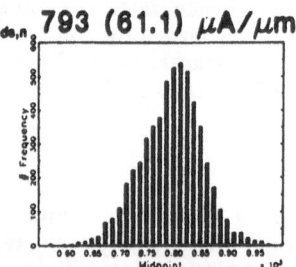

Fig. 3, 4, and Table 1: Simulated n-channel distributions for I$_{ds}$ and V$_t$, and analysis of manufacturing variation.

P-Channel Analysis of Manufacturing Variation

	V$_t$	g$_m$	g$_{ds}$	I$_{ds}$
typical	-0.516 V	9.85 μS/μm	108.5 μS/μm	-313 μA/μm
one sigma	0.0475	1.31	2.16	44.6
L$_s$	-5.76 %	3.19 %	-3.55 %	-3.30 %
L$_D$	1.01	-4.69	10.46	1.01
L	32.43	38.09	52.46	42.35
T$_{ox}$	13.52	9.80	16.97	3.48
W$_{ox}$	-0.33	3.70	-5.25	-2.92
D$_c$	57.81	53.67	55.95	57.08
Interactions	-1.34 %	3.76 %	27.04 %	-2.30 %

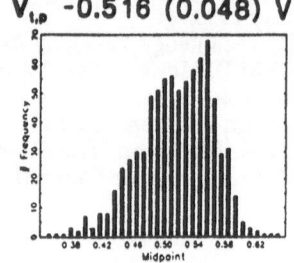

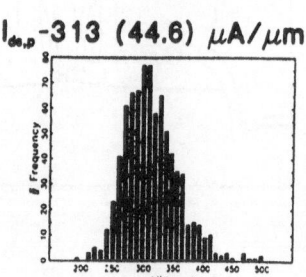

Fig. 5, 6, and Table 2: Simulated p-channel distributions for I$_{ds}$ and V$_t$, and analysis of manufacturing variation.

An Investigation of Coupled and Decoupled Iterative Algorithms for Energy Balance Calculations

Y. Apanovich, E. Lyumkis, B. Polsky, and P. Blakey

Algorithms Development Group, Silvaco International
4701 Patrick Henry Drive, Bldg. 3, Santa Clara, CA 95054, USA

Abstract

Coupled and decoupled iterative algorithms for the solution of nonlocal transport models are investigated. The decoupled scheme described in [1] exhibits excellent convergence properties except under strong breakdown conditions, where the full Newton method is required. Simulation results of snap-back curves for submicron BJT and MOS devices illustrate the potential of both algorithms.

1. Introduction

Deterministic nonlocal transport models are widely used for the analysis of transport phenomena in submicron devices. These models provide a description of effects that are neglected by the conventional drift-diffusion model (DDM). The effects include velocity overshoot and nonlocal impact ionization. These nonlocal models are commonly referred to as energy balance (EB), energy transport (ET) or hydrodynamic (HD) models. This class of models will be referred to here as 'energy balance' models.

From the numerical point of view device simulation using energy balance models is more complicated than simulation based on the DDM. Additional partial differential equations (PDE's) must be solved, and the system of PDE's exhibits greater nonlinearity. A natural decoupled iterative algorithm for energy balance models is as follows: the electron and hole continuity equations and Poisson's equation are solved by either Gummel or Newton methods with the carrier temperatures held constant; and then electron and hole temperatures are updated from the solution of the energy balance equations. This approach is added easily to existing general purpose device simulators, and is therefore widely used.

Meinerzhagen et. al. [1] have shown that the convergence rate of this algorithm is slow for high biases, and that for some cases convergence may not be obtained at all. They suggested a new algorithm that differs from the previous one at the second step. In the new algorithm the energy balance equation is solved simultaneously with the carrier continuity equation. A dramatic increase in convergence rate for MOS transistor calculations is obtained. At the moment this algorithm seems to be the best available for unipolar simulations. The work described here was performed to investigate the convergence rate of Meinerzhagen's method for bipolar simulations, especially in the strong breakdown region; and to establish whether it is necessary to also provide the full Newton algorithm in a truly general purpose device simulator.

2. Energy Balance Model

The variant of the energy balance model used in the present work follows [2], and was described in [3]. The impact ionization coefficient for electron was modeled as:

$$\alpha_n = a_n \exp(-\frac{b_n}{E_{eff}}),$$

with a_n and b_n taken from [4], and an effective electric field determined from the relationship:

$$E_{eff} = \frac{3}{2} \frac{kT_n}{q\lambda_n},$$

where λ_n is the energy relaxation length for electrons. A similar expression is used for α_p.

3. Methods

Meinerzhagen's method and the full Newton method for all five equations were implemented in SPISCES. Meinerzhagen's algorithm is implemented in the following form. For each outer loop, the electron and hole continuity equations and Poisson's equation are solved for either a specified number of iterations, or to convergence, using Newton iteration. One Newton iteration is then performed for the simultaneous solution of the electron continuity and electron temperature equations, after which one Newton iteration is performed for the simultaneous solution of the hole continuity and hole temperature equations. (This implementation differs slightly from [1] where only one Newton iteration was performed at the first step. This does not impact the asymptotic behavior of the convergence rate.) All linearized systems of equations are solved with a direct solver.

4. Results

Numerical experiments were performed to establish the behavior of the different algorithms for several situations. The first experiment was for a submicron bipolar transistor similar to the device considered in [5]. The emitter-base voltage was kept constant and equal to -0.75V. An external resistance of $1.0e5$ $\Omega \cdot \mu m$ was connected between the collector and a voltage source V_{cc}. Figure 1 shows the calculated collector current as a function of collector-base voltage. Figure 2 shows the error versus iteration number for different values of V_{cc}: curve 1 is for $V_{cc} = 8.5V$, curve 2 for $V_{cc} = 10.5V$, and curve 3 for $V_{cc} = 11V$, all calculated using Meinerzhagen's method with the equations solved to convergence in the first step. The corresponding collector-base voltages were 7.99, 9.54 and 9.87V respectively. Curve 4 is for $V_{cc} = 11V$, calculated using the full Newton method. Errors are measured from the fully converged solution obtained using Newton method with very tight convergence criteria. This figure shows that the rate of convergence of Meinerzhagen's algorithm is excellent for $V_{cc} = 8.5V$, but that it decreases rapidly near the snap-back voltage. The Newton algorithm exhibits similar behavior to that shown in Figure 2 for all biases.

A second experiment was for a conventional MOS device with a channel length of 0.45 μm. The gate, source and substrate voltages were grounded. An external resistance of $1.e10$ $\Omega \cdot \mu m$ was connected between the drain, and a voltage source V_{dd}. The drain current as a function of drain voltage is shown in Figure 3. Errors versus iteration number for different values of V_{dd} are shown in Figure 4. Curve 1 is for $V_{dd}=12.25V$, curve 2 is for $V_{dd}=12.4125V$, curve 3 is for $V_{dd}=12.425V$, and curve 4 is for $V_{dd}=12.425V$. The corresponding drain voltages were 12.246, 12.364 and 12.362V respectively. As before, the first three curves are calculated using Meinerzhagen's method with the equations solved to convergence in the first step, and curve 4 is obtained using full Newton iteration. The decoupled algorithm shows fast convergence for $V_{dd}=12.25V$, but fails near the snap-back voltage.

5. Conclusions

Meinerzhagen's decoupled algorithm [1] exhibits excellent convergence properties for pre-breakdown two carrier, two temperature simulations. The convergence rate decreases rapidly under strong breakdown conditions. The full Newton method is required to handle such conditions. Meinerzhagen's algorithm and the full Newton method will both be available in future releases of the SPISCES and BLAZE device simulators.

References

[1]   B. Meinerzhagen, K. H. Bach, I. Bork and W. L. Engl, "*A New Highly Efficient Nonlinear Relaxation Scheme for Hydrodynamic MOS Simulations*," NUPAD IV Abstracts, Seattle, USA, pp. 91-96, 1992.

[2]   R. Stratton, "*Diffusion of Hot and Cold Electrons in Semiconductor Barriers*," Phys. Rev., vol. 126, pp. 2002-2013, 1962.

[3]   Y. Apanovich, E. Lyumkis, B. Polsky, A. Shur, and P. Blakey, "*Numerical Simulation of Submicron Devices Using Energy Balance and Hydrodynamic Models in the General Purpose Device Simulator SPISCES-2B*," Abstracts of the International Workshop on Computational Electronics, University of Illinois, May 28-29, pp. 95-98, 1992.

[4]   R. Van Overstraeten and H. DeMan, "*Measurement of the Ionization Rates in Diffused Silicon p-n Junctions*," Solid-State Electronics, vol. 13, pp. 583-608, 1970.

[5]   E. F. Crabble, J. M. S. Stork, G. Baccarani, M. V. Fischetti, and S. E. Laux, "*The Impact of Non-equilibrium Transport on Breakdown and Transit Time in Bipolar Transistors*," IEDM Tech. Dig., pp. 463-466, 1990.

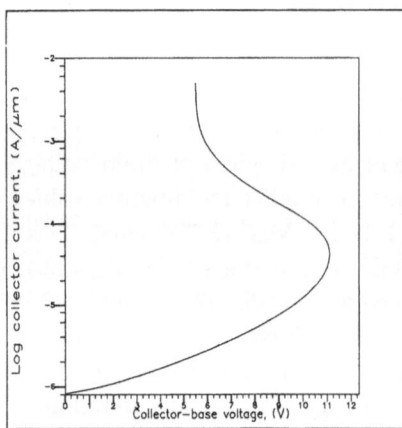

Figure 1.

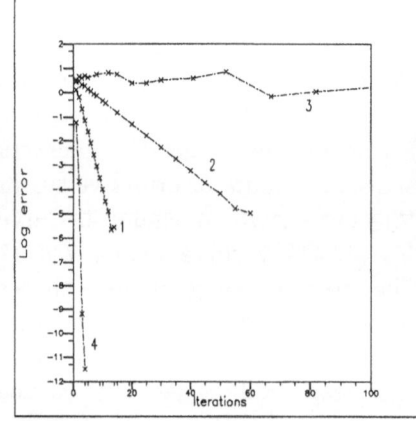

Figure 2.

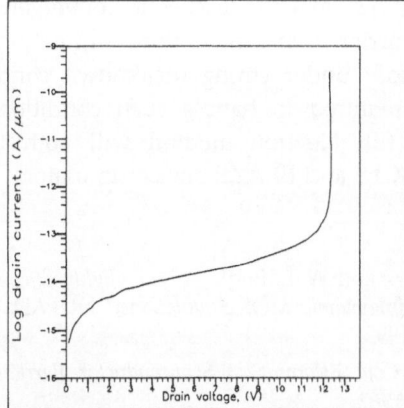

Figure 3.

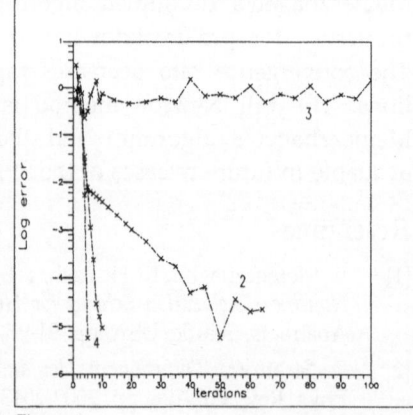

Figure 4.

SIMULATION OF SEMICONDUCTOR DEVICES AND PROCESSES Vol. 5
Edited by S. Selberherr, H. Stippel, E. Strasser – September 1993

On the Scharfetter-Gummel Box-Method

T. Kerkhoven

Department of Computer Science, University of Illinois at Urbana-Champaigne
1304 W. Springfield Avenue, Urbana, IL 61801, USA

Abstract

For a linear potential function one-dimensional constant current drift-diffusion
equations can be integrated in closed form, yielding the Scharfetter-Gummel
(SG) discretization. The box-method generalizes the insistence on exact current conservation to higher dimensions by imposing the exact balancing of
Scharfetter-Gummel fluxes through box-faces.

It has long been recognized that the one-dimensional SG discretization defines a finite element method that yields the exact solution by employing closed
form solutions as an approximant. Finite element analyses of the box-method
tend to employ piecewise linear approximating functions and fail to incorporate
the exact integration properties of the SG discretization.

Nevertheless, the current conservation validates for the SG box-method an
analytical coupling limitation for the differential drift-diffusion equations.

1. The Scharfetter-Gummel Discretization

In the discretization of the one-dimensional zero generation recombination drift-
diffusion equations, the Scharfetter-Gummel technique [9] reproduces exactly the
constant current. Let u denote the electrostatic potential ϕ in units of the thermal potential $U_T \equiv (k_B T)/q$, where k_B is Boltzmann's constant, T is the ambient
temperature, and q is the size of the electron charge. Then, under the assumption of
Einstein's relations the one-dimensional zero generation-recombination drift-diffusion
equation for the conduction electron density n is given by

$$[\mu_n(nu_x - n_x)]_x = 0. \tag{1.1}$$

The solution $n(x)$ to equation (1.1) can be expressed in terms of a closed form Green's
function, analogously to the procedure in [7]. To a piecewise linear potential function
$U_\mathbf{u}(x)$ with nodal values u_j at nodes x_j can be associated a vector $\mathbf{u}$. The Scharfetter-
Gummel expression for the one-dimensional electron current I_n on an interval $[x_{i-1}, x_i]$
with $x_i - x_{i-1} = h_i$ is expressed in terms of the Bernouilli function $B(x) = x/[\exp(x) -
1]$ as in

$$I_{n,j} = -\frac{1}{h_j}\mu_n(u_x)k_B T[B(u_{j+1} - u_j)n_{j+1} - B(u_j - u_{j+1})n_j]. \tag{1.2}$$

As observed in [4] and elsewhere, the SG approximation coincides with a finite element method in which the Slotboom variable $\nu(x) = \exp[-u(x)]n(x)$ is expanded in terms of nodal, $u(x)$-dependent basis-functions

$$\xi_j(x) = \begin{cases} (\int_{x_{j-1}}^x \mu_n \exp[-u(t)]dt)/(\int_{x_{j-1}}^{x_j} \mu_n \exp[-u(t)]dt) & \text{if } x \in [x_{j-1}, x_j], \\ (\int_x^{x_{j+1}} \mu_n \exp[-u(t)]dt)/(\int_{x_j}^{x_{j+1}} \mu_n \exp[-u(t)]dt) & \text{if } x \in [x_j, x_{j+1}], \quad (1.3) \\ 0 & \text{elsewhere.} \end{cases}$$

Let the vector with components n_j solve the SG discretization of (1.1). Then, the function

$$n(x) = \sum_j n_j \exp[u(x) - u_j]\xi_j(x) \tag{1.4}$$

also solves equation (1.1). Hence, solution of the SG discretization of (1.1) for a piecewise linear potential $U_u(x)$ yields the exact solution to (1.1). In terms of quasi-Fermi levels $n(x) = \exp[u(x) - v(x)]$, $p(x) = \exp[w(x) - u(x)]$.

2. Box Method Discretization

The prevalence in two and three-dimensional computational codes of the SG box-method, see e.g. [2, 10, 11, 3], is possibly due to the consistent handling of current conservation. Discretization by the box-method of the Slotboom variable equations

$$-\nabla \cdot [\mu_n \exp(u)\nabla\nu] = 0, \tag{2.1}$$
$$-\nabla \cdot [\mu_p \exp(-u)\nabla\omega] = 0, \tag{2.2}$$

is defined on a mesh of boxes B_k dual to the vertices x_k in a mesh of simplexes. Box-faces f_{jk} are planar. Even though the current is not equal to a constant in higher dimensions, in the Scharfetter-Gummel box-method fluxes through box-faces f_{ij} are approximated analogously to (1.2), yielding for the electron density vector **n**

$$\sum_{x_j \text{ adjacent } x_i} \frac{|f_{ij}|}{|e_{ij}|} \mu_n[B(u_j - u_i)n_j - B(u_i - u_j)n_i] = 0. \tag{2.3}$$

The nodal values of the Slotboom variable ν are then given by $\nu_j = n_j \exp[-u_j]$.

Both in the box-method, and in Galerkin's equations for a piecewise linear approximation V_h, one component ν_j of the solution vector ν corresponds to every vertex x_j of a simplicial mesh. In the sequel, the notation $V_\nu = \sum_j \nu_j \phi_j(x)$ will be employed for the piecewise linear interpolant of the vector of nodal values ν_j at the vertices x_j. With a vector **v** will also be associated the nodal piecewise polynomial function $V_{pp,v} = \sum_j v_j \psi_j(x)$. Finally, define piecewise constant box test-functions $\psi_{B_k}(x)$ that are equal to 1 in the interior of box B_k and 0 elsewhere.

The analysis of the box-method is simplified significantly by reducing (2.1) on each element S_r in the mesh to the Laplacean by replacing the coefficient $\mu_n \exp(u)$ by a function that assumes elemental average values $\overline{\mu_n \exp(u)}$. The boundary conditions are set piecewise linear. In [6] mild conditions are presented under which this simplified BVP approximates the original BVP (2.1) to sufficient accuracy.

Finite element analysis of the box-method commences with the observation (for two dimensions in Bank and Rose in [1], for N dimensions in Lemma 2.3 of [6]) that for box-faces f_{jk} perpendicular to edges e_{jk} in the finite element mesh the perpendicular bisector box-method Laplacean Element Matrix (LES) is identical to the Petrov-Galerkin LES for piecewise linear functions with box test-functions ψ_{B_k}.

The components of this perpendicular bisector box-method LES, E_{B,S_r}, are defined in terms of box-faces $f_{jk}^{(S_r)}$, normal to edges e_{jk} and delimited by the faces F_k of element S_r, by

$$E_{B,S_r,jk} = (|f_{jk}^{(S_r)}|/|e_{jk}|). \tag{2.4}$$

The components of the corresponding global Petrov-Galerkin stress-matrix are defined

$$A_{PG,ij} \equiv \sum_{S_r \text{ adjacent } e_{ij}} \overline{\mu_n \exp(u)}_{S_r} E_{B,S_r,ij}. \tag{2.5}$$

The analysis in [6] relies on piecewise linearity of the approximant in a generalization of the two-dimensional results of Bank and Rose in [1] (see also [5]). By the results in Lemmas 2.1 and 2.2 of [6] the Petrov-Galerkin LES E_{PG,S_r} for a linear approximation and test-functions ψ_i that assume on element faces F_k the average values $(\int_{F_k} \psi_i dx / \int_{F_k} dx) = p_{ik}$ can be expressed in terms of the piecewise linear E_{pl,S_r} and the differences $q_{ik} = p_{ik} - (1/N)$ of the face-averages p_{ik} of the test-functions ψ_i from the piecewise linear averages $\langle \phi_i \rangle_{F_k} = (1/N)$ as in

$$E_{PG,S_r} = [I - NQ_{S_r}]E_{pl,S_r}. \tag{2.6}$$

Here the matrix Q_{S_r} is defined by $Q_{S_r,il} = q_{il}$, the matrix of the deviations from the mean of the face-averages p_{il}. The equivalence $E_{B,S_r} = E_{PG,\psi_B,S_r}$ and equation (2.6) imply immediately (see Corollary 2.4 of [6]) that if boxes B_j partition equally all faces F_k of an N dimensional simplex S_r, then $E_{B,S_r} = E_{pl,S_r}$. This observation combined with equation (2.6) implies that in three dimensions $E_{B,S_r} = E_{pl,S_r}$ and $Q_{S_r} \equiv 0$ if and only if S_r is a regular tetrahedron.

The error analysis in [6] admits this difference in stress matrices subject to the following equivalences of energies defined by the piecewise linear Galerkin LES $E_{pl,S_r} = a_{S_r}(\phi_i, \phi_j)$ and the box-method LES E_{B,S_r} defined in (2.4) (here $\int_{S_r} \nabla f \cdot \nabla g dx = a_{S_r}(f,g)$ and $\mathbf{u}^t D_{S_r} \mathbf{u} \equiv a_{S_r}(V_{pp,u}, V_{pp,u})$.)

$$c_{S_r} \mathbf{u}^t E_{pl,S_r} \mathbf{u} \le \mathbf{u}^t E_{B,S_r} \mathbf{u}, \quad \mathbf{u}^t D_{S_r} \mathbf{u} \le C_{D,S_r} \mathbf{u}^t E_{B,S_r} \mathbf{u}. \tag{2.7}$$

Piecewise polynomial test-functions $\psi_{H,j}(x)$ that assume appropriate face averages are substituted for the $\psi_{B,j}$. Inequality (2.8), below, reflects a special case of Theorem 3.1 in [6].

If $c \le c_{S_r}$, and $C_D \ge C_{D,S_r}$ in (2.7) on all elements S_r. If $\tilde{\nu}$ solves the simplified version of (2.1), and the vector $\mathbf{v}$ solves the approximate box-method (2.5), then V_v realizes a piecewise linear order of accuracy because for all piecewise linear V_w that satisfy identical boundary conditions as V_ν

$$\sqrt{\int_G \overline{\mu_n \exp(u)} |\nabla(V_v - \tilde{\nu})|^2 dx} \le [1 + \sqrt{\frac{C_D}{c}}] \sqrt{\int_G \overline{\mu_n \exp(u)} |\nabla(V_w - \tilde{\nu})|^2 dx}. \tag{2.8}$$

Approximation results from [6] and inequality (2.8) yield for the function $n_\nu(x) = \sum_j n_j \exp[u(x) - u_j]\phi_{pl,j}(x)$, defined in terms of a piecewise linear Slotboom variable $\nu_{pl}(x)$ and the vector $\mathbf{n}$ solving the box-method (2.3), a bound similar to (2.8).

3. Equation Coupling and The Scharfetter-Gummel Box-method

The exact current conservation can be employed to validate for the SG box-method discretization an analogy of a simplified coupling limitation for the drift-diffusion equation for electrons from [8]. The mobilities μ_n and μ_p are assumed to be functions of the location x only. In terms of quasi-Fermi levels the system (2.1–2.2) is written

$$-\nabla \cdot [\mu_n \exp(u - v)\nabla v] = 0, \qquad (3.1)$$
$$-\nabla \cdot [\mu_p \exp(w - u)\nabla w] = 0. \qquad (3.2)$$

For $i = 1, 2$, let u_i be bounded with square-integrable derivative, let v_i be the solution to (3.1). We introduce the averages $\tilde{u} = \frac{1}{2}(u_1 + u_2)$, $\tilde{v} = \frac{1}{2}(v_1 + v_2)$, and the differences $\Delta u = u_2 - u_1$, $\Delta v = v_2 - v_1$. Then for (3.1)

$$\sqrt{\int_G \mu_n \exp(\tilde{u} - \tilde{v})|\nabla \Delta v|^2 dx} \leq \sqrt{\int_G \mu_n \exp(\tilde{u} - \tilde{v})|\nabla \Delta u|^2 dx} \qquad (3.3)$$

The SG box-method discretization of the drift-diffusion equations balances the sum of one-dimensional constant-current expressions for the fluxes through box-faces. The following inequality is valid for quasi-Fermi levels on edges e_{jk}, corresponding to the optimized expansion of the Slotboom variables (1.3).

For $i = 1, 2$, and the vectors u_i, let n_i solve the SG box-method equations (2.3). On each edge e_{ij} in the mesh let u be the linear interpolant of the nodal values u_i and u_j at the vertices x_i and x_j, and let v be the univariate quasi-Fermi level $v(t)$ corresponding to the conduction-electron density function (1.4). Then

$$\sqrt{\sum_{e_{jk}} \int_{e_{jk}} \exp[\tilde{u} - \tilde{v}]|\Delta u_t|^2 dt} \geq \sqrt{\sum_{e_{jk}} \int_{e_{jk}} \exp[\tilde{u} - \tilde{v}]|\Delta v_t|^2 dt}. \qquad (3.4)$$

References

[1] Randolph E. Bank and Donald J. Rose. Some Error Estimates for the Box Method. *SIAM J. on Numer. Anal.*, 24:777–787, 1987.

[2] Randolph E. Bank, Donald J. Rose, and Wolfgang Fichtner. Numerical Methods for Semiconductor Device Simulation. *SIAM J. on Scient. and Statist. Comp.*, 4(3):416–435, September 1983.

[3] E.M. Buturla, P.E. Cottrell, B.M. Grossman, and K.A. Salzburg. Finite-Element Analysis of Semiconductor Devices: The Fielday Program. *IBM J. Res. Develop.*, 25:218–231, July 1981.

[4] Walter L. Engl, Heinz K. Dirks, and Bernd Meinerzhagen. Device Modeling. *Proceedings of the IEEE*, 71(1):10–33, January 1983.

[5] W. Hackbusch. On First and Second Order Box Schemes. *Computing*, 41:277–296, 1989.

[6] Thomas Kerkhoven. Piecewise Linear Petrov-Galerkin Analysis of the Box-Method. *SIAM J. Numer. Anal.*, 20 pages, submitted.

[7] Thomas Kerkhoven. On the Effectiveness of Gummel's Method. *SIAM J. on Scient. & Statist. Comput.*, 9:48–60, January 1988.

[8] Thomas Kerkhoven. *A Computational Analysis Of The Steady State Drift-Diffusion Semiconductor Model.* SIAM, Philadelphia, 1993.

[9] D. Scharfetter and H.K. Gummel. Large signal analysis of a silicon read diode oscillator. *IEEE Trans. Electron Devices*, ED-20:64–77, 1969.

[10] Siegfried Selberherr. *Analysis and Simulation of Semiconductor Devices.* Springer-Verlag, New York, 1984.

[11] Jan W. Slotboom. Computer-Aided Two-Dimensional Analysis of Bipolar Transistors. *IEEE TRans. Electron. Dev.*, ED-20(8):669–679, August 1973.

A Spectral Method for the Numerical Simulation of Transit-Time Devices

M. Schlett

Institute for Applied Mathematics, University of Karlsruhe
Kaiserstraße 12, D-76128 Karlsruhe, GERMANY

Abstract

For the numerical simulation of semiconductor devices driven by a periodic voltage a new numerical approach is presented. The method is based on a temporal Fourier expansion to solve time-dependent nonlinear partial differential equations like the Drift-Diffusion Model. Disadvantages and problems of conventionally used time discretizations are avoided. To achieve high-accuracy results an interval based error-analysis is presented.

1. Introduction

The simulation of semiconductor devices with a periodic driven voltage requires the numerical solution of nonlinear partial differential equations. An often used model is the Drift-Diffusion Model (DDM). Typical applications are the simulation of IMPATT-, PIN- or BARITT-diodes, [1]. The usual way to approximate time derivatives is the discretization via Finite Differences. To obtain the periodic solution a decoupled one-step iteration is necessary, [2]. Besides of many required iteration cycles all known problems of Finite Difference Methods like low-order discretization are occuring. In order to avoid these problems a new method, the so-called t-Fourier Method (tFM), is presented here. Most theoretical aspects are common to the $\bar{x}$-Fourier Method [3]. With the formulated tFM all usual semiconductor equations describing time and space dependent transport phenomena of electrons and holes can be treated. For simplicity all subsequent details are performed for the DDM. For derivation, parameter selection and boundary values see [4]. The DDM can be written in the following manner

$$\frac{\partial}{\partial t} \begin{pmatrix} 0 \\ u_2 \\ u_3 \end{pmatrix} + F(u, \nabla u, \Delta u) = 0, \quad u = (u_1, u_2, u_3)^T \tag{1}$$

involving Poisson equation and continuity equations for electrons n and holes p. The electrostatic potential $u_1 = \varphi$ is determined at the contacts by the applied sinusoidal voltage $U(t) = U_0 + U_1 \sin(\omega t)$.

2. t-Fourier Method

The method presented is based on a temporal Fourier expansion of the solution vector

$$u(\vec{x}, t) \;=\; \sum_{k=-\infty}^{\infty} \hat{u}_k(\vec{x}) e^{-ik\omega t} \tag{2}$$

This approach fullfills the periodicity condition automatically. Inserting (2) into the semiconductor equations (1) and applying the inner product $< .,. >$ of the underlying Sobolev-space the equations are transformed to an infinite system of nonlinear differential equations for the space dependent Fourier coefficients $\hat{u}_k$. This can easily be seen by utilizing the main features of spectral series like linearity, convolution, and transformation of time-derivatives into algebraic terms. The Fourier Spectrum of the nonlinear transcendental functions $\mu_{n,p}, G, R$ can be calculated in t-space.

Boundary conditions have to be transformed analogously. Thus, we obtain the Fourier-Galerkin coefficients $F_k :=< F, e^{-ik\omega t} >, \quad k = -\infty, ..., \infty$ of the basic operator F. A solution vector $(\hat{u}_k)_{k=-\infty,...,\infty}$ now has to fullfill the following equations:

$$G_k := -ik\omega(0, \hat{u}_{2k}, \hat{u}_{3k})^T + F_k \;=\; 0, \quad k = -\infty, ...\infty \tag{3}$$

3. Numerical Solution Procedure

For numerical treatment the infinite sum (2) is approximated by the M-th partial Fourier sum $(k = -M, ..., M)$. Transcendental functions are calculated in t-space transformed with efficient FFT-techniques. Hence, one ends up solving a finite nonlinear system of differential equations. They can be discretized with well-known techniques like the Finite Element, Finite Difference or $\vec{x}$-Fourier Method. This leads to an algebraic nonlinear system G^M which is solvable with an efficient Newton algorithm:

$$u_0 \;:=\; (\tilde{u}_k, k = -M, ..., M) \tag{4}$$

$$u_{j+1} \;:=\; u_j - (\frac{dG^M}{du}(u_j))^{-1} G^M(u_j), \quad j > 0 \tag{5}$$

The Jacobian matrix dG^M/du depends on the $\vec{x}$-discretization used. The examples illustrated below were calculated with a classical Scharfetter-Gummel scheme which leads to a tridiagonal block matrix.

4. Examples

Some results for a Si-IMPATT diode are presented. The diode has a $N_D^+ N_D N_A N_A^+$ geometry (double drift) with $N_D = 10^{17}$ $1/cm^3$ (300 nm), $N_A = 1.25 \cdot 10^{17}$ $1/cm^3$ (300nm). The contacts were chosen to $N_D^+ = N_A^+ = 10^{18}$ $1/cm^3$. For simplicity external circuits are neglected and the DDM is used in an one-dimensional formulation. Fig.1-3 show the electric field E and the electron and hole concentration n and p at the 4 significant time-points $t_1 = 0, t_2 = T/4, t_3 = T/2, t_4 = 3T/4, T = 2\pi/\omega$. The parameters were chosen to $U_0 = 24V, U_1 = 10V, f = 1/T = 60GHz, J_{DC} = 28.4kA/cm^2$. The solid line represents the solution with $M = 32$, dots $M = 10$. The results illustrate the very fast convergency of the Fourier coefficients. The tFM is unconditionally stable. The Newton algorithm requires only few iterations and no damping technique.

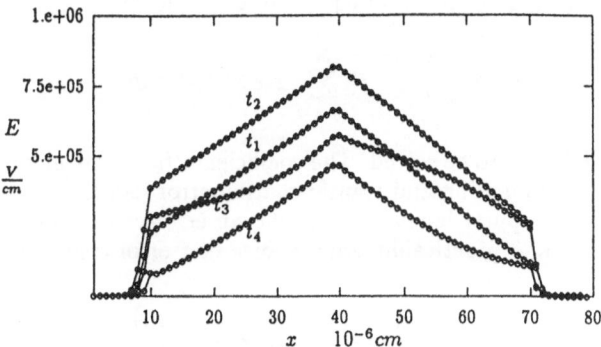

Figure 1: Electric Field $E(x, t_j)$

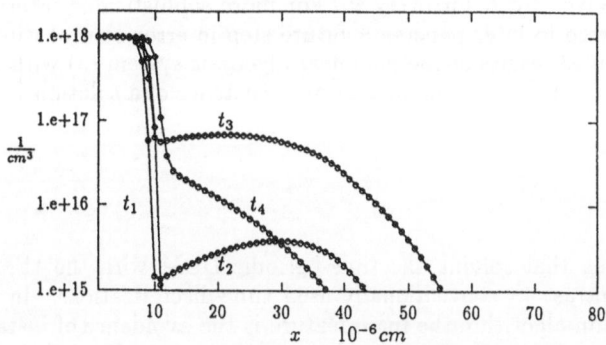

Figure 2: Electrons $n(x, t_j)$

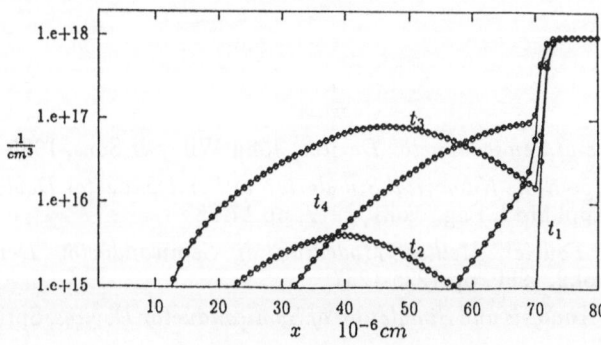

Figure 3: Holes $p(x, t_j)$

5. Interval based Error Control

The main advantage of Spectral Methods is the theoretical representation of the solution up to a very high accuracy. In the case of the semiconductor equations this attribute is impaired by rounding and transformation error when treating the transcendental functions. An easy way to control these errors is the usage of Intervals [5]. During the application of the tFM equations of the form

$$f_k = \frac{1}{T}\int_0^T f(\sum_{j=-M}^{M} a_k e^{-ij\omega t})e^{ik\omega t}\, dt \tag{6}$$

with known a_k and f have to be solved. The coefficients f_k are calculated numerically via FFT-techniques. Rounding and transformation error can lead to unpredictable results. With XSC-Computer languages, e.g. [6], an error-control with intervals can easily be implemented. A practicable formula for a first error control is given with

$$R := f([0,T]) - \sum_{k=-M}^{M} \widetilde{f_k} e^{-ik\omega[0,T]} \tag{7}$$

The error of the numerically calculated Fourier coefficients $\widetilde{f_k}$ and the exact ones lie between bounds given by R for every k. For more sophisticated interval methods the reader is referred to later papers. A future step in error control will also be the calculation of verified results of the nonlinear algebraic system (3) with an Interval-Newton algorithm to prove the uniqueness and existence of a solution within narrow interval bounds, [7].

6. Conclusion

It has been proven that solving the time-periodic DDM with the tFM has many advantages in contrast to conventionally used time-discretizations. In addition to the efficient Newton algorithm the main feature is the avoidance of instabilities and diffusion problems. The most impressive result is the very low number of Fourier coefficients required to represent the solution with sufficient accuracy. With the posibilities of interval computation in error analysis Spectral Methods can lead to a new dimension in accuracy. Hence, the tFM is also an interesting alternative in the transient case.

References

[1] S. Sze, *Physics of Semiconductor Devices*, John Wiley & Sons, 1985

[2] M. Reiser, *Large-Scale Numerical Simulation in Semiconductor Device Modelling*, Comp.Meth.Appl.Mech.Eng., vol.1, 1972, pp 17-38

[3] V. Axelrad, *Fourier Method Modelling of Semiconductor Devices*, IEEE Trans.CAD, vol.9, 1990, pp 1225-1237

[4] S. Selberherr, *Analysis and Simulation of Semiconductor Devices*, Springer-Verlag, 1984

[5] G. Alefeld, J. Herzberger, *Introduction to Interval Computations*, Academic Press, 1983

[6] R. Klatte et al. , *PASCAL-XSC, Language Reference*, Springer-Verlag, 1992

[7] E. Kaucher, W. Miranker, *Self-Validating Numerics for Function Space problems* Academic Press, 1984

SIMULATION OF SEMICONDUCTOR DEVICES AND PROCESSES Vol. 5
Edited by S. Selberherr, H. Stippel, E. Strasser – September 1993

PARDESIM — A Parallel Device Simulator on a Transputer Based MIMD-Machine

O. Kalz and D. Schröder

Technische Elektronik, TU Hamburg-Harburg
Eißendorfer Straße 38, D-21073 Hamburg, GERMANY

Abstract

In this paper we present a parallel 3D device simulator on a MIMD[1]-machine. We have investigated the efficiency of a parallel implementation, especially for the fine-grained parallelized iterative linear solvers like ILU-BCG[2]. As a result we have found that an efficient employment of some 10^1 processors is possible on a MIMD-computer with the same convergence behaviour as on an adequate large sequential computer.

1. Introduction

3D-semiconductor device simulation belongs to the field of numbercrunching data processing. This means that high computing performance and a large memory are necessary. Thus, vectorization and parallelism came early into the scope of interest for satisfying the increasing requests to solve the classical semiconductor equations on a discretization grid of $10\,000\ldots250\,000$ points.

First at all, efficient vector algorithms for the most CPU consuming iterative or direct linear solvers, like ILU-CG, ILU-BCG, or CGS, have been developed and applied [1]. But future requirements can be satisfied only, or at least in a much cheaper way, by massively parallel computers [2]. Traditional shared memory machines can neither solve the memory problem nor allow a fully scalable program, which means an efficiency nearly independent of the number of processors.

Unfortunately, distributed memory algorithms are harder to develop especially under usage of fine grain parallelism, because one has to organize the data exchange between the nodes with regard to the limited communication bandwidth and a noticeable communication setup time.

In consideration of this background we will present the parallelization strategy of the parallel device simulator PARDESIM, which is under development[3] on a SuperCluster SC-128 multiprocessor of Parsytec. Some parameters of the transputer based parallel computer are listed in Table 1.

---

[1]Multiple Instruction Multiple Data
[2]Incomplete LU Decomposition and Biconjugate Gradient Method
[3]This work was sponsored by the "Deutsche Forschungsgemeinschaft"

2. Implementation of the parallel algorithm

In order to concentrate on parallel algorithms, we restricted hitherto the simulator to a steady-state solver of the traditional semiconductor equation system with two carrier equations on a rectangular tensor mesh in 2 or 3 dimensions [3]. While the discrete nonlinear system is solved successively within the *Gummel* iteration [4], we chose preconditioned gradient methods like ILU-CG and ILU-BCG for the linear systems [5].

To use the advantage of distributed memory computers, we have to implement a domain decomposition algorithm that runs on some 10^1 processors with a high speedup S_p and a sufficient efficiency E_p, defined as follows:

$$S_p = \frac{t_{serial}}{t_{parallel}} \quad \text{on p processors (1)} \qquad E_p = \frac{S_p}{p} \qquad (2)$$

We have measured that the iterative solution of the linear systems takes about 80% of total CPU time, which in turn divides into 40% for the ILU preconditioner, 40% for the matrix vector multiplications, and 20% for local vector operations without communication effort. The hardest problem is to make the incomplete LU decomposition and forward and backward substitution parallel for sparse band matrices, because the algorithm seems to be inherently sequential. But the special matrix block structure (e.g. Fig. 1) caused by the discretization scheme gives us the chance for a time-displaced ILU decomposition (by one row) of each block in parallel. For that purpose the processors are arranged in a ring topology and the matrix blocks are mapped on them in a natural ordering (Fig. 2a).

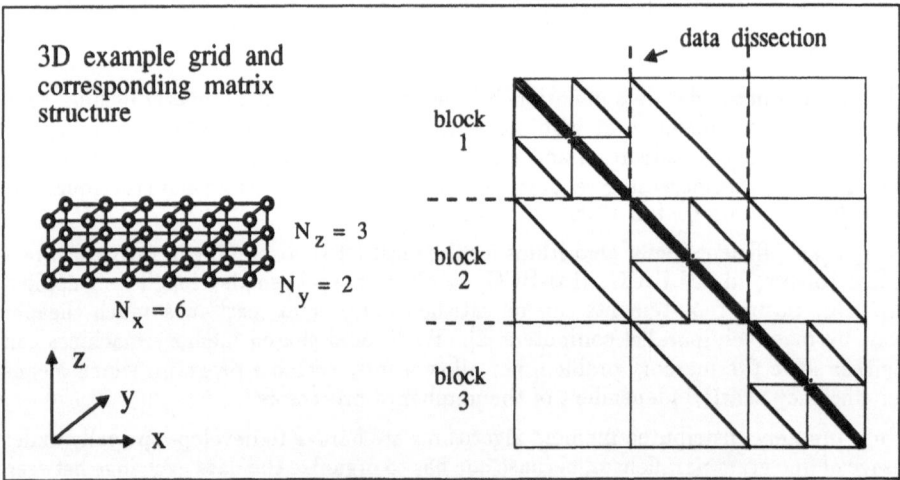

Figure 1: Nonzero matrix structure for an example grid

For the *incomplete* decomposition of the n-th row of one block only the results up to the (n-1)-th row of the *preceding* block and local block data are needed. The inclusion of the block coupling outer diagonals saves a good convergence behaviour but limits the degree of parallelism. As a theoretical maximum speedup resp. efficiency for ILU decomposition on N_B available processors without communication costs we find

$$S_{N_B,ILU} = \frac{N_B \cdot B_{Dim}}{B_{Dim} + N_B - 1} \quad (3) \qquad E_{N_B,ILU} = \frac{B_{Dim}}{B_{Dim} + N_B - 1} \quad (4)$$

where N_B is the number of blocks (N_z of a 3D-grid) and B_{Dim} is the block dimension ($N_x \cdot N_y$ of a 3D-grid).

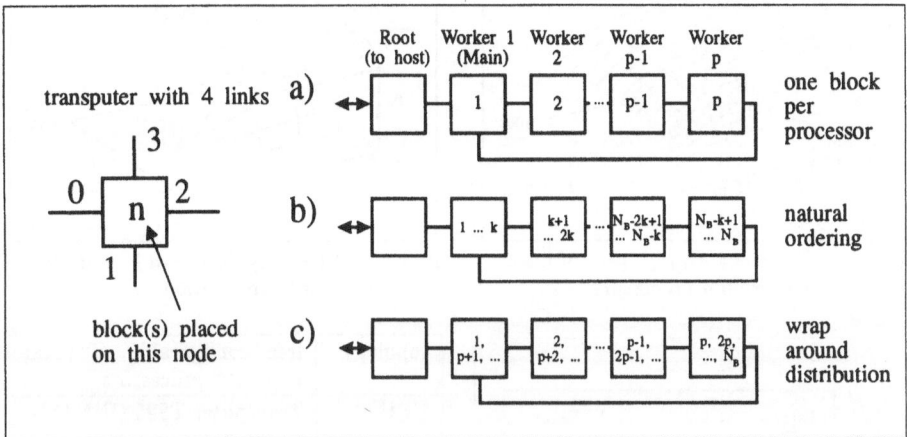

Figure 2: Topology of processor network and block mapping example

If there are more blocks than processors we use a mapping scheme that places neighbouring block-processes on the same processor (Fig. 2b), instead of a wrap around distribution (Fig. 2c) as usual for dense matrices. This profits from a partly drastic reduction of interprocessor communication effort and an increasing job size per processor. In this case, the estimations (3), (4) for $p = N_B$ have to be replaced by the following formulas for $p \leq N_B$, with k the number of blocks per processor.

$$S_{p,ILU} = \frac{B_{Dim} \cdot p}{B_{Dim} + p - 1} \quad (5) \qquad E_{p,ILU} = \frac{B_{Dim}}{B_{Dim} + p - 1} \quad (6)$$

From (6) one can see that the mapping of neighbouring blocks on the same processor does not decrease the ILU efficiency compared with the case of one block per processor (4). Nevertheless the maximum theoretical speedup can only be reached for a large block dimension B_{Dim} and is limited by N_B, i.e. the maximum number of grid coordinates in one space direction.

3. Numerical results

As a result we have found that the described parallelization strategy leads to a sufficient efficiency if the job size per transputer is large enough (Fig. 3). In Fig. 4 the measured efficiency of the critical forward backward substitution on 4 processors is compared with the theoretical maximum. Especially for $N_B = p = 4$, i.e. one block per processor, the estimation (6) is too optimistic because already few arithmetic operations require one communication step in this case. Finally Fig. 5 shows that a scalability of the complete linear solver on a different number of processors can be observed if the matrix size, i.e. the number of grid points, increases proportional to the number of used processors. Otherwise the efficiency decreases rapidly if the matrix size remains constant.

As a next implementation step the distributed matrix assembly and the nonlinear iteration scheme has to be implemented in parallel in order to achieve the aspired speedup for the complete simulator.

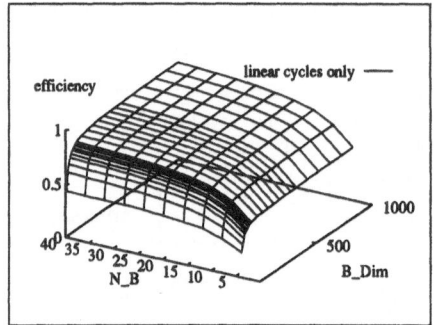

Figure 3: Efficiency $E_{4,ILU-BCG}$ of linear solver on 4 processors

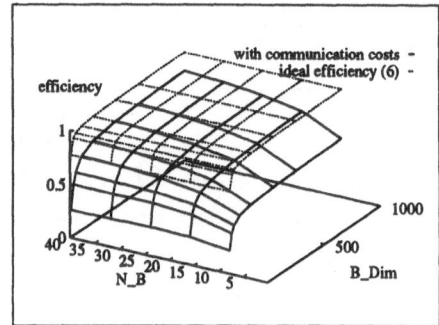

Figure 4: Efficiency $E_{4,ILU}$ of forward backward substitution

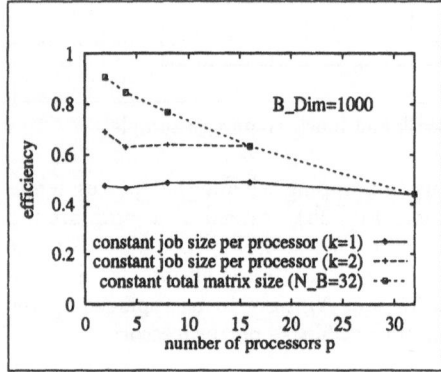

Figure 5: Efficiency $E_{p,ILU-BCG}$ of a variable number of processors

topology	free configurable of maximum 128 processors
CPU	Transputer T805 (INMOS) 32 bit RISC with FPU
clock	30 MHz
peak performance	about 2 MFLOPS per processor
memory	4 MByte + internal memory
communication	4 links per transputer (each 20 Mbit/s)
compiler	Par.C (C with parallel extensions)

Table 1: Parameters of multi processor system SuperCluster SC-128

References

[1] Heiser, G.; Pommerell, C.; Weis, J.; Fichtner, W.: Three-Dimensional Numerical Semiconductor Device Simulation: Algorithms, Architectures, Results, IEEE CAD-10 (1991), No. 10, p. 1218-1231

[2] Webber, D.; Tomacruz, E.; Guerrieri, R; Toyabe, T.; Sangiovanni-Vincentelli, A: A Massively Parallel Algorithm for Three-Dimensional Device Simulation, IEEE CAD-10 (1991), No. 9, p. 1201-1210

[3] Selberherr, S.: Analysis and Simulation of Semiconductor Devices, Springer-Verlag Wien - New York 1984

[4] Gummel, H.K.: A Self-Consistent Iterative Scheme for One-Dimensional Steady State Transistor Calculations, IEEE ED-11 (1964), p. 455-465

[5] van der Vorst, H.K.; Dekker, K.: Conjugate gradient methods and preconditioning, Journal of Computational and Applied Mathematics 24 (1988), North Holland

[6] Wu, K.-C.; Chin, G.R.; Dutton, R.W.: A STRIDE Towards Practical 3-D Device Simulation - Numerical and Visualisation Considerations, IEEE Trans. CAD 10 (1991), No. 9, 1132-1140

An Efficient and Accurate Method to Calculate the Two-Dimensional Scattering Rates in Heterostructure Semiconductors

A. Abou-Elnour and K. Schuenemann

Arbeitsbereich Hochfrequenztechnik, TU Hamburg-Harburg
Postfach 901052, D-21050 Hamburg 90, GERMANY

Abstract

An accurate and efficient method is investigated to solve Schrödinger's equation by using variational techniques. The electronic states inside heterostructure devices are determined by a self-consistent solution of Poisson's and Schrödinger's equations. The closed forms of the wave functions are used to calculate the two-dimensional scattering rates in these structures. The computational efficiency is compared to that of conventional finite difference models.

1. Introduction

The existence of the two-dimensional electron gas (2DEG) in the vicinity of a heterointerface inside heterostructure semiconductors requires an accurate method to calculate the scattering rates and to determine the carrier transport properties in these structures. We believe that the solution of Schrödinger-Poisson system of equations will become a common simulation tool for ultra-small and heterostructure devices.

A number of authors investigated different models [1-3] which were based on the finite difference approach to solve Poisson's and Schrödinger's equations self-consistently. We have solved Schrödinger's equation by using variational methods to obtain the wave functions in terms of a number of expansion functions [4]. In the present work, this method is applied to determine the electronic states in an $AlGaAs/GaAs$ single-well heterostructure by solving Schrödinger's and Poisson's equation self-consistently. The two-dimensional scattering rates are then advantageously calculated using the obtained closed forms of the wave functions. The computational speed of this method is compared to that of conventional numerical models.

2. Model

The effective mass, one-dimensional Schrödinger equation is given by

$$-\frac{\hbar^2}{2\,m^*}\frac{\partial^2\psi(x)}{\partial x^2} + V(x)\,\psi(x) = E\,\psi(x),\qquad(1)$$

where $V(x)$ means potential energy, E Eigenenergy, $\psi(x)$ wave function corresponds to the eigenenergy E, m^* effective mass, and $\hbar$ Planck's constant. For a semiconductor structure of width a, the eigenfunction satisfies the boundary conditions $\psi(0) = 0$, $\psi(a) = 0$. The wave functions can be expanded as

$$\psi_k = \sum_{n=1}^{N} a_{nk} \sin\left(\frac{n \pi x}{a}\right). \tag{2}$$

The accuracy of the solution depends on the number of Rayleigh-Ritz functions N. If N is infinite, the obtained wave functions are identical to the true ones. However, a finite N still leads to very good accuracy. The coefficients a_{nk} are obtained by means of variational integrals whose stationary values correspond to the true eigenvalues when the true eigenfunction are inserted in the integral. The variational integral for E is given by

$$E = \frac{\int_0^a \left(\frac{\hbar^2}{2m^*}\left(\frac{d\psi_k}{dx}\right)^2 + V(x)\,\psi_k^2\right)\,dx}{\int_0^a \psi_k^2\,dx} \tag{3}$$

The condition that (3) should be stationary is satisfied if the first-order variation in E vanishes for an arbitrary first-order variation $\delta\psi$ in ψ_k. Applying this condition, the following set of equations is obtained:

$$\sum_{n=1}^{N} a_{nk}\left(T_{ln} - \frac{2\,m^*}{\hbar^2}E_k\,\delta_{ln}\right) = 0 \qquad l = 1, 2, \ldots, N \tag{4}$$

$$T_{ln} = T_{nl} = \int_0^a \left(\frac{df_n}{dx}\frac{df_l}{dx} + \frac{2\,m^*}{\hbar^2}V(x)\,f_l\,f_n\right)\,dx \tag{5}$$

Solving these equations, the subband energies and the corresponding wave functions are determined. The electrostatic potential is then calculated by solving Poisson's equation. Knowing the electrostatic potential, the new potential energy function is calculated. For the next iteration, the effective potential energy function is expressed as a linear combination of its new and old values. The potential energy function is used to determine the wave functions which are then used to recalculate the carrier distribution. The procedure is repeated until initial and final values of $V(x)$, within the same iteration, differ by less than a specified error.

3. Two-dimensional scattering rates

The two dimensional scattering rates are calculated by defining the matrix element for scattering between the ith and the jth subbands according to

$$|M_{ij}|^2 = \int |M(Q, q)|^2\,|I_{ij}(q)|^2\,dq \tag{6}$$

where Q, q are the phonon wave vector components in parallel and normal to the hetero-interface, and $I_{ij}(q)$ means overlap integral

$$I_{ij}(q) = \int \psi_i(x)\,\psi_j(x)\,\exp(i\,q\,x)\,dx \tag{7}$$

$\psi(x)$ is the normalized envelope wavefunction. The closed form of the wavefunctions (2) decreases the required CPU time for calculating the 2D scattering rates compared to that of numerical integrations [1].

4. Application to an $Al_{0.7}Ga_{0.3}As/GaAs$ heterojuction

An $AlGaAs/GaAs$ heterojuction is considered with $N = 10^{21}\,cm^{-3}$ in a 0.09 μm $GaAs$ layer and $N = 10^{23}\,cm^{-3}$ in a 0.018 μm $AlGaAs$ layer. Both Rayleigh-Ritz and finite difference methods are applied to calculate the subband energies and the corresponding wavefunctions (Fig. 1). The numerical efficiency of the finite difference method is deteriorated by discretization and mesh size [3]. The wave functions are just numerically obtained so that any further application of these wave functions to calculate the scattering rates requires large CPU time because all quantities have to be calculated numerically [1]. Using Rayleigh-Ritz method, the required CPU time (Fig. 2b) to calculate the 2D scattering rates (Fig. 3) versus the number of subbands is nearly constant while it greatly changes using the finite difference method (Fig. 2a). This makes the application of the present method more practical in particular for device simulation.

5. Conclusion

An efficient variational method is applied to determine the electronic states inside heterostructure semiconductors by a self-consistent solution of Poisson's and Schrödinger's equation. Using this technique, the wave function is obtained in closed form which decreases the CPU time required for the calculation of the two-dimensional scattering rates. Moreover, the present method overcomes the limitations of the finite difference method which arise from mesh size and discretization.

Acknowledgement

The authors are thankful to Prof. Dr. A. S. Omar for fruitful discussions and to the Deutsche Forschungsgemeinschaft for financial support.

References

[1] K. Yokoyama, J. Appl. Phys. 63, 938 (1988).

[2] A. Cruz and H. Abreu Santos, J. Appl. Phys. 70, 2734 (1991).

[3] I.-H. Tan, G. Snider, L. Chang, and E. Hu, J. Appl. Phys. 68, 4071 (1990).

[4] A. Abou-Elnour and K. Schuenemann, will be published in Proc., MTT-S'93.

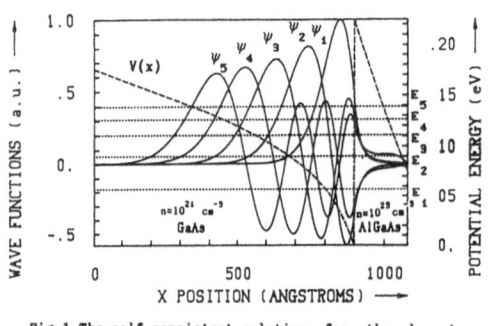

Fig.1 The self-consistent solution for the lowest five subband energies (dotted lines), the corresponding wave functions (solid lines), and the potential energy.

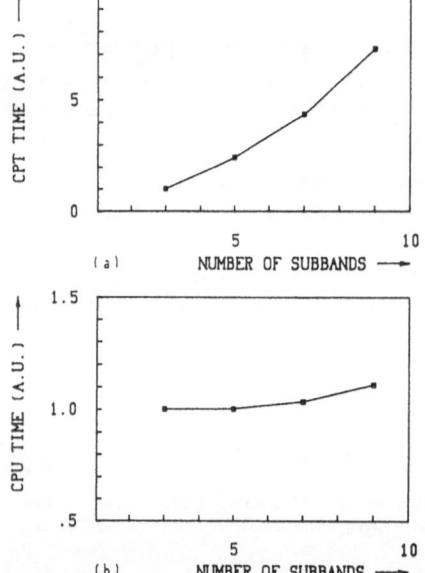

Fig.2 The required CPU time to calculate the 2D scattering rates versus the number of subbands.
a) finite difference method b) Rayleigh-Ritz method.

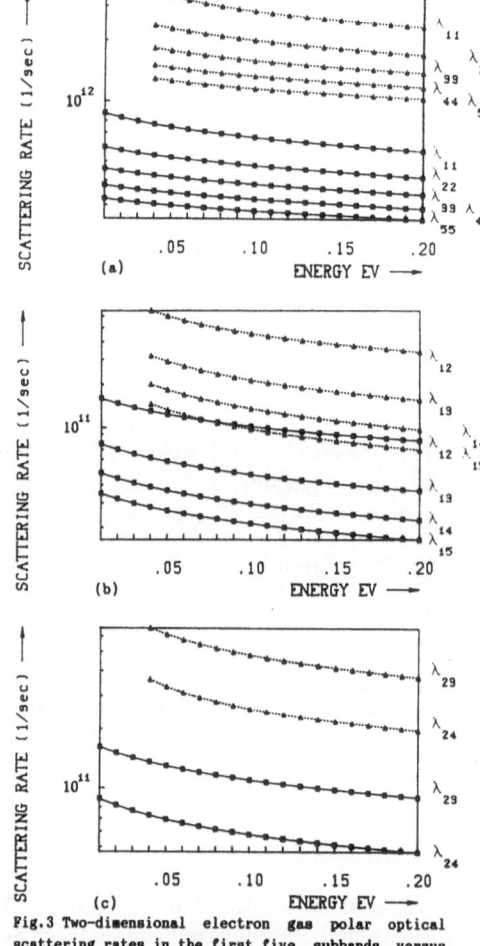

Fig.3 Two-dimensional electron gas polar optical scattering rates in the first five subbands versus electron energy. Solid lines stand for phonon absorption and dotted lines for phonon emission.
(a) intrasubband scattering.
(b) intersubband scattering for the first subband.
(c) intersubband scattering for the second subband.

Two-Dimensional Numerical Analysis on the Diffusion-Induced Degradation of AlGaAs/GaAs Heterojunction Bipolar Transistors

S. Hong, J. Kim, J. Lee, C. Park[†], J. Lee[†], and T. Won

Department of Electronic Materials and Devices, College of Engineering,
Inha University
253 Yonghyundong, Namgu, Inchun 402-751, KOREA
[†]Electronics and Telecommunications Research Institute
P.O. Box 8, Daeduck Science Town, Daejun 305-606, KOREA

Abstract

In this report, the effect of the beryllium *in situ* and *ex situ* outdiffusion into the emitter on the device characteristics is investigated by employing two-dimensional device simulators. It revealed that the current driving capability and RF performance are greatly affected by the beryllium diffusion due to thermal stress.

1. Introduction

Search for high speed electronic devices over the past few years has resulted in the demonstration of field-effect transistor (FETs) and heterojunction bipolar transistors (HBTs) with excellent microwave and millimeter-wave performance. As the circuit frequency limits of FETs are being approached, HBTs are gaining increasing attention in millimeter wave and high-speed digital applications because of their inherent high current handling capacity and high operation frequency. To achieve an outstanding power characteristics with high f_{MAX}, the base doping concentration of the typical AlGaAs/GaAs HBTs is made higher than 1×10^{19} cm^{-3} with extremely thin base thickness. To achieve this relatively high doping concentration for the base, Be are widely used as p-dopant for MBE and MOCVD growth, respectively [1]. However, the Be outdiffusion during epitaxial growth has been observed and was found to result in the redistribution of dopant profile at the heterointerface. In addition to the outdiffusion of Be during the growth, thermal stress at the high collector current densities can cause Be redistribution when the junction temperature is raised [2]. For the optimum design of the AlGaAs/GaAs HBTs in the microwave and millimeter-wave ranges, the influence of device characteristics on the Be redistribution due to thermal stress should be thoroughly investigated. In this study, we simulated device characteristics with redistributed Be diffusion profile in the base caused by the thermal stress.

2. Device Simulations

The structure of the simulated device consists of a 0.5-μm Si-doped GaAs subcollector, a 0.3-μm Si-doped collector with a doping concentration of 2×10^{17} cm^{-3}, 0.1-μm Be-doped GaAs base with a doping concentration of 1×10^{19} cm^{-3}, followed by a 0.2-μm Si-doped

AlGaAs emitter. On top of the emitter, a 0.2-μm GaAs layer with a doping concentration of 1×10^{19} cm^{-3} is assumed for ohmic contact. In addition, an undoped GaAs spacer of 300 Å is inserted between n-AlGaAs emitter and p-GaAs base.

In order to estimate device characteristics, a series of devices with simulated Be redistributed profile were employed. For instance, a sum of linear combinations of Gaussian profiles was employed to simulate diffused Be profile due to the thermal stress. The Be diffusion under the thermal stress shifts the junction into the AlGaAs region, which degrades the emitter injection efficiency of an HBT. The Be diffusion into the emitter region seems to form the potential barrier at the conduction band near the emitter-base junction. In this calculations, the assumed penetration depths of Be outdiffused into the emitter are 125, 250, 450, and 680 Å, respectively. The simulated device parameters of each structure are shown in Table 1.

Table 1. The parameters of the HBT simulated in this study influenced by Be diffusion at the emitter-base junction.

Devices	V_{BE} (V) (Ic=1μA)	V_{CEoff} (mV)	h_{fe} (Ic=0.1mA)	f_t (GHz) (Ic=0.1mA)	f_{max} (GHz) (Ic=0.1mA)
Structure #1	1.2	28	107	10	1
Structure #2	1.3	83	60	12	2
Structure #3	1.4	168	10	13	6
Structure #4	1.4	183	8	14	7

Two-dimensional device simulation revealed that one of the effects of Be outdiffusion into the emitter is the shift of the turn-on voltage at the emitter-base junction as shown in Table 1. The turn-on voltage changes from 1.2 to 1.4 V as the penetration depth varies from 125 to 680 Å due to the shift of the emitter-base junction into the AlGaAs region where the bandgap is the larger. Our calculation exhibits that a shift of turn-on voltage of 0.48 mV corresponds to approximately 1 Å of Be diffusion. In addition to the increase of the turn-on voltage of the emitter junction, the collector-emitter offset voltage in the common emitter output characteristics increases with the degree of Be outdiffusion. This seems to caused by the decrease of the reverse saturation current due to raised bandgap.

As shown in Fig. 1, the DC characteristics of AlGaAs/GaAs HBTs exhibits strong dependence on the redistributed Be diffusion profile. In other words, the deeper the Be dopant penetrates into the emitter, the lower common emitter current gain is obtained. This seems due to the decrease of the emitter injection efficiency. The calculated values of the common emitter current gain of the simulated devices varies by two orders of magnitude, as shown in Table 1. This implies the thickness of the spacer should be chosen such that the distance of Be outdiffusion is confined to the spacer.

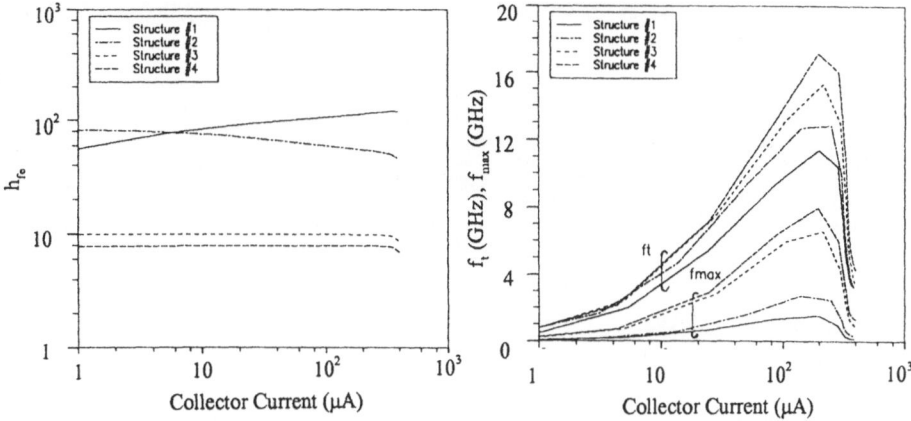

Fig. 1. A plot showing the dependence of the common emitter current gain on the redistributed Be diffusion profile under the thermal stress.

Fig. 2. A plot showing the dependence of cutoff frequency and maximum oscillation frequency on the Be diffusion.

Finally, the high-speed performance of the AlGaAs/GaAs HBTs is affected by the thermal stress. As shown in Fig. 2, the calculated cutoff frequencies changes from 10 to 14 GHz at the collector current of 0.1 mA as the penetration depth increases. Simultaneously, the maximum oscillation frequencies increase from 1 to 7 GHz with the penetration depth at the same collector current level. This seems to be due to the increase of the spike at the heterointerface, which results in the accelerated electrons launching into the base according to the calculation of the band diagram.

3. Conclusion

We report the study on the degradation of AlGaAs/GaAs HBT by thermal stress by employing two-dimensional simulators. Our simulation results revealed that the DC and RF characteristics are strongly affected by the *in situ* and *ex situ* thermal stress. The authors would like to express special thanks to Dr. Tim Crandle at Silvaco International for the encouragement and helpful discussions. This work was supported by ISRC-92-E-0023, Inha University, and ETRI.

References

[1] N. Jourdan, F. Alexandre, Chantal Dubon-Chevallier, Jean Dangla, and Y. Gao, "Heavily Doped GaAs(Be)/GaAlAs HBT's Grown by MBE with High Device Performances and High Thermal Stability," IEEE Trans. Electron Devices, vol.39, pp.767-770, 1992.

[2] Kazuhiro Mochizuki, Seiichi Isomae, Hiroshi Masuda, Tomonori Tanoue and Chuushiro Kusano, "Stress Effect on Current-Induced Degradation of Be-Doped AlGaAs/GaAs Heterojunction Bipolar Transistors," Jpn. J. Appl. Phys. vol.31, no.3, pp.751-756, 1992.

High Speed Performance of Si Homo- and $Si/Si_{1-x}Ge_x$ Heterojunction Bipolar Transistors

W. Molzer[‡†], T. F. Meister[‡], and S. Marksteiner[†]

[‡]Corporate Research and Development, Siemens AG
Otto-Hahn-Ring 6, D-81739 München, GERMANY
[†]Institut für Theoretische Physik, Leopold-Franzens-Universität Innsbruck
A-6020 Innsbruck, AUSTRIA

Abstract

One intention of our investigations was to compare devices with base profiles resulting from different types of base formation. The first type of base profiles used in the simulations (type I) is fabricated conventionally by ion implantation [1], the second by epitaxial growth of Si (type II). An epitaxially grown base offers the possibility of profiles which show a significantly increased gradient of the base dopant concentration towards the collector. This is the feature which we are mainly interested in. In terms of numbers this means 17 nm/decade in the case of an epitaxial base (even 10 nm/dec are possible) compared to a minimum of 40 nm/dec with an implanted one, as seen in Fig. 1.

A larger collector-side gradient of the base profile results in a thinner base necessary to reach a certain doping level starting from a given level at the collector-base junction, where we assumed a figure of $1.8 \cdot 10^{17}$ cm^{-3} for all our devices. A thinner base reduces base transit time. The grading also causes an accelerating electric field in the quasi-neutral base, which further reduces base transit time.

In fixing the value of the base sheet resistance, the base transit time τ_B can be reduced by raising the doping level in the base and simultaneously shrinking its width. This is explained by the fact that the base transit time depends much more strongly on the base

1. Introduction, Doping and Germanium Profiles

One intention of our investigations was to compare devices with base profiles resulting from different types of base formation. The first type of base profiles used in the simulations (type I) is fabricated conventionally by ion implantation [1], the second by epitaxial growth of Si (type II). An epitaxially grown base offers the possibility of profiles which show a significantly increased gradient of the base dopant concentration towards the collector. This is the feature which we are mainly interested in. In terms of numbers this means 17 nm/decade in the case of an epitaxial base (even 10 nm/dec are possible) compared to a minimum of 40 nm/dec with an implanted one, as seen in Fig. 1.

A larger collector-side gradient of the base profile results in a thinner base necessary to reach a certain doping level starting from a given level at the collector-base junction, where we assumed a figure of $1.8 \cdot 10^{17}$ cm^{-3} for all our devices. A thinner base reduces base transit time. The grading also causes an accelerating electric field in the quasi-neutral base, which further reduces base transit time.

In fixing the value of the base sheet resistance, the base transit time τ_B can be reduced by raising the doping level in the base and simultaneously shrinking its width. This is explained by the fact that the base transit time depends much more strongly on the base width than on the reduction of the diffusion constant, due to more dopant which alone would increase it. Yet the distance required to reach a certain base doping level is determined by the gradient of the base profile as stated above. If the base width is made smaller, the sheet resistance cannot be maintained. Other limits are given by the rise of EB-junction capacitance and reduction of EB-breakdown voltage with an increase of base doping level. We therefore consider profiles with a maximum doping level in the base which is not higher than $5 \cdot 10^{18}$ cm^{-3}.

For the epitaxially grown base, we discuss the improvements made possible by introducing linearly graded Ge into the base, reaching a maximum content of 20% (type III). The addition of Ge to the base gives rise to a much higher I_C at the same emitter base bias, resulting in a much lower emitter transit time τ_E. By grading the Ge content over the base layer, i.e. increasing its concentration from emitter to collector, an additional electric field in the base can also lower τ_B. The higher the Ge content at the emitter-sided edge of the base (or rather the increase over the emitter-base depletion region), the higher is I_C. But the maximum Ge content is limited by the permitted thickness of the strained SiGe-layer, for thicker layers might relax, leading to degradation of the device. Thus the higher the Ge content at the emitter-sided edge, the shallower the grading in the base. This results in the compromise of reducing both τ_E and τ_B in order to get opimal device performance for case III.

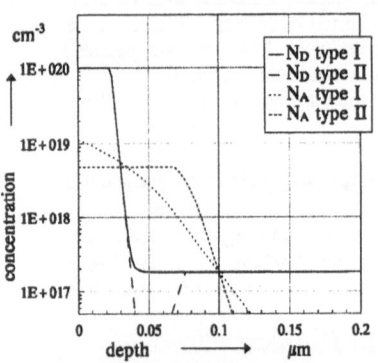 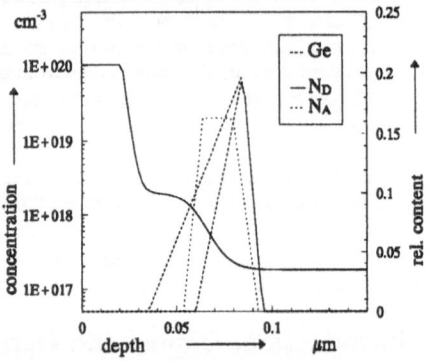

Fig 1: Doping profile of device types I and II

Fig 2: Doping and Ge profile of device type IV

In addition to these devices, we also present performance data of an advanced HBT with a lightly doped emitter layer at the emitter base junction (type IV), similar to the device presented in ref. [2]. With the aid of Ge in devices like thse of type III, the reduction of τ_E is much more noticable than that of τ_B. To make the most out of the addition of Ge, a device is to be bulit where τ_B is reduced to a comparable extent. This can be done by fully utilizing the possibilities of very steep and highly doped - almost box-like - base profiles, which can be fabricated by epitaxial growth. The very high doping of the base leads to base layers with very low sheet resistance - even when they are very shallow. This is another important characteristic of high-speed bipolar junction transistors. But the very thin space charge region of two highly doped layers succeeding one another (emitter and base in this case) leads to unwanted leakage currents [3]. To get rid of them a more lightly doped layer in the emitter can be introduced. In our studies we used a doping concentration of $2 \cdot 10^{18}$ cm^{-3} for this layer. Its width has an important influence on τ_E and thus on f_T. Therefore it will be made as small as permitted by the leakage currents [3].

With this type of profile f_T is much more sensitive to variations of the starting point of the Ge grading (Fig. 2), representing the inevitable compromise when trying to reduce both τ_E and τ_B.

The simulations are performed by running an extended version of a conventional one-dimensional drift-diffusion equation solver [4], which has been adapted for Si/SiGe heterostructures [5]. The device parameters are extracted from the quasistatic device internal steady state distributions of electrons and holes.

2. Results

For a device of type I holding the base sheet resistance at the value of 10 kΩ and taking the stated limits into consideration the base width be cannot reduced below 70 nm. This leads to a f_T of 43 GHz, compared to 32 GHz with a base width of 100 nm. The corresponding numbers for type II are 48 nm instead of 84 nm with an improvement in f_T from 33 to 54 GHz (Fig. 3), for a 10 nm/dec base profile gradient 63 GHz are reached.

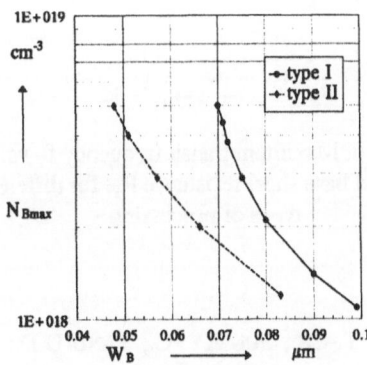

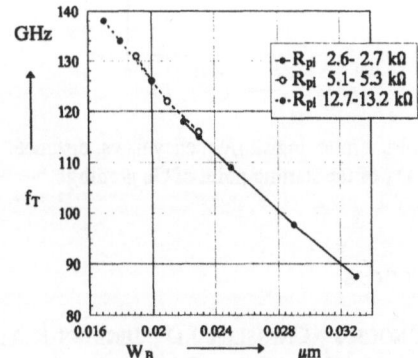

Fig. 3 : Boron doping concentration N_B vs. basewidth W_B at a fixed value of intrinsic base sheet resistance R_{pi} of 10 kΩ for device type I and II

Fig. 4: Maximum f_T vs. baswidth for various values of intrinsic base sheet resistance for device type IV

A similar study was also performed with device IV. For a base sheet resistance of about 2.6-2.7 kΩ a range for f_T can be covered which extends from as low as 87 GHz up to values of 120 GHz, representing doping values of the base ranging from $1.5 \cdot 10^{19}$ cm^{-3} to $5 \cdot 10^{19}$ cm^{-3}. The limits here are again set by the height of the maximum base doping which can be reached for a given base width. For about 13 kΩ sheet resistance even 140 GHz are possible (Fig 4) for $3.4 \cdot 10^{19}$ cm^{-3} base doping.

As mentioned above an important factor for device IV is the inevitable compromise of simultanously lowering τ_E and τ_B. The position in the emitter where the Ge grading starts is measured as the (negative) distance to the metallurgical emitter base junction. As can be seen in Fig. 5, an optimum of 95 GHz for this parameter at about -15 nm can be achieved, when $2 \cdot 10^{19}$ cm^{-3} is assumed as the doping level of the base.

Now the four types of profile can be compared by considering the variation of maximum transit frequency f_T with base sheet resistance R_{pi} (representing the changing of base width) for a maximum base doping of $5 \cdot 10^{18}$ cm^{-3} for technology I-III and $2 \cdot 10^{19}$ cm^{-3} for IV. The results are plotted in Fig. 6. As can be seen, a moderate improvement in f_T for a given R_{pi} is obtained due to the steeper slope of the epitaxial base compared to the implanted one. A significantly higher f_T can be achieved by grading Ge over the base of the transistor. This is further enhanced by the significantly shallower and higher doped base, accompanied by a steeper Ge concentration gradient in device IV.

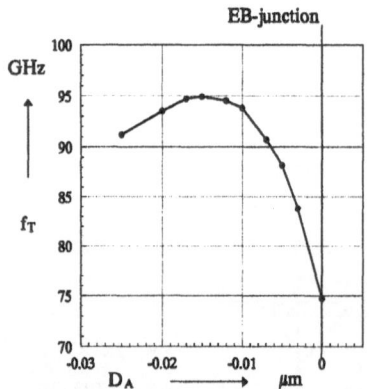

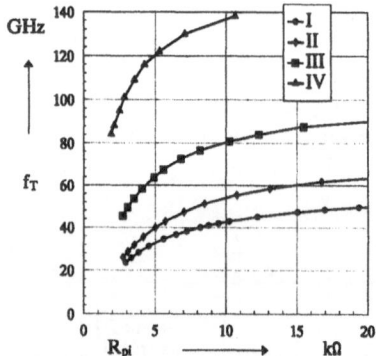

Fig 5: Maximum transit frequency f_T vs. distance D_A of the starting point of Ge grading

Fig. 6: Maximum transit frequency f_T vs. pinched base sheet resistance R_{pi} for different types of processing

References

1 WARNOCK J., CRESSLER J.D., JENKINS K.A:, CHEN T.-CH., SUN J.Y.-C., TANG D.D.: 50-GHz Self-Aligned Silicon Bipolar Transistors with Ion-Implanted Base Profiles,
IEEE **EDL-11**, pp. 475-477 (1990)

2 GRUHLE A., KIBBEL H., ERBEN U., KASPER E.: 91 GHz SiGe HBTs Grown by MBE,
Electron. Lett. **29**, pp. 415-417 (1993)

3 MATUTINOVIC-KRSTELJ Z., PRINZ E.J., SCHWARTZ P.V., STURM J.C.: Reduction of p$^+$-n$^+$ Junction Tunneling Current for Base Current Improvement in Si/SiGe/Si Heterojunction Bipolar Transistors,
IEEE **EDL-12**, pp. 163-165 (1991)

4 ENGL W. L., LAUR R., DIRKS H.: MEDUSA -- a simulator for modular circuits,
IEEE Trans., **CAD-1**, pp. 85-93 (1982)

5 MARKSTEINER S., FELDER A., MEISTER T.F.: Near Future Perspectives for Si and Si$_{1-v}$Ge$_v$ Bipolar Transistors,
Proc. ESSDERC 92, pp. 535-538 (1992)

Analysis of a CMOS-Compatible Vertical Bipolar Transistor

G. Schrom, S. Selberherr, F. Unterleitner[†], J. Trontelj[‡], and V. Kunc[‡]

Institute for Microelectronics, TU Vienna
Gußhausstraße 27-29, A-1040 Wien, AUSTRIA
[†]Austria Mikro Systeme GmbH
SchloßPremstätten, A-8141 Unterpremstätten, AUSTRIA
[‡]Laboratory for Microelectronics, University of Ljubljana
Tržaša 25, SLO-61000 Ljubljana, SLOVENIA

Abstract

A vertical npn bipolar transistor (BJT) which can be manufactured in a simple p-well CMOS process without additional process steps is described. The proposed BJT uses a p-well as base and an n^+ S/D doping as emitter. The collector consists of the n^- substrate and does not require an n^+ buried layer or a highly doped substrate. The device is especially suitable for high-voltage applications in electrically hostile environments such as automotive circuits.

1. Introduction

Four different BJT structures (Fig. 1) were designed and analyzed with the two-dimensional device simulator BAMBI 2.1 [1]. Using a high-voltage CMOS process, the S/D junction depth is $1.8\mu m$ and the p-well junction depth is about $6\mu m$, thus, the base thickness is $\approx 4\mu m$ (Fig. 2). The half-cell widths vary from $90\mu m$ (structure 1) down to $25\mu m$ (structure 4). The characteristics of the optimal structure (no. 4) are: $h_{FE} \approx 160$, $I_{KF} = 0.17A/cm$, $V_{AF} = 170V$, $\tau_F = 0.5ns$. Depending on the application, emitter current densities of more than $1kA/cm^2$ can be achieved (limited by the beta roll-off at high collector currents).

2. Collector Resistance

Due to the distributed collector resistance, the BJT exhibits a distinct quasi-saturation behavior which adversely affects the operation at low V_{CE} because of minority injection into the substrate. By optimizing the device, this effect can be kept sufficiently small even without a buried layer. The optimization was performed by reducing both the emitter stripe width and the distance between emitter and collector to a minimum which is determined by high-level injection in the base and the required V_{CEmax} respectively. The vertical doping profile was not changed so that the NMOS and PMOS transistors on the chip are not affected (in particular, the substrate resistivity is confined to $3–5\Omega cm$). In Fig. 5 and Fig. 7 it can be seen that the structures 3

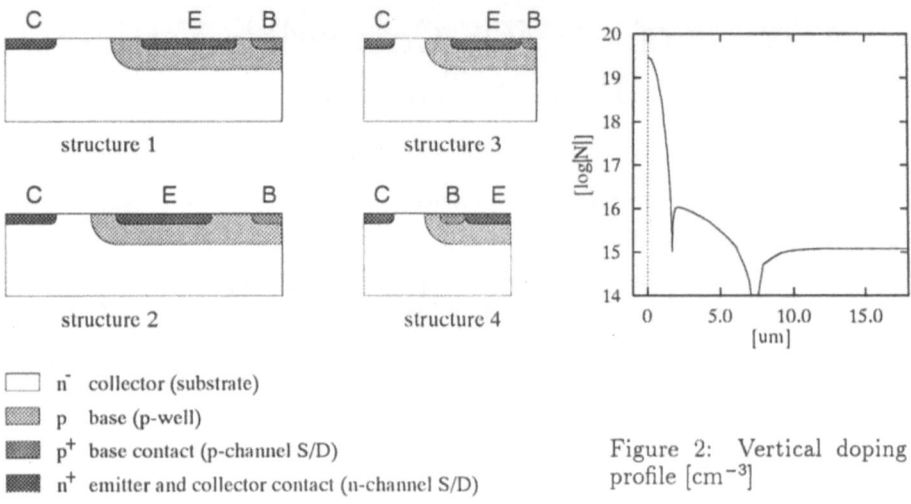

structure 1

structure 3

structure 2

structure 4

☐ n⁻ collector (substrate)

▨ p   base (p-well)

▦ p⁺ base contact (p-channel S/D)

▪ n⁺ emitter and collector contact (n-channel S/D)

Figure 2: Vertical doping profile [cm⁻³]

and 4 still exhibit some quasi-saturation which is, however, far smaller than that of the structures 1 and 2. The currents of the smaller structures 3 and 4 were scaled so that the comparison refers to the same total device area.

3. High-Voltage Capability

The placement of the base contact has an essential impact on the dynamic and high-voltage behavior: The lateral part of the transistor limits its overall high-voltage performance due to the higher electric field at the surface, which is responsible for punch-through, especially, when steep pulses are applied to the collector (structures 1–3). However, if the base contact is placed between collector and emitter (structure 4) the p^+ doping at the base contact virtually eliminates the lateral BJT, which allows for higher V_{CE} and reduces the avalanche multiplication current at the surface. Additionally, the effective base resistance is reduced which also enhances the dynamic and high-voltage behavior. Fig. 6 shows the output-characteristics of the four structures for high V_{CE}. The structures 1 and 2 suffer mainly from their high effective base resistance. The maximum V_{CE} of structure 3 is limited by the avalanche generation in the base-collector junction at the surface.

4. Applications

The BJT is used in a supply protection circuit for static and dynamic overvoltage protection. The circuit must handle a permanent overvoltage of 24V d.c. and pulses of up to 80V at $t_r = 1\mu s$ at a nominal load current of 300mA/cm. It consists essentially of a series regulator which limits the internal supply to $\approx$16V (Figs. 3, 4). The BJT must not saturate at all, otherwise it would be a bypass for steep overvoltage pulses. As the collector is contacted to the substrate, the circuit needs Zener diodes on its output too.

Fig. 8 shows the behavior of the four structures in the case of a steep V_{CE} pulse at the nominal load current ($R_B = 0$). The structures 3 and 4 can easily handle the pulse whereas the two wider structures 1 and 2 are overloaded due to saturation. A finite resistance R_B in the base branch has no detrimental effect because the additional voltage drop across R_B reduces the minority injection into the substrate so that the voltage overshoot is virtually not increased by the base resistor.

Further possible applications of the BJT which can also be merged with PMOS transistors include amplifiers, shunt regulators, buffers, solenoid drivers, and stepper motor drivers.

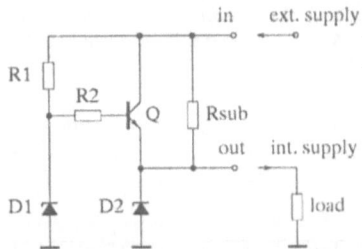

Figure 3: Series regulator as overvoltage protection circuit

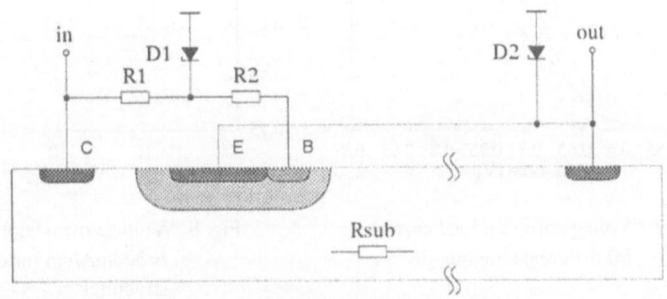

Figure 4: Integration of the protection circuit with structure 3

References

[1] W. Kausel, G. Nanz, S. Selberherr, H. Pötzl, *BAMBI – A transient two-dimensional device simulator using implicit backward Euler's method and a totally self-adaptive grid*, NUPAD II Workshop, May 9–10, 1988, San Diego, Ca., Digest No. 105/106.

[2] F. Berta, J. Fernandez et al., *A Simplified Low-Voltage Smart Power Technology*, IEEE Electron Device Lett., pp. 465-467, 1991.

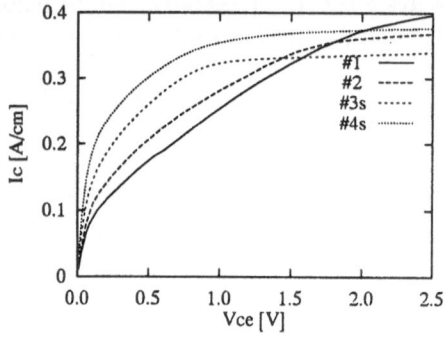

Fig. 5: Output-characteristics for different structures

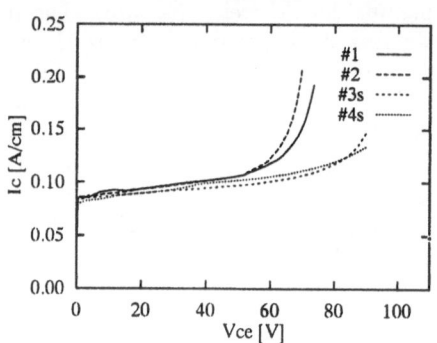

Fig. 6: Output-characteristics for different structures

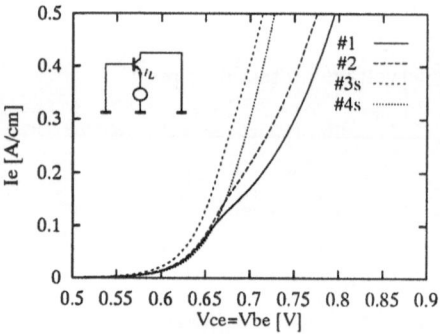

Fig. 7: Voltage drop vs. load current for different structures

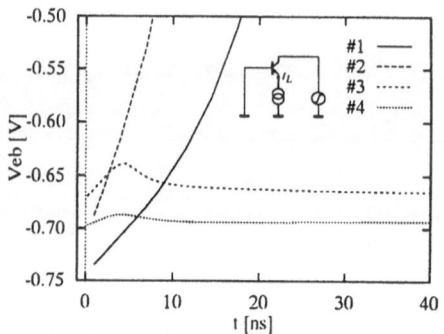

Fig. 8: Voltage overshoot at 80V/us, I=300mA/cm for different structures

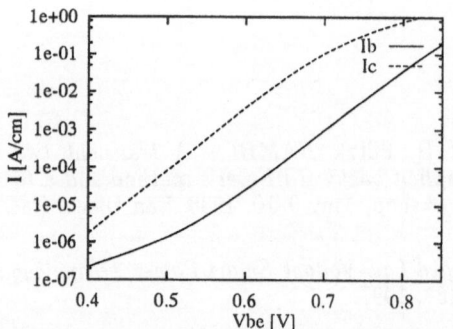

Fig. 9: Input-characteristics of structure 4

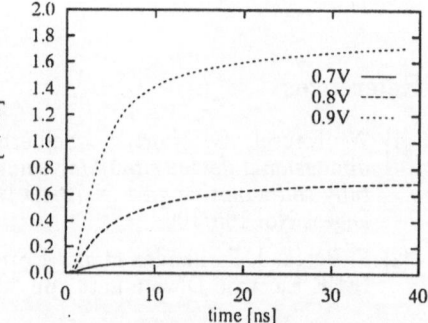

Fig. 10: Turn-on characteristics of structure 4 for different Vbe pulses (tr=1ns), Vce=10

Finite Element Simulation of Recess Gate MESFETs and HEMTs: The Simulator H2F

A. Asenov, D. Reid, J. R. Barker, N. Cameron, and S. P. Beaumont

Nanoelectronics Research Centre,
Department of Electronics and Electrical Engineering, The University of Glasgow
Glasgow, G12 8QQ, UNITED KINGDOM

Abstract

In this paper we present a new 2D finite element compound semiconductor device simulator H2F suited for simulation of the parasitic effects in recess gate MESFETs and HEMTS. Several simulation examples of real devices fabricated in the Nanoelectronics Research Centre at the University of Glasgow illustrates the usefulness of the adopted finite element approach.

1. Introduction

When dimensions of the modern MESFETs and HEMTs scale down to a few tenths of a micron, the device performance becomes strongly affected by device parasitics such as coupling capacitances and access resistances [1]. In recess gate devices these parasitics are critically affected by the shape and surface condition of the recess region. In addition the T-gate process designed to reduce the gate series resistance [2] may also reinforce the parasitic capacitances. Although Hydrodynamic [3] and Monte Carlo [4] simulation programs are making significant progress in properly describing the non equilibrium transport phenomena in compound FETs, the real shape of the gate recess is generally poorly modelled, assuming planar or rectangular simulation domains. Surface effects are also either neglected or modelled by fixing the surface potential or by increasing the surface doping [5]. Yet it is well known that these effects in many cases have a more profound impact on device DC characteristics and high frequency performance than the transport details in the 'intrinsic' region under the gate.

In this paper we report on a new Heterojunction 2D Finite element (H2F) device simulator which focuses on a precise description of the device's geometry and a realistic handling of surface effects and their influence on the device performance.

2. The program H2F

The program H2F is a 'classical' steady-state simulator, which self-consistently solves Poisson's and the current continuity equations in a drift-diffusion approximation. While this approach is unable to describe precisely the device's transport, in many cases it is justified by the need to accurately predict the device's parasitics. A great deal of attention has been paid to the proper handling of the surface effects in the simulation. For

Poisson's equation, the simulation domain includes the space above the semiconductor surface providing a proper interaction between the charge on the surface states and the spreading surface potential. A generalised surface trap model includes acceptor and donor like traps with an arbitrary energy position whose occupation depends on the quasi-Fermi level and the surface potential variation.

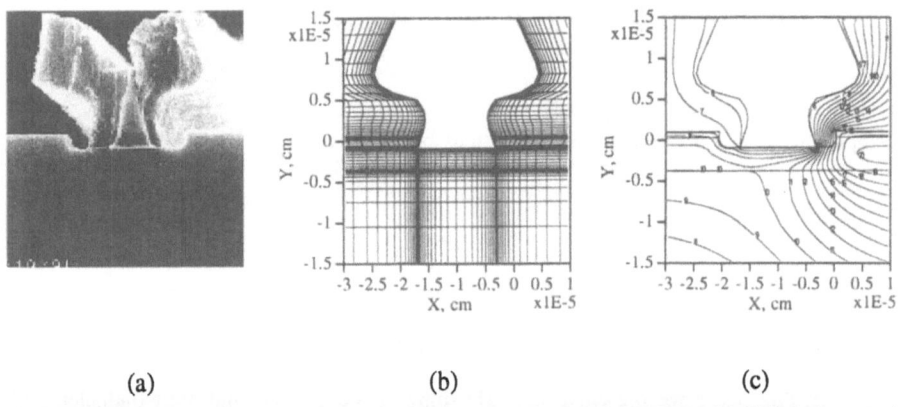

(a)   (b)   (c)

Figure 1. Finite element simulation of a 200 nm gate length MESFET. (a) SEM picture of the device cross sectional view (b) the corresponding H2F grid (c) Potential distribution at VG=-0.4 V and VD=2.5 V.

Quadrilateral finite elements have been used for discretization. The flexibility of the quadrilateral grid is illustrated in Fig. 1 where the cross sectional photograph of a 200 nm gate-length state of the art MESFET, fabricated in the Nanoelectronics Research Centre at the University of Glasgow [6], is compared with the corresponding H2F simulation domain. The grid is generated by appropriate deformation of originally rectangular sub domains. The Galerkin finite element method with a linear isoparametric mapping has been adapted to solve Poisson's equation. A control volume method has been developed for the discretization of the current-continuity equation [7]. In this approach each quadrilateral element is divided into four subelements and the discretization is carried out balancing the current flowing in and out of the subelements attached to a given condensation point. The current density is approximated by the standard Gummel-type expression. The growth functions involved in the derivation of this expression are also used for interpolation of the electron concentration along the sides of the element. It has been found that this discretization is stable for arbitrary shapes of the quadrilateral elements and do not leads to the spikes typical for obtuse triangles.

The grid generation preserves the number of grid points in lateral and vertical directions and leads to a regular nine diagonal matrix of the discretized equations. A Fast Incomplete LU factorisation Biconjugate Gradients (ILUBCG) solver is used for the numerically intensive iterations. The solution of the Poisson's equation involves only a few biconjugate gradient steps per Newton iteration that significantly reduces the total computation time. The convergence problems related to the strongly localised, potential dependent interface charge have been resolved by appropriate dumping. ILUBCG also solves without complication the discretized current continuity equation.

Although H2F is a 'serial' code, an universal pipeline fileserver have been developed to run the program on MIMD mashines like Parsytec Model 64 transputer system. Using this approach multiple copies of the program can calculate in parallel a separate set of input device data. This extends dramatically the capability of the simulator for real design work such as structure optimisation, sensitivity analysis and yield prediction where several hundred simulations are often carried out for a single investigation.

3. Simulation Examples

A set of examples illustrate the application of H2F simulating compound FETs with complex recess shapes. The influence of the position and the density of these surface states on the device's I_D-V_G characteristics for the 200nm MESFET shown in Fig. 1 is given in the Fig. 2 (a, b). The doping concentration in the 60 nm thick MESFET channel is 5×10^{17} cm^{-3}. A p-type buffer suppresses the electron penetration in the substrate. The experimental measurements are in good agreement with the expected position and states density $Ps=0.6$ eV and $Nit=2 \times 10^{12}$ cm^{-2} (Fig 5 (c)). The reduction of the drain current for gate voltages above 0 V and the presence of deep surface states is mainly due to the increase in the series resistance of the unprotected recess region.

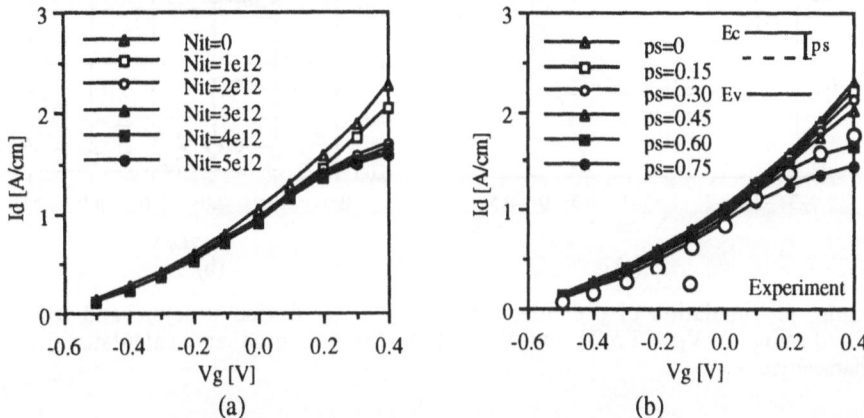

Figure 2. Simulated and measured ID-VG curves for the 200 nm gate length MESFET illustrated in Fig. 1. (a) influence of the acceptor type surface states position P_s (b) influence of the surface state density N_{it}.

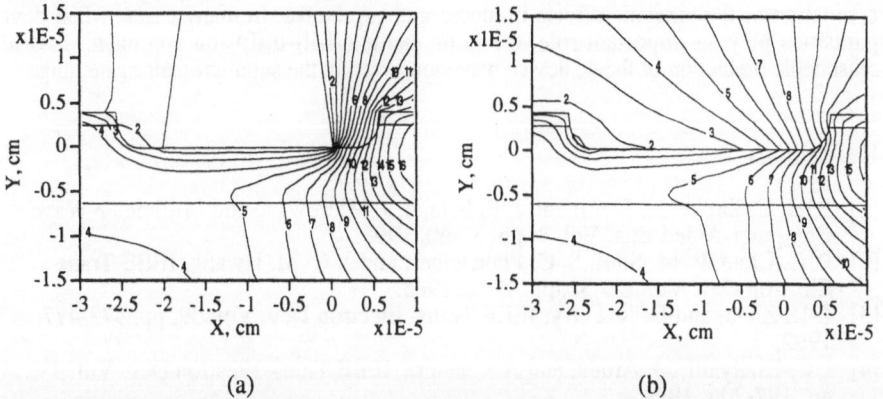

Figure 3. Potential distribution (a) in a gated MESFET (same as in Fig. 1) at $V_G=0$ V and $V_D=3.5$ V and (b) in the corresponding gate-less structure at $V_D=2.5$ V. In the both cases acceptor type surface states with density 4×10^{12} cm^{-2} and position 0.6 eV below the conducting band are assumed.

A practical technological problem is addressed in Fig 3 (a, b) where the potential distribution in gated and gate-less transistor structures are investigated. This reflect a part of the technology cycle as recess etching is often controlled by measuring the saturation current in the gate-less structure. The gateless structure shows approximately 160% more current than that of the gated transistor with zero gate bias. This is due to the change in the surface conditions and to the lateral penetration of the drain potential and the corresponding shortening of the effective channel length.

H2F is also suited to simulating parasitic effects in HEMTs and the results for a delta-doped pseudomorphic HEMT structure significantly influenced by the series resistances are presented in Fig. 4 (a, b). Although the drift diffusion approach underestimates the current, it has been found that by adjusting the saturation velocity in the mobility model (to 1.4×10^7 cm/s in this case) the measured characteristics can be acceptably matched.

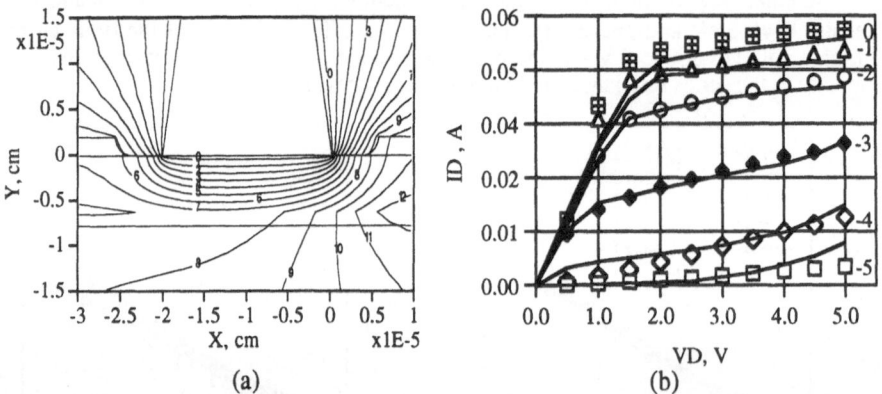

(a)   (b)

Figure 4. Simulation of pseudomorphic HEMT. (a) device structure and potential distribution at V_G=-3.5 V and V_D=2.5 V (b) measured and calculated I_D-V_G characteristics

Conclusions

A new finite element 2D simulator H2F has been developed. The quadrilateral elements used in the simulator provide the necessary flexibility for realistic description and proper estimation of the parasitic effects in recess gate structures. In many cases, when device parasitics play an important role, the implemented drift-diffusion approach leads to a reasonable prediction of the dc device behaviour even in the submicrometer gate range

References

[1] P. H. Ladbroke, A. J. Hill and J. P. Bridge, J. Microwave and Millimeter-Wave Computer-Aided Eng. Vol. 3, pp. 37-60, 1993.
[2] P. C. Chao, P. M. Smit, S. C. Plamaeteer, and J. C. M. Hwang, IEEE Trans. Electron Dev. Vol. ED-32, pp. 1042, 1985.
[3] J.-R. Zhou, and D. K. Ferry, IEEE Trans. Electron Dev. Vol. 39, pp. 473-477, 1992.
[4] I. C. Kizilyalli, M. Artaki, and A. Chandra, IEEE Trans. Electron Dev. Vol. 38, pp. 197-206, 1991.
[5] H.-F. Chau, D. Pavlidis and K. Tomizava, IEEE Trans. Electron Dev. Vol. 38, pp. 213-221, 1991.
[6] N. I. Cameron, G. Hopkins, I. G. Thayne, S. P. Beaumont, C. D. W. Wilkinson, M. Holland, A. H. Keanand and C. R. Stanley, J. Vac Sci. Technol. B9 3538 (1991).
[7] A. Asenov, and E. Stefanov, Proc. ISPPM Varna, pp. 272-   285, 1989.

SIMULATION OF SEMICONDUCTOR DEVICES AND PROCESSES Vol. 5
Edited by S. Selberherr, H. Stippel, E. Strasser – September 1993

A Smallsignal Databased HEMT Model for Nonlinear Time Domain Simulation

T. Felgentreff, G. Olbrich, and P. Russer[†]

Lehrstuhl für Hochfrequenztechnik, Technische Universität München
Arcisstraße 21, D-80333 München, GERMANY
[†]Ferdinand-Braun-Institut für Höchstfrequenztechnik
Rudower Chaussee 5, D-12489 Berlin, GERMANY

Abstract

Accurate and simple nonlinear modeling of semiconductor devices for use in nonlinear CAD is of significant importance, especially at higher frequencies. In this paper we present a nonlinear HEMT model, which is directly based on measured data. Therefore time consuming optimization techniques during the modeling process are eleminated. The values of the equivalent circuit elements are derived from DC and smallsignal RF measurements and are stored in a twodimensional lookup table. Using this model we calculate the harmonics generated by the nonlinear elements of the device. For the verification of this model we compare the results with measured data and results obtained by conventional empirical HEMT models [1], [2], [3]. The measured and calculated data deviate less than 1.5 dB.

1. Nonlinear Databased Model

The nonlinear model is based on DC and S-parameter measurements of the transistor at all bias points of interest. At operating points in the region near pinch-off a higher density of data points is choosen than in areas with much more linearity (Fig. 2). The data obtained from S-parameter measurements are deembeded from the extrinsic elements using techniques described in [4] and [5]. The next step is to calculate the value of all the intrinsic nonlinear elements at each bias point. Therefore the voltage drop at the extrinsic resisitive elements R_s and R_d has to be taken into acount. The values of the nonlinear elements are stored in a twodimensional lookup table as functions of the two independent controlling voltages V_{gs} and V_{ds} at the intrinsic HEMT. No optimization process is performed to calculate this table. Because there are no functional approximations to fit the measured data no information gets lost and the modeling process works quite fast.

The databased model is used in nonlinear time domain CAD programs. The values stored in the twodimensional lookup table are used directly by the simulation program. The description of the equivalent circuit of the transistor (Fig. 1) in the time domain is done by a system of differential equations.

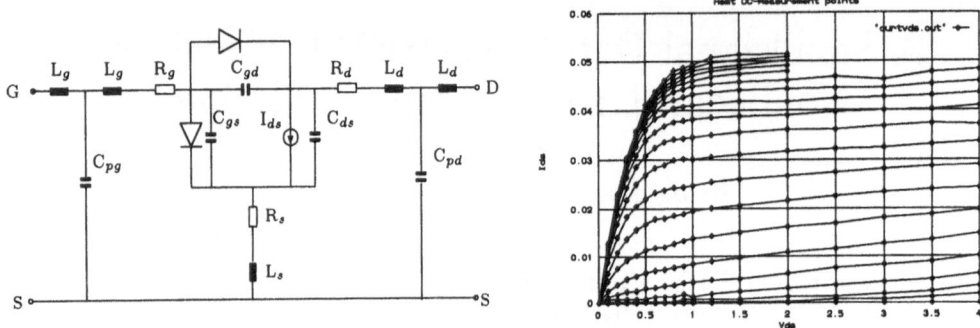

Figure 1: HEMT Equivalent Circuit   Figure 2: I_{ds} vs. V_{ds} (parameter V_{gs})

The values for untabulated points are obtained by interpolation. Two different interpolation methods -the bilinear interpolation and the bicubic spline interpolation- were tested for calculation. Both interpolation methods show little difference. So the faster bilinear interpolation was used. Two capacitors (C_{gs} , C_{gd}) and the nonlinear current source I_{ds} are considerd to depend on the bias voltages V_{gs} and V_{ds}. In addition the parameters of the two diodes were seperatly determined and applied to the model.

To verify the nonlinear model the output power at the fundamental frequency and its harmonics were measured. Measurements take place in a 50 Ω system using a wafer prober. The measured results are compared with calculations. A time domain simulator is used to compute the output power at the fundamental frequency and its harmonics.

The modeling process was performed on a standard GaAs/GaAlAs MODFET CFY 65. The semiconductor chip was mounted on a carrier Al_2O_3-substrate and bonded to coplanar waveguide interconnections. The S-parameters were measured in the frequency range 0.1 to 40 GHz at more than 200 bias points using a wafer prober and a computer controlled measurement setup. The verification of the model was done at a fundamental frequency of 5 GHz. In Figures 4 and 5 the empirical Curtice and Materka HEMT model given in [1] combined with the Meyer capacitance-voltage relationship [6] are compared with results from measurement and from the databased HEMT model. The results obtained from the different models agree within 2 dB; however the databased model describes best the measured results of the fundamental frequency and its higher harmonics compared to the measured data.

The nonlinear databased HEMT model was primarily developed for use in CAD software for the simulation of oscillators in the time domain and combined time/frequency domain using FATE [7]. As an example the simulated and measured results of a 16 GHz coplanar HEMT oscillator build on an Al_2O_3-substrate are presented. Figure 6 shows output power versus normalised frequency for different models and measured results with V_{gs} as parameter. Figure 7 shows the same quantities with V_{ds} as parameter. Only the databased model can describe the measured tuning characteristic. These results show the significant advantage of this model concerning especially the accurate modeling of the capacitances (Fig. 3).

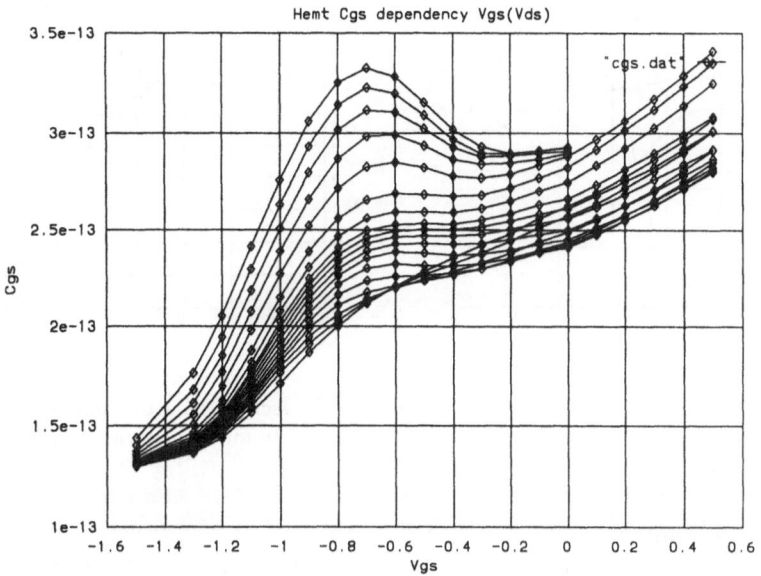

Figure 3: C_{gs} vs. V_{gs} (parameter V_{ds})

2. Conclusion

An accurate and simple nonlinear model for use in nonlinear CAD has been presented. The model was applied to a GaAs/GaAlAs-HEMT, but it is also applicable to MES-FEFTs. Because of the greater nonlinearities in HEMTs the use of the databased model is more advantageous. The model, that is directly based on measured data, is used for the time domain simulation of amplifiers and oscillators. The model has been proved to give more accurate results than conventional empirical models.

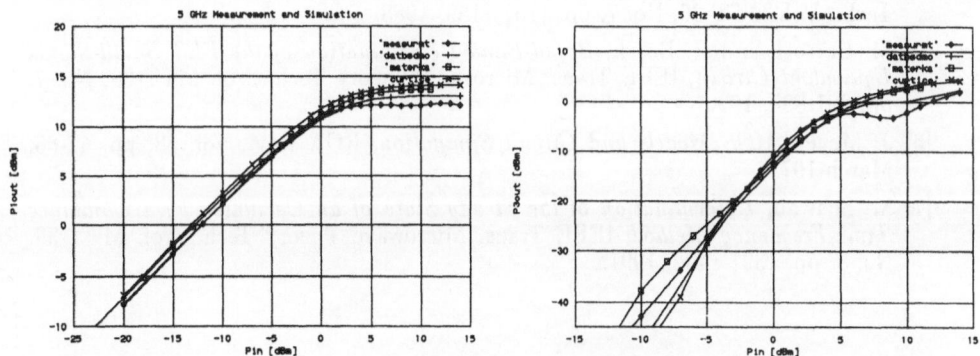

Figure 4: Fundamental Power   Figure 5: Third Harmonic Power

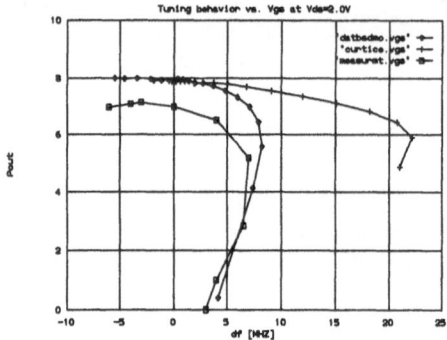

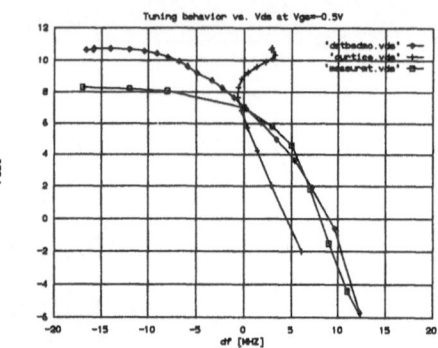

Figure 6: Oscillator output power vs. normalized frequency (parameter Vgs)

Figure 7: Oscillator output power vs. normalized frequency (parameter Vds)

Acknowledgement

This work has been financially supported by the Deutsche Forschungsgemeinschaft (DFG/SFB 348).

References

[1] J. M. Golio, *Microwave MESFETs and HEMTs*, Artech House, Boston, pp. 142-146, 1991.

[2] W. R. Curtice, *A MESFET Model for Use in Design of GaAs Integrated Circuits*, IEEE Trans. Microwave Theory Tech., Vol. MTT-28, No. 5, pp. 448-456, 1980.

[3] T. Kacprzak and A. Materka, *Compact dc Model of GaAs FETs for Large-Signal Computer Calculation*, IEEE Solid-State Circuits, Vol. SC-18, No. 4, pp. 211-213, 1983.

[4] G. Dambrine, A. Cappy, F. Heliodore, and E. Playez, *A New Method for Determining the FET Small-Signal Equivalent Circuit*, IEEE Trans. Microwave Theory Tech., Vol. MTT-36, No. 7, pp. 1151-1159, 1988.

[5] M. Berroth and R. Bosch, *Broad-Band Determination of the FET Small-Signal Equivalent Circuit*, IEEE Trans. Microwave Theory Tech., Vol. MTT-38, No. 7, pp. 891-895, 1990.

[6] J. Meyer, *MOS Models and Circuit Simulation*, RCA Rev., Vol. 32, pp. 42-63, March 1971.

[7] M. Schwab, *Determination of the Steady State of an Oscillator by a Combined Time-Frequency Method*, IEEE Trans. Microwave Theory Tech., Vol. MTT-39, No. 8, pp. 1391-1402, 1991.

Noise in $Si/Si_{1-x}Ge_x$ n-channel HEMTs and p-channel FETs

A. F. M. Anwar, K. W. Liu, and M. M. Jahan

Department of Electrical and Systems Engineering, The University of Connecticut
260 Glenbrook Road, Storrs, CN 06269, USA

Abstract

A theoretical model to evaluate noise in $SiGe/Si$ based n-channel MODFETs and p-channel MOSFETs is presented. The analysis is based on a self-consistent solution of Schrödinger and Poisson's equations. In this study, n-channel FETs exhibit a better noise performance than that of p-channel FETs. The influence of device parameters on noise properties for this class of devices are presented.

1. Introduction

Noise can be characterized by measuring the minimum noise figure F_{min} or the noise temperature. Extensive experimental and theoretical efforts have been directed to the study of noise figure in III-V HEMTs, however, no such study exists for $SiGe/Si$ based devices. In this paper a noise model is presented which is based on a self-consistent solution of Schrödinger and Poisson's equations [1-3]. The calculation of noise parameters follows the treatment presented by Anwar et. al. [1].

2. Model

The modeling of noise proceeds by first quantifying the quantum well (QW) behavior and identifying a realistic velocity-electric field $(v_d-\mathcal{E})$ characteristic for the carrier in the conducting channel. Instead of using a two-line or an exponential approximation the following $v_d - \mathcal{E}$ [1,3] is used in this calculation, $v_d = v_s\mathcal{E}/\sqrt{(v_s/\mu_0)^2 + \mathcal{E}^2}$, where v_s is the saturation velocity, μ_0 is the low field mobility and $\mathcal{E}$ is the applied electric field along the channel. This $v_d - \mathcal{E}$ characteristic makes the current-voltage and dc small signal parameters analytic in nature [4] and provides a better agreement with results obtained from Monte Carlo simulation. The QW parameters are calculated self-consistently [2] and introduced in the calculation of noise by recognizing the functional dependence of the average distance of the electron cloud x_{av} and the position of the Fermi level E_F on the 2DEG concentration n_s [2].

The analysis of noise is based on the identification of the different noise sources that are present in the conducting channel, namely (a) Johnson Noise in the ohmic region (unsaturated region), (b) noise associated with spontaneous generation of dipole layers in the saturation region, (c) gate noise due to elementary voltage fluctuations in the channel and (d) induced gate noise in the saturation region. The noise analysis is carried out by accounting for all these noise sources and their correlation coefficients [2]. The minimum noise figure, F_{min}, the noise conductance, g_n, the minimum noise temperature, T_{min}, and the optimized external generator source impedance, $Z_{s,opt}$ can be written as :

$$F_{min} = 1 + 2 \cdot g_n \cdot (R_c + \sqrt{R_c^2 + \frac{r_n}{g_n}}) \tag{1}$$

$$T_{min} = 2 \cdot T \cdot g_n \cdot [R_c + Z_{s,opt}] \tag{2}$$

$$g_n = g_m \cdot (\frac{f}{f_T})^2 \cdot [P + R - 2C\sqrt{PR}] \tag{3}$$

$$Z_{s,opt} = \sqrt{R_c^2 + \frac{r_n}{g_n}} \tag{4}$$

where r_n is noise resistance defined in Ref.[1], $Z_c = R_c + jX_c$, is the correlation impedance, R_s and R_g are the source and drain impedance, $R_i = \frac{L_g}{v_s C_{gs}}$, is the gate charging resistance. L_g represents the length of the gate. In several cases, the device is not matched for the minimum noise figure and the mismatch effect on the noise figure can be expressed as

$$F = F_{min} + \frac{g_n}{R_s} \cdot |Z_s - Z_{s,opt}|^2 \tag{5}$$

where $Z_s = R_s + jX_s$, the input termination, or source, impedance. Eqn. (5) shows that the mismatch effect is less sensitive for lower values of the noise conductance g_n.

3. Results and Discussion

In this paper, the n- and p-channel devices analyzed are based on the structures reported in Refs. [5] and [6]. In Fig.1, the calculated minimum noise figure, F_{min} (dB), is plotted as a function of the drain-source current for n- and p-channel FETs with temperature as a parameter. The calculations are performed at 15 GHz. The present noise model predicts "U" shape in minimum noise figure (F_{min}) versus saturation drain current I_{ds}. The minimum noise figure of n-channel devices are smaller than that of p-channel devices at both temperatures. At higher I_{ds}, the increase of noise figure for p-channel devices is faster than that of n-channel device. This may be due to the fact that n-channel FETs have higher transconductance g_m and unity current gain cut-off frequency f_T than those of p-channel FETs [4].

In Fig.2, F_{min} is plotted as a function of frequency for n- and p-channel FETs at various temperatures. For the p-channel MOSFET V_{gs} = -2.5V (-1.64V) and V_{ds} = -2.5V at 300K (82K), for the n-channel MODFET V_{gs} = -0.65V (-0.53V) and V_{ds} = 3.0V at 300K (77K). F_{min} increases with frequency for both families of FETs. A higher f_T for n-channel MODFETs results in a lower F_{min}. We have observed that F_{min} decreases with increasing QW width. This decrease in not so significant at the high frequency end. At 60 GHz a well width increase from 50Å to 300Å may result

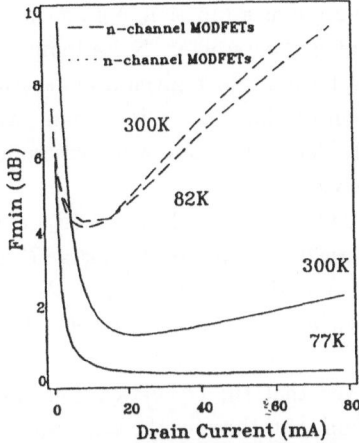

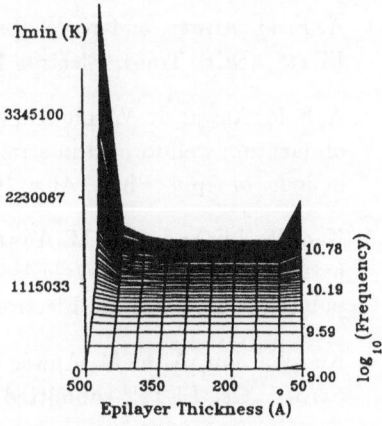

Fig.1 Minimum noise figure F_{min} is plotted as a function of drain current for $SiGe/Si$ based n- and p-channel FETs with temperature as a parameter.

Fig.2 Minimum noise figure F_{min} is plotted as a function of frequency for both n- and p-channel FETs with temperature as a parameter.

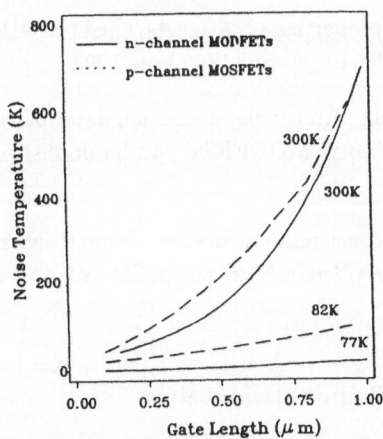

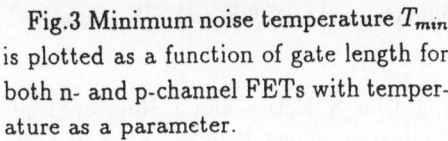

Fig.3 Minimum noise temperature T_{min} is plotted as a function of gate length for both n- and p-channel FETs with temperature as a parameter.

Fig.4 Minimum noise temperature T_{min} is plotted as a function of the doped epilayer thickness for an n-channel FET at 77K.

in a reduction of F_{min} by 1.5dB for n-channel FETs. F_{min} is less sensitive to QW width variation at cryogenic temperatures.

In Fig.3, minimum noise temperature T_{min} is plotted as a function of gate length for both n- and p-channel devices with temperature as a parameter. Noise temperature increases with increasing gate length for both devices. For a given drain-source voltage, the thermal noise induced from ohmic region will increase by increasing gate length (also increasing the length of ohmic region). Therefore, noise will increase by increasing gate length for both n- and p-channel devices.

In Fig.4, T_{min} is plotted as a function of the doped epilayer thickness (d_d) and frequency for an n-channel FET at 77K. As observed, a range of d_d exists (75Å-200Å) where T_{min} is a minimum ($T_{min} < 200$K at 60 GHz for 75Å$\leq d_d \leq$ 200Å). The reduction in T_{min} is prominent at higher frequencies. For p-channel devices T_{min} increases slightly with increasing d_d.

Based on the present noise model, it is observed that the calculated noise parameters are less sensitive to the donor concentration than those of the conventional $AlGaAs/GaAs$ based FETs.

4. Conclusion

A self-consistent model to calculate F_{min} and T_{min} in $SiGe/Si$ n-channel MOD-FETs and p-channel MOSFETs is presented. These predictions of noise parameters may lead one to optimize the noise performance for $SiGe/Si$ based FETs.

References

1. A. F. M. Anwar, and K. W. Liu, "Noise properties of $AlGaAs/GaAs$ MOD-FETs", IEEE. Trans. Electron Dev., vol. 40, no. 6, p. 1174, June 1993.

2. A. F. M. Anwar, K. W. Liu and R. D. Carroll, "An envelope function description of quantum well formed in strain layer $SiGe/Si$ MODFETs", to be published in Jour. of Appl. Phys., Aug. 1993.

3. Kuo-Wei Liu and A. F. M. Anwar, "A self-consistent calculation of small signal parameters for $AlGaAs/GaAs$ and $AlGaAs/InGaAs/GaAs$ HEMTs," to be published in Solid State Electronics.

4. Kuo-Wei Liu, A. F. M. Anwar and V. P. Kesan, "A self-consistent model for $Si/Si_{1-x}Ge_x$ FETs'" submitted to IEEE Electron Device Letter.

5. V. P. Kesan, S. Subbanna, P. J. Restle, M. J. Tejwanl, "High performance 0.25 μm p-MOSFETs with silicon-germanium channel for 300K and 77K operation", IEDM 91, p. 2.2.1-2.2.4, 1991.

6. K. Ismail, B. S. Meyerson, S. Rishton, J. Chu, S. Nelson and J. Nocera, "High-transconductance n-type $Si/SiGe$ modulation-doped field-effect transistors", IEEE Electron Dev. Lett., vol. 13, no. 5, p. 229-231, May, 1992.

SIMULATION OF SEMICONDUCTOR DEVICES AND PROCESSES Vol. 5
Edited by S. Selberherr, H. Stippel, E. Strasser – September 1993

Drift Velocities and Momentum Distributions of Hot Carriers in MOSFETs at Low Supply Voltages

C. C. C. Leung and P. A. Childs

School of Electronic and Electrical Engineering, The University of Birmingham
P.O. Box 363, Birmingham, B15 2TT, UNITED KINGDOM

Abstract : In this paper we determine the momentum distribution and drift velocity of hot carriers in silicon under the condition of finite supply energy. We show that the momentum distribution varies between the nearly isotropic in high field to strongly anisotropic when the carrier's energy is comparable to the supply energy.

1. Introduction

Hot Carriers have long been known to create serious reliability problems for short channel MOSFETs. It was believed that scaling of the supply voltage would virtually eliminate the problem. However, hot carrier currents and degradation have been observed at drain voltages of 3.3V and below [1]. Recent Monte Carlo simulations of hot electron transport in MOSFETs operating at low drain voltages have indicated that the high energy tail of the electron distribution, the most important quantity in determining reliability, is determined by the supply voltage rather than local electric fields [2]. In this paper we show how the momentum distribution and drift velocities of hot carriers in the high energy tail of the distribution relate to hot carrier phenomena.

The momentum distribution of hot carriers is obtained by solving the Boltzmann Transport equation using a modification of the Chambers' path integral solution [3]. This form of the Boltzmann transport equation can be solved efficiently by an iterative method developed by the authors [4,5]. At low drain voltage a one dimensional solution can be obtained in only seconds of CPU time on a SUN sparcstation.

2. Momentum Distributions and Drift Velocities

Earlier studies of electron distributions by Baraff [6] and Keldysh [7] provide accurate solutions to the Boltzmann transport equation in the case of uniform fields in infinitely long samples. However, these analyses are not applicable when the supply energy is finite. In this case the distributions vary between the nearly isotropic in high field and low energy regions to strongly anisotropic close to the supply energy. The high energy drift velocities are important when the electric field is in the direction of the potential barrier, for example in substrate hot electron injection, where they significantly modify the injection current. To illustrate the method the BTE is solved over a potential profile obtained from MINIMOS. The distribution functions obtained from the analytical solution of the BTE, derived by Keldysh for uniform fields [7], are used as boundary conditions in the low field region.

3. Results and Discussion

Figure 1 shows the hot carrier distribution obtained at the edge of the drain depletion region. The momentum distribution of hot electrons obtained at two points, one at the average energy and a second in the tail, is shown as a polar plot in figure 2. The magnitude of the vector represents the probability of finding an electron travelling at the angle made by the vector with respect to the electric field. At low energies the momentum distribution is nearly spherical symmetric, whilst in the tail it is strongly anisotropic. The effect of the momentum distribution on drift velocity is shown in figure 3. Around the average energy the velocity of the electrons in the field direction increases as $energy^{1/2}$ and the average drift velocity is saturated. At high energies the velocity in the field direction increases sharply as the distribution becomes increasingly anisotropic. Electrons in this region gain energy by making few collisions with phonons, effectively travelling by 'very lucky drift'. It is, therefore, apparent that although the probability of finding hot carriers falls off rapidly due to the finite supply energy, the drift velocity of these carriers will significantly increase the current density as measured in the direction of the electric field.

It is interesting to note the electron distribution developed at an energy above that of the supply. In this region the energy distribution is identical to that formed by the boundaries. However, the momentum distribution is modified by the field and remains highly anisotropic.

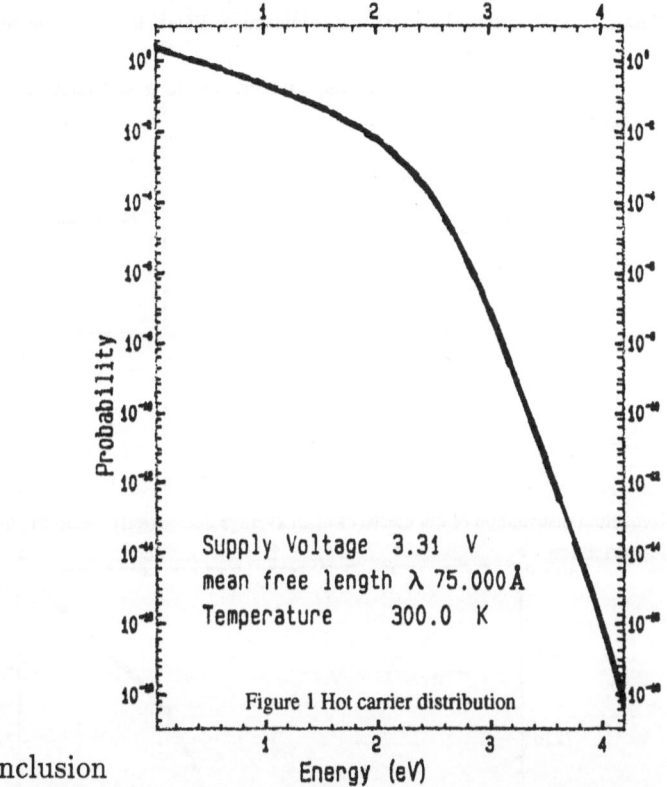

Figure 1 Hot carrier distribution

4. Conclusion

In summary the paper shows the importance of hot carrier momentum distributions and drift velocities on hot carrier currents at low supply voltages. In obtaining the hot carrier distributions the application of a new and highly efficient technique for solving the spatially dependent BTE has been demonstrated.

Acknowledgment: This work is part of the IED1939 project funded by the DTI and SERC

References

1. J E Chung, M-C Jeng, J E Moon, P-K Ko and C Hu, 'Low-Voltage Hot-Electron Currents and degradation in Deep-Submicron MOSFETs', IEEE Trans. Elec. Dev. **37** No. 7, p.1651 (1990)
2. F Venturi, E Sangiorgi, and B Ricco, 'The Impact of Voltage Scaling on Electron Heating and Device Performance of Submicrometer MOSFETs', IEEE Trans Elec. Dev., **38** No. 8, p. 1895 (1991)
3. R. G. Chambers, Proc. Phys. Soc. **A65**, 458 (1952)
4. C C C Leung and P A Childs, 'Hot Carrier Distributions in Silicon Determined by Path Integral Solution of the Boltzmann Transport Equation', Proc. IEDMS, Taiwan p.209 (1992)
5. C C C Leung and P A Childs, 'Spatially Transient Hot Electron Distributions in Silicon Determined from the Chambers Path Integral Solution of the Boltzmann Transport Equation', Solid State Electron, **36**, No. 7, 1001 (1993)

6.   G. A. Baraff, ' Distribution Functions and Ionization Rates for Hot Electrons in Semiconductors', Phys. Rev. **133**, A26 (1964)

7.   L. V. Keldysh, 'Concerning the theory of Impact Ionisation in Semiconductors', Soviet Physics JETP, **21**, 1135 (1965)

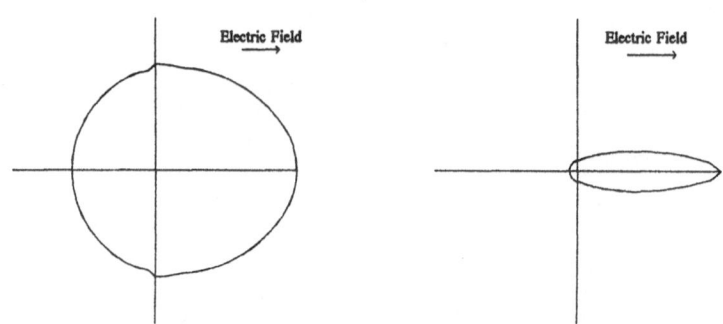

Figure 2 Momentum distribution of hot electrons at an average energy(left) and in the tail (right) of the distribution

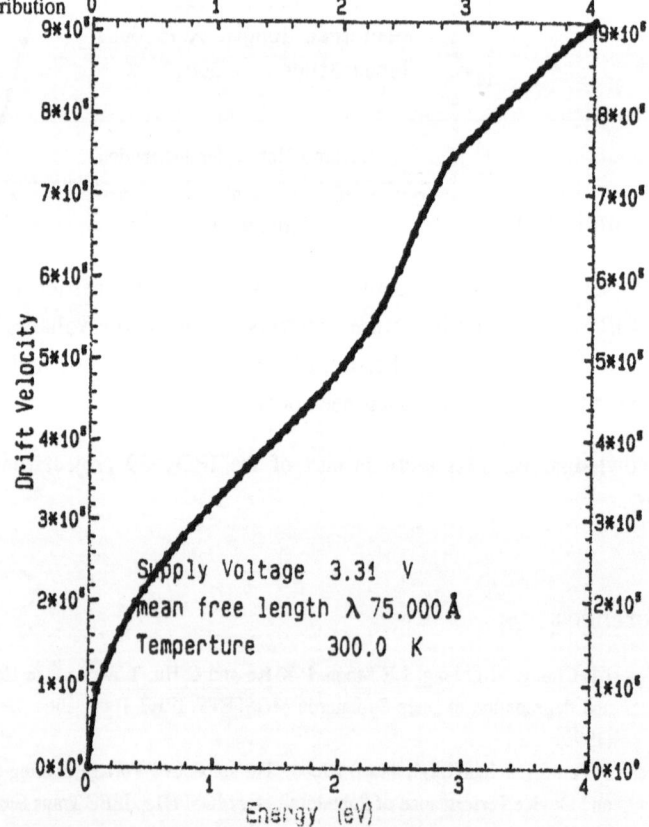

Figure 3. Drift velocity as a function of electron's energy

Numerical Simulation of MOSFETs Gate Capacitances for the Evaluation of Hot-Carrier Generated Interface States and Trapped Carriers

R. Ghodsi, Y. T. Yeow, and M. K. Alam[†]

Department of Electrical and Computer Engineering, The University of Queensland
Brisbane, Queensland, 4072, AUSTRALIA
[†]School of Microelectronic Engineering, Griffith University
Nathan, Queensland, 4111, AUSTRALIA

Abstract

The two dimensional device simulation program MINIMOS is used to calculate gate-to-drain capacitance in the presence of spatially localised interface states and trapped charge. The program has been modified to allow for the introduction of donor and acceptor interface states in different parts of the band gap and with arbitrary spatial distribution along the channel. Numerical simulation and measurement results for gate-to-drain capacitance before and after stress are presented and compared. It is concluded that numerical study of the influence of each parameter on the gate capacitances as well as changes in I-V characteristics would provide a better understanding of the charges induced by hot carriers in a device.

1. Introduction

Hot-carrier generated interface states and trapped charges are the main causes of the degradation of MOSFET's under electrical stressing. Much work has been reported in the literature on the evaluation of these charges via the I-V characterisation of the device[1-3]. The information obtainable from these measurements is generally considered as incomplete, for example it is still unclear if the degradation is due to charge in the interface states or to charge of carriers trapped within the gate oxide. In this paper we present the numerical simulation study of the effects of interface states and trapped charges on the small signal gate-to-drain capacitance. This will demonstrate how the comparison of numerically simulated gate capacitances with the measured gate capacitances would enhance the understanding of observed device degradation due to hot carriers. In particular it will be shown that by comparing the simulated and measured gate capacitance one can obtain an indication of the spatial distribution of interface states and trapped carriers along the channel as well as the distribution in the energy band gap for interface states.

2. Experimental Results

The gate-to-drain capacitance C_{gd} of n-channel MOSFET with L=1.7µm and W=50µm was measured using a Hewlett-Packard 4284A LCR meter as a function of the dc bias before and after electrical stressing of the drain[4]. Fig. 1 (a) shows the measured C_{gd} before and after stress. Increase in C_{gd} after stress in the weak inversion regime is attributed to the presence of positive charge in the oxide near the drain. Conversely the decrease in C_{gd} in the strong inversion is due to negative charge in the oxide at the same location. Positive charge decreases the local threshold voltage V_T and leads to lower channel resistance near the drian. Whereas negative charge increases the local V_T and leads to higher channel resistance.

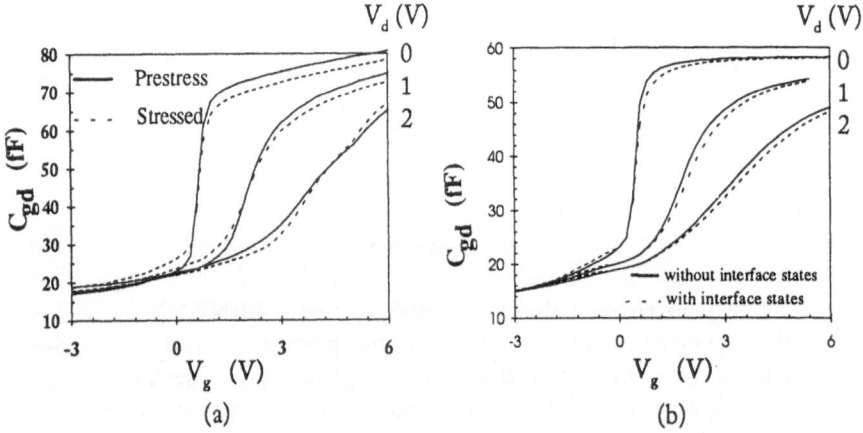

Fig. 1 (a) Measured C_{gd} versus V_g for n-MOSFET, stress condition: V_d=8 V, V_g=2 V, stress time=1800s. (b) Simulated C_{gd} versus V_g with and without interface states, see Fig. 3 for interface state distribution.

Fig. 2 (a), (b) and (c) show the variation of ac channel potential in response to an ac signal, V_{sig} applied to the drain, for (a) no charge, (b) positive charge and (c) negative charege near the drain. In each case C_{gd} is given by the shaded area of the graph:

$$C_{gd} = \frac{W \cdot C_{ox}}{V_{sig}} \cdot \int_{x=0}^{x=L} V_{ac}(x)\, dx$$

where C_{ox} is the gate oxide capacitance per unit area, W is the width of transistor, L is the length of transistor and x is the distance along the channel from the source to drain.

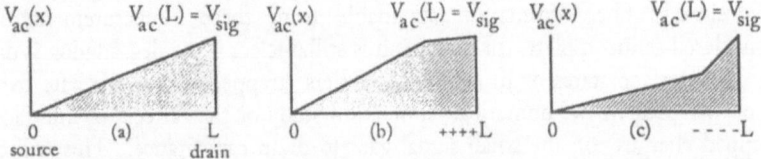

Fig. 2 AC channel potential profile showing the effect of charge in the oxide near drain: (a) with no charge, (b) with positive charge, (c) with negative charge.

As the sign of the charge is observed to vary with biasing condition a possible cause for this observation is the existence of both donor type and acceptor type interface states generated after electrical stressing located in different energy levels in the band gap. Due to the complexity of the problem the only way to examine this hypothesis is the use of numerical simulation.

3. Numerical Simulation Results

The two dimensional device simulator program MINIMOS is used to calculate C_{gd} in the presence of spatially localised interface states. The program has been modified to allow for the introduction of donor and acceptor interface states in different parts of the band gap as well as in different part of the channel.

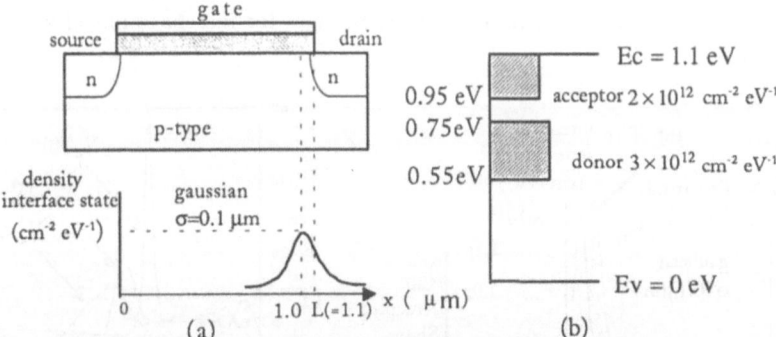

Fig. 3 (a)Assumed spatial distribution and (b)the distribution in the energy band gap of the interface states for the simulated device.

Fig. 1 (b) shows simulated C_{gd} versus gate voltage for different drain voltages with and without interface states. The assumed spatial distribution and the distribution in the energy band gap of the interface states is shown in Fig. 3. The simulation results confirm that the assumed interface state model does give rise to increased C_{gd} in low gate voltages and decreased C_{gd} in higher gate voltages (inversion).

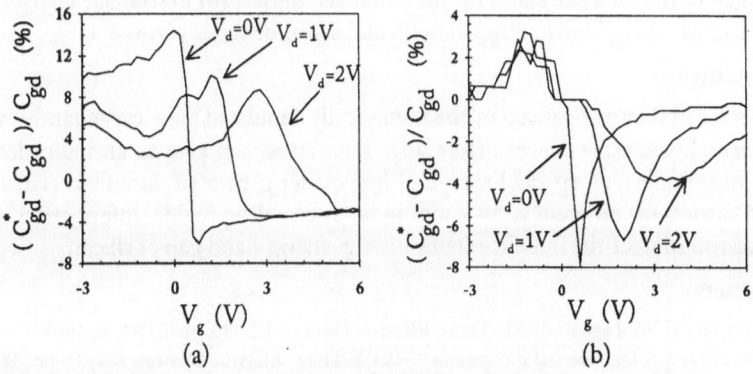

Fig. 4 (a) Percentage change in measured C_{gd} after stress versus V_g.
(b) Percentage change in simulated C_{gd} due to interface states versus V_g.

The fractional change in C_{gd} for measurement and simulation data is shown in Fig. 4. The agreement between measurement and simulation is not very good. In particular the positive peaks for the measurement case move to higher gate voltages as V_d is increased. This is not true for the simulated capacitances. This is due to the fact that the donor states are always occupied in the weak inversion regime regardless of the value of the drain voltage so their effect is not observable in the change in the C_{gd}.

It was found that an increase in C_{gd} in the weak inversion regimes after electrical stressing necessitate the existence of positive fixed charges or trapped holes near the drain junction. These positive fixed charges would then be compensated by the existence of the charges due to acceptor states in strong inversion. It was found that the interface states and fixed charge model shown in Fig. 5 (a) would produce best approximation to the observed change in C_{gd} before and after stress. The fractional change in C_{gd} for measurement and simulation data for the above model is shown in Fig. 5 (b).

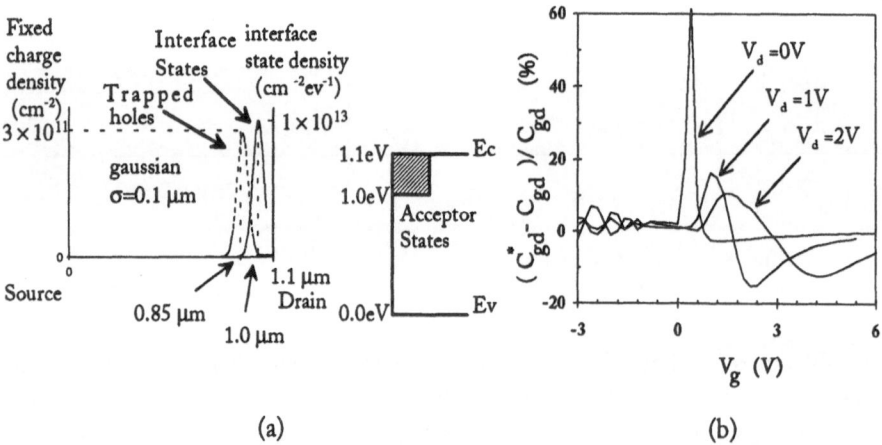

(a)   (b)

Fig. 5 (a) Assumed spatial distribution and the distribution in the energy band gap of the interface states for the simulated device. (b) Percentage change in simulated C_{gd} due to trapped holes and acceptor states versus V_g.

4. Conclusion

It is concluded that comparison of the numerically simulated gate capacitances with the experimental gate capacitances before and after stress can give us an indication of the nature, magnitude and spatial location of hot-carrier generated interface states and/or trapped carriers for different stress conditions. It can also provide information in regard to the distribution of the interface states in the energy band gap of silicon.

References

[1] K.K Ng and G.W. Taylor—IEEE Trans. Electron Devices, ED-30, pp. 871-876, 1983.
[2] T. Tsuchiya, T.Kobayashi and S.Nakajima —IEEE Trans. Electron Devices, ED-34, pp. 386-391, 1987.
[3] P. Heremans, R.Bellens, G.Groeseneken and H.E.Maes—IEEE Trans. Electron Devices, ED-35, pp. 2194-2209, 1988.
[4] Y.T. Yeow —IEEE Trans. Electron Devices, ED-34, pp. 2510-2520, 1988.

An Analytical Device Model Including Velocity Overshoot for Subquartermicrometer MOSFET

G. F. Niu, G. Ruan, and T. A. Tang

Institute of Microelectronics, Department of Electrical Engineering,
Fudan University
220 Handan Road, Shanghai 200433, CHINA

Abstract

A semi-empirical device model including velocity overshoot for subquarter-micrometer MOSFET's is presented, in which the velocity overshoot near source is attributed to the strong field here instead of the field gradient. Key features include successful modeling of the dependence of velocity overshoot on both channel length and doping, and closed form expressions for both drain current and transconductance. The proposed model is verified in comparison with experiments.

1. Introduction

Velocity overshoot was proved to enhance the drain current of subquarter-micrometer MOSFETs(SQM) by both experiments[1,2] and Monte Carlo simulation[3]. Now, analytical device models[4,5] including velocity overshoot have been developed to provide engineering tools for the development of SQM. In previous studies, the velocity overshoot near source, which is responsible for the current enhancement, is attributed to the gradient of chordal field, and the field at the source end of channel E(0) is neglected. In fact, the channel of SQM is so short that E(0) is comparable to the critical field E_s[6], and the value of E(0) derived from the equations in previous work[4,5] shows the same result. Therefore, the neglect of E(0) not only results in overestimation of the field gradient near source, but also leads to inconsistent modeling. In addition, the dependence of overshoot on channel doping cannot be modeled.

In this work, the velocity overshoot near source is attributed to the strong field here instead of its gradient, a semi-empirical overshoot factor model is proposed to incorporate drain current enhancement due to overshoot into compact device modeling.

2. Modeling

2.1. A simple model of the effect of overshoot on drain current

Although overshoot exists in the pinched-off region of even "long channel"(e.g. 2um) MOSFET, only when the channel length is down to below quarter-micrometer , is the current enhancement observable, since drain current is determined mainly by the velocity near source[7]. It is also well known that the major reason for the overshoot in the pinched-off region is the great field gradient here. However, as for the overshoot near source in the case of SQM, the field near source is quite gradual, as opposed to that in the pinched-off region, because the field E(0) increases correspondingly with the reduction of channel length, partly compensating for the increase of field gradient due to channel length reduction, the verification of this point is given later.

It is argued that the strong field near source is the very reason instead of the field gradient. The electrons are in equilibrium with lattice inside the source, then are injected cold into the channel existing strong field, the electrons gain energy from the field and the electron temperature rises. As a direct consequence of the rise of electron temperature, velocity overshoot occurs, which is very similar to the electron movement from a "cold" source into a constant field studied in [8]. The overshoot factor, defined as the average overshoot compared to the drift-diffusion value υ_{dd} near source that is responsible for the enhancement of drain current[5], is supposed to be directly proportional to the rise of electron temperature in the case of "cold" electron injection into a constant high field E(0), which is representative of the field near source as

$$r = 1 + \frac{\upsilon - \upsilon_{dd}}{\upsilon_{dd}} = 1 + \alpha \frac{T' - T_L}{T_L} \quad (1)$$

where r is overshoot factor, T_L is lattice temperature, T' is the final electron temperature after achieving balance with the field E(0), α is a coefficient determined by data fitting. With the fact that the diffusivity of electron is nearly independent of its temperature[9], T' can be obtained from the generalized Einstein relation as

$$\frac{kT_L}{q} \mu_{eff} = \frac{kT'}{q} \frac{\upsilon_{dd} (E(0))}{E(0)} \quad T' = T_L(1 + \frac{E(0)}{E_s}) \quad (2)$$

where μ_{eff} is low field effective mobility, and the following drift-diffusion velocity model is used[10]

$$\upsilon_{dd} = \frac{\mu_{eff} E}{1 + E/E_s} \text{ for } E < E_s \quad \upsilon_{dd} = \upsilon_s \text{ for } E > E_s \quad (E_s = 2\upsilon_s/\mu_{eff}) \quad (3)$$

where υ_s is the saturation velocity in silicon inversion layer, which is chosen as 7.0×10^6 cm/s[6], and the low field mobility model proposed in [10] is used. Substituting (2) and (3) into (1), one obtains

$$r = 1 + \frac{\alpha}{2} \frac{E(0)\mu_{eff}}{\upsilon_s} \quad (4)$$

The overshoot factor model proposed here is just for the modeling of enhancement of drain current due to the overshoot near source, in which the overshoot in the pinched-off region is neglected because of little contribution to drain current, as discussed above.

2.2. Drain current and transconductance model

With the proposed overshoot factor model, conventional charge control analysis in linear region, and the pseudo-two-dimensional analysis in saturation region[4], an analytical expression for drain current is obtained. An analytical expression for transconductance is obtained following the treatment in [10].

3. Results and discussions

The calculated drain current and transconductance are verified in comparison with Sai-Halasz et al.'s experiments[1,2](Fig.1 and Fig.2), where the determined coefficient α in (1) is 0.50.

The dependence of overshoot on channel doping is given in Fig.3. It can be seen that the current enhancement due to overshoot increases with the lowering of channel doping, because of improved low field mobility, which results from reduced bulk charge.

Transconductance and overshoot factor for different channel length as a function of gate voltage V_g are given in Fig.4 and Fig.5 respectively. Both of the two parameters increase with V_g first because of increased E(0), and then decrease gradually or saturate because of high gate voltage induced enhancement of surface roughness scattering and phonon scattering, which reduces the low field mobility.

For devices in Fig.2, as the channel length decreases from 0.3 to 0.1um, the field at the source end of channel E(0) increases from $0.1E_s$ to $0.7E_s$. Using the term $(E_s\text{-}E(0))/L$ as an approximation of the field gradient near source, where $E<E_s$, we see that there is little increase of the field gradient in the channel region near source. It follows that the increase of the field gradient near source is not the major reason for the increase of overshoot factor with the reduction of channel length, as discussed in the second part.

Another advantage of the proposed model is that the dependence of overshoot on both channel length and doping are accounted for. According to (4), the shorter the channel, the larger is E(0) and, therefore, the bigger is overshoot factor; similarly, lower channel doping gives less bulk charge, which in turn results in bigger overshoot factor.

4. Conclusions

The velocity overshoot near source, which is responsible for drain current enhancement, is attributed to the strong field near source instead of the field gradient. A semi-empirical model of overshoot factor is proposed and used in compact device modeling. The dependence of overshoot on both channel length and doping are successfully accounted for. Good agreement with experiments is obtained.

References

[1] G.A. Sai-Halasz et al., IEEE EDL, 8, p. 464, 1987
[2] G.A. Sai-Halasz et al., IEEE EDL, 9, p. 463, 1988
[3] S.E. Laux et al., Phys. Rev. B, 38, p. 9721, 1988
[4] K. Sonoda et al., IEEE T-ED, 38, p. 2662, 1992
[5] P.A. Blakey et al., IEEE T-ED, 39, p. 740, 1993
[6] C.G. Sodini et al., IEEE T-ED, 31, p. 1386, 1984
[7] D.A. Antoniadis et al., IEDM Tech. Dig., p. 20, 1990
[8] G. Baccarani et al., Solid State Electronics, 28, p. 407, 1985
[9] C. Jacoboni et al., Solid State Electronics, 20, p. 77, 1977
[10] K. Toh et al., IEEE J-SC, 23, p. 950, 1988

Fig.1 Calculated and measured drain current as function of drain voltage at 300K. Channel length 0.16um, acceptor concentration $3.0 \times 10^{17} \text{cm}^{-3}$, gate oxide thickness 4.5nm, junction depth 65nm, substrate bias 0.0V, gate voltage minimum 0.3V, gate voltage increment 0.1V.

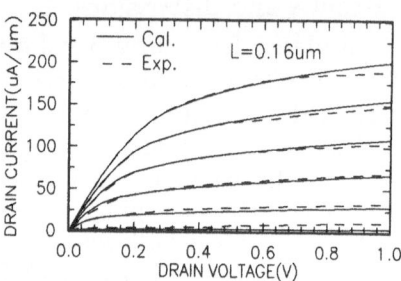

Fig.2 Calculated and measured transconductance as function of channel length at 300K. Solid curve and dashed curve are calculated results with and without velocity overshoot respectively. Transconductance are chosen at the gate voltage for maximum value at V_d=0.8V. Substrate bias is 0.0V. Device parameters are the same as that in Fig.1.

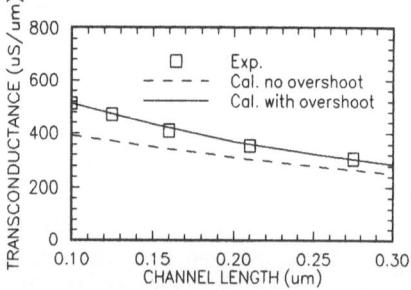

Fig.3 Current enhancement in terms of velocity overshoot factor as a function of channel length at the gate voltage for maximum transconductance with different channel doping level N_a. Device parameters and bias are the same as that in Fig.1.

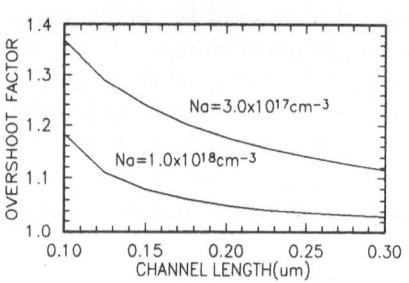

Fig.4 Transconductance as a function of gate voltage V_g-V_{th} for different channel length. Channel length minimum 0.1um, channel length increment 0.05um. Device parameters are the same as that in Fig.1. Substrate bias is 0.0V, drain bias is 0.5 V.

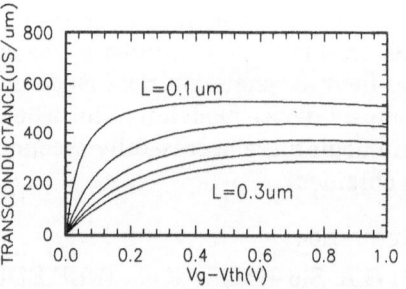

Fig.5 Overshoot factor r as a function of gate voltage V_g-V_{th} for different channel length. Channel length minimum 0.1um, channel length increment 0.05um. Device parameters and bias are the same as that in Fig.4.

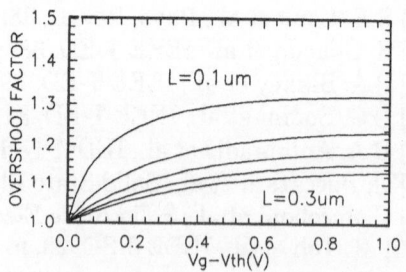

Numerical Modeling of Electrothermal Effects in Semiconductor Devices

Y. Apanovich, B. Cottle, B. Freydin, E. Lyumkis, B. Polsky, and P. Blakey

Algorithms Development Group, Silvaco International
4701 Patrick Henry Drive, Bldg. 3, Santa Clara, CA 95054, USA

Abstract

Nonisothermal steady-state and transient simulation capabilities have been implemented in the ATLAS general purpose 2D device simulator. The new capabilities support accurate simulation of silicon and heterostructure devices, taking into account lattice heating, heat sinks and other features of the thermal environment. Results of self-consistent simulations of HBT, HEMT and ESD protection devices are presented and compared with results obtained using isothermal simulation.

1. Introduction

Numerical simulation of electrothermal processes is important for the design and optimization of many semiconductor devices, including power devices, electrostatic discharge (ESD) protection devices, SOI MOSFET's, and hetero-junction devices. General electrothermal simulation capabilities have now been incorporated into the ATLAS [1] general purpose 2D device simulator. The ability of ATLAS to simulate electrothermal effects is illustrated by results obtained for steady-state simulation of AlGaAs/GaAs HBT and HEMT devices, and for transient simulation of a silicon ESD protection device.

2. Physical Model and Numerical Techniques

The electrothermal capabilities in ATLAS are based on Wachutka's thermody-namically rigorous model [2]. The current density equation includes terms that deal with variable band structure and thermal diffusion [3, 4]. The main heat source in the lattice heat equation is Joule heat. Generation/recombina-tion heat, and Thomson and Peltier terms are also included in Wachutka's model. These terms can be switched off in the ATLAS implementation to recover the 'intuitive' $(J_n \cdot \nabla E_c + J_p \cdot \nabla E_v)$ model. The dependencies of all physical parameters on lattice temperature are modeled. These dependecies lead to reduced mobilities and ionization coefficients with increasing temperature for a given electric field.

Very general thermal environments can be simulated. Realistic multi-layer heat sink structures may be specified. Thermal boundary conditions may be specified in terms of temperature, heat flux, or thermal impedance. The latter capability is useful for handling certain situations (e.g. thick substrates) in a computationally efficient manner.

The discrete nonlinear problem is solved using an iterative scheme in which Newton iteration or Gummel's method is used to solve Poisson's equation and the current continuity equations, and then the heat flow equation is solved. This approach is used for both steady-state and transient calculations.

3. Results

3.1. AlGaAs/GaAs HBT Example

Calculations that include electrothermal effects in a AlGaAs/GaAs HBT were simulated and compared to isothermal results calculated at T=300 K. Thermoisolation was assumed at all boundaries except for the substrate boundary which was modeled using a thermal resistor. Values for the thermal resistor of 0.0005 cm^2 K/W and 0.002 cm^2 K/W were considered. Figure 1 shows calculated common emitter I-V characteristics with Ib=0.5µA/µm and Ib=2µA/µm. The non-isothermal simulations predict lower currents than the isothermal calculations. The decrease is believed to be due to decreased mobility as the temperature increases. The temperature distribution in the device is shown in Figure 2 for Ib=2µA/µm, Vcol=4V, and Rth=0.002 cm^2 K/W. The temperature varies smoothly in the device with a peak value in the vicinity of the collector-base junction.

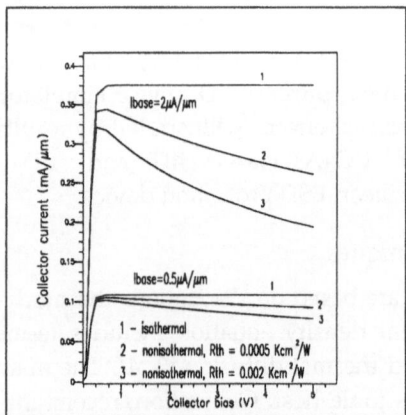

Figure 1. Common emitter I-V characteristics for the AlGaAs/GaAs HBT.

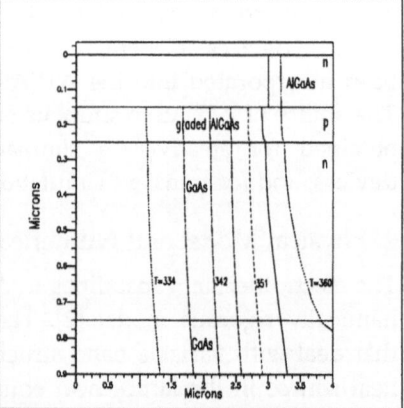

Figure 2. Temperature distribution in the HBT. Ib=2µA/µm, Vcol=4V, Rth=0.002cm^2 K/W.

3.2. AlGaAs/GaAs HEMT Example

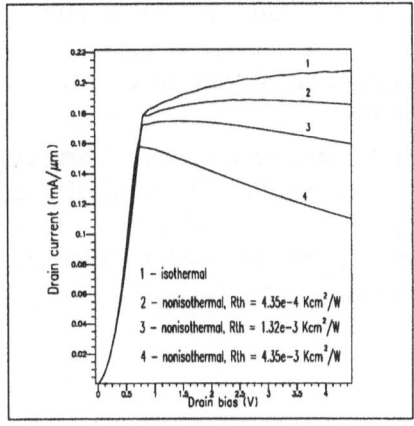

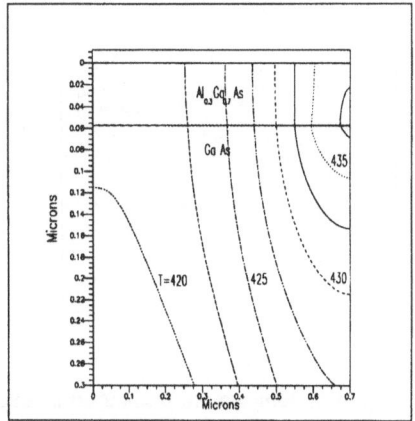

Figure 3. Id-Vd curves for the Al-GaAs/GaAs HEMT.

Figure 4. Temperature distribution in the HEMT for Vd=4V, Vg=0, Rth = 0.00132 cm^2 K/W.

A similar reduction of drain current in nonisothermal calculations, as compared with isothermal calculations, takes place in HEMT's. Figure 3 shows the calculated drain currents versus drain voltage for Vg=0. Three values of thermal resistance were used to model the effect of different heat sinks. As can be seen from Figure 3, a thermally-induced negative differential resistance may arise. The calculation with Rth=0.00435 cm^2 K/W models an extreme case with a poor heat sink. The calculated temperature distribution in the HEMT for Vd=4V, Rth=0.00132 cm^2 K/W is shown in Figure 4.

3.3. Transient Simulation of Si ESD Protection Device

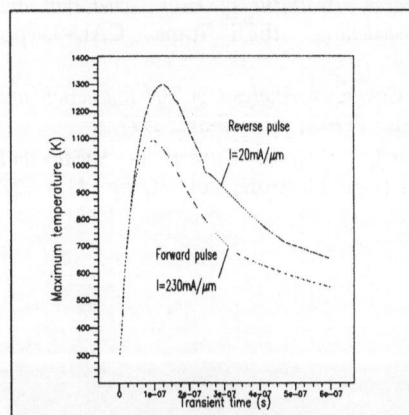

Figure 5. Maximum temperature in ESD protection device as function of time.

Electrothermal transients in a silicon ESD protection device were simulated under conditions of stress induced by an ESD human body model pulse. The current pulse was applied to the n-contact. A linear ramp of 10ns duration and a peak amplitude of 230 mA/μm for the forward pulse and 20 mA/μm for the reverse pulse was followed by exponential current decrease with a characteristic decay time of 150ns. The maximum temperature in the structure versus time is shown in Figure 5.

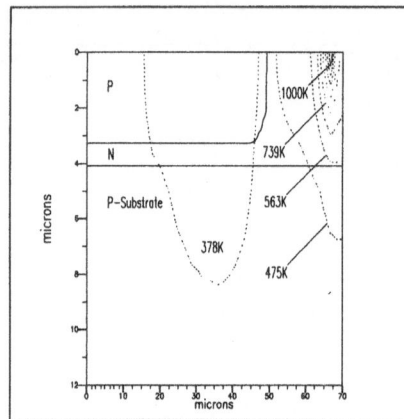

A temperature distribution for the case of a forward current pulse at t=92 ns is shown in Figure 6. The location of a "hot spot" is clearly visible.

Knowledge of the internal temperature distributions helps device designers to identify failure mechanisms, and evaluate alternative designs.

Figure 6. Temperature distribution in ESD protection device. Forward current pulse, t=92 ns.

4. Summary

A self-consistent electrothermal model of lattice heating has been incorporated into a general purpose 2D device simulator. Steady-state and transient capabilities are included, and heat sinks and general thermal boundary conditions are supported. Calculated results demonstrate the influence of thermal effects on the operating characteristics of HBT, HEMT, and ESD protection devices.

References

[1]   *ATLAS II Version 1.0 User Manual*, SILVACO International, Santa Clara, 1993.

[2]   G. K. Wachutka, "*Rigorous Thermodynamic Treatment of Heat Generation and Conduction in Semiconductor Device Modelling*," IEEE Trans., CAD-9, pp. 1141-1149, 1990.

[3]   J. E. Sutherland and J. R. Hauser, "*A Computer Analysis of Heterojunction and Graded Composition Solar Cells*," IEEE Trans., ED-24, pp. 363-372, 1977.

[4]   A. H. Marshak and K. M. van Vliet, "*Electrical Current in Solids with Position-Dependent Band Structure*," Solid State Electron., vol. 21, pp. 417- 427, 1978.

SIMULATION OF SEMICONDUCTOR DEVICES AND PROCESSES Vol. 5   293
Edited by S. Selberherr, H. Stippel, E. Strasser – September 1993

Analysis of Charge Storage in Polysilicon Contacts

D. Bardés and R. Alcubilla

Departament d'Enginyeria Electrònica, ETSETB - UPC
P.O.Box 30002, E-08034 Barcelona, SPAIN

Abstract

We present the use of a commercial 2D device simulator, MEDICI [1], to analyze a polysilicon emitter contact. By using a 2-box model we adjust simultaneously the effective recombination velocity and the stored charge in the polysilicon layer.

1. Introduction

Polysilicon emitter transistors have been a main topic in device research during the last years [2], [3]. The progressive scaling-down of both horizontal and vertical dimensions enhances the interest of an accurate simulation of the static and dynamic characteristics of this kind of contacts.

Simulation of these contacts is usually done [4] by using an effective recombination velocity as a boundary condition. The stored charge can be then considered through a correction in the emitter transit time τ_e [5].

$$\tau_e = \tau_{es} \cdot (1 + \frac{Q_{ep}}{Q_{es}})\tag{1}$$

In most 2D simulators as MEDICI, polysilicon contacts are simulated through the use of virtual semiconductor parameters in the polysilicon layer. This allows to fit the DC characteristics of the whole bipolar transistor [4].

In this work we propose a two layer model that can be easily implemented through MEDICI allowing to adjust the DC characteristics and at the same time to evaluate the stored charge in the polysilicon layer.

2. Simulation method

Classically, the simulation of polysilicon emitter contacts trough MEDICI is done providing an equivalent boundary condition [6] to the device, that considers an effective surface recombination velocity:

$$J = q \cdot S_{eff} \cdot p\tag{2}$$

S_{eff} is obtained by modifying the minority carrier mobility and lifetime in the region corresponding to the polysilicon. Solving the diffusion equation in this zone with these modified parameters leads to the desired effective recombination velocity.

$$S_{eff} = \frac{D_p}{L_p} \coth \frac{W_{Poly}}{L_p} \tag{3}$$

Where D_p is the modified diffusion coefficient for holes, L_p the modified diffusion length and W_{Poly} is the polysilicon layer width. Although this procedure can be useful for DC conditions, we have no mean to control the stored charge in the poly. As this charge may significantly modify the forward transit time in scaled down structures [5], it is worthy to have a realistic estimation for the stored charge in the poly, and, as a consequence, for τ_{poly}.

Our proposal, based on the box model [6], is to consider two regions to simulate the contact. One corresponding to the thin oxide layer (or a grain boundary, if there is no oxide present), and another to the rest of the contact. The properties of this oxide layer mostly determine the stored charge in the poly.

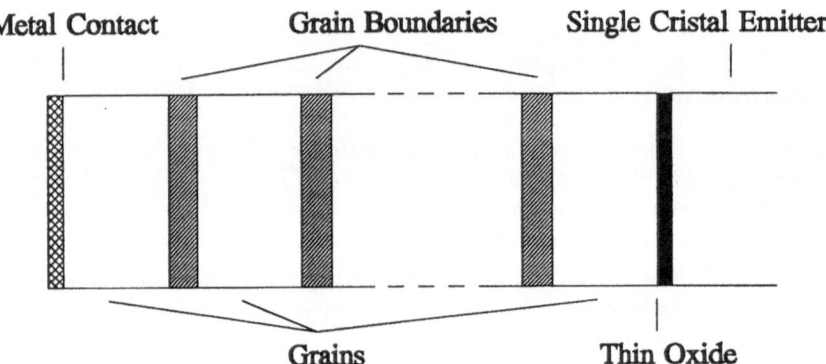

Figure 1: Scheme of a polysilicon emitter, box model.

The parameters describing this two regions, i.e. the minority carrier lifetimes and mobilities, are found from the equations and parameters given in [6] for the hole transport.

For the oxide layer, the effective diffusion coefficient, D_{eff}, and the effective diffusion length, L_{eff} are calculated through.

$$L_{eff} = \frac{d_{ox}}{\arg \cosh \frac{S_{ox}+T_{ox}}{T_{ox}}} \tag{4}$$

$$D_{eff} = T_{ox} \cdot L_{eff} \cdot \sinh \frac{d_{ox}}{L_{eff}} \tag{5}$$

T_{ox} is a coefficient related to the tunneling trough the oxide, and depends [6] on the hole potential barrier, χ_h and the hole effective mass in the oxide, m_h^*, S_{ox} a surface recombination velocity located at the two edges of the thin oxide layer, and d_{ox} is the oxide layer width.

The second region, which models the polysilicon, can be described classically through the modified mobility method mentioned above.

In this way good estimations for the charge in the poly can be obtained. If more accuracy is desired, it is easy to extend the method to consider the polysilicon grains and grain boundaries in different regions instead of joining them in a single region. The grain boundary zones are described in [6] through expressions (4) and (5) using a T_{gb} (a transport coefficient in the grain boundary), and S_{gb}, for recombination, instead of T_{ox} and S_{ox}.

3. Results

We have used the parameters shown in table I in equations (3) to (5) to simulate a polysilicon emitter with MEDICI, and to perform the analytical calculations. In Figures 2 and 3 we compare the obtained results with those found through analytical expressions.

Polysilicon layer thickness	W_{Poly}	$0.4\,\mu m$
Hole transport parameters in grain boundaries	T_{gb}	$2.5\ 10^5$ cm/s
	S_{gb}	$7.5\ 10^4$ cm/s
Grain boundaries thickness	d_{gb}	$20\,\text{Å}$
Recombination velocity at the oxide edges	S_{ox}	$1.5\ 10^3$ cm/s
Hole effective mass in the oxide	m_h^*	$0.42\ m_0$
Hole potential barrier in the oxide	χ_h	1 eV
Hole diffusion coefficient in the grains	D_p	3.28 cm^2/s
Hole diffusion length in the grains	L_p	$0.812\,\mu m$
Donor concentration in the polysilicon	N_{dpoly}	$7\ 10^{19}$ cm^{-3}

Table 1: Values used for the calculations

We can see in Figure 2 the stored charge in the poly as a function of the oxide layer thickness. Results obtained with the classical method and through the analytical expressions are shown for comparison.

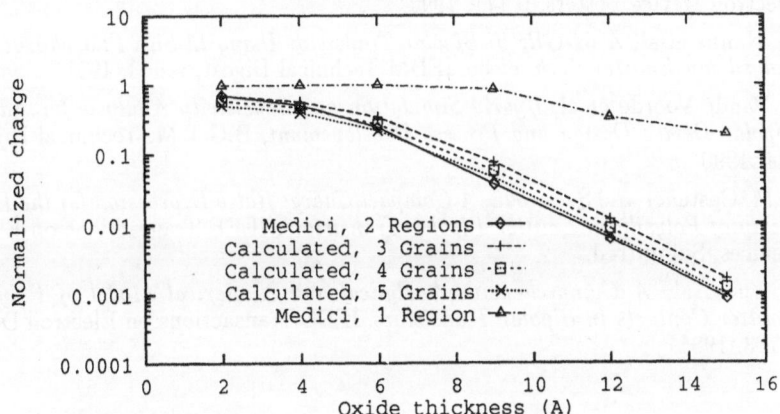

Figure 2: Integrated charge density as a function of oxide thickness.

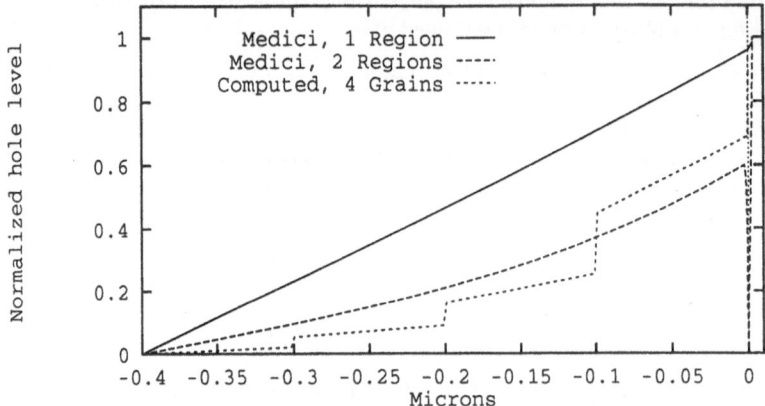

Figure 3: Hole concentration profile versus depth in polysilicon and oxide layers.

In Figure 3 the hole profile in the polysilicon layer is shown. Three cases are compared for an oxide thickness of 4 Å: the obtained analitically in the case of 4 grains [5], and those found with MEDICI, through the proposed method and by using the classical one.

4. Conclusions

This work uses a numerical simulator, MEDICI, to evaluate the charge storage in the polysilicon layer of a polysilicon emitter bipolar transistor. The used method provides an estimation for the stored charge that agrees with previous results. If more accuracy is needed, a straightforward extension of the method has been proposed.

References

[1] Technology Modelling Associates, Inc., *MEDICI User's Manual.* March 1992

[2] G. L. Patton et al, *75-GHz f_t SiGe-Base Heterojunction Bipolar Transistors*, IEEE Electron Device Letters, p 171. 1990

[3] M. Nanba et al, *A 64 GHz Si Bipolar Transistor Using In-Situ Phosphorus Doped Polysilicon Emitter Technology*, IEDM Technical Digest, p 443. 1991

[4] P. Vande Voorde et al, *Hybrid Simulation and Sensitivity Analysis for Advanced Bipolar Device Design and Process Developement*, B.C.T.M. Technical Digest, p 114. 1990

[5] L. M. Castañer and S. Sureda, *A Compact Charge Ratio Expression for the Emitter Delay of Polysilicon Emitter Bipolar Transistors*, IEEE Transactions on Electron Devices, Submitted.

[6] Z. Yu et al, *A Comprehensive Analytical and Numerical Model of Polysilicon Emitter Contacts in Bipolar Transistors*, IEEE Transactions on Electron Devices, p 773. 1984

Simulation of the Conduction Mechanisms in Polycrystalline Silicon Thin Film Transistors

S. Mottet and M. Kandouci

CNET Lannion B, France Telecom
Route de Trègastel, BP 40, F-22301 Lannion Cédex, FRANCE

Abstract

A simulator for the design of polycrystalline silicon thin film transistors, used for the command of flat panel liquid crystal display pixels, is discribed. This simulator is based on the plolycrystalline silicon specific physical equations. It permits to describe both passing and blocking operating modes.

1. Introduction

Polycrystalline silicon thin film transistors are being used for the command of flat panel liquid crystal display pixels [1]. A major point is to determine their blocking capability. Therefore the simulation needs to describe both passing and blocking characteristics. The transistors are made of a thin undoped polycrystalline silicon film deposited on a glass substrate (figure 1). The film thickness is about 500 Å. The monocrystalline grain thickness is the layer thickness. The grain size, in the film plane, is about 3000 Å. The drain and source contacts are made of n+ polycrystalline silicon.

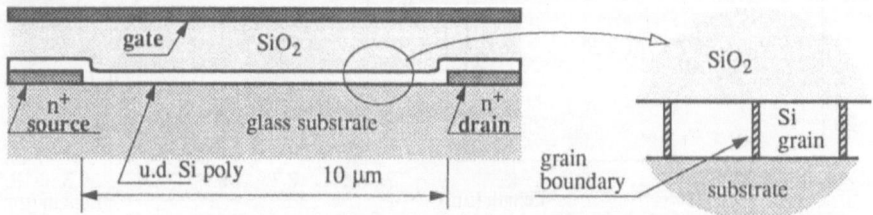

Figure 1 : Polycrystalline Silicon Thin Film Transistor structure, and detail of the grains.

The grain surface has a major influence on the conduction properties of the thin film. The surface lattice defect implies silicon dangling bonds and band tail densities which govern the surface charge density and the recombination-generation mechanisms. The grain boundary medium is a disordered material (amorphous) with a large band gap, comparable to an oxide one. The conduction through the grain boundary (about 10 Å thick) is governed by tunnel effect. The monocrystalline silicon grains are supposed to have the same energy band structure as bulk silicon, except at the surface where dangling bonds and band tails are added.

2. Physical model

The modelling of the layer supposes parallelepipedic monocrystalline grains. The physical model describing the steady state conduction in semiconductor material is composed of Poisson's equation

$$\nabla \cdot (\varepsilon \cdot \nabla \cdot \varphi) = -\rho \tag{1}$$

and electron and hole balance equations $-\dfrac{1}{q} \cdot \nabla \cdot J_n = -U$; $\dfrac{1}{q} \cdot \nabla \cdot J_p = -U$ (2)

2.1 Grain material

In the grains (bulk silicon), the carrier densities are expressed within Maxwell-Boltzmann statistics

$$n = N_c \cdot exp\left(-\frac{E_c - E_{Fn}}{k \cdot T}\right) \; ; \; p = N_v \cdot exp\left(\frac{E_v - E_{Fp}}{k \cdot T}\right) \tag{3}$$

and the charge density is $\rho = -q \cdot (n - p - dop)$ (4)
where *dop* is the residual doping density.

The drift-diffusion model is used for the electron and hole current densities (including a mobility law to take into account the high field velocity saturation) :

$$
\begin{aligned}
J_n &= n \cdot \mu_n \cdot \nabla E_{Fn} \\
J_p &= p \cdot \mu_p \cdot \nabla E_{Fp}
\end{aligned}
\quad \text{with } \mu_{n,p} = \frac{\mu_{On,p}}{1 + \dfrac{\mu_{On,p}}{v_{Sn,p}} \cdot \left|\dfrac{1}{q} \cdot \nabla E_{Fn,p}\right|}
\tag{5}
$$

The recombination-generation is governed by the Shockley-Hall-Read mechanism:

$$U_{SHR} = \frac{n \cdot p - n_i^2}{\tau_p \cdot (n + n_i) + \tau_n \cdot (p + n_i)} \tag{6}$$

In the blocking mode (hole accumulation), high fields appear in the space charge region close to the drain contact due to the reverse diode biased effect (figures 2, 3).

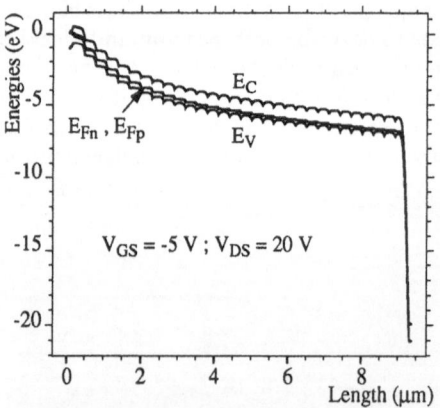
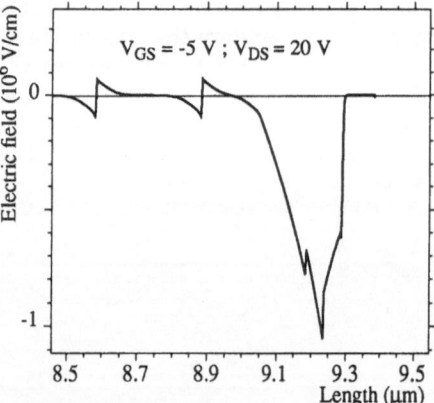

Figure 2: Band diagram along the channel in blocking mode showing both serial resistance effect and diode voltage drop on drain side.

Figure 3: Enhancement of the near drain region showing the very high electric field values (blocking mode).

This induces electron and hole generation by impact ionisation :

$$G_{In,p} = \frac{1}{q} \cdot \alpha_{n,p} \cdot |J_{n,p}| \quad \text{with } \alpha_{n,p} = \alpha_{On,p} \cdot exp\left(-\frac{E_{n,p}}{\left|\dfrac{1}{q} \cdot \nabla E_{Fn,p}\right|}\right) \tag{7}$$

So that the total recombination-generation term is $U = U_{SHR} - G_{In} - G_{Ip}$ (8)

2.2 Grain surface and grain boundary

Band tails and ionised dangling bonds densities govern the equilibrium of the grain surface. The carrier surfacic densities in the band tails are :

$$n_s = \int_{-\infty}^{\infty} \frac{E_c \, N_{0c} \cdot exp\left(-\dfrac{E_c - E}{E_{0c}}\right)}{1 + exp\left(\dfrac{E - E_{Fn}}{k \cdot T}\right)} \cdot dE \; ; \quad p_s = \int_{E_v}^{\infty} \frac{N_{0v} \cdot exp\left(\dfrac{E_v - E}{E_{0v}}\right)}{1 + exp\left(\dfrac{E_{Fp} - E}{k \cdot T}\right)} \cdot dE \qquad (9)$$

where N_{0c} and N_{0v} are the effective state densities in the conduction and valence band tails ;. E_{0c} and E_{0v} are the respective energy extrema of the band tails.
The dangling bonds of the silicon atoms make appear two deep energy levels E_{lp1} and E_{lp2}. The total dangling bond surfacic charge density is summation of the two ionised states of this amphoteric defect [2].

$$N_{lp}^- = N_{lp} \cdot \frac{n_s + p_{s1}}{(n_s + n_{s1}) + (p_s + p_{s1})} \; ; \; N_{lp}^+ = N_{lp} \cdot \frac{n_{s2} + p_s}{(n_s + n_{s2}) + (p_s + p_{s2})} \qquad (10)$$

where $n_{s(1,2)} = n_s \cdot exp\left(\dfrac{E_{lp(1,2)} - E_{Fn}}{k \cdot T}\right) \qquad P_{s(1,2)} = p_s \cdot exp\left(\dfrac{E_{Fp} - E_{lp(1,2)}}{k \cdot T}\right)$

the charge density is then $\rho_s = -q \cdot (n_s - p_s + N_{lp}^- - N_{lp}^+)$ (11)

The surfacic recombination is governed by the capture emission process between band tails and dangling bonds :

$$U_{lp} = N_{lp} \cdot C_{lp} \cdot \left[\frac{n_s \cdot p_s \cdot \left(1 - exp\left(\dfrac{E_{Fn} - E_{Fp}}{k \cdot T}\right)\right)}{(n_s + n_{s1}) + (p_s + p_{s1})} + \frac{n_s \cdot p_s \cdot \left(1 - exp\left(\dfrac{E_{Fn} - E_{Fp}}{k \cdot T}\right)\right)}{(n_s + n_{s2}) + (p_s + p_{s2})} \right] \qquad (12)$$

The conduction through the grain boundaries is described by electron and hole tunnel effects.

$$J_n = -q \cdot v_n \cdot \Pi \cdot n \cdot \left(1 - exp\left(\dfrac{\Delta E_{Fn}}{k \cdot T}\right)\right) \text{ and } J_p = q \cdot v_p \cdot \Pi \cdot p \cdot \left(1 - exp\left(\dfrac{\Delta E_{Fp}}{k \cdot T}\right)\right) \quad (13)$$

where ΔE_{Fn} and ΔE_{Fp} are the variations of the imref at the grain boundary ; n, p are the carrier concentrations at the surface of the grain. Π is the transmission factor of the intergrain medium and $v_{n,p}$ is the thermionic emission velocity [3].

The interband tunnel effect (Zener) which appears on the top of the drain contact for large gate biases in blocking mode is added to these mechanisms [4]. This mechanism depends directly on the electric field induced by the gate on the top of the n+ material.

3. Numerical method

The numerical model is composed of the equations 1 to 13 describing the different mechanisms. The discretisation scheme is a Box Method [5] type which allows to describe easily the surfacic terms. The internal behaviour of each grain and grain boundary has to be described. A full 2D simulation leads to a prohibitive number of nodes. In order to minimize the number of nodes the discretisation is simplified so as to have a 2D description of Poisson equation and a 1D description of the balance equations (see figure 4). So that the total equation set can be solved within a 1D direct coupled method.

4. Application

Figure 5 shows the characteristics of the structure described in figure 1. This figure shows both the passing and blocking characteristics. This classical device exhibits an excess of current under blocking operation which may penalize its use in flat liquid crystal display applications. The origin of the excess current is both impact ionisation and Zener effects. These mechanisms which increase rapidly with the fields are presently limited by the grain surface recombinations on the dangling bonds. No breakdown can be observed in the blocking characteristics because of this very efficient limiting mechanism. A new offset-gate structure has been optimized by means of the simulator [6]. This type of device does not exhibit any excess current. The study permitted to verify the quality of the simulator to predict the behaviour of new devices

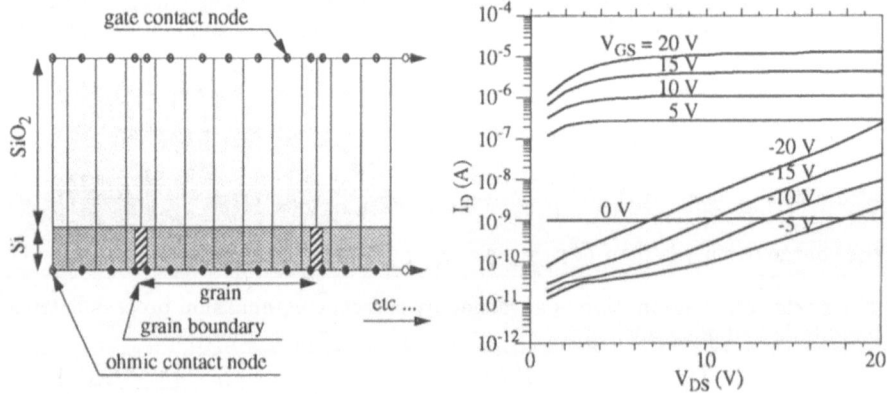

Figure 4 : Box method discretisation scheme used to simulate the transistor structure, including the grain boundaries. The unknown nodes are black dots.

Figure 5 : Simulation showing the dynamics between blocking ($V_{GS} < 0$) and passing ($V_{GS} > 0$) characteristics.

References :

[1] B. Loisel, L. Haji, P. Joubert and M. Guendouz, "Crystallized silicon films for actives devices", Springer Proceedings in Physics 35, pp. 283-288 (1988)

[2] J. Hubin, A.V. Shah and E. Sauvin,"Effects of dangling bonds on the recombination function in amorphous semiconductors," Philosophical Magazine Letters, Vol. 66, No. 3, pp. 115-125, (1992)

[3] S. Mottet and J.E. Viallet, "Thermionic emission in semiconductors," SISDEP, Vol.3, Ed. G. Baccarani, M. Rudan - Bologna (Italy), pp. 97-108, (1988)

[4] A. Schenk, "Rigorous theory and simplified model of the band to band tunneling in silicon," Solid-State Electronics, Vol. 36, No. 1, pp. 19-34, (1993)

[5] C. Simon, S. Mottet and J.E. Viallet, "Autoadaptative Mesh Refinement", SISDEP, Vol. 4, Ed. W. Fichner, D. Aemmer (Hartung-Gorre), pp. 225-233,(1991)

[6] M. Bonnel, N. Duhamel, M. Kandouci, B.Loisel and Y. Pelous, "Si Poly TFT's with low off-current for flat panel displays," Proceedings of EURODISPLAY' 93, Strasbourg (F), sept (1993)

SIMULATION OF SEMICONDUCTOR DEVICES AND PROCESSES Vol. 5
Edited by S. Selberherr, H. Stippel, E. Strasser – September 1993

Extraction of Parameters for Balance Equations from Monte-Carlo Simulations

M. Grabe and C. Peschke

Arbeitsbereich Hochfrequenztechnik, TU Hamburg-Harburg
Wallgraben 55, D-21073 Hamburg 90, GERMANY

Abstract

The Boltzmann transport equation (BTE) can be transformed in an equivalent balance equation system for an infinite number of moments. If this system is trunctated, an approximate balance equation system is obtained. The equations consist of moments, space and time derivatives of the moments, and coefficients being functions of the moments. We show that the coefficients for a three moment (hydrodynamic) model can easily be extracted from exact solutions of the BTE for homogeneous material. No approximations have to be made concerning the energy-momentum-relation or the shape of the distribution function. In principle, the method can be extended to higher order equations.

1. The Equivalent Moment Equations

The dynamics of the electron distribution function is described by the *Boltzmann-transport-equation* [1]:

$$\frac{\partial}{\partial t} f(\vec{x}, \vec{k}, t) = -\vec{u}\frac{\partial}{\partial \vec{x}}f - \dot{\vec{k}}\frac{\partial}{\partial \vec{k}}f - \hat{Q}[f] . \tag{1}$$

The electron distribution function may be expanded in any set of functions of the electron wave number $\vec{k}$ which is complete. Elements of such a series are designated with $\hat{m}_\alpha(\vec{k})$. Insertion of this expansion in the BTE, multiplication with the functions $\hat{m}_\alpha$ and integration over the k - space yields an equation system for the coeffiecients of this series. In the following, we first consider electron transport in one-dimensional varying fields. The extension to more than one dimension will be discussed later. The series for the distribution function may be written in the form [2]

$$\frac{\partial}{\partial t} n < \hat{m}_\alpha > = n \cdot \dot{k} < \frac{\partial}{\partial k}\hat{m}_\alpha > - \frac{\partial}{\partial x} n < u \cdot \hat{m}_\alpha > - n \cdot S^{(\alpha)} \tag{2}$$

with :

$$S^{(\alpha)} = \int d^3k \, \hat{m}_\alpha \cdot \hat{Q}(k) , \qquad < \hat{m}_\alpha > = \frac{\int d^3k \, \hat{m}_\alpha \cdot f}{\int d^3k \, f} = \frac{1}{n}\int d^3k \, \hat{m}_\alpha \cdot f . \tag{3}$$

The functions $n \cdot < \hat{m}_\alpha >$ are designated as *moments* of the distribution function. We can take the series (2) not to be an equation system for the coefficients of the series expansion but for the moments themselves. If this series is truncated, a finite number of equations describes electron transport. The equations can be solved, if all terms in (2) are known as functions of the moments.

The moments have to be chosen so that the physical behaviour of the system is described with as few parameters as possible. A natural choice for the moments of zeroth, first and second order are n, $n \cdot v$ and $n \cdot w$. Here, n is the electron density, $v = < u >$ the mean electron velocity and $w = < \epsilon >$ the mean electron energy. Equation sets of this kind are called *hydrodynamic* models.

2. Introducing Physical Parameters

In equation (2), the physical significance of the coefficients is not obvious, and we now introduce expressions relating the coefficients to relaxation times, mobilities, and electron temperature. Relaxation times are defined by

$$\tau^{(\alpha)} = \frac{< \hat{m}_\alpha > - < \hat{m}_\alpha >_0}{S^{(\alpha)}} \, .$$

Note that these relaxation times may depend on any other moments. With

$$\delta \hat{m}_\alpha = \hat{m}_\alpha - < \hat{m}_\alpha > \tag{4}$$

we define

$$\hat{C}^{(\alpha)} = < \delta u \cdot \delta \hat{m}_\alpha > \quad \text{and} \quad \hat{D}^{(\alpha)} = \tau^{(\alpha)} \cdot \hat{C}^{(\alpha)} \, . \tag{5}$$

Then (2) can be rewritten in the form

$$< \hat{m}_\alpha > = \hat{\mu}^{(\alpha)} \cdot k_{\gamma'} + \hat{D}^{(\alpha)} \frac{\partial}{\partial x} n + < \hat{m}_\alpha >_0 \, . \tag{6}$$

Here, $\hat{\mu}^{(\alpha)}$ is the generalized mobility

$$\hat{\mu}^{(\alpha)} = \tau^{(\alpha)} \cdot [< \frac{\partial}{\partial k} \hat{m}_\alpha > - < u > \frac{\partial}{\partial x} < \hat{m}_\alpha > + \frac{\partial}{\partial x} \hat{C}^{(\alpha)} + \frac{\partial}{\partial t} < \hat{m}_\alpha >] \tag{7}$$

as well as $\hat{D}^{(\alpha)}$ is the generalized diffusivity. The coefficients $\hat{C}^{(\alpha)}$ are related to the electron temperature which is given by

$$k_B T = < \delta u \cdot m^*(\epsilon) \cdot \delta u > \, . \tag{8}$$

For example, $\hat{C}^{(1)}$ can be rewritten as

$$\hat{C}^{(1)} = \frac{k_B}{< m^*(\epsilon) >} [T + \delta T^{(1)}] \tag{9}$$

with :

$$k_B \delta T^{(1)} = 2 \cdot < u > < m^*(\epsilon) \cdot \delta u > - < u \cdot \delta m^*(\epsilon) \cdot u > \, . \tag{10}$$

If $m^*(\epsilon) = \text{const.}$ one obtains $\delta T^{(1)} \Rightarrow 0$. Hence $\delta T^{(1)}$ is a correction term taking into account nonparabolicity . Similarly, $\hat{C}^{(2)}$ is given by

$$\hat{C}^{(2)} = < u > \cdot k_B [T + \delta T^{(2)}] \tag{11}$$

with :

$$k_B\,\delta T^{(2)} \;=\; \frac{1}{2}<u>\cdot(<\delta u\cdot m^*(\epsilon)> + \frac{<\delta u\cdot m^*(\epsilon)\cdot\delta u^2>}{<u>^2}) \,. \qquad (12)$$

The correction $\delta T^{(2)}$ disappears if the distribution function is symmetric with respect to $<u>$. If this symmetry is not given, a contribution of $\delta T^{(2)}$ to heat conduction results [3].

To obtain a conventional hydrodynamic model, several of the following approximations usually are made (e. g. [4], [5]):

- The electrons have a constant effective mass

- The distribution function is symmetric with respect to $<u>$

- The relaxation times depend only on the mean electron energy and are always given by their stationary values

3. Extraction of the Coefficients from Monte-Carlo Simulations

Usually, the coefficients in (2) are determined using some approximations yielding the v- and w-dependence. However, in principle these coefficients can be determined exactly by performing appropriate simulations. Here we simulate the reaction of an electron ensemble to square pulses of the electric field strength and store the coefficients occuring in (2) as functions of v and w. Fig. 1 shows that no ambiguities arise since the curves $w(v)$ belonging to different values of the height of the pulse do not intersect. Furthermore, any relevant combination of v and w occurs, thus yielding coeffients for any combination of v and w occuring during a simulation. The coefficients do not depend on n as far as electron-electron-interaction and Pauli-exclusion-principle are not taken into account.

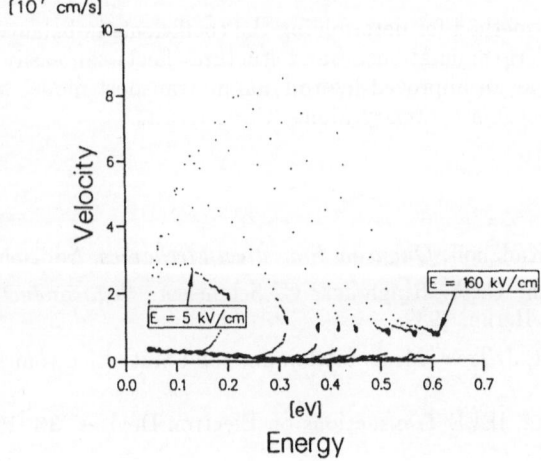

Figure 1: Mean electron velocity as a function of mean electron energy.

Fig. 2 shows the energy relaxation as function of electron energy and electron velocity $S^{(2)}(v, w)$. Usually, this function is taken to be a function of energy only $S^{(2)}(w_{stat})$. Hence a systematic error occurs if strongly nonstationary processes are simulated. The values tabulated in simulations of the one-dimensional case can also be used to perform two- or three-dimensional calculations. In these cases, one has to assume that the distribution function as a function of $|v|$ and w has the same shape as in the one-dimensional case. If heat conduction caused by gradients of the temperature is to be taken into account, higher order equations have to be studied. In principle, parameters can be calculated in a similar manner as described above.

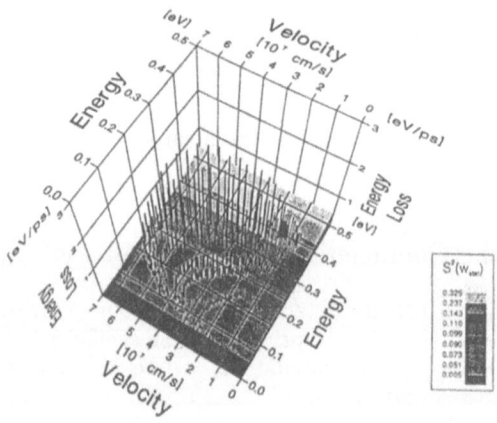

Figure 2: Mean electron energy loss $S^{(2)}(v, w)$ versus $S^{(2)}(w_{stat})$

4. Conclusions

We have developed a method for determining the coefficients in balance equations by appropriate Monte-Carlo simulations. Band stucture effects can easily be taken into account. We arrived at an improved hydrodynamic transport model which yields in specific cases the exact nonstationary Monte-Carlo results.

References

[1]   G. Baym, L. P. Kadanoff, *Quantum Statistical Mechanics*, Addison-Wesley, 1986

[2]   P. A. Markowich, C. A. Ringhofer, C. Schmeiser, *Semiconductor Equations*, Springer Verlag, Berlin, 1989

[3]   D. L. Woolard, R. J. Trew, M. A. Littlejohn, Solid State Electronics, **31**, pp. 571-574

[4]   Y. Feng, A. Hintz, IEEE Transactions on Electron Devices, **33**, 1988, pp. 1419-1431

[5]   M. Nekovee, B. J. Geurts, H. M. J. Boots, M. F. H. Schuurmanns, Physical Review B, **45**, 1992, pp. 6643-6651

SIMULATION OF SEMICONDUCTOR DEVICES AND PROCESSES Vol. 5   305
Edited by S. Selberherr, H. Stippel, E. Strasser – September 1993

The Influence of Technological Parameters on Ultra-Short Gate Si-NMOS Transistor Performances

M. Charef, F. Dessenne, J. L. Thobel, L. Baudry, and R. Fauquembergue

IEMN, UMR 9929, Université des Sciences et Technologies de Lille
Bât P3, F-59655 Villeneuve d'Ascq, FRANCE

Abstract

We have studied the effect of technological parameters on the performances of an ultra-short Si-NMOS transistor, at room temperature, using 2D Monte Carlo simulation. Results obtained for a device with source-drain extensions are compared with experiments. Doping profiles of Gaussian form are assumed. The main physical phenomena involved in the device behaviour are described. Additionally, electrical characteristics are analysed (transconductance, threshold voltage, cut-off frequency). Transconductance of 600mS/mm and intrinsic cut-off frequency close to 300Ghz are obtained for a NMOS structure with optimized source drain extensions and 0.07μm gate length.

1. Introduction

We used a 2D Monte Carlo method [1, 2] to study the influence of technological parameters on the performances of ultra-short gate NMOS transistor including source-drain extension (SDE) [3] in order to define an optimized structure. The structure, sketched in figure 1, has a gate oxide thickness of 45Å, a junction depth of 0.1μm. The SDE doping profile is Gaussian with maximum concentration at a projected range Rp, a parallel standard deviation ΔRp and a transverse standard deviation ΔRt close to 800Å and 400Å respectively. The channel doping profile is also Gaussian. The doping level is 10^{19}cm^{-3} for the heavily doped region. A uniform substrate doping concentration of 10^{16} cm^{-3} is used. The Si/SiO$_2$ interface presents uniform surface states Qss close to 3* 10^{10}cm^{-2} and fixed oxide charges Qox [4]. The substrate and source are assumed to be grounded. We will focus on the physical information and the performances of device extracted from the simulation and their comparison with experimental data at room temperature [3].

2. Influence of Source-Drain Extensions

In this section, we assume channel profile parameters: DI=2*10^{12}cm^{-2}; Rp=100Å; and Qox= 3*10^{10}cm^{-2}.

2.1. Influence of Source-Drain Length Lz

We have presented in figure 2, the dependence of the drain current on gate voltage, for two different values of the source drain extension length Lz (0.12μm and 0.07μm), with SDE doping profile parameters: DI=10^{14}cm^{-2}, and Rp=100Å. It should be noted that the gate charge control decreases if the extended zone length increases. As the length of the SDE increases, it introduces series resistance to source and drain, raising the electric field. This increased field is undesirable because it creates energetic carriers that degrade the velocity overshoot resulting in a limitation of the maximum voltage that can be used. Through a set of simulations, we found that the optimal value of the SDE length is close to 0.07μm.

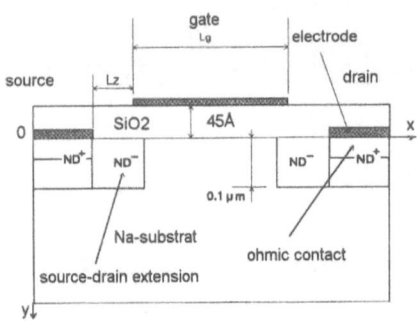

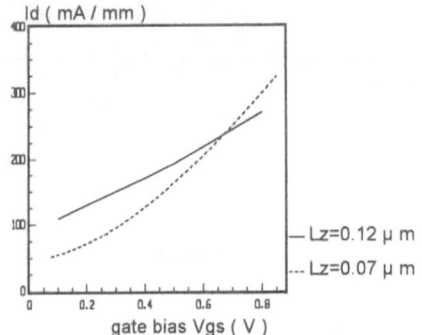

Fig 1 Schematic cross section showing simulated device structure with ultra-short gate.

Fig.2 Transfer characteristics for two values of source drain extension length: 0.12μm   and   0.07μm   (Lg=0.1μm; Vds=1V).

2.2. Source-Drain Extensions Doping Profile

Now, we assume Lz parameter is 0.07μm. Figure 3 shows the evolution of the transconductance as a function of the applied gate voltage Vgs, for two implantation doses (10^{14}cm^{-2} and $5*10^{13}$cm^{-2}). Projected range Rp of the SDE is 100Å. We observe a decrease of the transconductance with reducing doping level. This behaviour is caused by mobile charges deficiency in SDE zones and thus in the transistor channel. Figure 4 shows transconductance evolution with gate voltage for two values of the projected range: Rp=100Å and Rp=1000Å. The implantation dose of the SDE is 10^{14}cm^{-2}. We observe that the conductance decreases when the distance between the maximum doping and the interface increases. This reduction is attributed to the spreading of carrier distribution into the semiconductor as the projected range increase.

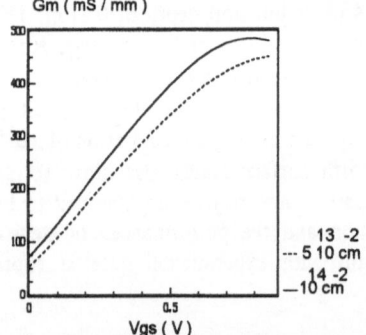

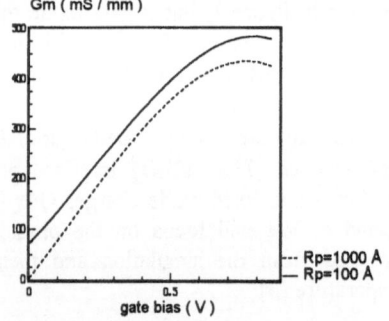

Fig.3: Intrinsic transconductance versus gate bias for two values of implantation ion doses in SDE: 10^{14}cm^{-2} and $5*10^{13}$cm^{-2}(Lg=0.1μm Vds=1V).

Fig.4: Intrinsic transconductance versus gate bias with two values of projected range: 100Å and 1000Å (Lg=0.1μm; ΔRp=800Å; ΔRt=400Å and Vds=1V)

In the following study, the parameters of SDE are: DI=10^{14}cm^{-2}, Rp=100Å, Lz=0.07μm.

3. Treshold Shift

For n-channel MOSFET structure with positively biased gate a negative space charge appears in the semiconductor beneath the oxide. If the electron density at the surface is equal to holes density, the potential applied to the gate is called reversal threshold voltage VT. This parameter is very important in MOSFET's digital applications. We assume $Qox=3*10^{10}cm^{-2}$ and study the threshold shift. Figure 5 shows the evolution of threshold voltage as a function of the channel implantation dose DI with Rp=100Å. The tension VT is proportional to DI. This behaviour is attributed to the diffusion of the channel electrons into the semiconductor as the implantation dose DI increases. Figure 6 shows the evolution of tension VT in relation to the projected range Rp of channel profile with a channel implantation dose of $2*10^{12}cm^{-2}$. We observe that the threshold voltage is proportional to Rp because the channel carriers density is spread into the device when the projected range increases. Now, we assume channel doping parameters of $DI=2*10^{12}cm^{-2}$; Rp=100Å and study the effect of the oxide charges on threshold voltage. Figure 7 displays the evolution of threshold voltage as a function of oxide charge assuming that $Qss=3*10^{10}cm^{-2}$. We observe a linear decrease of VT versus Qox. This is produced by positive charges which attract more and more electrons to the surface, resulting in an earlier conduction in the channel.

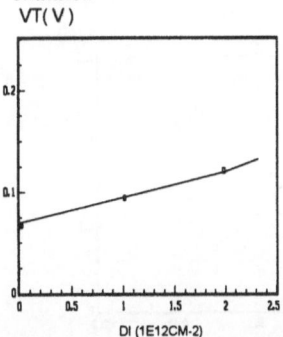

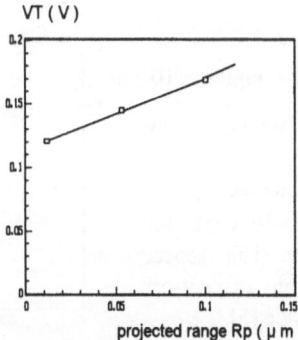

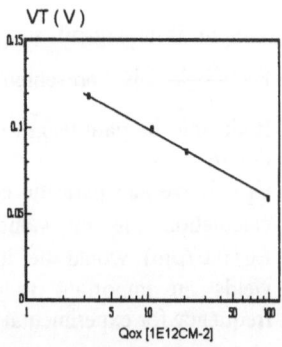

Fig.5 Threshold voltage versus channel implantation dose (Lg=0.1μm; Vds=1V; Qox=3*10^{10}cm^{-2}; channel doping parameters are Rp=100Å and ΔRp=800Å).

Fig.6 Threshold voltage versus projected range Rp of doping profile in the channel. (Lg=0.1μm; Vds=1V; Qox=3 *10^{10}cm^{-2}; DI=2*10^{12}cm^{-2}).

Fig.7 Threshold voltage versu oxide charges. (Lg=0.1μm; Vds=1V).

4. Influence of the Gate Length

In this paragraph, we consider that the channel doping profile parameters are: $DI=2*10^{12}cm^{-2}$, Rp=100Å and $Qox=3*10^{10}cm^{-2}$.

Figure 8a shows the evolution of VT in relation to the gate length. We notice that, when the gate length decreases, threshold tension decreases also because channel conduction is favoured by carrier diffusion between source and drain. In figure 8b is presented the evolution, with gate length, of the maximum transconductance obtained both theoretically (MC and Drift Diffusion (DD) simulation) and experimentally [1]. For gate length larger than 0.07μm, the MC results are in a good agreement with experiments. The differences between MC and DD results increases with decreasing gate length. This is due to the non-stationary transport which is not accounted for in the DD model. In ultra-short gate length devices this is the dominant mechanism. For gate length shorter than 0.07μm, electrons are not able to reach their maximal non-stationary velocity (figure 9). This results in a reduced gate control and Gm decreases with decreasing gate length. Thus the optimized gate length should be close to 0.07μm and transconductance of 600mS/mm could be achieved.

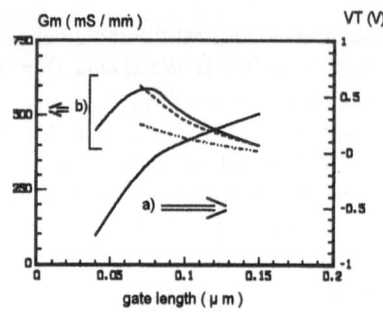

Fig. 8 a):Threshold voltage versus gate length (DI=2*10^{12}cm^{-2}; Qox=3*10^{10}cm^{-2}). b): Intrinsic maximum transconductance versus gate length (Vds=1V). ---- exp. Sai-H. -.-.- drift diffusion ___ monte carlo

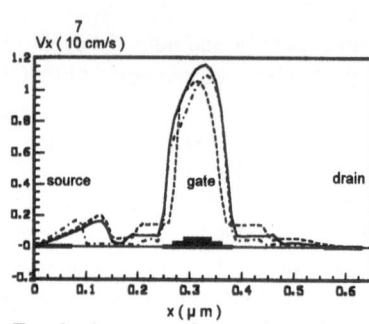

Fig. 9: Average velocity along the channel for three values of gate length: --- Lg=0.04μm; Lg=0.07μm; -.-.- Lg=0.1μm (Vds=1V; Vgs=0.8V). Average velocity defined by: $<v_x>_y = <v_x * n>_y / <n>_y$; where $<\bullet>_y = \int \bullet dy$

The intrinsic cut-off frequency defined by: $Fc = \dfrac{Gm}{2\pi Cgs}$ is presented in Figure 10. Reducing the gate length increase Fc, mainly due to a drastic reduction of gate capacitance Cgs. If we add parasitic capacitances to our calculation the Fc values (≈300GHz for Lg=0.07μm) would be lower. This aspect yields an important reduction of cut-off frequency for experimental device [5].

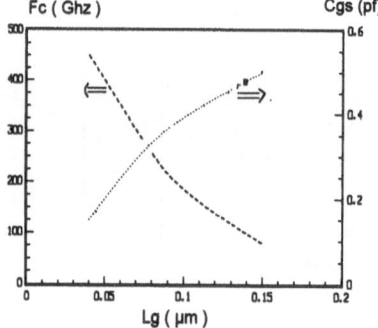

Fig.10: Intrinsic cut-off frequency and gate capacitance versus gate length (Vds=0.8V).

5. Conclusion

We have used a Monte Carlo program to investigate the ultra-short gate Si-NMOS transistor performances. The results obtained by this model are in good agreement with experimental ones. Transconductance of 600 mS/mm and intrinsic Fc close to 300 Ghz could be obtained for a NMOS structure with optimized SDE region and 0.07μm gate length.

References

[1] M.V. Fischetti and S.E. Laux, "Monte carlo analysis of electron transport in small semiconductor devices including band-structure and space-charge effects", Phys. Rev. B, Vol.38, pp. 9721-9745, 1988.
[2] R. Fauquembergue, "Computer simulation of III-V MESFET's, MODFET's and Mis-Like FET's", Computer Physics Communications, N°67, pp.63-72, North-Holland, 1991.
[3] G. A. Sai-Halasz, et al. "Design and experimental technology for 0.1μm gate length low-temperature operation FET's", IEEE Electron Device Lett., Vol. EDL-8, N°10, pp. 463-466, 1987.
[4] S.M. Sze, "Physics of semiconductor devices", 2nd Edition, J. Wiley & Sons, New York, 1981.
[5] R.H. Yan, et al., "89 Ghz fT Room-Temperature Silicon MOSFET's", IEEE Electron Device Lett., Vol. EDL-13, N°5, pp. 256-258, 1992.

Semianalytical Universal Simulation of the Electrical Properties of the Permeable Base Transistor

P. Chenevier, G. Kamarinos, and G. Pananakakis

LPCS - URA CNRS 840 - Enserg
23 rue des Martyrs, BP 257, F-38016 Grenoble Cédex, FRANCE

Abstract

Using only a few numerical calculations, we give the analytical current-voltage and charge-voltage characteristics valid for any PBT. The highest unity current gain frequency (f_τ) corresponding to the current technology is on the order of 30 GHz ; nevertheless, the oscillation frequency can be higher than 100 GHz.

1. Introduction

There are two types of PBT : the buried base PBT, and the etched groove PBT [1].
The PBT is essentially a two dimensional structure ; it is so impossible to find *ab initio* analytical expressions for its characteristics. The optimization of the device needs time consuming 2-D programs, and in particular, concerning its frequency limits.
In this paper, we show, for the first time, that the modelization of the PBT do not require to have a continuous recourse to a 2-D numerical simulation program, and we give a definitive answer to the high frequency performances of the device. To reach these results : (i) we have worked in the buried base PBT, and (ii) we have utilized the 2-D numerical program (TITAN+JUPIN) of CNET/CNS. Evidently, the PBT is a short channel MESFET. So, we have studied the half period of the structure $\left(L = \dfrac{a+d}{2}\; ; \text{Fig. 1}\right)$, and we have supposed that the device is limited in its active zone : $W_E + W_C$; the device bias are then applied on the limits of these zones (V'_{BE} and V'_{CE}).

2. Similitude Laws for Low Collector Bias

We suppose here that in the electron velocity expression : $\vec{V}_n = \mu_n \vec{E}$, the mobility is constant. Besides, we adopt the classical hypothesis : (i) SCR totally depleted, (ii) at SCR boundaries, $\dfrac{\partial \phi}{\partial \bar{n}} = 0$, and (iii) into the channel $n \approx N_D$.

The study of Poisson equation with its boundary conditions shows that, if one takes $\dfrac{d}{2}$ as the length unity, and $\dfrac{qN_D}{\varepsilon}\dfrac{d^2}{4}$ as the potential unity, the potential $\phi(X,Y)$ depends only on the two dimensionless parameters $X = \dfrac{2W_E}{d}; Y = \dfrac{2W_C}{d}$, where W_E and W_C are the lengths of SCR around the base :

$$W_E = \left[\frac{2\varepsilon}{qN_D}(V_D - V'_{BE})\right]^{1/2} ; W_C = \left[\frac{2\varepsilon}{qN_D}(V_D - V'_{BE} + V'_{CE})\right]^{1/2} \tag{1}$$

We show then that the transistor current can be written as :

$$I_C = -\frac{q^2 N_D^2 \mu_n Z d^2}{4\varepsilon} \ f\left(\frac{2W_E}{d}, \frac{2W_C}{d}\right) \tag{2}$$

To determine the function $f(X,Y)$, it is sufficient to plot the $I_C(V'_{BE}, V'_{CE})$ characteristics for only one set of technological parameters $\left(\dfrac{d}{2}, N_D, \mu_n\right)$. Therefore, using the 2-D numerical program for calculating the current characteristics of only one (non particular) PBT, we can calculate analytically the characteristics of any other device.

The threshold voltage of PBT (for $V_{CE} = 0$) can also be calculated :

$$l(0) = H(0)W_E ; V_{BET} = V_T \ \text{for} \ \frac{d}{2} - l(0) = 0, \ \text{so,} \ V_T = V_D - \frac{q}{2\varepsilon H^2} N_D d^2 \tag{3}$$

The simulated characteristics give : $H(0) \approx 0,7$.

3. High Collector Bias Regime

The used model takes $\vec{V}_n = V_s \dfrac{\vec{E}}{E + E_c}$ with $V_s = 1.04 \times 10^7 \, cm \cdot s^{-1}; E_c = 1.04 \times 10^4 \, Vcm^{-1}$

The current I_C can be written :

$$I_C = -qN_D ZV_m\left[\left(\frac{d}{2} - h\right)\right] \tag{4}$$

where V_m is an average velocity ; V_m and h are, *a priori*, dependent from bias, doping (N_D) and $\dfrac{d}{2}$. Using the 2-D program, we show that : (i) V_m and h are independent from $\dfrac{d}{2}$, (ii) V_m is independent from V'_{BE}, (iii) we can write : $h = H(N_D, V'_{CE})W_E$,

(iv) $H = 0.705 - 0.0525 V'_{CE}$ (5)

so H is, in a first approximation, independent from N_D,

(v) satisfactory analytical expression for V_m is :

$$V_m = \frac{V'_{CE}}{V'_{CE} + E_c\left(W_{E_0} + W_{C_0}\right)} \tag{6}$$

W_{E_0} and W_{C_0} being the values for $V'_{BE} = 0$.

We establish so a universal expression for I_C :

$$I_C = -qN_D ZV_m\left[\frac{d}{2} - HW_E\right] \tag{7}$$

with V_m and H given by (6) and (5).

These analytical expressions fit very satisfactorily the 2-D simulations (Fig. 2 for example, where the dots correspond to the analytically calculated current). The 2-D simulations allow us to find also an analytical expression for the total charge in the SCR :

$$Q_{SC} = qN_D ZS = qN_D Z\left(W_E + W_C\right)\left[\frac{a}{2} + \alpha_0 \frac{W_E\left(W_E + W_E\right)}{3W_E + W_C}\right] \tag{8}$$

we find $\alpha_0 = 0.77$.

4. Small Signal Parameters; Frequency Limits

Using the analytical expressions of current (7) and charge (8), we can calculate : the conductance g_D, the transconductance g_m and the interelectrode capacitances C_{BE} and C_{BC}. The unity current gain (transition) frequency is then calculated :

$$f_T = \frac{g_m}{2\pi\left(C_{BE}\left(C_{BE} + 2C_{BC}\right)\right)^{1/2}} \tag{9}$$

$$g_m = Z\frac{\varepsilon}{W_E}V_m H$$

$$C_{BE} = Z\frac{\varepsilon}{W_E}\left[\left(\frac{a}{2}\right)+\alpha_0 W_C\frac{(1+\gamma)}{(1+3\gamma)^2}\left(6\gamma^2 + 3\gamma + 1\right)\right] \quad)$$

$$C_{BC} = Z\frac{\varepsilon}{W_E}\left[\left(\frac{a}{2}\right)\gamma+\alpha_0 W_C\frac{(1+\gamma)}{(1+3\gamma)^2}\gamma^2\left(5\gamma + 1\right)\right] \quad) \quad (10)$$

$$\gamma = \frac{W_E}{W_C} = \left[\frac{V_D - V'_{BE}}{V_D - V'_{BE} + V'_{CE}}\right]^{1/2} \quad)$$

5. Conclusion and Remarks

The f_T, corresponding to the Fig. 2 parameters, is not higher than 30 GHz (Fig. 3). This result is confirmed by all the published experimental works [2,3,4,5,6,7] ; the PBT on GaAs has the same limits (V_s is nearly the same for Si and GaAs) ; the etched groove PBT can be slightly better (C_{BE} is lower) ; the maximum frequency oscillations, according to our estimations, must be higher than 100 GHz ; so, the PBT, as it can deliver an important power, can be an interesting device for high frequency power amplification [8].

References

[1] B.J. VOJAC, G.D. ALLEY - IEEE Trans. ED-30, 877 (1983)
[2] B.O. BOZLER - Surf. Science, 174, 487 (1986)
[3] K.B. NICHOLS et al - IEEE Trans. ED-35, 2246 (1988)
[4] D.D. RATHMAN, W.K. NIBLACK - MTT Int. Microw. Symp., Dig.1, 357 (1988)
[5] L.J. KUSHER - Microw. J., 33, 87 (1990)
[6] T. OSHIMA et al - IEDM Proceed. (IEEE ed.) 33 (1991)
[7] A. GRUHLE et al - ESSDERC'91 Proceed., Elsevier ed., 27 (1991)
[8] P. CHENEVIER, G. PANANAKAKIS, G. KAMARINOS, F. STEINHAGEN
       Superl. & Microst., 8, 269 (1990)

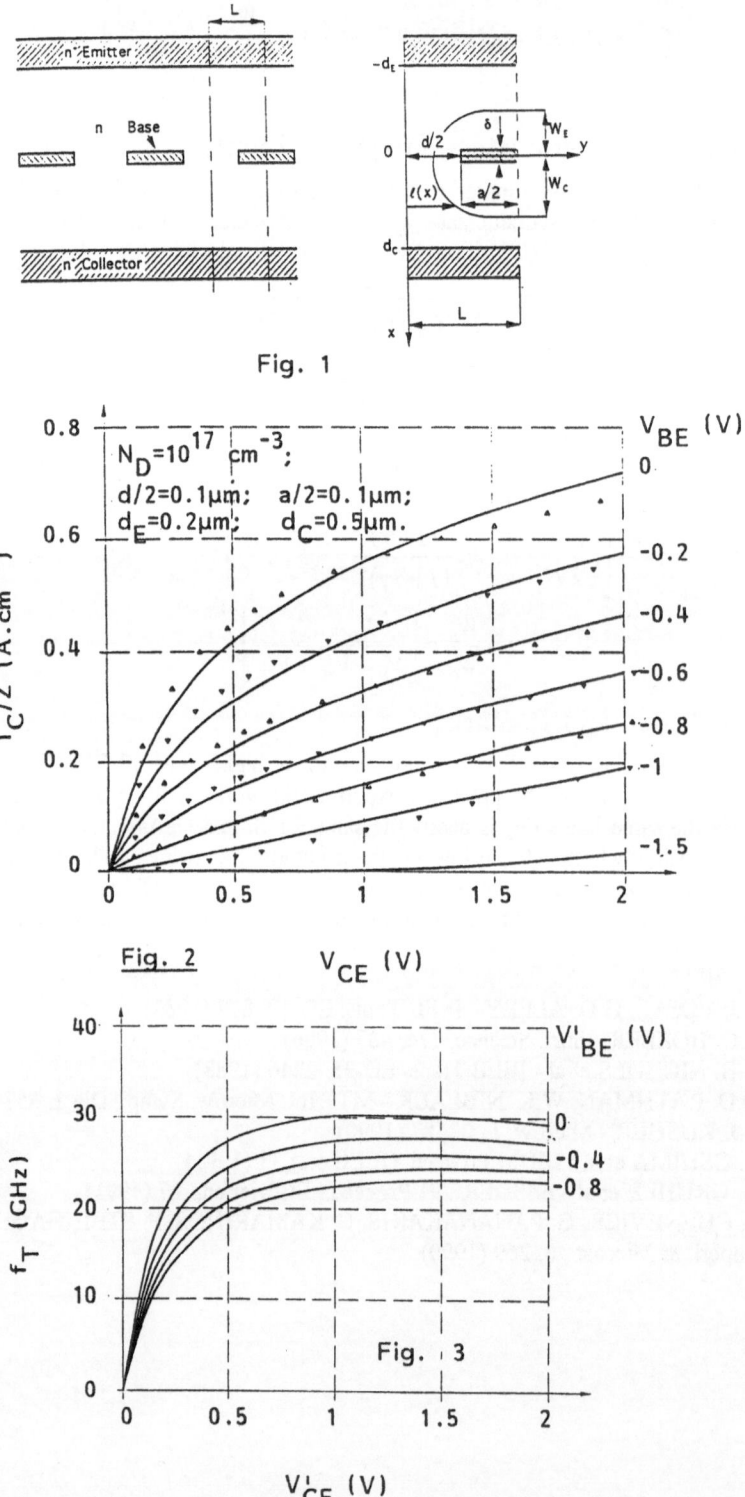

Fig. 1

$N_D = 10^{17}$ cm^{-3};
$d/2 = 0.1\mu m$; $a/2 = 0.1\mu m$;
$d_E = 0.2\mu m$; $d_C = 0.5\mu m$.

Fig. 2

Fig. 3

The Static and Dynamic Behaviour of NPT-IGBTs with Different p+-Anode Designs

W. Feiler, W. Gerlach, and U. Wiese

Institut für Werkstoffe der Elektrotechnik, TU Berlin
Jebensstraße 1, D-10623 Berlin, GERMANY

Abstract

Two analytical models for the NPT-IGBT are presented. The first model explains the static carrier distribution in the on-state and gives approximate values for the emitter efficiency. The second model describes the turn-off behaviour under inductive load conditions. The results of the analytical models are compared with 2D-simulations obtained with the device simulator TMA-MEDICI. Moreover the influence of a shorted anode on the behaviour of the IGBT is investigated by numerical simulations.

1. The static model for the NPT-IGBT

In this model it is shown, that the carrier distribution in the NPT-IGBT and the emitter efficiency can be calculated based on the two-dimensional current flow equations within the device. For this purpose the IGBT is devided into two one-dimensional sections (Fig.1), which are coupled via the lateral electron current below the p+-cathode-well. In both regions of the model the current flows one-dimensionally in positive x-direction between $-w_0 \leq x \leq -d_1$, where d_1 is chosen as half the cell width. Due to the electric field in the space charge region (SCR) the electron current cannot enter the p+-cathode in the pnp-transistor part (section II). Therefore the electron current has to flow laterally into the pn-diode part (section I). It is assumed, that this lateral electron current is constant between $-d_1 < x < 0$. Furthermore a lateral hole current is neglected. The electron current grows therefore linearly in this area of section I, whereas for $x \geq 0$ it flows undisturbed into the accumulation layer below the gate. Because the hole current is zero in the accumulation layer, it must change its direction between $0 \leq x < d$ and has to flow laterally into the p+-cathode. For simplicity it is assumed, that here the lateral hole current density is constant. Neglecting recombination, the carrier distri-

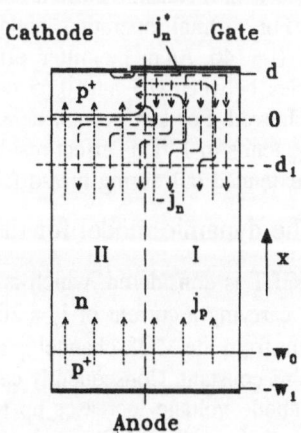

Fig. 1: The two sections of an IGBT
Part I: pn-diode   Part II: pnp-transistor

bution within the two sections of the IGBT can be calculated for high injection conditions in the base region by simply solving the current equations in the ambipolar form

$$j_n = \frac{\mu_n}{\mu_n + \mu_p} \cdot j + q \cdot D \cdot \frac{dp}{dx} \qquad\qquad j_p = \frac{\mu_p}{\mu_n + \mu_p} \cdot j - q \cdot D \cdot \frac{dp}{dx} \qquad (1)$$

The solution of this calculation is shown in Fig.2 together with 2D-simulation results, which are extracted along the cuts AA′ and BB′. Both curves show rather good agreement. Due to the neglected recombination the carrier distribution falls off linearly between $-w_0 \le x \le -d_1$, whereas the constant lateral current density is responsible for a parabolic decrease below the p$^+$-cathode and a parabolic increase under the gate. As the slope of the carrier density can be deduced from the model, it is possible to derive an analytical expression for the emitter efficiency γ_p:

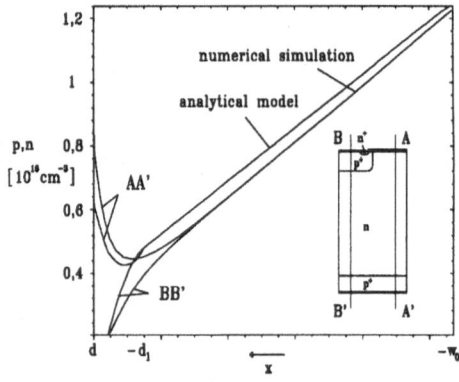

Fig. 2 : Carrier concentration in the IGBT

$$\frac{\gamma_p}{\gamma_p + 1} = \frac{\mu_p}{\mu_n} \cdot \left(1 - \frac{d_1}{2 w_0} + \frac{2 \cdot q}{w_0} \cdot \sqrt{\frac{(w_1 - w_0) \cdot D_n \cdot N_A}{q \cdot j_n}} \right) \qquad (2)$$

In Fig.3 the emitter efficiency according to (2) is compared with the values extracted from the MEDICI-simulation. The dependence of the analytical emitter efficiency on the current density (solid lines) is in a good agreement with the results derived by MEDICI (dashed lines). Generally the emitter efficiency decreases with increasing current density. For normal operating conditions ($j \approx 40$ A/cm^2) emitter efficiencies between 0,3 and 0,65 can be achieved. For low current densities γ_p tends to 1. This effect results in a extended tail during turn-off.

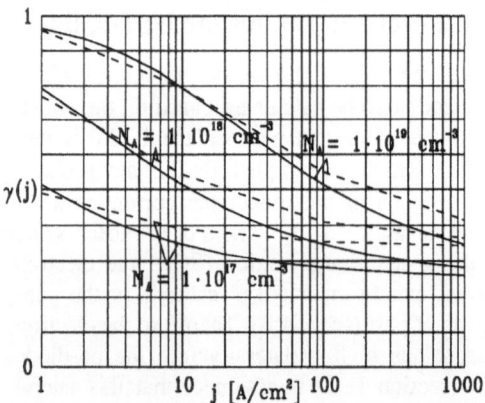

Fig. 3 : Emitter efficiency of uniform p$^+$-emitters

2. The dynamic model for the NPT-IGBT

An IGBT is considered, which is driven in a chopper-circuit with a clamped inductive load, carrying a current of $j_0 = 20$ A/cm^2. When the IGBT is switched off, the electron current from the MOS-channel is interrupted. Due to the inductive load the anode current remains constant. Consequently carriers are extracted from the base, the SCR is growing. The anode voltage increases up to the external supply voltage. Then the extraction of electrons from the SCR periphery stops and the anode current begins to fall. As it is known from the numerical simulation the first phase of this decrease is characterized by an increase of the ratio j_n/j_p. For this case the behaviour of the total current and the extraction of the carriers from the base is described by the developed dynamic model. For this purpose the ambipolar diffusion equation

$$D \cdot \frac{\partial p^2}{\partial x^2} = \frac{\partial p}{\partial t} - \frac{p}{\tau} \tag{3}$$

is solved taking advantage of the Laplace-transform method. For that purpose the initial carrier distribution is approximated by two linear functions as it is shown in Fig. 4 :

$$p_I(x,0) = \frac{j_0}{2\,q\,D_p} \cdot x = c_1 \cdot x \tag{4}$$

$$p_{II}(x,0) = c_1 x_1 + \Delta p \left(\frac{x - x_1}{w - x_1} \right) \tag{5}$$

The boundary conditions are given by

$$p(0,t) = 0 \tag{6}$$

$$\left. \frac{\partial p}{\partial x} \right|_{x=0} = k \cdot \left. \frac{\partial p}{\partial x} \right|_{x=w} \tag{7}$$

The parameter k in (7) is a measure for the ratio j_n/j_p at the anode x = w. For simplicity a vanishing hole current density at the anode is assumed. For this case k = -3. This yields the following expression for the hole density :

$$p(x,t) = 2 \cdot w \cdot e^{-\frac{t}{\tau}} \cdot \sum_{n=-\infty}^{\infty} \frac{\sin\left((\varphi + 2 \cdot \pi \cdot n) \cdot \frac{x}{w} \right)}{(\varphi + 2 \cdot \pi \cdot n)^2 \cdot \sin(\varphi + 2 \cdot \pi \cdot n)}$$

$$\cdot \left\{ \frac{c_1}{k} - c_2 + (c_2 - c_1) \cdot \cos\left((\varphi + 2 \cdot \pi \cdot n) \cdot \left[1 - \frac{x_1}{w} \right] \right) \right\} \cdot e^{-\left(\frac{\varphi + 2 \cdot \pi \cdot n}{w/L} \right)^2 \cdot \frac{t}{\tau}} \tag{8}$$

where the value φ is determinated from $\varphi = \arctan\left(\sqrt{k^2 - 1} \right)$. At x = 0 the total current density j is given by the hole current only. The current can therefore be derived from the hole density (8) according to

$$j(t) = -2 \cdot q \cdot D_p \cdot \left. \frac{\partial p}{\partial x} \right|_{x=0} \tag{9}$$

Fig. 5 shows the carrier profiles according to (8) and Fig. 4 visiualizes the total current density (9) in comparison with the results of MEDICI. It can be seen, that the accordance of the results from the analytical model and the numerical simulations are in a remarkably good agreement. Only for the very first time interval the profiles are different, due to the simplified initial condition (4), (5). Nevertheless the model describes the time-dependence

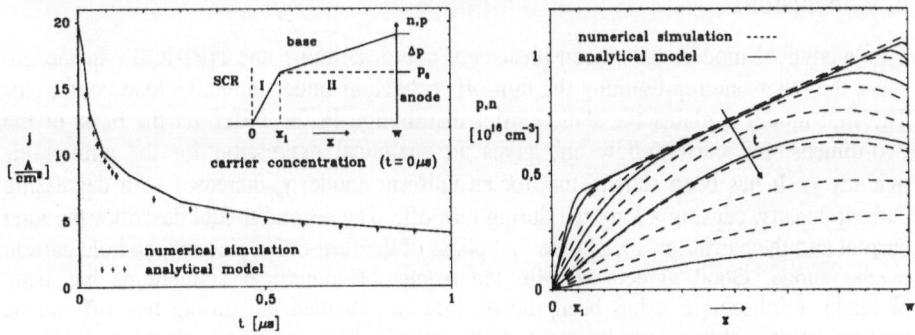

Fig. 4 : Total current density j during turn-off   Fig. 5 : Carrier profiles in the neutral base

of the total current very well, until the current-ratio j_n/j_p decreases and the assumption (7) is violated. This is the case in the exponential phase of the tail, where the current-ratio tends to zero

3. The NPT-IGBT with a shorted anode

In order to improve the dynamic behaviour of the IGBT in the exponential phase of the tail, it is recommended to introduce an anode, which has a low emitter efficiency for low current levels. It is shown in the first section of this paper, that this cannot be achieved by lowering the doping concentration in the anode. However by introducing a shorted anode this aim can be accomplished (Fig. 6). The doping concentration in the p$^+$-section of the anode is chosen quite high ($N_A = 10^{20}$cm^{-3}) to ensure a high conductivity modulation in the base during the on-state; the narrow n$^+$-shorts have a doping concentration of $N_D = 10^{17}$ cm^{-3}. Fig. 6 shows the resulting static IV-characteristic for the IGBT with a shorted anode (KS-IGBT) and two devices with uniform anodes, which have doping concentrations of $N_A = 10^{17}$ cm^{-3} (IGBT17) and $N_A = 10^{20}$ cm^{-3} (IGBT20). For high current levels the KS-IGBT is superior to the IGBT17, but it doesn't achieve the low forward voltage of the IGBT20. The advantage of the shorted anode becomes evident in Fig. 7, where the simulation data for the total current density in the IGBTs during turn-off is plotted. The KS-IGBT has only a short tail, resulting in a small turn-off loss of 2,5 mWs/cm^{-2}. In contrast the IGBT17 has a turn-off loss of 3,9 mWs/cm^{-2} and the IGBT20 of 12,5 mWs/cm^{-2}.

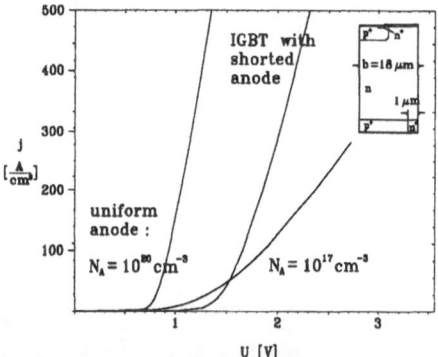

Fig. 6 : IV-charactristic of IGBTs with different anode designs

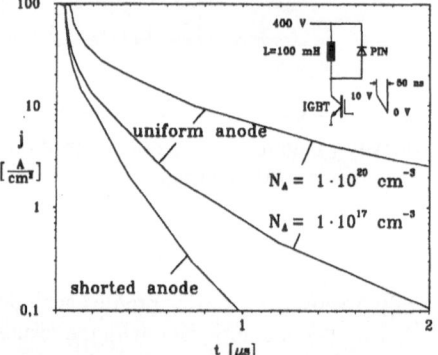

Fig. 7 : Current density during turn-off

4. Conclusions

Two analytical models have been presented, one describing the NPT-IGBT in the on-state, the other one representing the turn-off behaviour under inductive load conditions. The first model explains the static carrier distribution in the IGBT on the basis of the two-dimensional current flow and gives an analytical expression for the emitter efficiency γ_p. It has been shown, that for an uniform anode, γ_p increases with decreasing current density, causing a long tail during turn-off. The second model describes the total current and the carrier profiles in the first phase of the turn-off neglecting the hole current in the anode. Good agreement with the results of numerical simulations has been obtained. Furthermore it has been shown, that an extended tail during turn-off can be avoided using a shorted anode.

Numerical Modeling and Simulation of Multi-Electrode Semiconductor Optical Amplifiers

J. L. Pleumeekers[‡†], T. Mercier[‡], S. Mottet[‡], and F. Clérot[‡]

[‡]CNET Lannion B, France Telecom
Route de Trégastel, BP 40, F-22301 Lannion Cédex, FRANCE
[†]Department of Electrical Engineering, TU Delft
Mekelweg 4, P.O. Box 5031, NL-2600 GA Delft, THE NETHERLANDS

Abstract

The purpose of our study was to build a two-dimensional time-dependent simulator for multi-electrode semiconductor optical amplifiers with bulk active layers. This simulator was used to design an InGaAsP-InP external modulator at 1.55 µm for fast optical fibre transmission applications.

1. Introduction

For optical fibre applications, a new generation of multifunctional devices, based on semiconductor optical amplifiers (SOAs), is arising. The main studied functions are optical amplification, modulation, detection and wavelength conversion. A main studied point is the capability of integration of different functions in one device. The integration leads to multielectrode devices. To study and design these devices, a time-dependent simulator for multielectrode multifunction SOA has been developped. An SOA has anti-reflection coatings, so the travelling wave approximation (no facet reflectivities) can be used. Moreover, its structure (fig. 1) permits to solve the optical problem only along the light propagation z-axis. On the other hand, the electrical problem has to be solved within the xz plane : the main conduction takes place in the x direction but inhomogeneities, due to the internal optical mode amplification and the different biases of the top electrodes, are present in the z-direction.

2. Equation set

The electrical and optical behaviour of SOAs is described by the following five equations:

Poisson equation:
$$\nabla \bullet (\varepsilon \cdot \nabla \varphi) = q \cdot (n - p - dop) \qquad (1)$$

Electron continuity equation:
$$\frac{\partial n}{\partial t} - \nabla \bullet (n \cdot \mu_n \cdot \nabla \varphi_n) = -U(I) \qquad (2)$$

Hole continuity equation:
$$\frac{\partial p}{\partial t} + \nabla \bullet (p \cdot \mu_p \cdot \nabla \varphi_p) = -U(I) \qquad (3)$$

Light intensity propagation equation: $\quad \dfrac{1}{v_p} \cdot \dfrac{\partial I}{\partial t} + \dfrac{\partial I}{\partial z} = [\Gamma \cdot g(\varphi, \varphi_n, \varphi_p) - \alpha_s] \cdot I \quad (4)$

Light phase propagation equation: $\quad \dfrac{1}{v_g} \cdot \dfrac{\partial \phi}{\partial t} + \dfrac{\partial \phi}{\partial z} = \alpha_H \cdot \Gamma \cdot g(\varphi, \varphi_n, \varphi_p) \quad\quad (5)$

where φ_n and φ_p are the Fermi-potentials, Γ is the optical confinement factor and α_H is the linewidth enhancement factor [1].

3. Solving method

After discretisation in time with a coefficient of implicitness θ [2], equations 1 to 3 can be written at time $t + \Delta t$ as:

$$\nabla \bullet (\varepsilon \cdot \nabla \varphi) = q \cdot (n - p - dop) \qquad (6)$$

$$\nabla \bullet (n \cdot \mu_n \cdot \nabla \varphi_n) = U + \frac{n}{\theta \cdot \Delta t} - \frac{n^0}{\theta \cdot \Delta t} - \frac{1 - \theta}{\theta} \cdot \nabla \bullet (n^0 \cdot \mu_n^0 \cdot \nabla \varphi_n^0) + \frac{1 - \theta}{\theta} \cdot U^0 \qquad (7)$$

$$\nabla \bullet (p \cdot \mu_p \cdot \nabla \varphi_p) = -U - \frac{p}{\theta \cdot \Delta t} + \frac{p^0}{\theta \cdot \Delta t} - \frac{1 - \theta}{\theta} \cdot \nabla \bullet (p^0 \cdot \mu_p^0 \cdot \nabla \varphi_p^0) - \frac{1 - \theta}{\theta} \cdot U^0 \qquad (8)$$

where the superscript means the former solution at time t. These equations will be solved with as unknowns the potentials φ, φ_n and φ_p. They are all expressed in volt, so scaling or transformation of the variables is not necessary. The stationary solution is directly obtained by setting $1/\Delta t = 0$ and $\theta = 1$. Thus, the same solving method is used for both stationary and transient simulations. These equations are then discretised in a two-dimensional xz-space by a box method [3].

The optical equations, 4 and 5 , are solved only in the z-direction (decentred on the same mesh). The light intensity at time $t + \Delta t$ at mesh point i can be computed recursively after discretisation in time and z-space as:

$$I_i = \frac{\theta \cdot \dfrac{I_{i-1}}{H_{z_{i-1}}} + \dfrac{I_i^0}{v_p \cdot \Delta t} + (1 - \theta) \cdot (\Gamma \cdot g_i^0 - \alpha_s) \cdot I_i^0 - (1 - \theta) \cdot \dfrac{I_i^0 - I_{i-1}^0}{H_{z_{i-1}}}}{\dfrac{1}{v_p \cdot \Delta t} - \theta \cdot (\Gamma \cdot g_i - \alpha_s) + \dfrac{\theta}{H_{z_{i-1}}}} \qquad (9)$$

where the superscript means the former solution at time t. The boundary condition is the incident light power I_0. In the same way, the light phase at time $t + \Delta t$ at mesh point i is computed as:

$$\phi_i = \frac{\theta \cdot \dfrac{\phi_{i-1}}{H_{z_{i-1}}} + \dfrac{\phi_i^0}{v_g \cdot \Delta t} + \theta \cdot \alpha_H \cdot \Gamma \cdot g_i - (1 - \theta) \cdot \dfrac{\phi_i^0 - \phi_{i-1}^0}{H_{z_{i-1}}} + (1 - \theta) \cdot \alpha_H \cdot \Gamma \cdot g_i^0}{\dfrac{1}{v_g \cdot \Delta t} + \dfrac{\theta}{H_{z_{i-1}}}} \qquad (10)$$

The full set of equations is solved within a direct coupled Newton iterative scheme [4].

The SOAs are driven by modulated current sources, so transient current boundary conditions have to be implemented within the two-dimensional direct coupled

Newton method. This condition can be written for each contact as:

$$\sum_{nodes} (J_n + J_p + J_d) \bullet S_{box} = I_{bias} \tag{11}$$

where J_d is the displacement current density.

4. Applications

Among the different possible applications, the simulator was used to design an InGaAsP-InP external modulator at 1.55 µm for fast optical fibre transmission applications. It allows to access the internal and external behaviour of a SOA, under stationary and transient conditions together with the performances in system operation. As an example, the electron density is shown in figure 2 for an input power of -5 dBm and currents of 20 mA and 50 mA on the two contacts which gives a total optical gain of 11 dB. The inhomogeneity in the z-direction can be seen on that figure. One important parameter is the electrodes isolation under operation. The current density in the longitudinal z-direction for the etched region between the two contacts is shown in figure 3. It permits to evaluate the interelectrode leakage current under operation. This current appears in the active layer under operation, due to internal equilibrium and cannot be measured directly. It is a penalising effect which has to be taken into account for the design of the external current drivers. In figure 4 the optical output power time response is shown under current modulation of one of the electrodes. Owing to system performance requierements, the optimisations of the SOA showed that a counterphase current modulation [5] had to be used to obtain a fast modulation of the optical power together with an acceptable variation of the phase. The typical output power and phase under operation are shown in figure 5.

These two-dimensional transient simulations required a mesh of typically 3000 nodes together with 40 time steps.

References

[1] C.H. Henry, "Theory of the Linewidth of Semiconductor Lasers," IEEE J. Quantum Electron., Vol. QE-18, No. 2, pp. 259-264, 1982.
[2] J.E. Viallet and S. Mottet, "Transient Simulation of Heterostructures," NASECODE IV Conf., J.J.H. Miller (Ed.), Boole Press, Dublin, pp. 536-542, 1985.
[3] C. Simon, S. Mottet and J.E. Viallet, "Autoadaptive Mesh Refinement", SISDEP 1991, Vol. 4, W. Fichtner and D. Aemmer (Ed.), Hartung-Gorre, pp. 225-233, 1991.
[4] C. Simon, "Numerical Simulation of the Electronic Conduction in Semiconductor Devices: Discretisation and Auto-adaptive Meshing", Ph.D. thesis, university of Rennes, 1990.
[5] T.N. Nielsen, U. Gliese, T. Durhuus, K.E. Stubkjaer, B. Fernier, P. Garabedian, E. Derouin and F. Leblond, "Chirp-free 2.5 Gbit/s Amplitude Modulation / Gating in Two-electrode Semiconductor Optical Amplifiers", ECOC 1992.

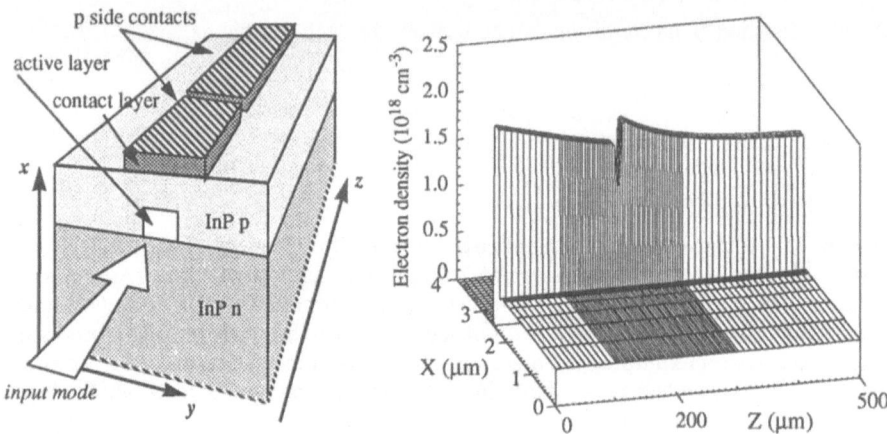

Figure 1: Sketch of a bielectrode SOA, with the different axes.

Figure 2: Electron density in the device. The inhomogeneity in the z-direction is clearly seen.

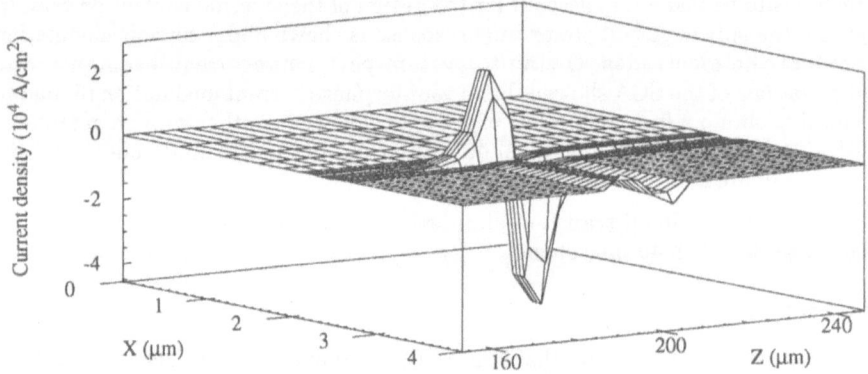

Figure 3: The z-component of the current density vector.

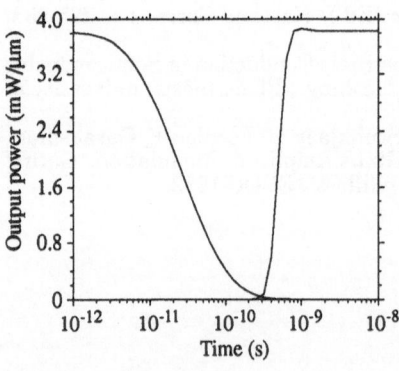

Figure 4: The optical response of the device.

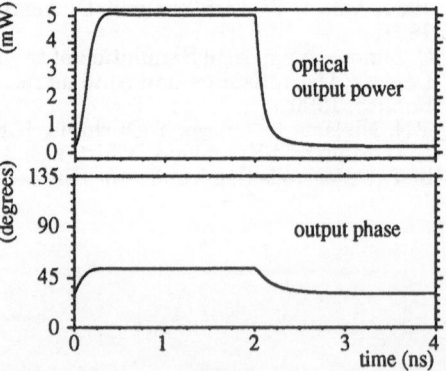

Figure 5: The optical response of the device under counterphase modulation of the electrodes.

SIMULATION OF SEMICONDUCTOR DEVICES AND PROCESSES Vol. 5
Edited by S. Selberherr, H. Stippel, E. Strasser – September 1993

321

Simulation of Carrier Heating Induced Picosecond Operation of GaInAsP/InP Laser Diode

S. V. Polyakov and V. I. Tolstikhin

Institute of Radio Engineering & Electronics, Russian Academy of Sciences
11 Mokhovaya, RU-103907 Moscow, RUSSIA

Abstract

A self-consistent model of a single-frequency GaInAsP/InP laser diode under carrier heating conditions is presented. The problem is formulated using the rate equations approach to the coupled carrier-phonon-photon system and includes a priori all important interaction processes. The carried out simulation shows the efficiency of a carrier heating induced high-speed modulation.

In the recent years it has been well established that carrier heating effects play an important role in determining the behaviour of laser diodes (LD) and especially their picosecond dynamics. This is due to a strong gain and losses dependences on the carrier temperature [1], while the latter usually relaxates at the picosecond time scale [2]. The phenomenon is of particular significance in the case of single-frequency long-wavelength GaInAsP/InP LD [3], those seem now to be most promising light sources for long-haul optical fiber communications [4]. Here we present the self-consistent model of a such LD and use it for investigation of their high-speed operation induced by carrier heating.

Very fast (during a time less than 0.1 ps [2]) thermalization of a dense electron-hole plasma (EHP) into the active region (AR) of a LD allows to suppose the energy distribution $f(\varepsilon)$ for each sort of carriers as a quasi-equilibrium at any actual time scale, i. e. to write: $f(\varepsilon) = [\exp(\varepsilon/T_e - \xi) + 1]^{-1}$, where T_e is the common effective temperature (measured in energetic units); ε and ξ are the energy and the chemical potential (normalized to T_e). When this distributions is postulated, the objective is to give the consistent discription of a LD's behaviour in terms of ξ and T_e. To solve the problem we use the model [3,5...7] based on the treatment of a LD as a coupled system of nonequilibrium carriers, LO-phonons and guided photons. Under the homogeneous approximation for EHP into the AR it reduces to four rate equations describing the temporal evolution of 1) EHP concentration N_e; 2) EHP energy density E_e; 3) LO-phonon occupation number N_{LO} and 4) effective photon concentration of an excited mode (with a frequency ω) N_ω:

$$\frac{dN_e}{dt} = \frac{J}{e \cdot d} - R_S - R_A - \gamma \cdot v_\omega \cdot N_\omega \,, \tag{1}$$

$$\frac{dE_e}{dt} = \frac{Q}{d} - W_{LO}\{N_{LO}\} - W_S + W_A + \eta_\omega \cdot v_\omega \cdot N_\omega \,, \tag{2}$$

$$\frac{dN_{LO}}{dt} = \nu_e \cdot (N_{LO,e} - N_{LO}) - \nu_d \cdot (N_{LO} - N_{LO,o}) \,, \tag{3}$$

$$\frac{dN_\omega}{dt} = \beta_S \cdot R_S - (\alpha_\omega - \gamma_\omega) \cdot v_\omega \cdot N_\omega . \tag{4}$$

Here, J is the pumping rate per unit volume, Q is the energy flux density flowing into (out of) the AR, R_S and R_A are spontaneous and Auger recombination rates, W_S and W_A are changing rates of the EHP energy due to these processes, W_{LO} is LO-phonon induced energy relaxation rate, $N_{LO,e}$ and $N_{LO,o}$ are LO-phonon Plank's functions with temperature equal to the EHP effective temperature T_e and equilibrium temperature T_o, ν_e and ν_d are LO-phonon emission and decay frequencies depending on the EPH parameters and LA-phonon temperature (equal to T_o) respectively, β_S is the spontaneous emission factor, γ_ω and α_ω are modal gain and common losses coefficients, v_ω is the group velocity of an excited mode, factor η_ω determines the include of an excited mode into the EHP energy changing due to electron-photon interactions.

Basic assumption and detailes of the model one can find elsewhere [3,5..7]. Here we only emphasize, that these rate equations result from the Boltzmann-like equations approach to a coupled carrier-phonon-photon system of a LD, which includes a priori any interaction processes into the AR. To describe these processes we employ the four-band Kane's model and the usual quasi-particle scattering theory.

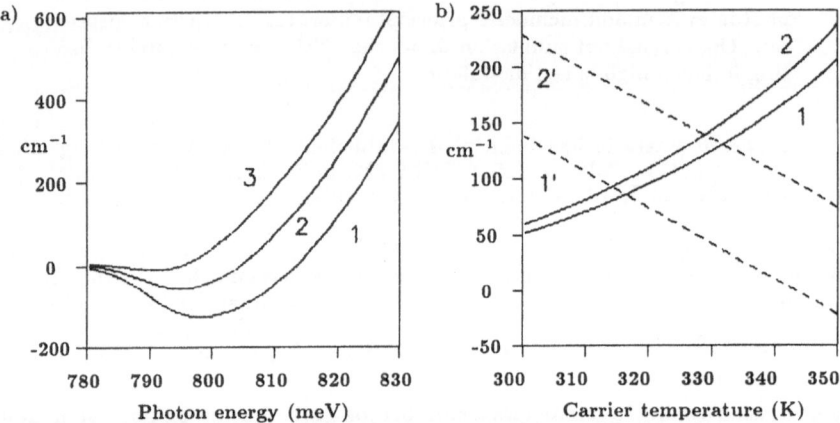

Figure 1: Absorption/gain coefficients. a) - Interband absorption/gain spectra. $N_e = 1.05 \cdot 10^{18} cm^{-3}$. T_e = 300K (1); 325K (2); 350K (3). b) - Peak gain (dashed) and free-carrier absorption (solid). $N_e = 1.05 \cdot 10^{18} cm^{-3}$ (1,1'); $1.20 \cdot 10^{18} cm^{-3}$ (2,2').

First, we examine the carrier heating affect on CW operation of a GaInAsP/InP LD (all presented numerical results correspond to a single-frequency 1.55 μm laser with d = 0.1 μm, $\beta_S = 2 \cdot 10^{-4}$ and material parameters adopted from [8]; the only one free parameter is the LO-phonon decay time $\tau_{LO,d} \sim$ 10 ps [9]). As an example of the calculated data the interband absorption/gain and intraband absorption coefficients are shown in Fig. 1. As it may bee seen from those results the main effects induced by carrier heating are: interband gain suppression (due to degeneracy decreasing) and intraband free-carrier losses (due to intervalence band absorption) activation. Both of them lead to saturation of a light-current curve (see Fig. 2 a)) and give rise to a pump dependence of EHP parameters over the lasing threshold. It must be mentioned, that the latter phenomenon is impossible in the case of a LD with isothermal EHP, but leads to essentially nonlinear behavior of lasing under carrier heating conditions. In particular, it causes a significant CW wavelength chirp of a single-frequency DFB laser (see Fig. 2, b)).

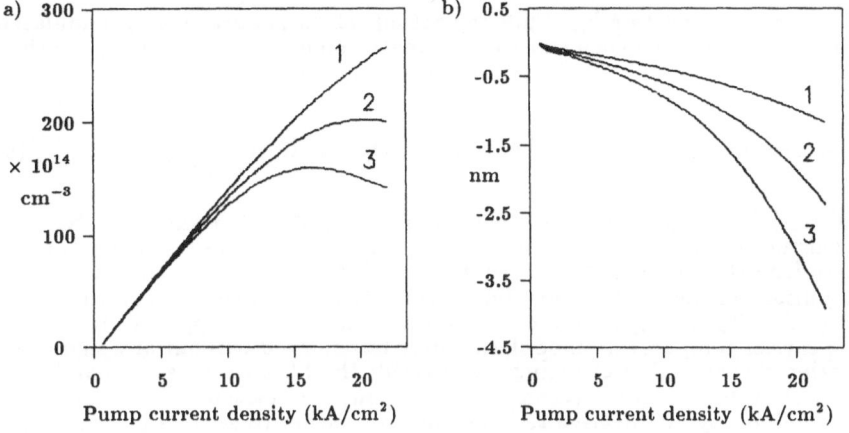

Figure 2: CW characterization. $\tau_{LO,d} = 3$ ps (1); 6 ps (2); 9 ps (3).
a) - Effective photon density N_w. b) - Wavelength of generation λ.

Second, we determine the small-signal modulation response of a such laser. There are two different methods for direct modulation under carrier heating conditions. We can do this changing (i) the pump rate J and/or (ii) the energy flux Q coming into an AR. Taking $J = J_0 + \delta J(t)$, $Q = Q_0 + \delta Q(t)$, where $|\delta J| \ll J_0$, $|\delta Q| \ll Q_0$, we'll obtain in a usual way two types of a frequency f response for any modulated parameter A, $\mathcal{M}_J^A(f)$ and $\mathcal{M}_Q^A(f)$:

$$\mathcal{M}_J^A(f) = \frac{\delta A(f)/\delta J(f)}{\delta A(0)/\delta J(0)} , \quad \mathcal{M}_Q^A(f) = \frac{\delta A(f)/\delta Q(f)}{\delta A(0)/\delta Q(0)} . \tag{5}$$

Typical calculated results, presented the intensity and wavelength responses of a GaInAsP/InP single-frequency DFB laser, are shown in Fig. 3. It's clear from these data (see Fig. 3, a)) the bandwidth expansion and nonmonotonous pump dependence of the relaxation-oscillation peak, all caused by carrier heating. Also the strong carrier temperature sensitivity of a DFB laser generation wavelength λ may be seen if one compares $\mathcal{M}_J^\lambda(f)$ (solid lines) and $\mathcal{M}_Q^\lambda(f)$ (dashed lines) curves, presented in the Fig. 3, b).

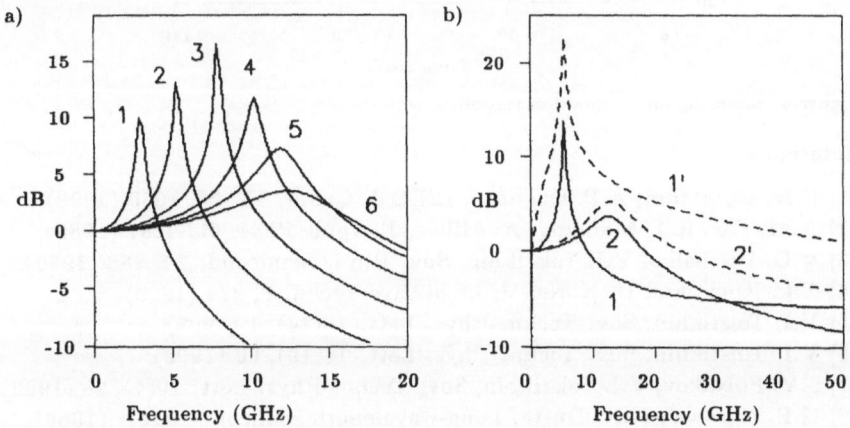

Figure 3: Small-signal modulation response. $\tau_{LO,d} = 6$ ps. a) - Intensity response. $J_0 = 1.1$ kA/cm^2 (1); 1.7 kA/cm^2 (2); 3.1 kA/cm^2 (3); 5.0 kA/cm^2 (4); 7.3 kA/cm^2 (5); 10.2 kA/cm^2 (6). b) - Wavelength response. $J_0 = 1.7$ kA/cm^2 (1,1'); 10.2 kA/cm^2 (2,2').

Third, we investigate the large-signal modulation response of a considered LD. In particular, the dynamics under pumping by a sequence of a very short pulses with a duration less than any actual relaxation processes is simulated. Since the shape of a such short pulses makes no differences we suppose them to be δ-shaped and write temporal dependence of a pumping rate $J(t)$ and energy flux $Q(t)$ as:

$$J(t) = J_0 + ed \cdot \mathcal{N}_e \cdot \sum_{(i)} \delta(t - t_i) \ , \quad Q(t) = Q_0 + d \cdot \Delta \cdot \mathcal{N}_e \cdot \sum_{(i)} \delta(t - t_i) \ . \qquad (6)$$

Here $\mathcal{N}_e$ is a single pulse induced jump of the EHP concentration N_e, Δ is a surplus energy introduced by a single injected electron-hole pare, t_i is the time for i-th pulse supply (interpulse time is supposed to be large compared to pulses duration). The most interesting for the high-speed operation is the case $\Delta \gg T_e$, i.e. generation of hot carriers, when an appropriate level of the carrier temperature modulation may be achieved without any significant carrier concentration changing. As a result the LD's response is mainly due to carrier temperature (than concentration) modulation and has a speed determined by an effective temperature relaxation time. In Fig. 4 we show the calculated time-dependent response of a GaInAsP/InP single-frequency LD excited by a periodic sequence of a δ-shape optical heating pulses like (6) with a period $\tau = 5$ ps and a surplus energy $\Delta \sim 500$ meV. One can see, that after a few pulses the LD comes to a regime when intensity response with a modulation depth ~ 10 dB repeats with a picosecond sequence of heating pulses.

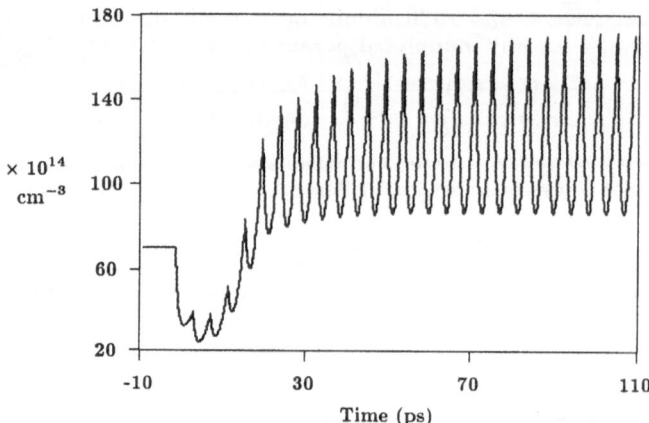

Figure 4: Large-signal modulation response.

References

[1] B.N. Gomatam, A.P. DeFonzo, IEEE J. Quant. El. 26, 1689 (1990).
[2] A.M. Fox, R.J. Manning, A. Miller, J. Appl. Phys. 65, 4287 (1989).
[3] V.D. Pischalko, V.I. Tolstikhin, Sov. Phys. Semicond. 24, 288 (1990).
[4] T.L. Koch, and U. Koren, J. Lightwave Techn. 8, 274 (1990).
[5] V.I. Tolstikhin, Sov. Techn. Phys. Lett. 18(14), 1 (1992).
[6] V.I. Tolstikhin, Sov. Techn. Phys. Lett. 18(19), 50 (1992).
[7] S.V. Polyakov, V.I. Tolstikhin, Sov. Techn. Phys. Lett. 19(4), 28 (1993).
[8] G.P. Agrawal, N.K. Dutta, Long-wavelength semicond. lasers (1986).
[9] W. Potz, P. Kocevar, Phys. Rev. B. B28, 7040 (1983).

SIMULATION OF SEMICONDUCTOR DEVICES AND PROCESSES Vol. 5
Edited by S. Selberherr, H. Stippel, E. Strasser – September 1993

Computer Simulation of Inverse Problems of Crystal Growth and Photoconductivity of Graded Band Gap Semiconductors

S. V. Kletskii

Institute for Semiconductor Physics, Academy of Sciences of Ukraine
Prospekt Nauki, 45, SU-252028 Kiev, UKRAINE

Abstract

Some problems of semiconductor science and technology in conformity with graded band gap materials are studied by methods of complex computer simulation. The main goal of the study is analysis and optimization of optical characteristics of infrared photodetectors based on narrow gap semiconductors such as $Cd_xMn_yHg_{1-x-y}Te$.

1. Introduction

This paper is a brief review of our results in a few directions which are so related that make possible to calculate the nonlinear composition profiles of graded band gap semiconductors with predeterminated optical characteristics. These directions are: phase equilibria in multicomponent systems, Stefan problems of crystal growth, interdiffusion in heterostructures, electron–hole transport in graded band gap semiconductors under illumination and contact phenomena ($n^+ - n$ and $p^+ - p$ contacts). In the case under consideration the narrow gap photoresistor is studied only.

2. Phase equilibria in multicomponent systems

From the practical point of view the first step in the using of concrete semiconductor consists in plotting of phase diagram. Among a number of relatively new semiconductor alloys is of great interest semimagnetic semiconductor $Mn_xHg_{1-x}Te$. The first detailed phase diagram of the ternary $Hg - Mn - Te$ system was plotted [1]. On the base of regular associated solutions theory were calculated solidus lines ($0 < x < 0.35$) and liquidus ones for Te-rich corner $0.6 < y < 0.85$.

3. Crystal growth of multicomponent systems and Stefan problems

Crystallization of multicomponent systems in inhomogeneous temperature field leads to inhomogeneous compositional distribution over the thickness of single crystals and epitaxial layers. In narrow gap semiconductors the energy gap depends, to a large extent, on the composition. Thus nonlinear statement of the inverse Stefan problem, associated with the determination of the temperature variation law at the boundary in time is very interesting in the connection with construction of solid-state structures posssesing predeterminated characteristics. Special modification of swipping method was developed for numerical solution of such type problems in regions with unknown moving boundary. The computer simulation has made it possible to determine initial compositions of growth solutions, a required degree of their oversaturation and the law of the controlling temperature variation of a growth sell in time for growing epitaxial layers with predeterminated compositional profiles. Crystal growth of $Cd_xHg_{1-x}Te$ under liquid phase epitaxy conditions was considered in [2], $Mn_xHg_{1-x}Te$ - in [3].

4. Interdiffusion in heterostructures

At high temperatures diffusion coefficient of Cd in solid is high and close to one in liquid, so under crystal growth conditions interdiffusion changes planned compositional profiles, band parameters and optical characteristics. Hence, determination of interdiffusion coefficient is a nessesary stage in the calculation of real graded structures. For example, it was calculated the nonlinear interdiffusion coefficient for $Cd_xMn_yHg_{1-x-y}Te/CdTe$ in Arrhenius form $D(x,T) = 300exp(-8.4x - 1.89/kT)$ sm^2/s by special procedure of successive approximations [4]. Boltzmann-Matano procedure [5] is not applicable becouse the thickness of epitaxial layers is very small.

The numerical solution of nonlinear interdiffusion equation in a region with moving boundary is connected with high level of truncation errors. The special modification of iterative implicit swipping method with control of mass integral was developed for these Stefan type problems. This method makes it possible not only to calculate real diffusional profiles under crystal growth conditions but to determine the real time of structure homogenezation on the whole. This homogenezation for vapour phase epitaxy conditions is shown in Fig.1.

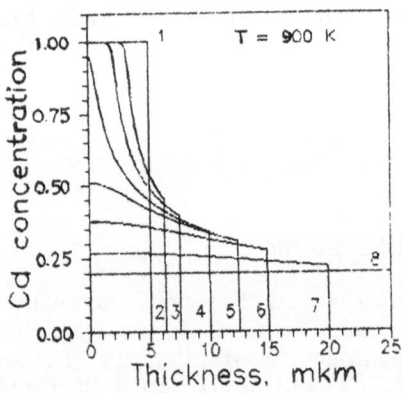
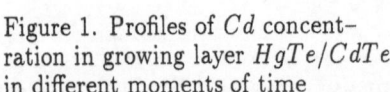
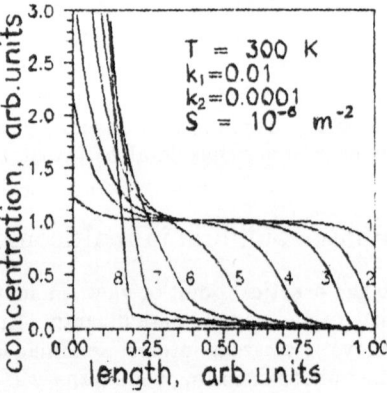

Figure 1. Profiles of Cd concentration in growing layer $HgTe/CdTe$
in different moments of time

Figure 2. Profiles of electron
concentration in the sample for
different values of current

5. Contact phenomena

It is necessary to consider phenomena in contacts $n^+ - n$ or $p^+ - p$, becouse one does not know any way to use semiconductor structure without outside devices. Following [6], longitudinal exclusion-accumulation model was developed for few special cases and numerical procedure for its investigation was constructed. The idea of obtaining at 300 K the electron subsystem parameters that may usually be obtained only at liquid nitrogen temperature seems to be very attractive [7]. The typical distribu- tions of carriers over structure thickness are shown in Fig.2. V-A-characteristics were calculated too.

6. Electron-hole transport in graded band gap semiconductors with nonlinear profile of composition under illumination

Quasielectric field in graded band gap semiconductors is of great influence on the distribution of nonequilibrium carriers and in fact determines device spectral char- acteristics. The analysis of the problem shows that construction of devices based on this principle is very perspective. It is often convenient to approximate experimental compositional data by polynoms of high degree and to solve the direct problem of graded structure photoconductivity for determination of final spectral characteristics. But more convenient this approch is for solution of inverce problem. The degree of polynom is determined by number of reper points. By choosing the number and coor- dinates of reper points one can construct the distribution of nonequilibrium carriers and spectral characteristics needed in special cases [8, 9].

Some results of similar computer construction in the case of n-type graded structure under monochromatic light illumination one can see in Fig.3. Fig.3-a shows given types of spectral characteristics, Fig.3-b - compositional profiles needed for arrival at data shown in Fig.3-a with same curve numbers.

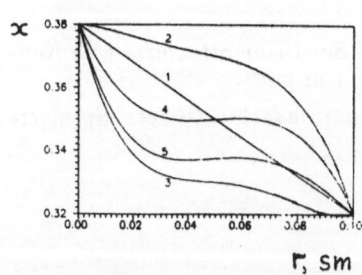

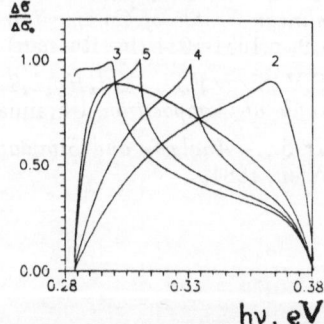

Figure 3–a. Compositional profiles in graded band gap structure $Cd_xHg_{1-x}Te$

Figure 3–b. Spectral characteristics of these structures with the same numbers of curves on Fig.3–a

7. Conclusion

Of course, it is not any discussion on formulas in this paper. A more concrete statements can be found in references, a description of numerical methods – in more fundamental books, for example in [10]. The most part of our calculations were tested experimentally and by special methods (comparison with exact solutions, conservation of mass integral, steady-state regimes, Runge principle and so on). Theoretical models under consideration are so cumbersome that it is hard to link them mathematically. But particular parts of our simulation can be easily generalized to the more complicated cases for in-depth study and optimization of concrete semiconductor devices. A few program products can demonstrate the approach possibilities in the computer construction of new graded band gap structures.

References

[1] Zhovnir G.I.,Kletskii S.V.,Sochinskii N.V., Frasunak V.M., - *Phase Equilibria in Mn-Hg-Te System*, Izv.AN SSSR, Neorg. mater., 1989, v.25, p.1216-1218 (in Russian).

[2] Zhovnir G.I.,Kletskii S.V., - *Numerical Solution of Inverse Stefan Problem*, Kristallografia, 1988, v.33, No. 5, p.1271-1273 (in Russian).

[3] Zhovnir G.I.,Kletskii S.V.,Sochinskii N.V., - *Computer Simulation of the Inverse Stefan Problem in Conformity with the Liquid Phase Epitaxy of* $Mn_xHg_{1-x}Te$, Phys.stat.sol.(a), 1989, v.115, No. 1, p.K31-K34.

[4] Kletskii S.V.,Sochinskii N.V.,Zhovnir G.I., - *Interdiffusion in Heterostructures* $Cd_xMn_yHg_{1-x-y}Te/CdTe$, Ukrainian Phys.J. - in press.

[5] Shewmon P.G., - *Diffusion in Solids*, McGraw-Hill Book Company, N.-Y., 1964.

[6] White A.M., - *The Characteristics of Minority-carrier Exclusion in Narrow Direct Gap Semiconductors*, Infrared Phys., 1985, v.25, No. 6, pp.729-741.

[7] Ashley T.,Elliot C.T., Harker A.T., - *Non-equilibrium Modes of Operation for Infrared Detectors*, Infrared Phys., 1986, v.26, No.6, pp.303-315.

[8] Kletskii S.V., - *Spectral Characteristics of Graded Band Gap Semiconductors with Nonlinear Profile of Composition*, Fizika i tehnika poluprovodnikov, 1992, v.26, No. 9, p.1631-1634 (in Russian).

[9] Kletskii S.V.,Sizov F.F., - $Cd_xHg_{1-x}Te$-*Graded Band Gap Structures with Nonlinear Profile of Composition*, Ukrainian Phys.J. - in press.

[10] Selberherr S., - *Analysis and Simulation of Semiconductor Devices*, Springer-Verlag, Wien, 1984.

SIMULATION OF SEMICONDUCTOR DEVICES AND PROCESSES Vol. 5   329
Edited by S. Selberherr, H. Stippel, E. Strasser – September 1993

Charge Distribution and Capacitance of Double Barrier Resonant Tunneling Diodes

P. Mounaix, X. Wallart, J. M. Libberecht, and D. Lippens

Institut d'Electronique et de Microélectronique du Nord,
Département Hyperfréquences et Semiconducteurs, UMR - CNRS 9929
Université des Sciences et Technologies de Lille
F-59655 Villeneuve D'Ascq Cédex, FRANCE

Abstract

Charge and potential profiles are self-consistently calculated in double-barrier heterostructures to derive the capacitance of resonant tunnelling devices. We show that the dipole charge integrated over the accumulation or the depletion side of the device is the result of a complex arrangement of the mobile charge dragged and drifted when a bias is applied. Excellent agreement is found with capacitance measurements carried out on high performance resonant tunnelling diodes up to 40 GHz .

1. Introduction

The resonant tunnelling effect through double barrier heterostructure is an intrinsically fast process which is currently used in the proposal of new high speed devices. For this application view point, it is essential to predict how the charges are distributed within the heterostructure, notably to derive the tunnelling conductance and the intrinsic capacitance of the device and hence its frequency capabilities. Basically, several interdependent processes are involved in the arrangement of the mobile charge. This includes diffusion phenomena from the highly doped contact regions and the accumulation effects of carriers in the emitter and in the well regions respectively. In addition, the situation is complicated by the fact that the structure consists of quantum zones with a two-dimensional density of states $D(E)$ and of semi-classical regions acting as reservoirs of electrons with monotonous variations of $D(E)$.

In this paper, the problem of the charge distribution in resonant double barrier heterostructure is more especially addressed. After discussing the various effects involved in a generic sample, we report the conduction band and the internal electric field profiles we calculated in the Thomas-Fermi approximation, i.e. carriers concentrations determinated assuming Fermi-Dirac statistics and constant Fermi levels within each contacts. Theses results were validated by comparison with more sophisticated formalisms. Lastly, to test the model, we designed and fabricated Double Barrier Heterostructures (D.B.H.) with very thin barriers. Their capacitance variations measured up to 40 GHz are in very good agreement with those predicted by simulations .

2. Charge arrangement

In figure 1, we reported the typical band bending we got for a resonant tunnelling diode which consists of AlAs/In$_{0.53}$Ga$_{0.47}$As/AlAs D.B.H. (structure A). Moreover , this tunnelling structure is cladded between two spacer layers with a stepped doping profile to provide electron reservoirs and to make possible the formation of low resistance contacts. Except for the diffusion potential due to the gradient of doping, one can note

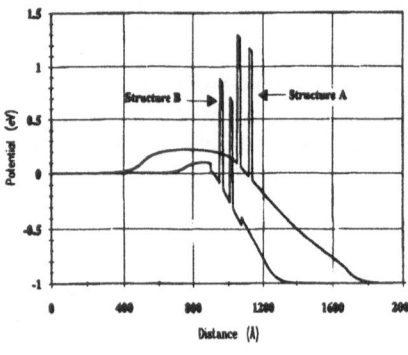

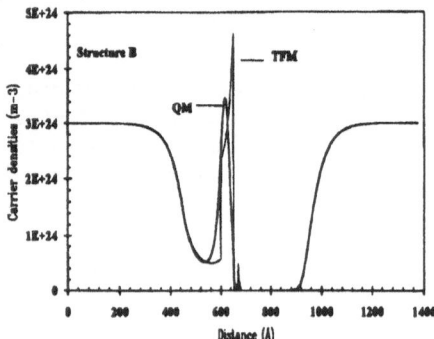

Figure 1 : Conduction Band Edges at 1 V   Figure 2 : Carriers densities under bias with 2
for structure A and B   models : TFM and QM

that an accumulation zone forms in the front of the D.B.H . In contrast , at the right hand side, the spacer layer is partly depleted . In practice, the calculation of such a profile raises a number of fundamental issues i.e. how the charge is shared between the different regions.

First of all, we have to stress that the amount of charge trapped in the well is small because we used symmetric ultra-thin barriers for increasing the current density , a welcome feature for high speed applications. It results from this that the carrier lifetime on the ground state is very short and hence the sheet carrier density n_s trapped in the well. Confidence in such an assertion can be found through Wigner distribution function simulation which gives a direct evaluation of n_s [1]. A maximum value of 5 10^{10} cm^{-2} was thus found at the peak voltage for a peak current density of 60 kA/cm^2 in agreement with aforementioned arguments [2]. Concerning the emitter-accumulation layer , the situation is not so straightforward. In fact, the problem that faces us is that the net charge is the superposition of a 3D contribution due to the extended states and a 2D component due to the electrons in the triangular-shaped potential. This 2D injection is exemplified in figure 1 which also gives the band bending for a strained InGaAs/GaAs/AlAs heterostructure (structure B). Such a pseudomorphic structure enforces the two-dimensional (2D) character of the carrier injection due to the presence of a pre-well prior to the first barrier [3], whereas, the undoped prewell controls the escape process and the postwell is grown to preserve symmetry .

At this stage, it is interesting to compare the carrier distribution obtained by a self-consistent calculation between the Poisson equation and the carrier concentration n(z) calculated classically (Thomas-Fermi Model) and by the Schrödinger equation for the two dimensional states (Quantum Model) . Such a comparison is made in figure 2 . Despite the fact that a Thomas-Fermi model gives rise to unphysical jumps in the carrier distribution , it was found that the variation against voltage of the integrated charge over distance is conserved and hence the capacitance of the device. From the conduction profile given above , we can also derive the internal electric field F and determine the peak values of F and the screening length. The results we obtained for structure A at various bias voltage are reported in figure 3. At the interface between the $4 10^{18}$ cm^{-3}

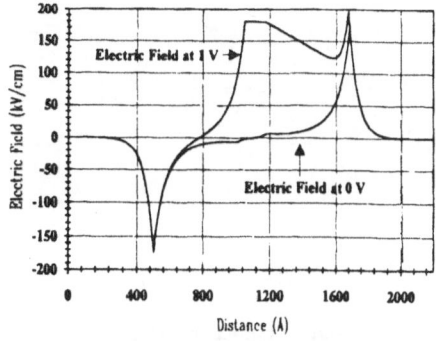

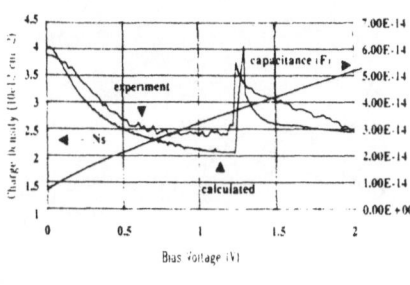

Figure 3 : Electric Field Profiles at 0 and 1 V
for structure A

Figure 4 : Charge density and calculated
capacitance versus bias voltage

doped layer and the spacer layer with a doping concentration of 10^{17} cm^{-3} , the electrical field reaches values as high as 170 kV/cm. As a general rule, this field is screened over relatively long distance comparable to the thickness of spacer layers. Under bias , the situation appears more complex with a close interdependence of the diffusion field and of the internal field in the quantum zone. This lack of separation between the different physical processes complicates notably the numerical procedure. Here , the simulation code we used is a double sweep routine (Choleski code) which reveals very efficient for realistic potential profile .

3. Capacitance

From the conduction band profile, the electrical properties (tunnelling conductance and capacitance) can be directly evaluated by solving the Shrödinger equation throughout the structure and thus get the tunnelling transmission probabilities [4] and by summing the charge either in the emitter or the collector side [5]. Here, we focus our attention on the second issue. Figure 4 gives the variation of the integrated charge up to the first hetero-interface leading to the potential profile given in figure 1. The calculated capacitance values, $C = A \partial Q / \partial V$, with A the diode area , is also given . For A = 20µm^2 , corresponding to the samples we fabricated, the capacitance values Cmax is 60 fF. For comparison, a simple parallel plate capacitor with a dielectric thickness limited to the double barrier will lead to a value exceeding Cmax by a ratio of 4. At increasing bias, the sheet density monotonously increased with a quasi-linear variation above 0.5 Volt. The value Cmax/Cmin obtained in that case is of about 2. The calculations reported before were made assuming that no current will flows through the device. One can expect that the introduction of the mobile charge forming the current will drastically affect the capacitance, when the diode is-in or out-of resonance. Such an effect can be simply introduce using the conductance of the diode and a mean velocity [6]. The results obtained in that case are shown in figure 4 .

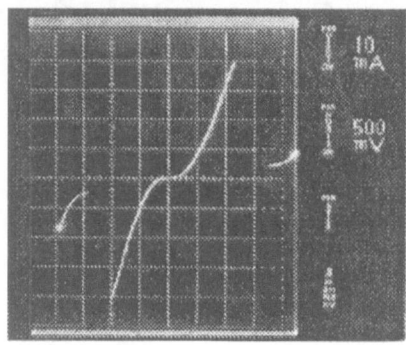
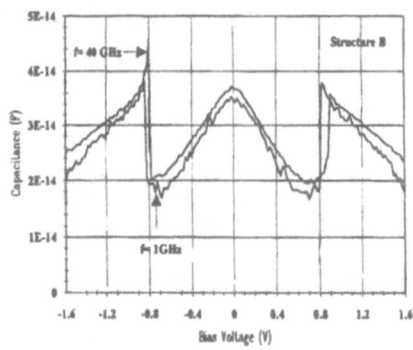

Figure 5 : Typical I(V) curve at 300K (A=100μm2)
for structure A

Figure 6 : Capacitance values of
Structure B (A=20μm2) at 1 and 40 GHz

4. Comparison with experiment

Structures A (AlAs/GaInAs) and B (GaInAs/AlAs/GaAs) were grown by molecular beam epitaxy . In order to have a direct evidence of the intrinsic capacitance effect, the device were fabricated in a low parasitic technology we developed previously. Details of fabrication procedure can be found elsewhere [7]. The devices exhibit excellent D-C performances with a nominal peak to valley ratio Jp/Jv of 9:1 and a peak current Jp of 30kA/cm^2 for structure A whereas Jp/Jv and Jp were typically 7 and 50 kA/cm^2 for structure B. For example , figure 5 gives a typical I-V curve measured at room temperature from which the tunnelling conductance can be deduced. We reported in figure 6 the variations of the capacitance measured at 1 and 40 GHz respectively for structure B. Similar C(V) variations are obtained proving the inherent fast response of the structure. Besides , by comparing the numerical and experimental data of figure 4, a good agreement is now obtained over the voltage range investigated.

5. Conclusion

In summary, the capacitance-voltage relationship of resonant tunnelling diode was successfully obtained by performing calculations based on the self consistent approach of the charge and potential distribution. The Cmax and Cmax/Cmin values are due to the variation of the charge density and the sudden increase near peak voltage is attributed to the tunnelling current flow.

[1] W.R. Frensley , Solid States Electronics , Vol 32 , N°12 , pp 1235-1239 , 1989 .
[2]O. Vanbésien , Private communication .
[3]O. Vanbésien and al , to be published in proceedings of 20Th GaAs & Related
        Compounds , Freiburg , Sept 1993 .
[4]P. Mounaix and al , Appl Phys Lett 57 (15) , pp 1517 , 1990 .
[5]J.P. Sun and al , J. Appl. Phys . 72 (6) , pp 2340 , 1992 .
[6]O. Borlé and al , Int. J. of Inf and Millimeter Waves , Vol 13 , N°13 , pp 799,1992
[7]E. Lheurette and al , Electronics Letters , Vol 28 , N°10 , pp 937 , 1992

SIMULATION OF SEMICONDUCTOR DEVICES AND PROCESSES Vol. 5
Edited by S. Selberherr, H. Stippel, E. Strasser – September 1993

333

Generation and Amplification of Microwave Power in Submicron n^+nn^+ Diodes

V. Gružinskis, E. Starikov, P. Shiktorov, L. Reggiani[†], M. Saraniti[†], and L. Varani[†]

Semiconductor Physics Institute
A. Goštauto 11, LT-2600 Vilnius, LITHUANIA
[†]Dipartimento di Fisica ed Istituto Nazionale di Fisica della Materia,
Universitá di Modena
Via Campi 213/A, I-41100 Modena, ITALY

Abstract

The current voltage characteristics and the spectra of the: small-signal response, current self-oscillations, and microwave power generation are analyzed by means of a theoretical simulation of near micron InP diodes. To this purpose, both the hydrodynamic and Monte Carlo approaches are used and compared. Good agreement with available experimental data is obtained.

1. Introduction

Wide-band generation of microwave power in the frequency range $100 < f < 200\ GHz$ has been recently observed experimentally in near-micron n^+nn^+ InP diodes [1,2]. Because of the short active-length involved, non-local effects such as velocity overshoot become dominant and the mechanism of generation is associated with the transit of an accumulation layer across the diode rather than the propagation of a dipole domain. The aim of this work is to present a theoretical analysis of the transport characteristics which are exhibited by the diode under a stable generation of microwave power.

2. Theory and results

The diode characteristics are simulated on the basis of both a kinetic and hydrodynamic (HD) approach. The model warrants a one-dimensional geometry, therefore macroscopic quantities normalized to unit cross-sectional area are considered in the following. The kinetic approach makes use of a numerical solution of the Boltzmann equation through an ensemble Monte Carlo (MC) method which is coupled self-consistently with a Poisson solver. We take a nonparabolic conduction band structure with Γ-L-X valley ordering, and account for impurity, acoustic and polar optical phonon scattering in each valley as well as intervalley scattering between each couple of equivalent and non-equivalent valleys with spherical equienergetic surfaces. The HD approach makes use of the set of conservation equations for number, velocity and energy in the single-electron gas model and it is coupled with a

Poisson solver [3]. Input parameters of this approach are the energy dependence of the average moments (up to the third) and scattering rates of the bulk material as provided by MC calculations under stationary and homogeneous conditions. Abrupt homojunctions are assumed and the parameters of the InP diode are taken close to those of the experiments [1,2] as: $n^+ = 10^{18}\ cm^{-3}$, $n = 2 \times 10^{16}\ cm^{-3}$, the cathode, n-region and anode lengths respectively of 0.1, 1.0 and 0.3 μm. Since a simulation of the same physical situation with the HD approach has the advantage of requiring a considerably simpler and faster computer enviroment when compared with the kinetic approach, we use the former to perform all calculations and the latter only for validation purposes.

Calculations performed at room temperature are summarized in Figs. 1 to 5. Figure 1 reports the current-voltage characteristics of the unloaded diode as calculated by the HD (lines) and kinetic (full circles) approaches for a constant voltage U_d applied to a diode. Both approaches predict the onset of a nonvanishing self-oscillation of the current I in the narrow region of values $2 \leq U_d \leq 5\ V$. As an example, the insert in Fig. 1 shows the time dependence of the current in the presence of self-oscillations as calculated by the kinetic approach for $U_d = 3\ V$. The current self-oscillations are caused by a periodic transit across the diode of an accumulation layer which is formed near the center of the n-region due to the joint action of velocity overshoot and Gunn effects. The amplitude of the self-oscillations obtained by the HD approach is shown by the dashed lines. In the region $2V \leq U_d \leq 5\ V$ the solid line and the circles show the mean values of I averaged over the period of the self-oscillations. For diodes with the same length of the n-region, the starting of the self-oscillations in the low-voltage region is found to be almost independent of n. On the contrary, the ending in the high-voltage region is found to depend on n as follows. By increasing n it shifts to higher voltages and by decreasing n it shifts to lower voltages until the region of self-oscillations disappears for $n \leq 1 \times 10^{16}\ cm^{-3}$. The appearance of current self-oscillations evidences that the diode can be used for microwave generation. Therefore, to obtain detailed information we have calculated the frequency dependence of the small-signal admittance and impedance, $Y_d(f)$ and $Z_d(f)$, respectively. For this purpose, we use a small signal analysis previously developed [3]. In the general case, a negative sign of the real part of $Z_d(f)$, $Re[Z_d(f)]$, represents the necessary condition for a small-signal amplification at the given frequency. In the case of interest this condition is fulfilled in a rather wide region of f and U_d. (As an example, Fig. 2 shows the negative part of $Re[Z_d]$ (continuous line) calculated for $U_d = 3\ V$.) The frequency regions corresponding to negative values of $Re[Z_d]$ are plotted in Fig. 3 as a function of the applied voltages. These regions, whose boundaries are shown by the continuous lines, define the amplification bands in the (f, U_d)-plane. In each band the dashed line corresponds to the frequency at which $Re[Z_d]$ is minimum. It should be remarked that at the highest voltages an additional amplification band is found. The frequency region where $Re[Z_d]$ is negative ranges from 90 up to 240 GHz for $4.5 \leq U_d \leq 6\ V$, and agrees well with the experimental values for microwave generation [1,2]. This region is much wider than that corresponding to self-oscillations, as can be seen by the full circles in Fig. 3 which refer to the $f - U_d$ region where self-oscillations are present. The imaginary part of Z_d, $Im[Z_d]$, determines the resonance condition which, for the series resonant circuit consisting of the external inductance L and the load resistance R, takes the form $L = -Im[Z_d]/(2\pi f)$. A positive value of L corresponds to the inductance of the external circuit. The frequency region of positive L for $U_d = 5\ V$ is presented in Fig. 4 by the dot-dashed curve. However, in the region of self-oscillations, L is negative. This indicates that the diode can generate without an external circuit. The frequency region of negative L is presented in Fig. 2 by the dotted curve. Here, L is found to become negative only around the maximum frequency for amplification which, in turn, corresponds to the shortest transit-time of the accumulation layer. This is the frequency region of the self-oscillations which is presented in Fig. 3 by full circles. Thus, the linear analysis provides an estimate for the values of the parameters of the external resonant circuit which are needed for generation. For instance, we find that R must be less than 1×10^{-8} Ωm^2 (which from Fig. 2 is the maximum value of $|Re[Z_d]|$) and L must be less than 4×10^{-20} $H m^2$ (see Fig. 4). By using these estimates, within the HD approach we have simulated the diode performances in the series resonant circuit for three sets of values of the total voltage applied to the circuit U and R, namely: (i) $U = 5.5$ V, $R = 5 \times 10^{-10}$ Ωm^2; (ii) $U = 7.5$ V, $R = 2.5 \times 10^{-9}$ Ωm^2; (iii) $U = 10$ V, $R = 5 \times 10^{-9}$ Ωm^2. In all cases, U_d is found to be of about 5 V. The frequency variation in each set is obtained by changing L from 10^{-22} to 3×10^{-20} $H m^2$. Such a variation of the generation frequency is very similar to that obtained experimentally by changing only one parameter, that is the length of the resonance cavity in which the diode is placed [1,2]. For the sake of providing a comparison with the linear theory, these results are summarized in Fig. 4. Here the largest difference is obtained for small values of the resistance, when the deviations from the linear theory are most pronounced (i.e. the current oscillations exhibit an amplitude greater than the average value). Nevertheless, the predictions coming from the linear theory serve as good estimates of the inductance necessary for generation. The efficiency of the generation is finally presented in Fig. 5 where full circles refer to a comparative kinetic calculation. The good agreement we have found between the results of the HD and the kinetic approach strongly supports the physical reliability of our modeling.

3. Conclusions

We have presented a theoretical analysis of linear and nonlinear response characteristics of microwave generators made with near-micron $n^+ n n^+$ InP diodes. The comparison between a kinetic and a hydrodynamic approach supports the reliability of the second approach which has the advantage of requiring a simpler and faster computer enviroment. The frequency range for generation and the estimated efficiency are found to agree well with experiments [1,2]. We would emphasize that by extending the above analysis to the case of submicron GaAs and InP diodes we have found a maximum increase of the generation frequency up to values in the range $600 - 800$ GHz when the length of the n-region is reduced to sizes within $0.3 - 0.2$ μm.

This work has been partially supported by the Commision of European Comunity CIPA 3510PL921499 contract and Ministero della Universitá e Ricerca Scientifica e Tecnologica.

References

[1] A. Rydberg, Int. J. IR and MM Waves, **11**, 383 (1990).

[2] A. Rydberg, IEEE Electron. Dev. Lett., **11**, 439 (1990).

[3] V. Gružinskis, E. Starikov, P. Shiktorov, L. Reggiani, M. Saraniti and L. Varani, Appl. Phys. Lett. **61**, 1456 (1992).

Figure 1: Current-voltage characteristics of the n^+nn^+ InP diode calculated with the HD (lines) and the kinetic (full circles) approaches. Dashed
lines show the amplitude of self-
oscillations. The insert shows the current self-oscillations within the kinetic
approach for $U_d = 3\ V$. Diode parameters are: $n^+ = 10^{18}\ cm^{-3}$, $n =
2 \times 10^{16}\ cm^{-3}$, $l_1^+ = 0.1$, $l_n = 1.0$,
$l_2^+ = 0.3\ \mu m$.

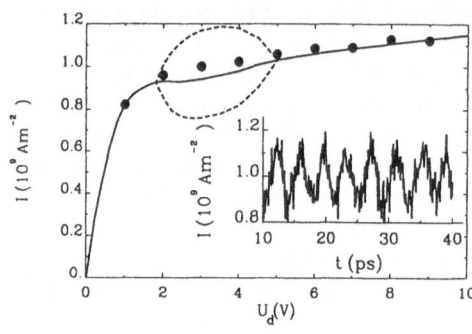

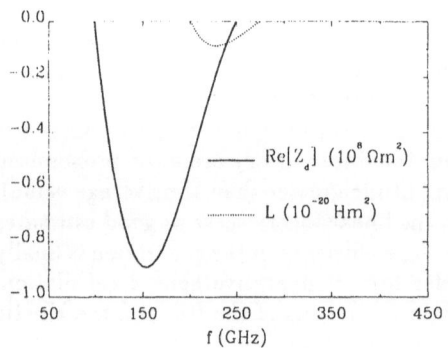

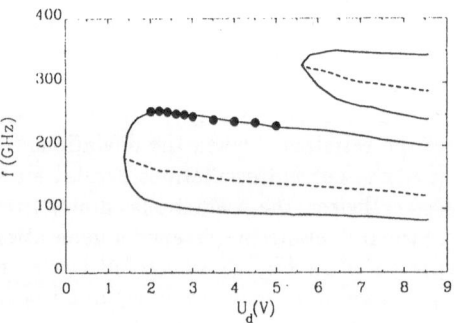

Figure 2: Spectra of the negative parts
of $Re[Z_d]$ and L calculated with the
HD approach for $U_d = 3\ V$.

Figure 3: Amplification band diagram.
Dashed lines refer to $min\{Re[Z_d]\}$.
Points to the self-oscillations region.

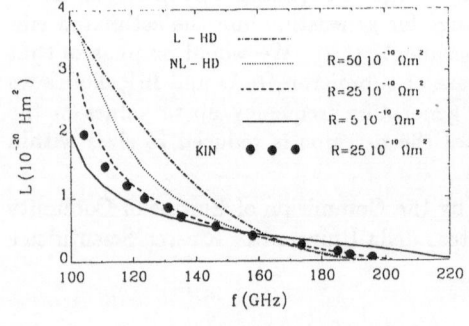

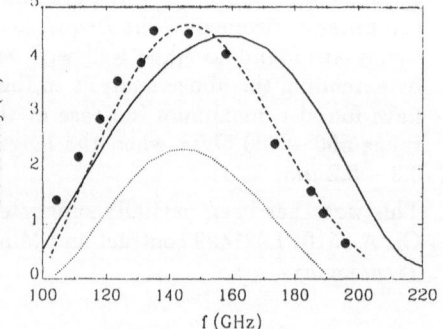

Figure 4: Inductance spectra. L-HD,
NL-HD, KIN refer to linear HD, nonlinear HD, and kinetic approaches.

Figure 5: Efficiency of the microwave
power generation. The notation is the
same of Fig. 4.

Modelling of IGBTs and LIGBTs for Power Circuit Simulation

B. Fatemizadeh, R. Constapel[†], and D. Silber

Institut für Mikroelektronik und Bauelemente der Elektrotechnik,
Universität Bremen
Kufsteiner Straße, Postfach 330440, D-28334 Bremen, GERMANY
[†]Forschungsinstitut AEG, Daimler-Benz AG
Goldsteinstr. 235, D-60528 Frankfurt, GERMANY

Abstract

A network model for the insulated gate bipolar transistor (IGBT) is presented
for circuit analysis programs, such as PSPICE and SABER. The model
contains a fast and rather accurate description of internal plasma dynamics,
and accounts for shorted-anode lateral IGBT structures on Dielectric Isolated
substrates. Good agreement has been obtained between measured data and
simulation results.

1. Introduction

The insulated gate bipolar transistor is a MOS-gate controlled power switching device,
which presently replaces the power bipolar junction transistor in many applications.
Similarly to thyristors, the on-state characteristics are determined by ambipolar current
transport and conductivity modulation.
Basically there are vertical IGBT structures (VIGBT), which are mainly used as discrete
devices in high-power applications, and lateral IGBT structures (LIGBT), which serve as
output devices integrated with MOS logic in power IC's. Lateral devices with dielectric
isolation on SOI-substrates have been manufactured, with and without anode shorts. The
cross sections of the considered device structures are shown in Fig.1.

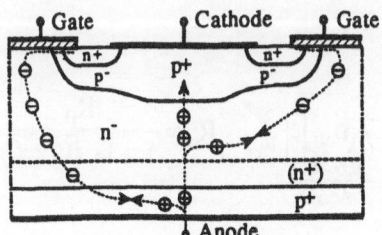
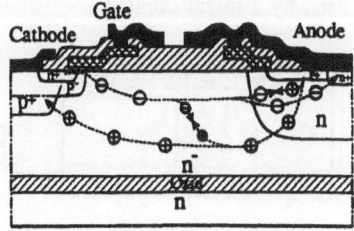

Figure 1: Cross section of VIGBT structure and the shorted-anode LIGBT structure

2. Static Analysis

Structurally, the IGBT works as a bipolar transistor that is supplied base current by a MOSFET. But standard models of bipolar transistors like Gummel-Poon are not suitable or applicable for the modelling of IGBT, as they do not consider appropriately the high injection-effects and the recombination in the base region. Our network model for the LIGBT is based on a "vertical" model, which has been presented in detail in [1]. The modified network model for the LIGBT structure is shown in Fig .2. The parameter b is the ambipolar mobility ratio. The material-parameters k, C1, C2, C3 depend on the base high-level lifetime, the base doping concentration, carrier mobilities, emitter parameter and the base width.

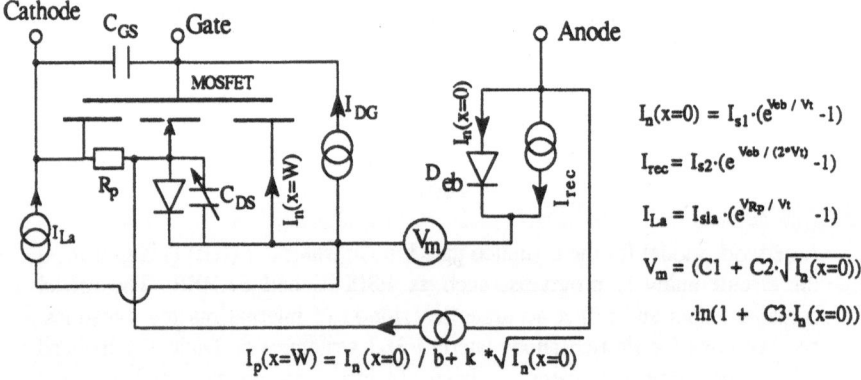

$$I_n(x=0) = I_{s1} \cdot (e^{V_{eb}/V_t} - 1)$$

$$I_{rec} = I_{s2} \cdot (e^{V_{eb}/(2 \cdot V_t)} - 1)$$

$$I_{La} = I_{sla} \cdot (e^{V_{Rp}/V_t} - 1)$$

$$V_m = (C1 + C2 \cdot \sqrt{I_n(x=0)})$$
$$\cdot \ln(1 + C3 \cdot I_n(x=0))$$

$$I_p(x=W) = I_n(x=0) / b + k * \sqrt{I_n(x=0)}$$

Figure 2: The electrical network model of the LIGBT

Since the IGBT operates under high-injection (conductivity modulated) conditions in the n'-base region (p ≈ n >> n'), the current transport in the base can be described by the one-dimensional ambipolar transport equation. Using the quasi-equilibrium simplification (p•n-product at emitter-base junction), assuming high level injection in the base, the solution of the ambipolar diffusion, the continuity and the current density lead to the proposed network model for IGBT.

2.1. Modelling of the shorted anode emitter

The anode short structure, combined with an n⁺-buffer, has been proposed to reduce turn-off switching power loss. The critical difference between this structure and a conventional LIGBT is the incorporation of an n⁺-drain diffusion together with the p⁺-anode that is shorted by a metal contact.

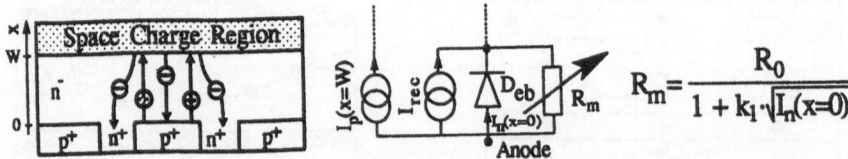

$$R_m = \frac{R_0}{1 + k_1 \cdot \sqrt{I_n(x=0)}}$$

Figure 3: representation of the physical mechanism in the anode short structure

The p$^+$-emitter layer provides conductivity modulation of the drift region, whereas the n$^+$-layer provides an electron extraction path during device turn-off. Fig. 3 illustrates the physical mechanism of the anode short structure.

The "anode short" effect is in the network model represented by the emitter-base-diode shunted by the shorting resistance. The shorting resistance is modulated by the injected excess carrier concentration in the base.

2.2. Modelling of DMOS-Transistor in the IGBT-structure

The device operation of the inherent DMOS in the IGBT-structure is very different from the standard MOSFETs, due to the laterally varying doping concentration in the channel region [2]. The electrons reach the saturation velocity first on the source side with increasing source drain voltage and drift with that velocity across the channel. Therefore this device shows no pinch-off behavior in the classical sense. On the basis of Kim-Fossums-DMOS-model, a simplified version is introduced that easily can be implemented in the circuit simulator. The current flowing throughout the channel in the triode region ($V_G > V_{th}$ and $0 < V_{ch} < V_{sat}$) can be derived as

$$I_{ch} = \frac{\mu_{n0} \cdot c_{ox} \cdot (w/L)}{1 + \theta (V_G - V_{th})} \cdot \frac{\left(V_G - V_{th} - \frac{a}{2} V_{ch} \right) \cdot V_{ch}}{1 + V_{ch}/V_c} \quad ; \qquad V_c = L \cdot E_c \qquad (1)$$

Where L is the the gate length, E_C the critical field, μ_{n0} the low field inversion layer mobility, V_{ch} the drain-source-voltage on the channel and θ the normal field dependence parameter. The depletion charge and the nonuniform doping effect can be considered by the parameters a and V_{th}. A realistic assumption for the saturation of the DMOS ($V_G > V_{th}$ and $V_{ch} \geq V_{sat}$) is, that near the source the channel current is limited by velocity saturation, when the drain voltage goes to saturation. We assume the following model for the saturation current in the channel.

$$I_{ch} = \frac{1}{2a} \cdot \frac{\mu_{n0} \cdot c_{ox} \cdot (w/L)}{1 + \theta (V_G - V_{th})} \cdot \frac{V_c \cdot (V_G - V_{th})^2}{V_c + (V_G - V_{th})} \quad ; \qquad V_{sat} = \frac{V_G - V_{th}}{a} \cdot \frac{1 + 0.5\lambda}{1 + \lambda} \cdot \left(1 - \sqrt{1 - \frac{1 + \lambda}{(1 + 0.5\lambda)^2}} \right) \quad , \quad \lambda = \frac{V_G - V_{th}}{V_c} \qquad (2)$$

3. Transient Analysis

For quantitative simulations of IGBT turn-off behaviour, the time dependent plasma distributions and the moving space charge layer boundary have to be considered.

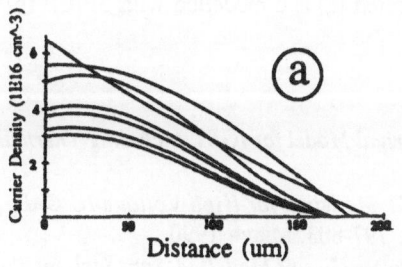
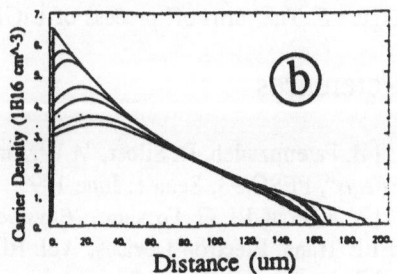

Figure 4: Calculated carrier distribution in the base region by
a) the Galerkin method b) a device simulator

The time-dependent carrier transport in the base of the IGBT is described by the ambipolar diffusion equation

$$\frac{\partial p(x,t)}{\partial t} = D_a \cdot \frac{\partial^2 p(x,t)}{\partial x^2} + \frac{p(x,t)}{\tau_H} \tag{3}$$

Boundary conditions for the ambipolar diffusion equation are a function of the time and are represented in the form

$$\left.\frac{\partial p}{\partial x}\right|_{x=0} - \left.\frac{\partial p}{\partial x}\right|_{x=W(t)} = \frac{h_p \cdot p^2(x=0,t) + p(x=W(t)) \cdot \partial W/\partial t}{D_a} \; ; \qquad \frac{\partial p(x=W(t))}{\partial t} = \frac{\partial W(t)}{\partial t} \cdot \left.\frac{\partial p}{\partial x}\right|_{x=W(t)} \tag{4}$$

Where h_p is the emitter parameter, D_a the ambipolar diffusivity, τ_H the base high-level lifetime. Conventional charge control models cannot cover a wide range of switching conditions with reasonable accuracy. Finite difference methods which have been used in diodes and GTO modelling [3] are very exact but require much computation time. We have found that a reasonable way for fast and rather accurate modelling is based on the use of linear superpositions of suitable basis functions, which is optimized using the methods of weighted residuals. This has recently been proposed for BJT modelling [4]. Divergent from that paper we have used the Galerkin method to optimise superpositions of linear and sine-wave-type functions. The method is especially useful if very different p-emitter properties are considered, as is required to simulate advanced IGBT devices. Implementation of the corresponding device model is performed in SPICE and SABER. In order to illustrate the model accuracy in a transient analysis, in figure 5 the comparision between the experimental and simulated output characteristics for a manufactured LIGBT with shorted-anode is shown.

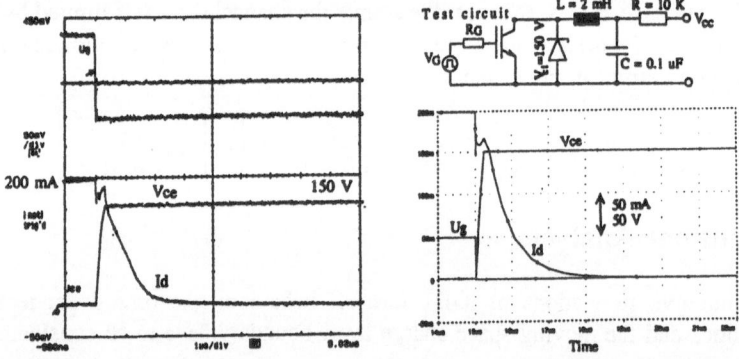

Figure 5: The turn- off process of LIGBT measured (a) and modelled with SPICE (b)

References

[1] B. Fatemizadeh, D. Silber, "*A Versatile Electrical Model for IGBT Including Thermal Effects*", PESC'93, Seattle, June 1993

[2] Y. Kim and J. G. Fossum , "*Physical DMOST Modeling for High-Voltage IC CAD*", IEEE Trans. Electron Devices, vol. ED-37, pp. 797-803, March 1990

[3] T. Vogler and D. Schröder, "*A New and Accurate Circuit-Modeling Approach for the Power-Diode*", PESC'92, Toledo, June 1992, pp. 870-876

[4] B. Allard, H. Morel and J. P. Chante, "*State-Variable Modeling of High-Level Injection Regions in Power Devices*", PESC'92, Toledo, June 1992, pp. 885-892

SIMULATION OF SEMICONDUCTOR DEVICES AND PROCESSES Vol. 5   341
Edited by S. Selberherr, H. Stippel, E. Strasser – September 1993

Physical IGBT Model for Circuit Simulations

M. Andersson, M. Grönlund, and P. Kuivalainen

IC Design, Semiconductor Laboratory, Technical Research Center of Finland
Olarinluoma 9, SF-02200 Espoo, FINLAND

Abstract

An improved analytical model for IGB power transistors has been developed for
circuit simulators. Special attention is paid to the physical modeling of the short
channel in the DMOS branch and the buffer layer near the anode in IGBT device
structures. Good agreement between simulated and measured data has been
obtained.

1. Introduction

Recent advances in power and high-voltage integrated circuits have allowed the integration
of new emerging power devices with high performance MOS logic. One of the most
promising power devices in terms of integration possibilities is the lateral insulated gate
bipolar transistor (LIGBT) [1]. Recently, the integration of a vertical IGBT with a control
circuit has been demonstrated [2]. The IGBT's have unique electrical characteristics which
combine the low on-state resistance of the bipolar power transistor and the low driving power
requirements of the MOS gate.

Computer-aided design of analog circuits including power IGBT's is critically dependent on
reliable device models implemented in circuit simulators, e.g. SPICE. Due to deficient
MOSFET and BJT models in SPICE, simulations in which these conventional power device
models are used to describe new devices cannot accurately account for the special features
of these new devices. In the present work we have developed a physical IGBT model for
circuit simulations to overcome this problem. Special attention is paid to the modeling of the
DMOS channel region and the n^+-buffer layer close to the pnp-emitter region. In the
previously published IGBT models [3-5] for circuit simulations, the treatment of these
regions has been neglected. In the MOS channel region model, mobility degradation due to
high electric fields and the effect of doping gradient on the threshold voltage are taken into
account. For the low gain, high injection bipolar pnp-transistor part of the IGBT structure,
ambipolar transport theory is applied, extending the treatment to the case of a heavily doped
buffer layer close to emitter.

2. Model

The basic structure of a vertical IGBT having a n^+-buffer is shown in Fig. 1. The wide base pnp bipolar branch is driven by an enhancement-type n-MOSFET. With a positive gate voltage the electrons flow towards the anode through the inverted surface channel in the p-well. In the forward conduction mode the anode (or drain) injects minority carriers (holes) through a buffer layer into the lightly doped n-type base. This injection greatly enhances the conductivity of the wide base region, which in turn results in an additional lowering of the on-state resistance.

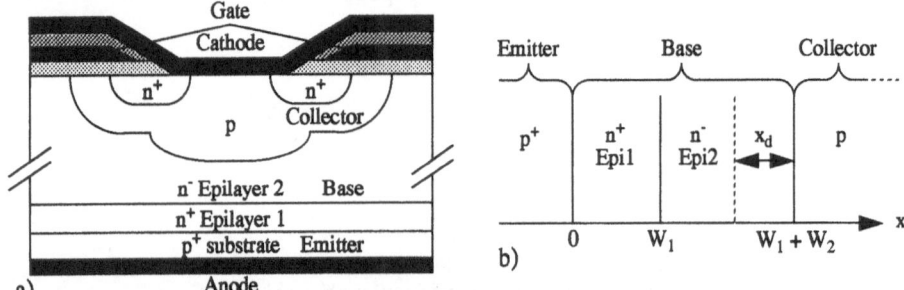

Figure 1. (a) Cross-section of a vertical IGBT. (b) Coordinate system used in the model.

2.1 Buffered wide base BJT model

We now extend the previously published [3-5] physical model equations for the pnp BJT by taking into account the buffer layer close to the emitter (anode), Fig. 1. This layer has been added to the IGBT structure since it increases the forward punch-through voltage and speeds up the turn-off process. The latch-up resistance is also improved by the buffer layer.

Base transport is described by the ambipolar transport equation in both epilayers of the base

$$\frac{d^2 p_i(x)}{dx^2} - \frac{1}{L_{pi}^2} \cdot p_i(x) = 0 \tag{1}$$

where $L_{pi} = \sqrt{D_{Ai} \tau_{Hi}}$ is the ambipolar diffusion length and the index i(=1,2) refers to the epilayer regions shown in Fig. 1 (b). The solution to (1) is a linear combination of exponential functions and the coefficients can be determined from proper boundary conditions obtained from the continuity of the hole current and concentration at W_1. Furthermore, we know that $p_2(x=W_1+W_2-x_d) = 0$, at the end of the neutral base region near the base-collector junction, since this junction is always reverse biased. Applying these boundary conditions we obtain the one-dimensional hole distribution from (1)

$$p_i(x) = \frac{p(0)}{A} \left[\cosh\left(\frac{x-W_1}{L_{pi}}\right) - \left(\frac{D_{A2}L_{p1}}{D_{A1}L_{p2}}\right)^{2-i} \coth\left(\frac{W_2-x_d}{L_{p2}}\right) \sinh\left(\frac{x-W_1}{L_{pi}}\right) \right] \tag{2}$$

with

$$A = \frac{D_{A2}L_{p1}}{D_{A1}L_{p2}} \coth\left(\frac{W_2-x_d}{L_{p2}}\right) \sinh\left(\frac{W_1}{L_{p1}}\right) + \cosh\left(\frac{W_1}{L_{p1}}\right) \tag{3}$$

$p(0)$ is the hole density at the edge of the emitter-base (p^+-n^+) junction space charge region, and by using the quasi-equilibrium assumption it can be expressed in terms of the emitter-base-voltage V_{EB}, $p(0) = [n_i^2/N_{epil}] \exp(qV_{EB}/k_BT)$. x_d is the depletion region width of the collector-base junction.

With the aid of the hole densities (2), the emitter and base current densities can be expressed as

$$J_E = \frac{b+1}{b} \{J_{no} \left[\frac{p_1(0)}{n_i}\right]^2 - qD_A \frac{dp_1}{dx}(x=0)\} \tag{4}$$

$$J_B = J_{no} \left[\frac{p_1(0)}{n_i}\right]^2 + q \int_0^{W_1} \frac{p_1(x)\,dx}{\tau_{H1}} + q \int_{W_1}^{W_1+W_2-x_d} \frac{p_2(x)\,dx}{\tau_{H2}} \tag{5}$$

where J_{no} is the saturation current density that defines the back injection of electrons into the emitter region, b the ratio between the electron and hole mobilities, and τ_{Hi} (i=1,2) is the high injection carrier life-time. The collector current density J_C is expressed as J_E-J_B. By using the hole and current densities (2) - (5) the voltage drop in the neutral base region can be calculated as in [3].

Finally, in the transient modeling of the IGBT [3], we need the depletion charge density at the base collector junction $Q_{BC} = qA_C N_{epi2} x_d$, and the base charge

$$Q_B = qA_E \int_0^{W_{M1}} p_1(x)\,dx + qA_E \int_{W_1}^{W_{M2}} p_2(x)\,dx \tag{6}$$

where W_{Mi} is the point where $p_i(x)$ is equal to the doping density in the epilayer i.

2.2 Channel region model

A physical channel region model for IGBT has been modified from the SPICE level 3 MOSFET model (MOS3)[6]. The modifications include an improved treatment of charge carrier mobility degradation phenomena according to experimental measurements and the effect of the doping concentration gradient on the threshold voltage. The final equations are the same as in MOS3 model excluding the effective mobility expression μ_{eff} and the threshold voltage V_{TH}:

$$\mu_{eff} = \frac{\mu_s}{\left[1 + \left(\frac{V_{DS}\mu_s}{VMAX \cdot L}\right)^p\right]^{1/p}} \tag{7}$$

$$V_{TH} = V_{FB} + PHI + \frac{2}{\eta}(1 - e^{-\eta/2})(GAMMA \cdot F_S \sqrt{PHI}) \tag{8}$$

Here the model parameters are the same as in the original MOS3 model excluding a new parameter p which is related to the electric field dependence of the carrier velocity and the parameter η for the doping concentration gradient. An exponential position dependence was assumed for the acceptor doping in the channel region of a n-type DMOS structure, $N_A(x) = N^0_A \exp(-\eta x/L)$. The new parameter p improves the accuracy of the model and it allows the use of a physical value for the saturation velocity parameter VMAX. Our modified MOSFET model reduces to the SPICE MOS model in the case $\eta=0$ and $p=1$.

3. Simulation results

The device physics based IGBT model, which suits both lateral and vertical structures, has been implemented in the APLAC circuit simulator [6]. An advantage of the APLAC is that it provides optimization capabilities for the model parameter extraction. Fig.2 (a) compares the results from the simulations and the measurements of the dc electrical characteristics of a lateral IGBT fabricated in our laboratory. Fig. 2 (b) shows the transient behaviour of the same device. In both cases the agreement between the simulations results and the measured data is good.

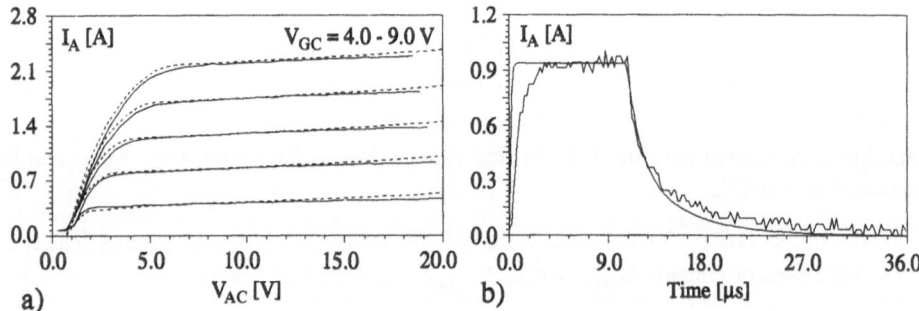

a) V_{AC} [V]   b) Time [µs]

Figure 2. (a) Measured and simulated (dotted lines) LIGBT anode current versus anode voltage for gate voltages in the range 5.0-9.0 V. (b) Measured and simulated current versus time for turn-on and turn-off transients in a LIGBT. Risetime discrepancy is due to unknown parasitic inductance. The LIGBT has been processed in our laboratory, and its gate dimensions are W/L = 0.03 m / 8 µm.

References

[1] M.N.Darwish, M.A.Shibib, *Lateral MOS-Gated Power Devices- A Unified View*, IEEE Trans.Electron Dev. 38 (1991) 1600.

[2] T.Mizoguchi, T.Shirasawa, M.Mori, Y.Sugawara, *600 V, 25 A Dielectrically-Isolated Power IC with Vertical IGBT*, IEEE Proceedings of the 3rd International Symposium on Power Semiconductor Devices and ICs, 1991, 40.

[3] J.G.Fossum, R.J.McDonald, M.A.Shibib, *Network Representations of LIGBT Structures for CAD of Power Integrated Circuits*, IEEE Trans. Electron Dev. 35 (1988) 507.

[4] S.Eränen, M.Grönlund, M.Blomberg, J.Kiihamäki, *The Kirk Effect in the LIGBT Devices*, IEEE Trans.Electron Dev. 38 (1991) 1919.

[5] Z.Shen, T.P.Chow, *An Analytical IGBT Model for Power Circuit Simulation*, IEEE Proceedings of the 3rd International Symposium on Power Semiconductor Devices and ICs, 1991, 79.

[6] M. Valtonen et al., Helsinki University of Technology, Circuit Theory Laboratory & Nokia Research Center, Hardware Design Technology, *APLAC, An Object-Oriented Analog Circuit Simulator and Design Tool*, 6.1 User's Manual & 6.1 Reference Manual, December 1992.

Model-Independent Distortion Analysis in SPICE Realized for Complex Modelled Bipolar and MOS Transistors

E. Klostermeier, J. Wilk, and R. Laur

Institut für Mikroelektronik und Bauelemente der Elektrotechnik,
Universität Bremen
Kufsteiner Straße, Postfach 330440, D-28334 Bremen, GERMANY

Abstract

Small–signal distortion analysis requires the second–order and third–order derivatives of the model equations. The use of numerical differentiation schemes enables a fast and model–independent derivation. We propose here two efficient algorithms. Both use a central differential quotient. One algorithm is based on a Romberg scheme and finds iteratively an optimal step size. The other algorithm was developed especially for bipolar elements. Here, a *logarithmus naturalis* transforms the equations to the *ln*–plane, where they behave nearly linear. This makes a step–size control for the differentiation superfluous. The methods have been realized in SPICE for complex modelled bipolar and MOS transistors.

1. Introduction

Small–signal distortion analysis in SPICE uses the perturbation approach. This means, that the characteristic of each nonlinear circuit element is expanded about its dc operating point by a third–order Taylor series. The nonlinear components are lumped together and are represented by a nonlinear voltage-controlled current source, the distortion current source.

The Taylor–series representation of the nonlinear element requires the second–order and third–order derivatives of its characteristical equation. If they are evaluated symbolically, as it is done in SPICE, the following disadvantages occur: The evaluation of the derivatives is very time-consuming, even for a moderately complex model as the Gummel–Poon transistor model. Furthermore, the evaluation must be done for each nonlinear device, and after each modification of the equations.

A numerical approximation of the derivatives circumvents these disadvantages. We propose here two algorithms for approximation of the derivatives. Both solve the problem of finding an appropiate step size for a numerical differentiation. One algorithm uses a Romberg scheme [3] to find iteratively an optimal step size. It requires only topological information on the device because it uses the dc analysis routine. The other algorithm was developed especially for bipolar devices. A *logarithmus naturalis* of the equations has nearly linear characteristics. This makes a step–size control for the differentiation superfluous.

We have applied the methods in SPICE to complex modelled bipolar and MOS transistors. They can be adapted for other models of nonlinear devices with little effort.

2. Model–independent numerical approximation scheme

We use a Romberg scheme on the first–order derivatives of the model equations. They can be found as the Jacobi elements, calculated in the dc analysis routine. Fig. 1 shows a pseudo code of the method. The voltages are varied at the operating

0. Initialize: $i = 0$; h_0: starting step size

1. CALL DC$(V_{OP} \pm h_0)$ $\Rightarrow$ $f'(V_{OP} \pm h_0)$

2. $f''_{h_0} = (f'(V_{OP} + h_0) - f'(V_{OP} - h_0))/2h_0$

3. $i = i + 1$; $h_i = h_{i-1}/2$

4. CALL DC$(V_{OP} \pm h_i)$ $\Rightarrow$ $f'(V_{OP} \pm h_i)$

5. $f''_{h_i} = (f'(V_{OP} + h_i) - f'(V_{OP} - h_i))/2h_i$

6. Romberg linear combination$(f''_{h,j}; \quad j = 0(1)i)$ $\Rightarrow$ f''_i

7. Continue 3. ... 6. until $|f''_i - f''_{i-1}| < \varepsilon$ $\Rightarrow$ h_i is optimal step size h_{opt}

8. $f''' \approx (f'(V_{OP} + h_{opt}) - 2f'(V_{OP}) + f'(V_{OP} - h_{opt}))/h_{opt}^2$

Figure 1: Model–independent differentiation scheme

point (OP) to a step size h_i. The dc analysis routine returns the first–order derivatives, $f'(V \pm h_i)$. A differential quotient approximates the second–order derivatives, f''_{h_i}. A linear combination of the approximations of the second–order derivatives to different step sizes yields a second–order derivative with smaller truncation error, f''_i. Decreasing of the step size and linear combination of the approximations is continued, until the second–order derivatives reach convergency. The last step size is considered optimal. With this step size, we approximate the third–order derivatives, f'''.

3. Numerical approximation scheme for bipolar devices

The differentiation scheme for bipolar devices exploits the exponential characteristic of the device equations. A *logarithmus naturalis* of the device equations obtains linear characteristics. The *ln*–function is used on the equations at the operating point (OP) and at a variation to a constant step size around the OP. Then a differential quotient approximates the second–order and third–order derivatives in the *ln*–plane. The inverse transformation leads to the second–order derivatives in the original plane. Eq. 1 shows that a multiplication of the *logarithmical differential quotients* with the original equations corresponds to the inverse transformation to the original plane.

$$\frac{d\ln(f(V_{OP}))}{dV} = f'(x) \cdot \frac{1}{f(V_{OP})} \Rightarrow$$

$$f'(V_{OP}) \approx \frac{\Delta \ln(f(V_{OP}))}{\Delta V} \cdot f(V_{OP}). \tag{1}$$

To increase the accuracy of the numerical differentiation, we split the first–order derivatives into summands of exponential functions. Fig. 2 illustrates the method.

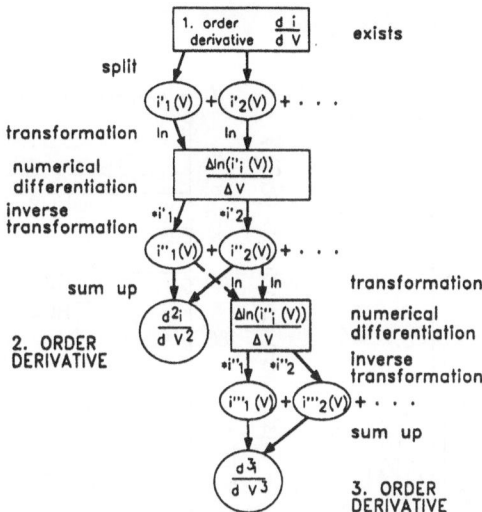

Figure 2: Differentiation scheme for bipolar devices

4. Examples

Fig. 3 shows the small signal distortion equivalent circuit of the MOS transistor model with emphasis on analog applications. The distortion current sources are indicated with an asterisk. Fig. 4 shows the harmonic and intermodulation products of a simple MOST circuit.

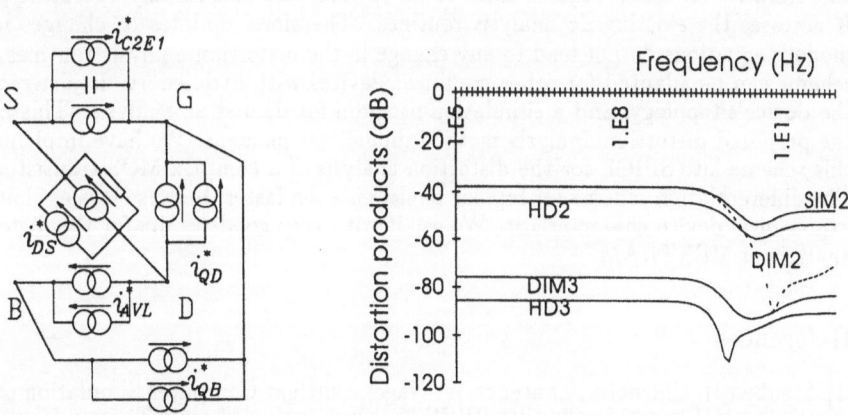

Figure 3: MOST equivalent circuit for distortion analysis

Figure 4: Distortion analysis of a MOST circuit

The "Most EXquisite TRAnsistor Model" (MEXTRAM) is described in [2]. Topologically, it is an extension of the Gummel–Poon type model with extra internal nodes. It incorporates many physical effects, for example the quasi–saturation is extensively modelled. Fig. 5 shows its equivalent circuit for distortion analysis. Fig. 6 shows the harmonic and intermodulation products of a MEXTRAM circuit.

For both devices, a comparison of the harmonic distortion products with the Fourier analysis following a transient excitation shows very good agreement for both the second–order and the third–order harmonics.

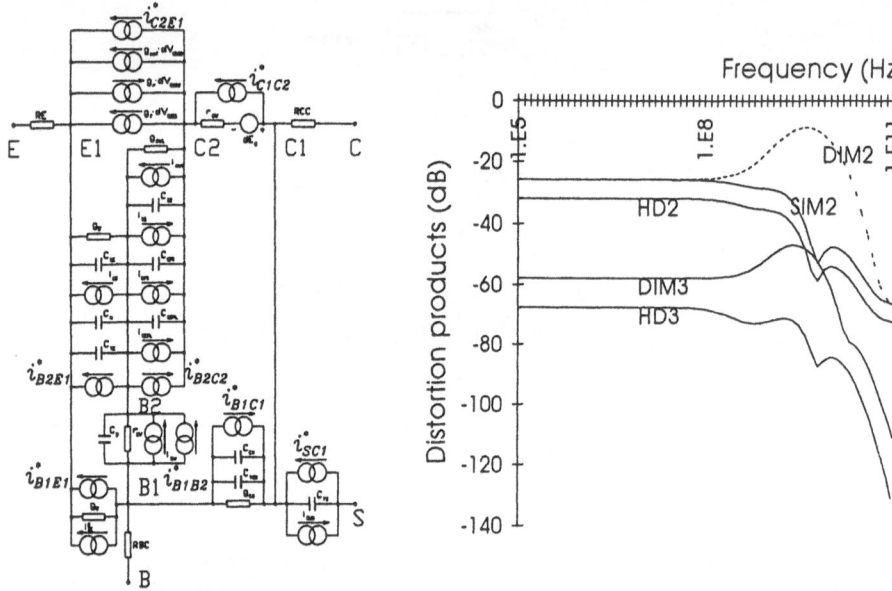

Figure 5: MEXTRAM equivalent
circuit for distortion analysis

Figure 6: Distortion analysis of a
MEXTRAM circuit

5. Discussion

The iterative Romberg scheme leads to an optimal step size for any operating point.
It accesses the existing dc analysis routines. Therefore, updates or changes in the
model's equations do not lead to any change in the distortion analysis routines. The
scheme can be adapted for other nonlinear devices with little effort. It just requires
the device's topology, and a simulation program for dc and ac analysis. This makes
the proposed distortion analysis method model–independent. We have implemented
this scheme into SPICE for the distortion analysis of a complex MOS transistor.
The differentiation scheme for bipolar devices is even faster, because it is exploits the
exponential device characteristic. We use it with very good results for the distortion
analysis of MEXTRAM.

References

[1] Stephen H. Chisholm, Laurence W. Nagel, Efficient Computer Simulation of Dis-
    tortion in Electronic Circuits, IEEE Transactions on Circuit Theory. Nov. 1973,
    pp. 742–745

[2] H. C. DeGraaff, F. M. Klaassen, *Compact Transistor Modelling for Circuit Design*,
    Springer Verlag Wien New York. 1990, pp. 114ff

[3] G. Engeln–Müllges, F. Reutter, *Numerische Mathematik für Ingenieure*, BI Wis-
    senschaftsverlag Mannheim/Wien/Zürich. 1988, pp. 357–360

A Newly Proposed Delay Improvement on CMOS/SOI Future Technology

M. Lee[‡†] and K. Asada[‡]

[‡]Department of Electronic Engineering, The University of Tokyo
7-3-1 Hongo, Bunkyo-ku, Tokyo 113, JAPAN
[†]Nippon Motorola Ltd.
3-20-1 Minami-azabu, Minato-ku, Tokyo 106, JAPAN

Abstract

Contribution of gate fringing capacitance to CMOS/SIMOX inverter time delay in deep sub-micrometer gate is propounded. Measurements of the fifty-one stage ring oscillator's time delay are completed for comparison with analytical model. Propagation Delay Times(TPD) by reducing Poly-Si gate thickness were improved up to two times in deep-submicron CMOS/SIMOX inverters. It is concluded that SOI technology is promising for a *high speed* by reducing gate fringing capacitance which is correlated to the poly-Si gate thickness.

1. Introduction

Deep sub-micrometer gate SOI devices built in ultra-thin Si film have attracted much attention for their high performance. It is generally acknowledged that the gate oxide capacitance strongly affects on the TPD [1][2][3][7], however parasitic capacitances other than the gate oxide capacitance become a major factor limiting device speed with device miniaturization [4] [5]. In this paper, we studied on the contribution of gate fringing capacitance to the time delay. Reduction of poly-Si gate thickness proportionally decreases the gate fringing capacitance as shown in Fig. 1 and Fig. 2, resulting in high speed of fabricated SOI CMOS inverters. It is indicated that a method for reducing the poly-Si gate thickness is proposed to improve the TPD as in future SOI technology.

2. Formulation for Model

It has been reported[5] that gate capacitances of CMOS/SOI inverter are composed of gate oxide capacitance, C'_{ox}, gate fringing capacitance, C_f and wiring capacitance, C_w, in nMOS and pMOS devices. A suggestion for high speed has also been raised by reducing parasitic capacitances other than intrinsic gate oxide capacitance using effective time-dependent capacitance model [6]. In this study we focused on the contribution of the C_f to the delay time, reducing poly-Si gate thicknes (t_m). It is found that gate fringing capacitance (C_f) is significantly decreased by reducing the t_m with assumption of invariable resistance for poly-Si gate as $t_m = 350$ (nm) whose assumption can be done by salicidation technique. The formulation of C_f is separated into three components at various distances from the edge which involves transforming a rectangular electrode and ground plane into the complex potential plane [6]. Introducing the parallel plate model for C_{ox} and summarizing the gate fringing capacitance, C_f, in the CMOS/SOI inverter as follows:

$$C_f = \epsilon_{ox} \times \gamma \times (W_g + L_g),$$ (1)

where L_g is gate length, W_g is gate width, ϵ_{ox} is permittivity of oxide, ϵ_{ni} is permittivity of nitride, and

$$\gamma = \frac{\left(0.613 + \ln(\frac{u}{a}) + (\frac{\epsilon_{ni}}{\epsilon_{ox}})\ln(a)\right)}{\pi},$$ (2)

while constants for u and a are determined by the following equations: $a = -1 + 2K^2 + 2K(K^2 - 1)^{1/2}$ and $u = (R^2 a - 1)/(R^2 - 1)$, where K and R are calculated as follows:

$$K = 1 + t_m/t_{ox},$$ (3)

and nonlinear expression on R as

$$\frac{\pi}{2} \cdot \frac{L_G}{t_{ox}} = \frac{a-1}{a^{1/2}} \cdot \frac{R}{(R^2-1)} + \ln\left(\frac{a^{1/2}R+1}{a^{1/2}R-1}\right) - \frac{a+1}{2a^{1/2}} \cdot \ln\left(\frac{R+1}{R-1}\right).$$ (4)

We analyzed the influence of the C_f via varying the poly-Si gate thickness, t_m, appeared in eq. (3). The constant, R, should first be solved numerically to evaluate the fringing parameter, γ, in Fig. 1(a) and corresponding C_f in Fig. 2(a) varying the poly-Si gate thickness, t_m. Fig. 2(b) shows the calculated results for the conventional gate oxide capacitance, C_{ox}, using the parallel-plate model, $C_g(t)$ by new model, gate fringing capacitance, C_f, in eq. (1) by the above sequence and drain-substrate capacitance, C_d, using the parallel-plate model for the buried oxide. It is noted that the gate capacitance was formulated using a Time-Dependent Gate Capacitance(TDGC) model, $C_{out}(t)$ [6] as follows:

$$C_{out}(t) = C_g(t) + C_f^{(n)} + C_f^{(p)} + C_w,$$ (5)

where $C_f^{(n)}$ and $C_f^{(p)}$ are gate fringing capacitances of nMOS and pMOS, respectively, C_w comprises both of metal wiring stray capacitance C_{metal} and drain-substrate capacitance C_d. However, the C_{metal} is negligible for SOI circuits. $C_g(t)$ at average of nMOS pull-down and pMOS pull-up circuit is shown in Fig. 2(b). Details for the $C_g(t)$, TDGC model, are expressed [6]. The output level $V_{out}(t)$ of an inverter during low-to-high transient process of the inverter gate potential can be represented as

$$\frac{d}{dt}\{C_{out}(t) \times V_{out}(t)\} = -I_{nMOS}(V_{out}(t)),$$ (6)

where I_{nMOS} is the nMOS drain current and $C_{out}(t)$ is output capacitance in eq. (5). The eq.(6) was solved numerically with dynamic change of the I_{nMOS}. The program sequence for the Propagation delay time(TPD) is depicted in Fig. 1(b). The TPD is defined as an average of a time for 50 % V_{dd} of nMOS pull-down circuit and pMOS pull-up circuit [5][6]. Details for numerical technique by multistep methods using Milne's *Predictor-Corrector*, MPC, algorithm are explained in [5]. It was found that the MPC appears to be somewhat more favorable than other techniques such as Runge-Kutta 5th order method [8] and Adams-Moulton method [8] such as better accuracy by the iterations, economing the number of corrections, automatic error estimate at each step which will minimize the computing time [9][10][11].

3. Results and Discussion

To provide experimental support for the analysis in section 2, we measured DC characteristics and TPD at various operating frequencies and power supply voltages at three-terminal SOI MOSFET's and CMOS inverters, respectively. The devices were fabricated at Nippon Telegraph and Telephone(NTT) with poly-Si gate thickness(t_m), 350(nm), same as [6]. Typical case TPDs are clearly shown as solid circles in Fig. 3 at $t_m = 350$(nm) for comparison with experimental results. Fig. 3 also shows other simulated results at different gate thicknesses and supply voltages of 1.5V and 2.0V. It is found that approximately two-time decreased speeds to the typical case TPDs were observed as the gate thickness was approached to zero. It is implied the reduction of C_f was obtained by decreasing the t_m and was significant to improve the propagation delay times for next generation in integrated circuit, i. e., *high speed* as in SOI future technology.

4. Conclusion

Contributions of gate fringing capacitance on the propagation delay times are studied. It is concluded that reduction of the poly-Si gate thickness(t_{n_i}) decreases the gate fringing capacitance(C_f), resulting in improving the propagation delays up to about two times than typically simulated TPDs. This is promising for high speed CMOS/SOI devices by reducing the t_m with current SOI technology. Thinning the t_m for high performance is proposed in CMOS/SIMOX technology. It is also predicted that this proposed reduction of the gate height leads to low power as next generation in integrated circuit.

References

[1] T. Douseki K. Aoyama and Y. Omura, "Dependence of CMOS/SIMOX inverter delay time on gate overlap capacitance," *IEICE Trans.*, ED92-62(176):39–44, Aug. 1992

[2] T. Sakurai and A. R. Newton, "Alpha-Power Law MOSFET Model and its Applications to CMOS Inverter Delay and Other Formulas," *IEEE J. Solid-State Circuits*, SC-25(2):584–594, Apr. 1990

[3] T. Sakurai and A. R. Newton, "Delay Analysis of Series-Connected MOSFET Circuits," *IEEE J. Solid-State Circuits*, SC-26(2):122–131, Feb. 1991

[4] M. Fujishima, M. Ikeda, K. Asada, Y. Omura and K. Izumi, "Analytical Modeling of Dynamic Performance of Deep Sub-micron SOI/SIMOX based on Current-Delay Product," *IEICE Trans. Electron.*, E75-C(12):1506–1514, Dec. 1992

[5] M. Lee and K. Asada, "Deep Sub-micron CMOS/SIMOX delay modeling by Time-Dependent Capacitance Model," *IEEE Trans. on Electron Devices*, To be published, July, 1993

[6] M. Lee and K. Asada, "Sub-100nm CMOS/SIMOX Delay Modeling by Time-Dependent Gate Capacitance Model," *1993 International Symposium on VLSI TSA*, Presented, Taipei, Taiwan, May 1993

[7] S. R. Vemuru and A. R. Thorbjornsen, "A MODEL FOR DELAY EVALUATION OF A CMOS INVERTER," *1990 IEEE Int. Sym. on CAS*, ISCAS-90(1):89–92, May. 1990

[8] S. D. Conte and C. Boor, *Elementary Numerical Analysis: An Algorithmic Approach*, 3rd ed., London; McGRAW-Hill:120–127, 1980

[9] R. E. Bank and D. J. Rose, "Global approximate Newton Methods," *Numer. Math.*, (37):279–295, 1981

[10] L. A. Hageman and D. M. Young, "Applied Iterative Methods," New York; Academic Press, 1981

[11] D. Luenberger, *Linear and Nonlinear Programming*, 2nd ed., Philippines; Addison-Wesley, 1984.

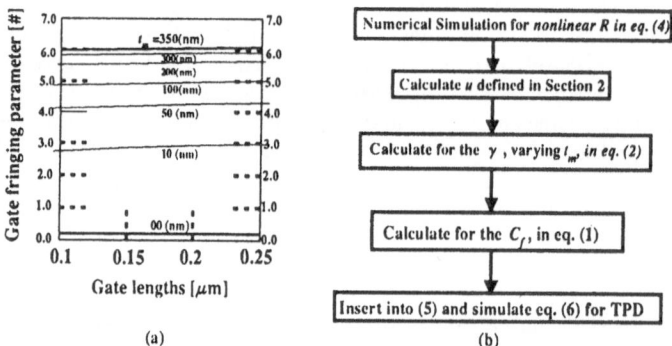

Fig. 1 (a) Simulation results for gate fringing parameter, γ, reducing poly-Si gate thickness (t_m).

(b) Program sequence to obtain the gate fringing capacitance (C_f) for the delay time.

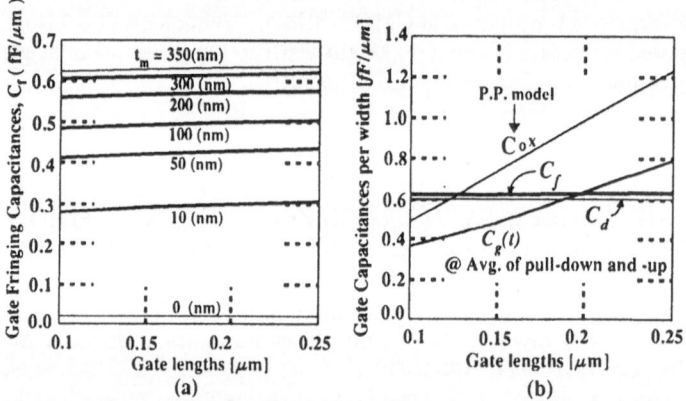

Fig. 2 (a) Normalized C_f by gate width, depending upon poly-Si gate thickness (t_m).

(b) Normalized gate capacitances by the gate width based on measured parameters.

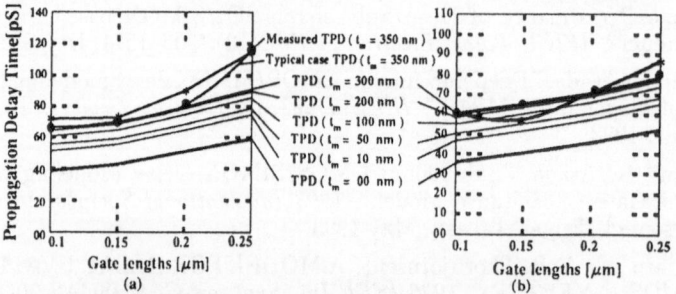

Fig. 3. The Simulated TPDs with the Measured data at (a) V_{dd} = 1.5v; (b) V_{dd} = 2.0v.

* : Measured TPDs
Solid line with filled circles : Typical case TPDs by TDGC with C_f @t_m=350nm.
Other Solid lines : TPDs with C_f @ different t_m = 300nm to 00 nm.

CMOS inverters; W_g of pMOSFET is twice larger than W_g of nMOSFET. Back-gate bias of 0.0V and

Supply voltages are V_{dd} as the above. Horizontal axis represents the gate lengths as micrometers.

SIMULATION OF SEMICONDUCTOR DEVICES AND PROCESSES Vol. 5   353

Edited by S. Selberherr, H. Stippel, E. Strasser – September 1993

Simulation of a Novel Scheme for 700-1000 V Wiring Applications

A. F. J. Murray and W. A. Lane

Institute of Advanced Microelectronics, National Microelectronics Research Centre
University College, Prospect Row, Lee Maltings, Cork, IRELAND

Abstract

This paper investigates an isolation and wiring scheme for a typical 700–1000-V, junction isolated (JI), high voltage integrated circuit (HVIC), process. The 2-D device simulator S-PISCES2B [1] is used to simulate the behaviour of the isolation, and that of a novel wiring technique used to run high voltage wires over the isolation.

1. Introduction

A common problem in today's high voltage integrated circuit (HVIC) technology is the ability to route metal wires around the chip. A 200-V wiring scheme has already been investigated by the authors [2]. Studies have also been carried out by other authors in the 400-V range [3]. However, I.C.'s in the 700–1000-V range are now been fabricated using junction isolation [4], and these are particularly vulnerable to parasitics induced by high voltage interconnect on chip. If figure 1 is examined, ignoring the field plates and p- extension, a general wiring problem is illustrated. The n+ drain of a high voltage device (e.g. an LDMOS), must be connected to other circuit elements, requiring a high voltage wire to run across the isolation. This paper uses the 2-D device simulator S-PISCES2B [1] to investigate isolation and a novel wiring scheme for a typical 700–1000-V HVIC that might be used in a.c. off-line applications.

2. Simulation structure

The general simulation structure is shown in figure 1. The epi layer is represented by uniform doping and the profiles by 1-D gaussians rotated by a lateral ratio of 0.7. The wire was simulated by placing an aluminium contact on top of a $3\mu m$ oxide. Field plates fp1–fp3 represent biased polysilicon field plates connected by biasing resistors. For simulation purposes, the resistors were ignored, fp1–fp3 were defined as n-type poly contacts and the required potential was placed directly on the field plate. The structure in layout is shown in figure 2, without the field plates. Simulation has been carried out using the 2-D device simulator S-PISCES2B [1]. Wiring has been investigated for the case of a $20\mu m$, $8 \times 10^{14} cm^{-3}$ n-type epi layer on a $1 \times 10^{14} cm^{-3}$ p-type substrate. Breakdown of the p+ isolation/n-epi diode was modelled using the full two carrier impact ionisation model with ionisation coefficients of Van Overstraeten and De Man [5]. A rectangular mesh was used to eliminate the generation of obtuse triangles by the regrid statement, and to keep the number of nodes at a minimum. The total number of nodes used was 1476, with a spacing ratio of $0.667 < x < 1.5$.

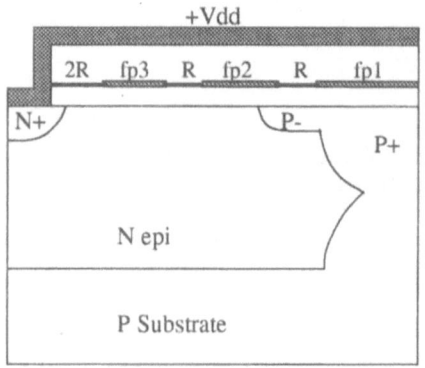

Figure 1: Field plate scheme.

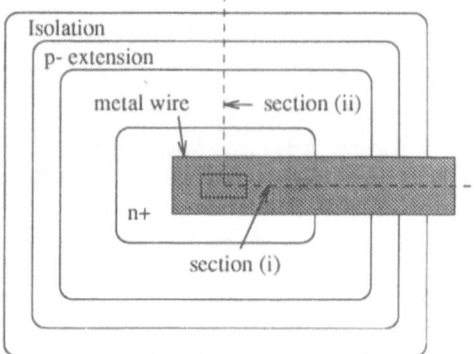

Figure 2: Layout of wiring structure.

3. Simple isolation structure

The breakdown of the p+ isolation/n-epi junction was simulated as 681-V without the wire. When the n+ contact was extended to run over the isolation, it was found that field crowding at the surface near the p+ edge caused the breakdown to drop to 229-V. This compares well with a previously published value of 180-V [6]. A p-extension can be used to reduce the field crowding at the p+ isolation edge [6]. This is shown clearly in figure 3, where we can see the potential contours moved from the p+ edge and distributed evenly across the p- extension.

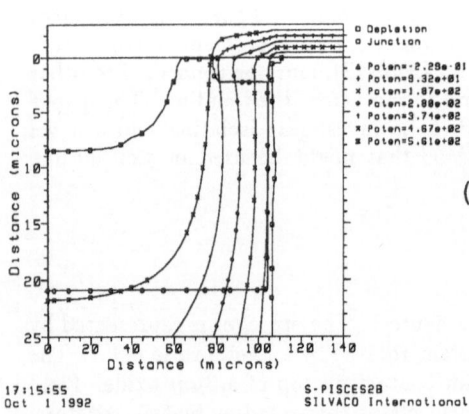

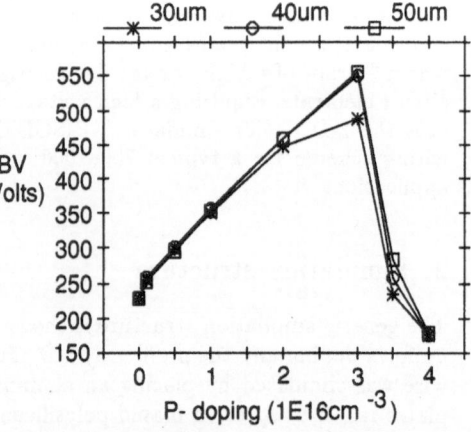

Figure 3: P- extension effectively prevents potential crowding at p+ edge. Wire is at 560-V.

Figure 4: Breakdown vs p- doping for three different extensions. Extension is the drawn distance between the p-/p+ edges.

Breakdown versus p- doping for three different extensions is shown in figure 4, the optimum being a drawn extension of $40\mu m$ at a peak doping of $3 \times 10^{16} cm^{-3}$. At low dopings, the extension depletes too quickly, and the breakdown approaches that of the p+ value. At high dopings, the extension depletes slowly, and breakdown moves from the p+ region to the edge of the p-. The optimum extension gives a breakdown of 550-V with the wire.

4. Wiring structure

In order to obtain 1000-V wiring, biased polysilicon field plates are used to redistribute the electric field along the surface. In a real structure, the field plates would be inserted as rings between the p- extension and the n+, and be biased as a function of V_{DD} using a voltage division scheme. The structure is shown in figure 1. For simulation purposes, the resistors are ignored and the required potential is placed directly on the field plates. The voltage on the electrodes is stepped until the required potential for each one is reached. Stepping each electrode as a multiple of V_{DD} would have been more realistic, however this was too cpu intensive, as the algorithm would only allow estimates from one previous solution.

Since the field plate over the isolation (fp1) is grounded, the inter-layer dielectric needs to sustain the full drain voltage. Here the simulation just monitors the oxide field. The field plate fp3 is biased at approximately $V_{DD}/2$ and fp2 at approximately $V_{DD}/4$. The distribution of the potential contours is shown in figure 5, with the wire V_{DD} at 880-V, fp3 at 480-V, fp2 at 240-V and fp1 at 0-V. Also shown in figure 5 are impact ionisation contours, illustrating areas of maximum field crowding, and total current vectors, showing the flow of carriers generated due to impact ionisation. We can see that breakdown occurs at the surface. A p- extension of $5 \times 10^{15} cm^{-3}$ was used. The breakdown voltage of the non-wire part of the structure, (illustrated by section (ii) of figure 2), was also simulated. Using the same p- extension of $5 \times 10^{15} cm^{-3}$, the breakdown voltage was increased from 681-V to 1340-V. Potential contours for this structure are shown in figure 6.

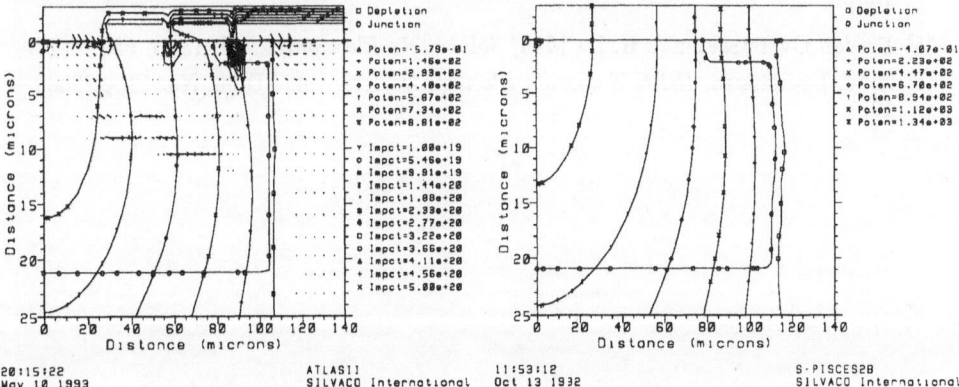

Figure 5: Potential and impact ionisation contours in field plate structure at 880-V. Also shown are total current vectors. Corresponds to section (i) of figure 2.

Figure 6: Potential contours of structure without wire and with a p- extension of $5 \times 10^{15} cm^{-3}$ at 1340-V. Corresponds to section (ii) of figure 2.

5. Limitations of simulation

As the field plate voltages were not stepped as multiples of V_{DD}, the potential distributions in going from 0-V to breakdown would be different in a real device.

The simulation only monitored the oxide field, and did not account for failure mechanisms that may occur in a real oxide.

Also, the simulation only modelled a 1-D field plate, as shown by section (i) of figure 2. The wire was assumed to have infinite lateral geometry whereas a real wire would have a finite dimension. 2-D edge effects were therefore ignored. However, since the width of the wire in a real structure is small in comparison to the overall size of the device, it may be the case that the electric fields under a real wire may not be as dominant as those simulated here.

6. Conclusions

A simple optimised p- extension can increase the breakdown of an isolation structure with a wire running over it from 229-V to 550-V. To obtain isolation greater than 800-V, biased polysilicon field plates can be used to redistribute the critical electric fields. The p- extension used here can increase the breakdown of the non-wired part of the structure from 681-V to 1340-V.

References

[1] Silvaco International., *S-PISCES2B User's Manual*, version 5.2, 1992.

[2] A.F.J. Murray & W.A. Lane *Proc. ESSDERC'91, Microelectronic Engineering*, vol. 15, p. 377, 1991.

[3] N. Fujishima & H. Takeda, *Proc. International Symposium on Power Semiconductor Devices & I.C.'s*, paper 3.3.2, p. 91, 1990.

[4] A.W. Ludikhuize, *IEEE Trans. on Electron Devices*, vol. ED-38, No. 7, p. 1582, 1991.

[5] R. Van Overstraeten & H. De Man, *Solid-State Electronics*, vol. 13, p. 583, 1970.

[6] A.W. Ludikhuize, *IEEE Trans. on Electron Devices*, vol. ED-33, No. 12, p. 2008, 1986.

A General Simulation Method for Etching and Deposition Processes

E. Strasser and S. Selberherr

Institute for Microelectronics, TU Vienna
Gußhausstraße 27-29, A-1040 Wien, AUSTRIA

Abstract

A new method for simulation of etching and deposition processes has been developed. This method is based on a cellular material representation and on morphological filter operations for surface movement. In this paper we describe theory and application of our approach. Simulation results both in two and three dimensions demonstrate the simulation capabilities of this new method.

1. Introduction

Accurate simulation of pattern transfer processes such as wet and dry etching, chemical vapor deposition, evaporation and sputtering requires three-dimensional models and algorithms for wafer topography evaluation. A variety of surface evolution algorithms has been studied to build three-dimensional topography simulators. Among them many algorithms have been reported for lithography simulation [1], [2], [3]. A few methods have been applied to three-dimensional simulation of etching and deposition processes [4], [5]. Surface advancement algorithms offer highly accurate results but with potential topological instabilities such as erroneous surface loops resulting from a growing or etching surface intersecting with itself. Cell removal methods allow the simulation of arbitrary structures, but suffer from inherent inaccuracy. Our new method is based on spatial adaptive filter operations for advancing the etch front. These filter operations are based on Minkowski algebra which makes it possible to simulate topography processes by use of the fundamental morphological operations of erosion and dilation, as they are termed in image processing [6]. Our method allows the accurate and absolutely stable simulation of three-dimensional arbitrary structures.

2. Simulation Method

The morphological approach is a general method in image processing used for many purposes including edge detection and segmentation of images. Several fundamental morphological operations provide a well defined methodology for altering an given image in terms of some predetermined geometric shape [6]. If we consider the simulation geometry as a black and white image (material or vacuum) these operations can be applied to the etching and deposition problem. We use an array of square or

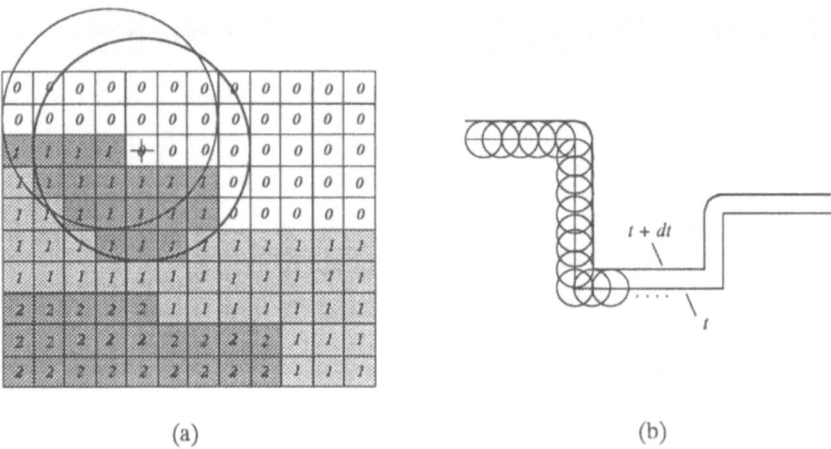

(a)   (b)

Figure 1: Simulation method

cubic cells for geometry representation, where each cell is characterized as etched or unetched. A material identifier is defined for each cell, therefore material boundaries need not be explicitly represented. To advance the etch front adaptive spatial filter operations are performed along the surface boundary as shown in Fig. 1. During etching, all cells within a filter are etched away, while cells outside stay unchanged. In general, for anisotropic simulation filters are ellipsoids, for isotropic movement of surface points filters are spheres. The spatial filter dimension is related to the simulation time step and to the local etch rates. The etch front at a given time step is obtained by the envelope of filtered cells. With our method we avoid the inherent inaccuracy of the original cell algorithm [7] which in two dimensions produces an octagon instead of a circle during uniform etching from a single point [8].

Filter operations at material boundaries are performed using composite filters. In general, interfaces lead to an abrupt change in etch rates. For this reason, on both sides of the interface a filter operation has to be performed selectively to the actual material. Filters which extend over a material boundary demand an additional filter operation for this time step. The etch rate for those filters depend on the etch rates on both sides of the interface and on how far a filter reaches into the other material.

3. Simulation Results

Fig. 2 shows a deposition from a hemispherical vapor source. The growth rate of the evaporated film at each point depends strongly on the surface topology. The growth rate varies along the surface as a result of shadowing. The side wall deposition depends on the solid angle visibility of surface points. Fig. 3 shows the simulation of reactive ion etching. The flux arrives with an angle of twenty degrees. This etching process is modeled regarding an isotropic and a directional etch rate component. The isotropic component models the chemically reactive gas. It produces profiles with large undercut and circular cross sections. The directional component correspond to the ion-enhanced surface etching effects due to ionic species. The flux can be shadowed and has a local $cos\phi$ rate dependences (where ϕ denotes the angle between incident flux and local surface normal).

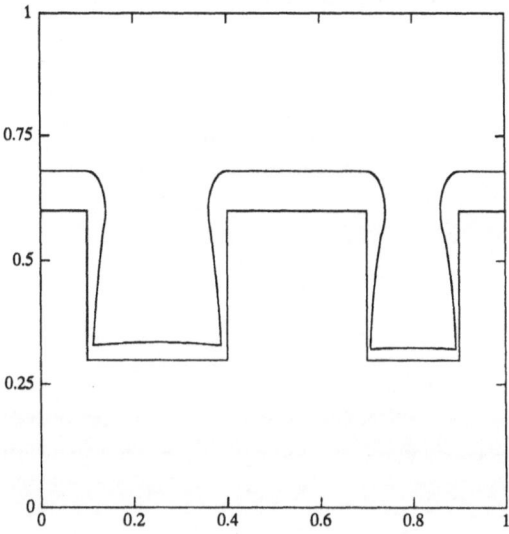

Figure 2: Sputter deposition simulation

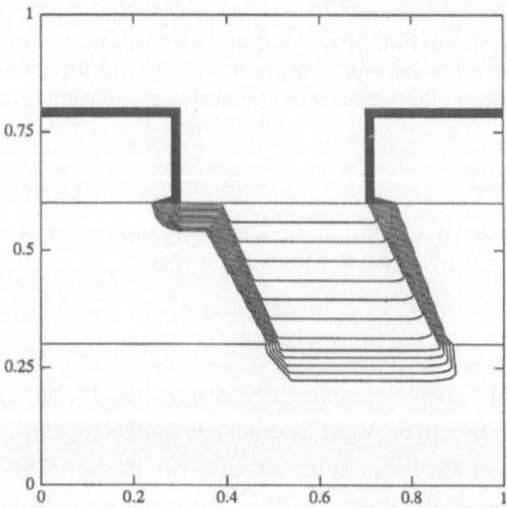

Figure 3: Simulation of reactive ion etching

Fig. 4 shows the result of sequential etching processes to simulate contact hole etching. The simulation for this example starts with a circular mask opening of 1 μm diameter. The first isotropic etching process etches the substrate to a depth of 0.5 μm. This was followed by a directional etch simulation for 0.5 μm additional material removal.

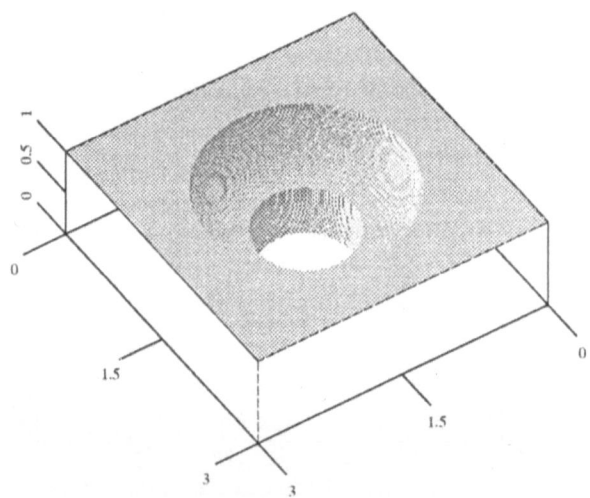

Figure 4: Simulation of contact hole etching

4. Conclusion

A new method for simulation of etching and deposition processes was introduced. Based on a cellular material representation and spatial filter operations for surface movement this method allows accurate and stable simulation of arbitrary structures.

Acknowledgements

Our work is significantly supported by Digital Equipment Corporation at Hudson, USA; and Siemens Corporation at Munich, Germany.

References

[1] W. Henke et al., Microelectronic Engineering, Vol. 14, pp. 283-297, 1991.

[2] Y. Hirai et al., Symp. on VLSI Technology, pp. 15-16, 1987.

[3] T. Matsuzawa et al., IEEE Trans. on ED, Vol. 32, pp. 1781-1783, 1985.

[4] E. W. Scheckler et al., Symp. on VLSI Technology, pp. 97-98, 1991.

[5] M. Fujinaga et al., IEDM Technical Digest pp. 905-908, 1990.

[6] C. R. Giardina and E. R. Dougherty, Prentice-Hall, New Jersey, 1988.

[7] F. H. Dill et al., IEEE Trans. on ED, Vol. 22, pp. 456-464, 1975.

[8] E. Strasser et al., Proceedings of VPAD, pp. 54-55, 1993.

SIMULATION OF SEMICONDUCTOR DEVICES AND PROCESSES Vol. 5
Edited by S. Selberherr, H. Stippel, E. Strasser – September 1993

361

Two Dimensional Monte Carlo Simulation of Ion Implantation in Crystalline Silicon Considering Damage Formation

A. Simionescu and G. Hobler

Institut für Allgemeine Elektrotechnik und Elektronik, TU Vienna
Gußhausstraße 27-29, A-1040 Wien, AUSTRIA

Abstract

One- and two-dimensional concentration profiles of high-dose boron implants have been computed using our Monte Carlo simulator IMSIL. In the 1-D case they have been compared with experiments obtaining very good agreement. The lattice damage formation is taken into account according to the modified Kinchin-Pease model. A factor of 1/8 was found to properly describe the self-annealing process reducing the amount of lattice defects. In the 2-D case it is observed that the channeling in directions leading below the mask edge is not suppressed in contrast to vertical channeling.

1. Introduction

In recent years considerable research activities on modeling 1-D implantation profiles in crystalline silicon have been carried out. 2-D simulations of ion implantation in crystalline silicon have been rare and have neglected damage formation [1], [2]. Souce/drain profiles of MOSFET's, however, require high implantation doses, and therefore considering damage formation and its influence on the channeling and dechanneling of ions is mandatory. Furthermore, the knowledge of the damage profile resulting from the implantation process is useful in modeling diffusion phenomena.

In this paper we have extended our Monte Carlo simulator [2] in order to calculate 2-D dopant profiles considering damage formation. We have called the program IMSIL (IMplantation in SILicon). In the following section we describe the physical models implemented in IMSIL. Afterwards we present the results of our simulations and finally we discuss them.

2. Models Implemented in the Monte-Carlo Simulator

In our studies we use an extended version of the Monte Carlo simulator described in [2]. The version includes an improved model of the electronic stopping power [3], a modified Debye temperature of 490K [4] for the description of lattice vibrations, and the damage model to be discribed below. Furthermore, the present version is not restricted to point responses, but allows to perform the implantation in the whole simulation area.

This is important because the superposition of point responses cannot be used to get the dopant distribution at a mask edge, if implantation damage is taken into account. The increasing lattice damage during the implantation is dynamically recorded as a function of the space coordinates, so we have permanent knowledge of the actual defect distribution. The lattice damage encountered by an ion along its trajectory is considered by carrying out amorphous collisions with a probability proportional to the local number of displaced atoms. This presumes as a physical background the existence of quasi-amorphous zones embedded in the crystal lattice.

During every collision the ion transfers energy to the lattice atom. If this energy exceeds the displacement energy, the target atom leaves its lattice site and may produce further damage on its way through the crystal. Restricting ourselves to boron implantations, we may neglect the recoil range. The amount of damage is taken into account using the modified Kinchin-Pease model [5]. Although self-annealing is a complicated process, we found it sufficient for our purpose to use a factor to reduce the amount of created damage resulting from the modified Kinchin-Pease model [5]. Comparing many simulated implantation profiles with experimental results we found a value of 1/8 for this factor at a temperature of 300K.

3. Results and Discussions

In all simulations boron has been implanted into amorphous or (100) silicon with a 5A native oxide layer. Figure 1 shows the dose dependence of the 1-D concentration profile of a boron implantation at 35keV in random direction (7° tilt, 15° rotation from the [110]-direction). Comparing the profiles we see that, at a dose of 10^{16} cm^{-2} (curve 1), we obtain a shallower profile than in the simulation of the same implantation without taking damage formation into account (curve 3).

At a dose of 10^{14} cm^{-2} (curve 2), the implantation profiles with and without damage are almost the same, so lattice amorphization can be neglected. We remark the paralellism of the profile calculated for the dose of 10^{16} cm^{-2} without taking damage formation into account (curve 3) and the profile for the dose of 10^{14} cm^{-2} (curve 2). This is due to the fact, that in the absence of damage only lattice vibrations and the native oxide layer [6] influence the channeling and dechanneling of implanted ions regardless of the dose, while at a dose of 10^{16} cm^{-2} lattice amorphization becomes the major dechanneling mechanism. For both doses there is excellent agreement between simulation (curves 1,2), and experiment [7].

In Figures 2,3, and 4 we see the 2-D dopant profiles resulting from 5keV boron implantations near a mask edge in amorphous target, in crystalline target without damage consideration and in crystalline target with taking lattice amorphization into account, respectively. The simulations into crystalline silicon were performed with 7° tilt and 45° rotation from the [110]-direction.

At the low implantation energy used, channeled ions have a far larger range than non-channeled ions. Therefore, the profiles resulting from the implantation in the crystalline target are much deeper than in the amorphous case.

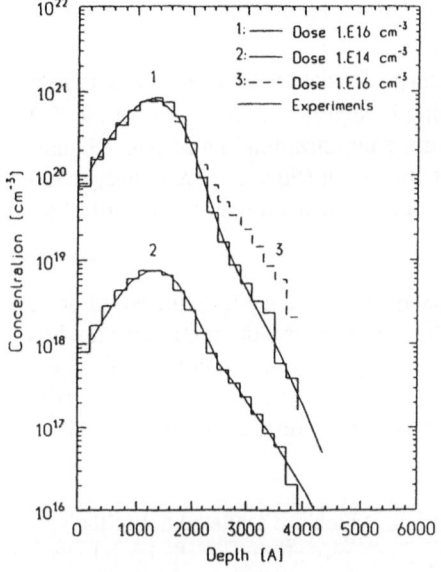

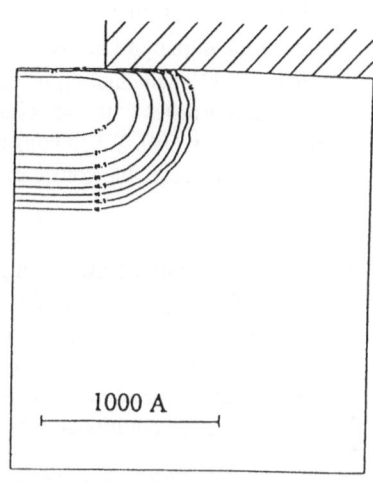

Figure 1: 35 keV boron implantation into (100)-silicon with 7° tilt and 30° rotation.

Figure 2: 5keV boron implantation into amorphous silicon near a mask edge.

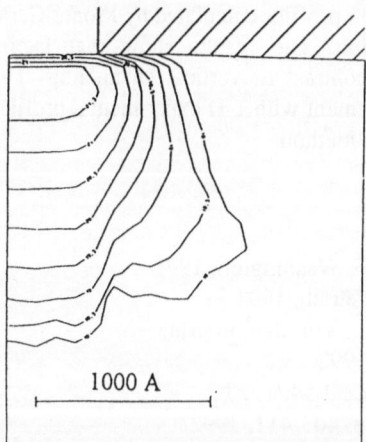

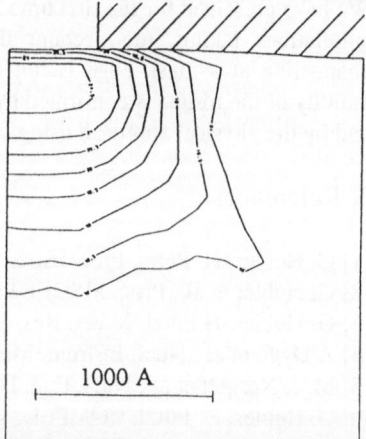

Figure 3: 5keV boron implantation into (100)-silicon near a mask edge without considering lattice amorphization.

Figure 4: 5keV boron implantation into (100)-silicon near a mask edge considering lattice amorphization

The 2-D simulations were performed with 7° tilt and 0° rotation and an implantation dose of $2 \cdot 10^{16}$ cm^{-2}. There are two contour lines per decade of concentration.

Comparing Figure 2 on the one hand and Figures 3 and 4 on the other hand, we remark the major effect of channeling on the dopant profile in the crystalline case. At the considered high implantation dose ($2 \cdot 10^{16}$ cm^{-2}) damage formation becomes the dominant channeling and dechanneling mechanism, resulting in significant differences of the 2-D implantation profiles with and without taking lattice amorphization into account (Figures 3 and 4 respectively). When considering damage formation (Figure 4) we notice, analogous to the 1-D case, the considerable suppression of the channeling effect in vertical direction.

In directions leading below the mask edge the channeling tendency is favorized. In order to explain this behaviour, we first notice that the region below the mask contains less damage than the silicon region which is not covered by the mask. However, this is not true when comparing the region below the mask (where lateral channeling occurs) with the deeper part of the region not covered by the mask (where vertical channeling occurs). Therefore we need additional explanations.

The first reason of the observed behaviour is that ions which are scattered out of the vertical channel by lattice damage are available with some probability for the lateral channels, while the reverse process, i. e. the scattering out of a lateral channel into the vertical channel, is less likely due to the smaller number of ions channeled lateraly as compared to those channeled vertically. The second reason is that a displaced lattice atom is more effective in scattering the ion into a channel than a regular lattice atom. This effect is similar to the enhancement of the channeling probability by a screening oxide [6].

4. Conclusions

We have presented for the first time 2-D implantation profiles calculated by Monte-Carlo simulations taking into account damage formation. The results show that lateral channeling at a mask edge is not suppressed in contrast to vertical channeling. The validity of the results is confirmed by the good agreement with 1-D experimental profiles and by the physical approach using the Monte-Carlo method.

5. References

[1] G. Hobler, H. Pötzl, Proc. IEDM 91, pp. 693-696, Washington, 1991
[2] G. Hobler et al., Proc. SISDEP 91, pp.389-398, Zürich, 1991
[3] G. Hobler, H.Pötzl, Mater. Res. Soc. Symp. Proc., Vol. 279, in print
[4] A.Dygo et al., Nucl. Instrum. Meth. B, 64, 701, 1992
[5] M. J. Norgett et al., Nucl. Eng. Des., Vol. 33, pp. 50-54, 1975
[6] G. Hobler, H. Pötzl, COMPEL, Vol. 11, No. 4,pp. 403-411, 1992
[7] Al F. Tasch et al., J. Electrochem. Soc., Vol 136, No. 3, pp. 810-814, 1989

SIMULATION OF SEMICONDUCTOR DEVICES AND PROCESSES Vol. 5   365
Edited by S. Selberherr, H. Stippel, E. Strasser – September 1993

MOSFET Two-Dimensional Doping Profile Determination

N. Khalil and J. Faricelli

Advanced Semiconductor Development, Digital Equipment Corporation
Hudson, MA 01749, USA

Abstract

Direct experimental measurement techniques have had limited success in the determination of the two-dimensional (2D) doping profile of a MOSFET. In this paper, we describe an alternative methodology that uses source/drain (S/D) diode and gate overlap capacitance measurements to determine the 2D profile by inverse modeling [1]. Our approach is based on the optimized tensor product spline (TPS) representation of the profile. We use nonlinear multiple outputs, least squares optimization to extract the values of the B-splines coefficients.

1. Introduction

The determination of the two-dimensional (2D) doping profile of a MOSFET is a crucial factor for device characterization and a very important input for device simulation. Direct experimental measurements of 2D doping profiles have met with limited success on very shallow junction devices, or rely on heroic experimental preparation techniques that require sophisticated data reduction or complicated angle lap and stain techniques [2,3,4]. In this paper, we describe a method for determining the 2D profile from MOSFET and S/D diode electrical measurements by inverse modeling [1]. We rely mainly on measuring various capacitances under different bias conditions that "probe" various portions of the doping profile by depleting or accumulating that region of carriers. In addition, we use one-dimensional profiling information from secondary ion mass spectroscopy (SIMS) or deep depletion C-V measurements as the "boundary conditions" of our extracted 2D profile. An initial guess is derived from these 1D measurements. We then perform an optimization using device simulation to reproduce the measurements numerically, and adjust the doping to minimize the difference between measurement and simulation.

2. Profile Representation

We parameterize 2D doping profiles by representing their logarithm as a tensor product spline (TPS) [5]. This representation greatly decreases the amount of computation as compared to the hundreds of parameters required by a straightforward mesh representation. It is more general than other analytically based approxima-tion functions. Moreover, the local approximation properties of the B-splines makes it suitable for the extraction procedure. Using the TPS representation a 2D doping profile is defined by its knots sequences t_x and t_y and the values of its coefficients c_{ij} and is given by:

$$f(x,y) = \sum_{i=1}^{nx}\sum_{j=1}^{ny} c_{ij} * B_{i,k_x,t_x}(x) * B_{j,k_y,t_y}(y) \tag{1}$$

Where $B_{i,k,t}$ is the ith B-spline of order k for the knot sequence t, nx and ny the number of knots in the X and Y direction respectively.

We represent the donors and the acceptors in a device by two TPS with a common t_x sequence but different t_y sequences. For fixed orders k_x, k_y, and t_y sequences, we can write the net doping as:

$$net(x,y) = F(t_x, \alpha_{ij}, \delta_{ij}) \tag{2}$$

Where α_{ij} and δ_{ij} are the splines coefficients for the acceptors and donors.

3. 2D Extraction

We use three different capacitances in our extraction: The gate to S/D overlap capacitance C_{ov}, and the depletion capacitance per unit area Ca, and per unit perimeter C_p of the reverse S/D junction diode. We calculate the capacitances by integrating the appropriate MINIMOS [6] charges. We avoid numerical problems by resorting to grid refinement and careful handling of rounding and integration errors. As done experimentally, we determine the values of C_a and C_p from the S/D diode capacitance values of two devices with different S/D area sizes.

The capacitances are a nonlinear function of the net doping. Using (2) we can write this functional relationship as:

$$\mathbf{C} = F(\mathbf{V}, t_x, \alpha_{ij}, \delta_{ij}) \tag{3}$$

Where $\mathbf{C}$ is the vector of capacitance values: C_a, C_p, and C_{ov}; and $\mathbf{V}$ is the input vector of the device bias voltages.

The extraction problem reduces to determining the values of the t_x sequence and the sets of coefficients α_{ij} and δ_{ij}. We use a nonlinear least square, multiple outputs, optimizer based on the well-known Levenberg-Marquardt algorithm. The Jacobian is calculated by finite differences with usage of the Broyden update scheme [7] to reduce the computational load. The overall figure of merit of the optimization can be formulated as follows:

$$\sum_{i=1}^{m} \left(\frac{C_a^{exp} - C_a^{sim}}{C_a^{exp}}\right)^2 + \left(\frac{C_p^{exp} - C_p^{sim}}{C_p^{exp}}\right)^2 + \left(\frac{C_{ov}^{exp} - C_{ov}^{sim}}{C_{ov}^{exp}}\right)^2 \tag{4}$$

Starting from 1D profile information for the channel and the S/D regions, we per-form a curve fitting optimization to generate their 1D spline representations. We assume the 1D channel profile is spread uniformly in the horizontal dimension and the S/D profile to have a certain lateral subdifusion, and we generate the initial TPS representations accordingly.

We then extract from the capacitance data values for the the t_x knots sequence, and the coefficients δ_{ij} and α_{ij}. The extraction is done in multiple iterations using different sets of capacitances data. To avoid redundancy, we only optimize the coefficients that are of the same sign as the net doping at the knot location.

4. Results

To validate the proposed extraction methodology, we show the results of of applying our extraction to two examples. In example 1, the original profile consisted of two TPS profiles for donors and acceptors. We also assumed the knot locations to be known. For example 2, the original profile was formed by rotating 1D profiles for both donors and acceptors.

As shown in figures 1 and 2, we obtain good agreement between the original and extracted lateral surface doping. This is due to the strong dependence of the gate overlap data in accumulation on the surface doping. However the resolution of the method deteriorates as the distance from the surface increases. This is apparent in figures 3 and 4 where the lateral doping at $0.1~\mu$ in depth is plotted. For both examples, figures 5 and 6 show that the spatial delineation of the 2D junction is good. As expected, the accuracy of the extraction is higher in example 1 where there are no errors due to knots placement and 1D spline approximation.

5. Conclusion

We have presented a 2D MOSFET profiling method based on inverse modeling of capacitance data. The method is non-destructive and provide important information for device characterization. We plan on further improving the method by using MOSFET threshold and drain current data in our extraction.

References

[1] G. J. L. Ouwerling, *Nondestructive One- And Two-Dimensional Doping Profiling By Inverse Methods*, Delft University of Technology, UMI, Ann Arbor MI, 1989.

[2] S. H. Goodwin-Johansson, R. Subrahmanyan , C. E. Floyd and H. Z. Massoud, "Two-Dimensional impurity profiling with emission computed tomography techniques", *IEEE-TCAD*, CAD-8(4),1989.

[3] R. Subrahmanyan, H. Z. Massoud and R. B. Fair, "Experimental characterization of two-dimensional dopant profiles in silicon using chemical staining", *Applied Physics Letters*, 52(25), 1988.

[4] S. Kordic, E. Van Leonen, D. Dijkkamp, A. Hoeven and H. Moraal, "Scanning Tunneling Microscopy on Cleaved Silicon PN Junctions", *IEDM Technical Digest*, 1989, pp.277-280.

[5] Carl De Boor, *A Practical Guide to Splines* , Springer-Verlag New York Inc., 1978.

[6] S. Selberherr, A. Schultz and H. W. Protzl, "MINIMOS — A Two-Dimensional MOS Transistor Analyser", *IEEE Transactions on Electron Devices*, ED-27 (8), August 1980.

[7] C. G. Broyden, in *Numerical Methods for Unconstrained Optimization*, W. Murray, Ed., Academic Press, 1972.

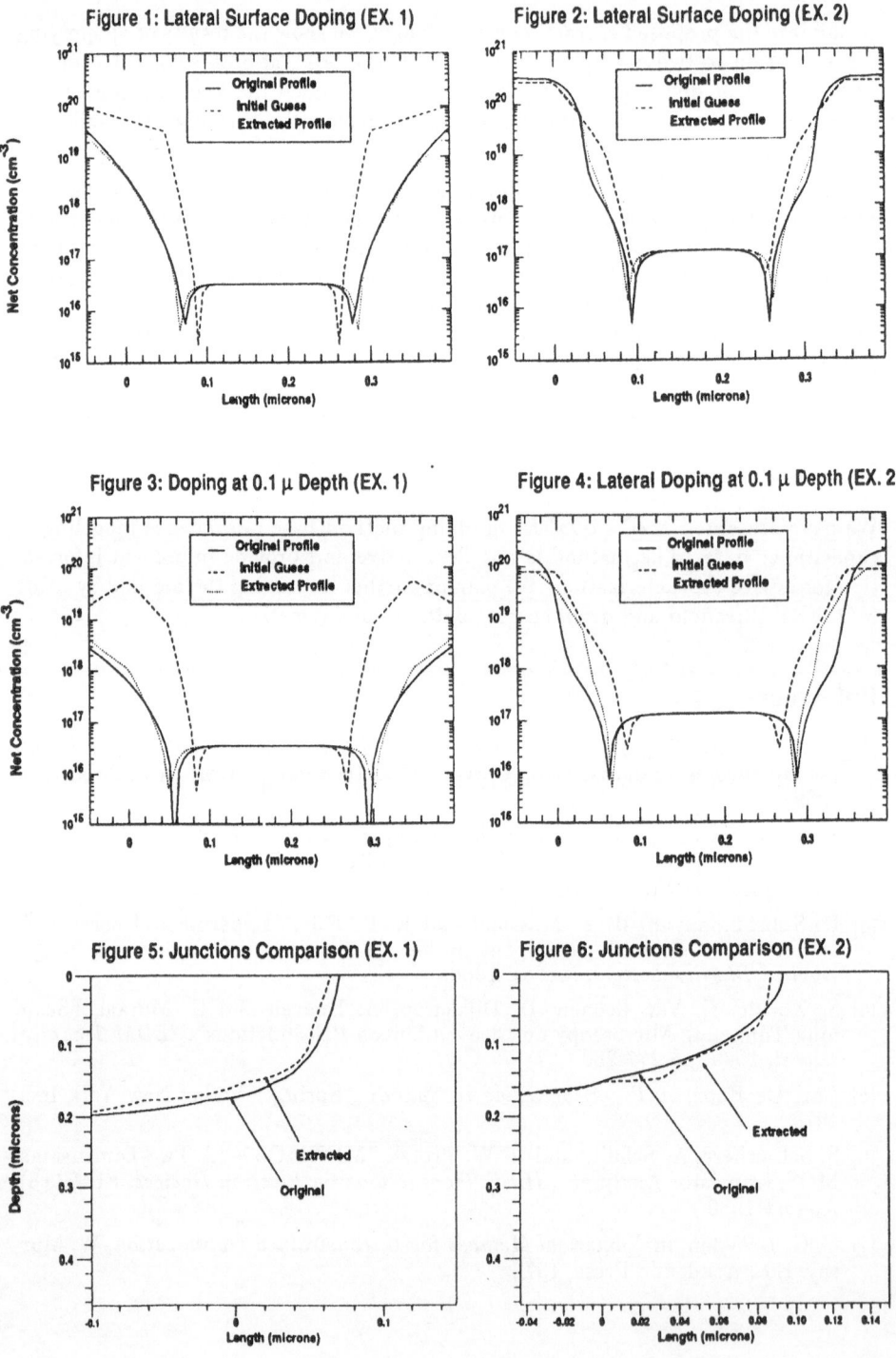

Figure 1: Lateral Surface Doping (EX. 1)

Figure 2: Lateral Surface Doping (EX. 2)

Figure 3: Doping at 0.1 μ Depth (EX. 1)

Figure 4: Lateral Doping at 0.1 μ Depth (EX. 2)

Figure 5: Junctions Comparison (EX. 1)

Figure 6: Junctions Comparison (EX. 2)

Determination of EBIC Response by Two-Dimensional Device Simulation

A. Erlebach, R. Stephan, and G. Dallmann[†]

Fraunhofer-Institut für Mikroelektronische Schaltungen und Systeme
Grenzstraße 28, D-01109 Dresden, GERMANY
[†]Institut Fresenius, Angewandte Festkörperelektronik GmbH
Königsbrücker Landstraße 159, D-01109 Dresden, GERMANY

Abstract

The carrier generation as a result of electron beam irradiation of semiconductor devices has been incorporated into the device simulator ToSCA. The simulator extended with this new capability makes feasible the investigation of complex device structures under conditions which are equivalent to those of EBIC measurements. Simulation results are compared with experimental data.

1. Introduction

Computer simulations of processes and devices have become significant in the design of novel device structures and manufacturing processes. Better approaches for simulation embedded in process and device development are created. Investigations of device behavior, analysis of sensitivity to different process or design parameters and vizualisation of non-measurable parameters are well-known. In the following we describe an extension and new application which aim at supporting the development of new high-tech measurement methods.

The parameters of microelectronic devices are defined by dopant profiles. Therefore the lateral extension of profiles gains importance. Extensive investigations are necessary providing methods for lateral features like effective channel length and channel asymmetry. The use of electron beam induced current (EBIC) [1] for quantitative determination of two-dimensional profiles appear to be possible. This method needs extensive interpretation and numerical treatment.

In the present work the carrier generation as a result of electron beam irradiation is incorporated into a general purpose device simulator. Simulation has been done with and without electron beam incidence during EBIC measurement for calculation of two-dimensional electric field and carrier distributions. These results are applied for a better interpretation of EBIC images, where the comparison of simulated and measured EBIC profiles gives information about the lateral dopant distribution.

2. Principles of the EBIC method

The electron beam of a scanning electron microscope (SEM) generates electron-hole pairs in a semiconductor sample. If the mobile excess minority carriers reach by diffusion an electric field (e. g. of a pn-junction), they give an external current in a low impedance amplifier connected to both sides of the pn-junction. The electric field can be caused by dopant gradients, external voltages and Schottky barriers. One can obtain EBIC images or profiles (line scans) by scanning the focused beam over the sample. In the past, analytical calculations of the EBIC signal have been performed. Most of the models divide the sample in a space charge region (SCR) and a field free diffusion region. The EBIC signal is calculated by using the collection probability of the excess carriers at a given distance of the SCR. One-dimensional numerical calculations of the EBIC signal have been published [2,3].

In modern semiconductor devices the SCR width, the depth of doped regions and the generation volume have comparable dimensions. In this case a numerical simulation of the EBIC signal using the real two-dimensional dopant distribution of the device is necessary.

3. Two-Dimensional Simulation of EBIC Profiles

The EBIC is determined by introducing an additional generation rate for electrons and holes into the continuity equations which is due to the primary electrons entering the semiconductor. The entire system of semiconductor equations takes into account the extended continuity equation and is solved with the advanced two-dimensional device simulation program ToSCA[1]. The generation rate is described by a gaussian function [4] characterized by the standard deviations σ_x, σ_y and the depth of the generation centre y_0 in the following manner:

$$G(x,y) = G_0 \cdot \exp(-(x-x_0)^2/(2 \cdot \sigma_x^2)) \cdot \exp(-(y-y_0)^2/(2 \cdot \sigma_y^2)) \tag{1}$$

where

$$G_0 = \frac{(1-f) \cdot W_e}{E_{eh} \cdot 2 \cdot \pi \cdot \sigma_x \cdot \sigma_y} .$$

W_e is the power of the electron beam, f characterizes the rescattering of electrons, E_{eh} is the energy required to produce an electron hole pair and x_0 is the location of the electron beam on the x-axis.

Essential effects influencing the EBIC signal are surface recombination velocity, additional surface charges, electron and hole life times and mobilities. The surface recombination velocity is increased by perturbations caused by surface treatment. Therefore the diffusion length of electrons and holes decreases near the surface. The bulk life time of the semiconductor material, the dopant profile and trap density distribution have a comparatively smaller influence on the diffusion length of electrons and holes. Careful determination of the parameters of generation and recombination processes mentioned above has to be carried out in order to obtain accurate simulation results.

4. Results

This new simulation method is applied for EBIC signal calculation of an 1.2 μm n-channel MOS transistor. All process steps influencing the dopant profile and the substrate surface contours are simulated with high accuracy. This is important for the accurate calculation of electric field.

To calculate the EBIC profile the electron beam is scanned over the device surface with a

---

[1] development of the Institute of Applied Analysis and Stochastics Berlin

step width of 100 nm. For each distinct irradiation step the EBIC signal is calculated by solving the non-stationary problem from the beginning of electron irradiation to stationary conditions achieved after about 10^{-7}s. Figure 3 shows the electron density distribution in the n-channel transistor during stationary electron irradiation. The generation rate produced by an electron beam (3 keV and 120 pA) is assumed to have a depth of the generation centre of 60 nm and standard deviations of 90 - 150 nm. The EBIC profile represents the sum of drain and source current for each irradiation position.

If the electron irradiation causes a small perturbation of the electric field only, there is a strong correlation between pn-junctions, undisturbed field strength distribution (fig. 2) and the EBIC image (fig. 1). Figure 4 shows that the maximum of the EBIC signal is located at the same position as the maximum of the electric field and the pn-junction. The influence of the dopant profile asymmetry and the surface inversion in the channel region can be clearly seen in the simulation data. The EBIC profiles in figure 4 (obtained at a depth of 50nm) show the good agreement between measurement and simulation. Differences occured in the tails of the profile are caused by deviations in the life time near surface of high doping areas. The EBIC plateau in the channel region found in measurements of short n-channel transistors is reproduced by the simulation too.

The special sample treatment (e.g. beveled sample) influences the surface recombination velocity and surface charge density strongly and thus the electrical properties. The surface charge density can produce an inversion layer and an additional electric field in the transistor channel enhancing the collection of electrons for the EBIC signal. For our calculations we use a surface recombination velocity of 10^6 - 10^7 cm/s and a surface charge density of 10^{12} e/cm^{-2}. The width of the EBIC peak is found to decrease by increasing of surface recombination. Very high values of recombination velocity can even provide two maxima in the EBIC profile instead of a plateau. It is possible to distinguish the several space charge regions of a n-channel transistor in these cases.

Acknowledgments

The authors would like to thank Prof. H. Gajewski for providing of the code ToSCA, Dr. H. Kück for helpful discussion and F. Richter for the technical assistance.

References

[1] J. D. Schick: " Electron Beam Induced Current Analysis of Integrated Circuits", Scanning Electron Microscopy, I, p. 295-304, 1981

[2] H. W. Marten and O. Hildebrand: "EBIC Profiling of Bevelled Samples: A Precise Method to Determine the Position of PN-Junctions and Doping Gradients", ESSDERC, 1984

[3] H. W. Marten and O. Hildebrand: "Computer Simulations of Electron Beam Induced Current (EBIC) Linescans across PN-Junctions", Scanning Electron Microscopy, III, p. 1197-1209, 1983

[4] S. Valkealahti, J. Schou and R. M. Nieminen: "Energy Deposition of keV Electrons in light Elements", J. Appl. Phys. 65(6), 2258-2266, 1989

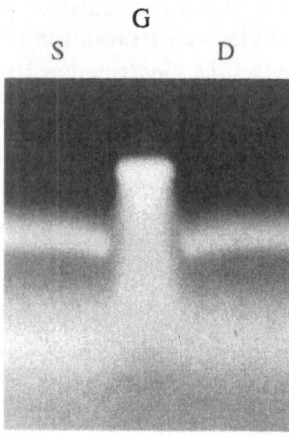

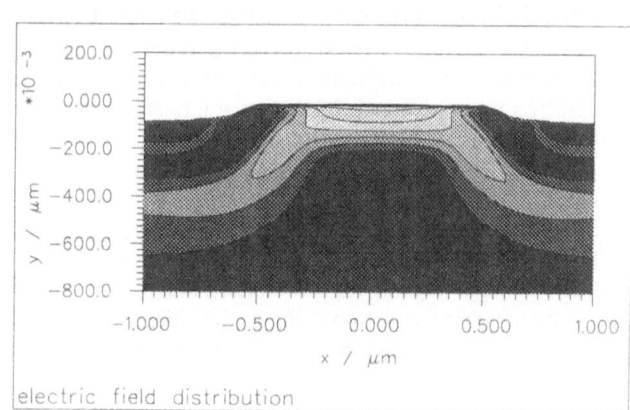

Fig. 1.   EBIC image of an   Fig. 2. Distribution of electric field strength in the n-channel
1.2 μm n-channel transistor   transistor without electron beam irradiation

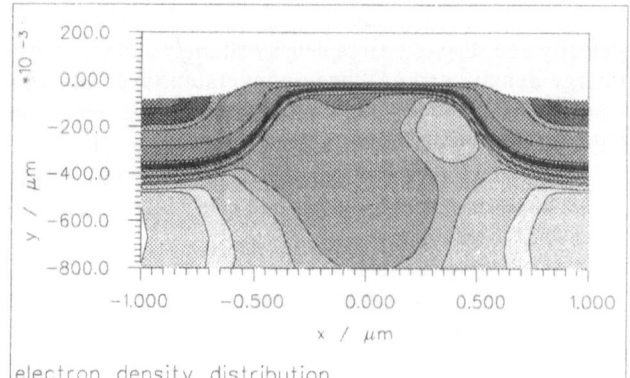

Fig. 3. Electron density distribution in the n-channel transistor   during   stationary
electron irradiation

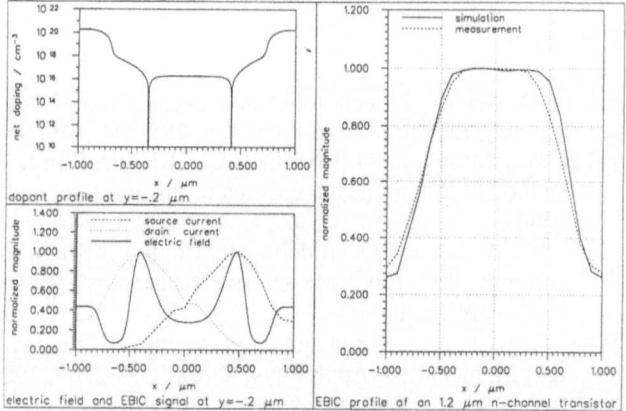

Fig. 4. Simulated lateral doping
profile, electric field and EBIC
signal   and   comparison   of
measured and simulated EBIC
profile

Radiative Heat Transfer with Quasi Monte Carlo Methods

A. Kersch and W. Morokoff[†]

Corporate Research and Development, Siemens AG
Otto-Hahn-Ring 6, D-81739 München, GERMANY
[†]IMA, University of Minnesota
206 Church Street S.E., Minneapolis, MN 55455, USA

Abstract

Monte Carlo simulation is often used to solve radiative transfer problems where complex physical phenomena and geometries must be handled. Slow convergence is a well known disadvantage of this method. In this paper we demonstrate that a significant improvement in computation time can be achieved by using Quasi-Monte Carlo (QMC) methods to simulate Rapid Thermal Processing.

1. Introduction

Monte Carlo simulation is an indispensable tool for modeling Rapid Thermal Processes (RTP) with an accuracy of a few degrees. Only this method allows very detailed modeling of the physical processes and geometrical complications which arise in real applications involving radiation transport. A well known disadvantage of this method is slow convergence, which has prohibited wide spread usage. In this paper we demonstrate that a significant improvement in computation time can be achieved by using Quasi-Monte Carlo (QMC) methods to simulate radiative heat transfer.

The QMC method [1] involves using deterministic, quasi-random sequences in place of random numbers in a Monte Carlo calculation. One well known example of such a sequence is the Halton sequence. While the idea of quasi-Monte Carlo is almost as old as the Monte Carlo approach itself, these methods have rarely been used in real world engineering problems. This may be attributed to the fact that many Monte Carlo problems are high dimensional, and quasi-random sequences tend to lose their advantage as dimension increases [2][3]. Also, great care must be taken to avoid problems arising from the fact that the quasi-random numbers are not independent.

As the following results show, however, a considerable advantage can be won by applying quasi-random sequences to the modeling of the radiative heat transfer from the heater to the wafer inside Rapid Thermal Processing equipment. Figure 1 shows the draft of the cylinder symmetric projection of a typical single wafer reactor.

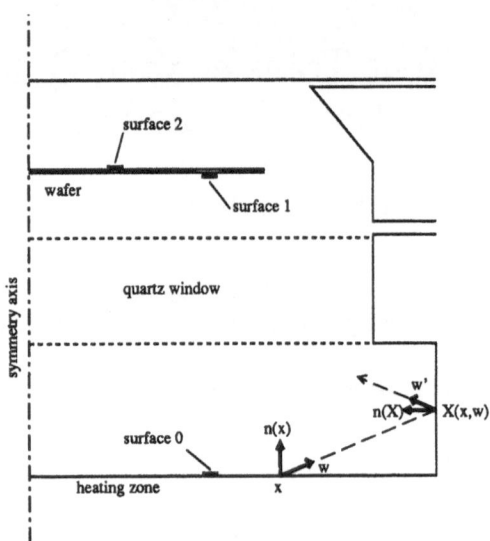

Figure 1: Geometry of the reactor and position of specific surface elements.

2. Description of Radiation Exchange

The radiative energy transfer inside the reactor can be described mathematically by an integral equation for the spectral radiation intensity. The solution of the integral equation leads to a problem of multidimensional integration.

Let D be the interior surface of the reactor, x be a point on D and n_x the interior normal. Let ω be the direction of a radiation bundle and Ω^+ be the half sphere of directions oriented toward the reactor interior. The optical surface properties are described by $a_{\lambda,T}(x,\omega)$, the spectral absorptivity ($=$ emissivity), $r_{\lambda,T}(x)$ the spectral reflectivity and $P_x(\omega \leftarrow \omega')$ the (normalized) probability that a bundle coming from direction ω is reflected into direction ω'. $T(x)$ is the temperature distribution of the surface. $X(x,\omega)$ describes the target of a bundle on the surface starting at x with direction ω. We restrict ourselves to opaque surfaces.

The iterative solution of the integral equation leads to a series expression for the distribution E_λ of radiation intensity impinging on the surface, $E_\lambda = \sum_{k=1}^{\infty} E_\lambda^{(k)}$, where

$$E_\lambda^{(k)}(x) = \int_{[D]^k} \int_{[\Omega^+]^k} a_{\lambda,T}(x,\omega) r_{\lambda,T}(y_{k-1}) \ldots r_{\lambda,T}(y_1) P_{y_{k-1}}(\omega \leftarrow \omega_{k-1}) \ldots P_{y_1}(\omega_1 \leftarrow \omega_0)$$

$$\delta(x - X(y_{k-1},\omega_{k-1})) \ldots \delta(y_1 - X(y_0,\omega_0)) a_{\lambda,T}(y_0,\omega_0) I_{\lambda,T}^{BB}(y_0) (d\omega)^k (dy)^k$$

The series may be truncated such that the remaining terms are smaller than the desired accuracy. The dimension of integration is $2k + 2 + 1$ in the k-th step (2 dimensions for the initial position, one dimension for the spectral distribution). In case of specular reflection, $P_x(\omega \leftarrow \omega') = \delta(\omega - \omega' + 2(n_x \cdot \omega')n_x)$ and the path is determined completely by the initial direction. The dimension of integration in this case is only $2+2+1$. The above procedure is the fractional absorption method. When a probability for a complete absorption is introduced, we can call it a discrete absorption method. This method introduces one more dimension of integration for every discrete process.

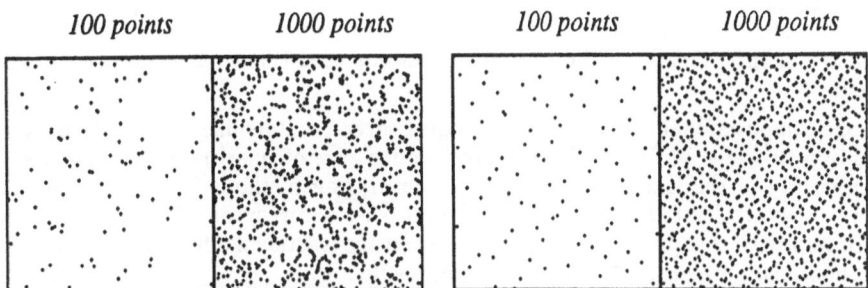

100 points   *1000 points*   *100 points*   *1000 points*

Figure 2: Points with pseudo-random sequence, with quasi-random sequence

3. The Computational Experiment

The goal of the experiments was to determine the accuracy of the amount of heat transfer between surface elements as a function of computation time. We show here the results for specularly reflecting surfaces. The first step towards this goal was to compute error size as a function of N, the number of rays emitted from the source. This error size was determined by calculating results for an ensemble of 30 experiments for every choice of N and taking the standard deviation. In a typical experiment, N rays were emitted from a surface or point source located at *surface 0*. The initial direction was sampled using Lambert's Law, which means that the elevation angle was sampled from a cosine distribution, while the azimuthal angle was sampled from a uniform distribution on $[0, 2\pi]$. The sampling was done using one point from a multi-dimensional quasi- or pseudo-random sequence, such that two angles were assigned separate dimensions. A further dimension was used to sample the initial energy of the ray from Planck's black body distribution.

In the standard Monte Carlo method in which a pseudo-random sequence is used as the source of integration nodes, the expectation of the integration error for the integral $\int f(\mathbf{x})d\mathbf{x}$ is

$$\epsilon(f, N) = \sigma(f)N^{-\beta} \tag{1}$$

with $\beta = 0.5$. For many problems, quasi-random sequences [4] significantly outperform random sequences in the range of practical N. The convergence rate is generally between $\beta = 0.5$ and $\beta = 1.0$. The best way to predict performance for a specific type of problem is to analyze the results for a test problem, as is done below. Figure 2 shows the coverage of the 2-D cube with points generated from a pseudo-random sequence and a Halton sequence.

4. Results

In Figure 3 we show results using the fractional energy absorption method and specularly reflecting surfaces.

Figure 3 shows the expected relative error $\epsilon(N)$ as a function of the number of emitted rays N for the Halton sequence and a pseudo-random sequence using the fractional absorption method with a constant surface absorptivity of 0.4. Results are given for *surface 1* and *surface 2* on the wafer. The plotted points are the calculated errors for various N (averaged over 30 runs), while the lines are a least squares fit of the data to the functional form (1).

Figure 3 illustrates a clear advantage of using a quasi-random sequence over a pseudo-random sequence in the calculation, both in error size and in convergence rate (i.e., β). The error in calculating the energy transfer to *surface 1* with $N = 100000$ is over a factor of three smaller if the Halton sequence is used than if a random sequence is used.

In our computational experiments we studied several factors like surface absorptivity, position of wafer surface to heat source, and choice of quasi-random sequence. A comparison of the fractional and discrete absorption methods was also made. Results show accelerated convergence and improved accuracy of QMC over the standard Monte Carlo approach, and indicate when the fractional and discrete absorption methods should be used to obtain optimal results.

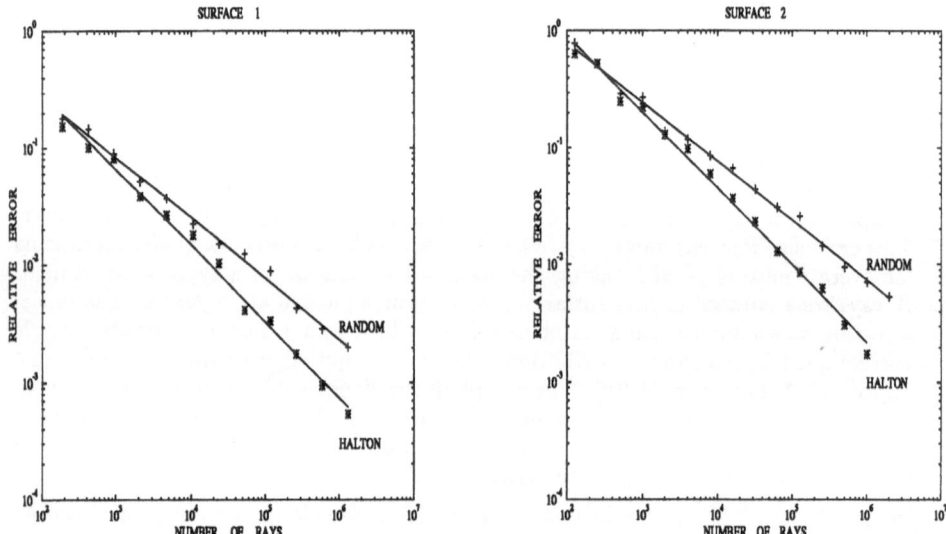

Figure 3: Comparison of random and quasi-random sequences using the fractional absorption method with absorptivity = 0.4.

References

[1] J.H. Halton, *On the efficiency of certain quasi-random sequences of points in evaluating multi-dimensional integrals*, Numer. Math. 2, 84-90, 1960.

[2] W.J. Morokoff and R.E. Caflisch, *Quasi-random sequences and discrepancy*, SIAM Sci. Comp. (to appear).

[3] W.J. Morokoff and R.E. Caflisch, *Quasi-Monte Carlo Integration*, submitted to J. Comp. Phys. (1991).

[4] H. Niederreiter, *Random number generation and quasi-Monte Carlo methods*, SIAM Regional Conference Series in Applied Mathematics, CBMS-NSF 63, 1992.

Numerical Simulation of Piezo-Hall Effects in n-Doped Silicon Magnetic Sensors

W. Allegretto, A. Nathan[†], and T. Manku[†]

Department of Mathematics, University of Alberta
Edmonton, Alberta, T6G 2G7, CANADA
[†]Electrical and Computer Engineering, University of Waterloo
Waterloo, Ontario, N2L 3G1, CANADA

Abstract

The output response of n-type Hall sensors are presented for various device (or current flow) orientations in the presence of both stress and magnetic field. The resulting distributions of potential and terminal characteristics are based on a numerical solution of the piezo-Hall transport equation where the conductivity and Hall coefficient are tensors.

1. Introduction

The encapsulation of magnetic sensors induces significant mechanical stresses (due to the piezoresistance effect [1]) affecting accuracy and long term stability of the output response. The stress on the device induces an offset which is highly undesireable when detecting low frequency fields since the offset voltage (or current) and the useful magnetic output signal cannot be distinguished. In addition to the piezoresistance effect, the effects of stress coupled with the magnetic field can potentially affect the magnetic response of Hall devices due to the piezo-Hall effect, which describes the stress-induced modulation of the Hall coefficient [2].

In this paper, we present simulation results of the stress-dependent magnetic response in n-type Hall and split-electrode geometries (Fig. 1) for various device (or current flow) orientations. Unlike previous analytical approaches which are restricted to certain limiting structures [3], the approach presented here is based on a finite element scheme and hence allows simulation of carrier transport and subsequent device optimization for arbitrary device geometries and structures.

2. Modeling Approach

The magnetic field dependent current density in the presence of homogeneous stress reads

$$\mathbf{J_n} - \sigma_n \cdot [(\mathbf{R_H} \cdot \mathbf{B}) \times \mathbf{J_n}] = \sigma_n \cdot \mathbf{E} \tag{1}$$

where, the conductivity σ_n is a symmetric second rank tensor. The resistivity (σ_n^{-1}) for not too large stress levels, is linearly related to stress, T *via* the well known fourth rank tensor of piezoresistance coefficients, viz.,

$$(\Delta\rho)_{ij}/\rho_0 = \Sigma_{k,l}\ \pi_{ijkl}T_{kl}. \tag{2}$$

The Hall coefficient tensor, $\mathbf{R_H}$ carries a similar form but it is related to the fourth rank tensor of piezo-Hall coefficents,

$$(\Delta R_H)_{ij}/R_{Ho} = \Sigma_{k,l}\ P_{ijkl}T_{kl}. \tag{3}$$

values of which have been measured only recently [4]. In the absence of stress, the conductivity and Hall coefficient become scalars, and the original form of the equation in terms of the electron Hall mobility, μ_{Hn} (= $|\sigma_n R_H|$) is recovered [5].

We transform eqn. (1) from the cubic crystallographic axes to the Cartesian system of arbitrary orientation. The transformation of vectors J_n and E lead to $J_n' = \alpha J_n$ and E' = αE, where α is the transformation matrix expressed in terms of Euler's angles [6]. The reduced form of equation (1) reads

$$J_n' = A . E' , \tag{4}$$

where, A is an asymmetric second rank tensor which contains terms in conductivity ($\sigma_n' = \alpha \sigma_n \alpha^{-1}$) and Hall coefficient ($R_H' = \alpha R_H \alpha^{-1}$) both being functions of stress ($X = \alpha T \alpha^{-1}$), and the magnetic field ($B' = \alpha B$). The matrix becomes skew symmetric in the absence of stress and becomes symmetric in the absence of both stress and magnetic field.

Consider a two-dimensional system with the orientations of current, magnetic field, and uniaxial stress as depicted in Fig. 2. Assuming negligible generation or recombination in the device, the divergence of the current density in (4) under steady state conditions can be cast into the following form:

$$\Sigma_{i,j=1,2} \quad D_i (a_{ij} D_j \psi) = 0 \tag{5}$$

where $D_1 = \partial/\partial x$, $D_2 = \partial/\partial y$, and the entries a_{ij}, which carry a cumbersome form, will not be shown for convenience. The boundary conditions for (5) consist of Dirichlet conditions (prescribed by the applied potential) at the current electrodes and the Neumann condition at insulating boundaries, $J_n'.n = 0$, where n denotes the outward normal vector. We observe that the latter is a natural condition associated with (5) and is influenced by both the stress and the magnetic field. At the Hall probe regions in the Hall geometry, we impose the condition, $\int_{probe} (J_n'.n) \, d(probe) = 0$, since we are only interested in the open circuit Hall voltage. Equation (5), together with the boundary conditions, is numerically solved based on a standard Galerkin finite element discretization procedure, using an adaptively generated grid. Since the conditions at the insulating boundaries are natural to (5), the discretization procedure is simplified. The numerical scheme employed here is based along the lines discussed in [5].

3. Results and Discussion

The samples considered in the analysis are n-type of <100> and <110> crystallographic orientations with doping concentrations in the range 10^{14} - 10^{16} cm^{-3}. The values of piezo-Hall coefficients used in the computations are based on the measurement data reported in [4]. In practical device structures, the stress on the device is induced primarily by encapsulation (or packaging), and arises mainly due to the difference in thermal expansion coefficients between the silicon die and the epoxy employed in the bonding process. To obtain practical estimates of the stress on the die surface, the general purpose finite element software package ANSYS [7] was used. The simulated stress distribution on the die surface is shown in Fig. 2.

Using a stress value, X = 10^8 dyne/cm^2, the output response for both (100) and (110) Hall and split-drain devices is computed as a function of device orientation. Here, we rotate the current density, J about a given direction by an angle ψ, keeping the directions of magnetic field, B and stress, X fixed (Fig. 3). For example, with no rotation, ψ = 0'; J || X|| <010>, B || <100> for (100) devices, and J || X || <001>, B || <110> for the (110) counterparts.

The output voltage for the Hall geometry, calculated at B = 0 and at B = 100 mT is shown in Fig. 4 for $J = 2\times10^4$ A/cm^2. Here, we see that the offset is zero when the device is oriented such that the current flow is parallel or transverse to the uniaxial stress. Any variation from these directions gives rise to an offset, and in particular, at angles 45° from these directions, the offset is seen to be a maximum. As expected, stresses acting on the <100> sample result in higher output voltages compared with the <110> orientation due to its relatively larger piezoresistance coefficient. Although not shown, the same behaviour is noted for the output current (I_1-I_2) of the split-electrode device geometry.

No significant nonlinear coupling between the stress and the magnetic field was noted at fields, B ≤1 Tesla. This can be noted from the Hall voltage response in Fig. 4. The output Hall voltage at 100 mT field is simply the zero field response shifted by an amount which corresponds to the "unstressed" Hall voltage. At relatively larger fields (B = 2 Tesla), a very weak (hardly noticeable) interaction can be observed in the potential distribution (see Fig. 5). This coupling, however, is not significant for practical values of field strengths.

Conclusions

In this paper, we have presented two-dimensional numerical solutions to the piezo-Hall current density equation which accounts for the effects of stress on galvanomagnetic carrier transport. The interaction between the stress and magnetic field was found to be very weak and evident only at large field strengths.

References

[1]   C. S. Smith, Phys. Rev. **94** (1954) 42.
[2]   R. W. Keyes, Solid State Physics **11** (1960) 149.
[3]   A. Nathan, T. Manku, Appl. Phys. Lett. **62**, No. 23 (1993) in press.
[4]   B. Halg, *J. Appl. Phys.* **64** (1990) 276-282.
[5]   W. Allegretto, A. Nathan, H. Baltes, *IEEE Trans. CAD* **10** (1991) 501-511.
[6]   Y. Kanda, IEEE Trans. Electron Devices ED-29 (1984) 64.
[7]   User's Manual, ANSYS Engineering Analysis System, Houston: Swanson Analysis Systems Incorporated.

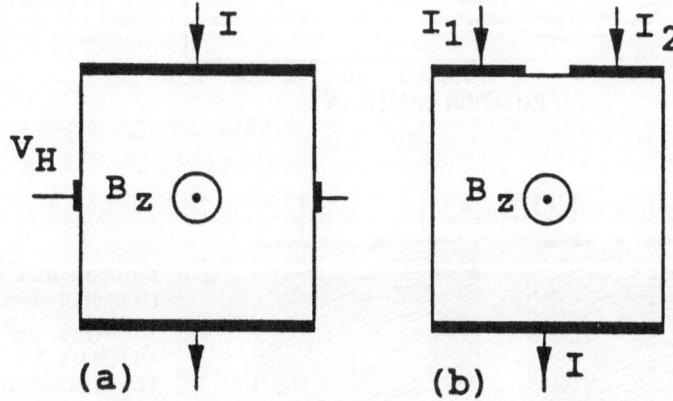

Fig. 1 (a) Hall and (b) split-electrode devices considered in analysis.

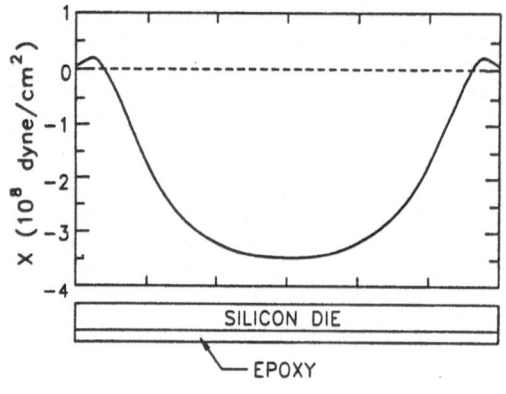

Fig. 2 Simulated stress distribution on die surface.

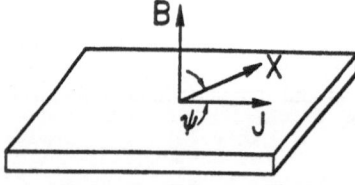

Fig. 3 Orientations of current density, magnetic field, and uniaxial stress.

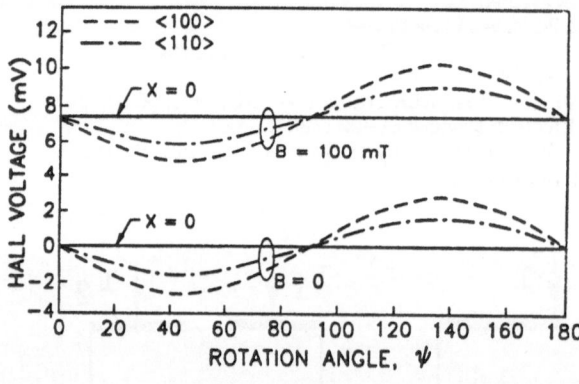

Fig. 4 Output response of (100) and (110) Hall geometries as a function of device orientation under uniaxial stress of 10^8 dynes/cm^2.

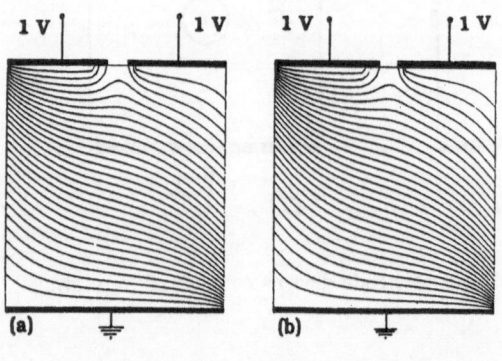

Fig. 5 Equipotential lines in the (100) split-electrode under 10^8 dynes/cm^2 for (a) B = 0 T and (b) B = 2 T. The orientations are B∥ <100>, X∥ <010>, and J∥ <011>. The rotation angle is ψ = 45°.

SIMULATION OF SEMICONDUCTOR DEVICES AND PROCESSES Vol. 5
Edited by S. Selberherr, H. Stippel, E. Strasser – September 1993

381

Electro-Elastic Simulation of Piezoresistive Pressure Sensor

P. Ciampolini, A. Rossi[†], A. Pierantoni[†], and M. Rudan[†]

Istituto di Elettronica, Università di Perugia
I-06131 Santa Lucia-Canetola, Perugia, ITALY
[†]Dipartimento di Elettronica, Informatica e Sistemistica, Università di Bologna
Viale Risorgimento 2, I-40136 Bologna, ITALY

Abstract

Transport properties of the silicon crystal are sensitive, to some extent, to mechanical perturbations: this allows for integrating mechanical sensors, together with the sensing circuitry, within a silicon chip by using an almost standard IC technology. In this paper, the numerical simulation of a silicon pressure sensor, based on the piezoresistive effect, is described. The simulated transducer is made of a thin silicon diaphragm, on top of which a four-resistor bridge is diffused. For a given pressure, the distribution of the stress components over the diaphragm is fed to the program, which then computes the sensor response depending on its geometrical and physical features. To this purpose, an anisotropical, stress-dependent, mobility model has been introduced into the device simulator HFIELDS-3D.

1. Introduction

The progress in silicon technology makes it possible to fabricate sophisticated microsensors where sensing part, transducer, and signal amplifier are all integrated within the same chip. At the same time, the miniaturization makes the analytical simplifications describing the sensors' performance less and less reliable. The purpose of this abstract is to present preliminary results of the numerical analysis of a piezoresistive pressure sensor. More specifically, the attention will focus here on the sensing part and transducer which, as shown below, are combined in a single device. In this respect, a few analysis codes have already been implemented elsewhere to simulate mechanical deformation and stress. On the other hand, the underlying physics of the electro-elastic interrelations in semiconductor crystals has been investigated and well understood. This makes it possible to take the further step of incorporating the electrical aspects into the simulation.

2. Theory

From the microscopical point of view, deformations in the lattice of a semiconductor crystal (induced, e.g., by an external force) modify the shape of the energy function $E(\mathbf{k})$. Due to this, an extrinsic anisotropy caused by the deformation adds to the intrinsic one already possessed by the energy function in the non-deformed case. Hence, when a semiconductor crystal subject to a deformation is brought to a non-equilibrium state by forcing a current flow in the material, its macroscopic transport properties must inherit to some degree the effects of the deformation. This concept can better be illustrated with reference to a specific semiconductor material, e.g., silicon, in the simple case where the deformations are kept within the linear regime and a near-equilibrium condition holds as for the electric behavior. Thanks to the latter assumption, the current transport is well described by the drift-diffusion model. Taking by way of example the conduction band, the electron current density $\mathbf{J}_n$ is obtained by adding the contributions of six valleys, each of them giving rise to a 3×3 mobility tensor. In the non-deformed case the six mobility tensors happen to combine in such a way that the resulting (macroscopic) mobility is a scalar. This, however, is no longer true when the crystal undergoes a deformation: in this case, in fact, the macroscopic mobility retains a tensor nature, which is a direct effect of the deformation's anisotropy and can be exploited for measuring purposes. Letting $\mathcal{M}$, μ_0 be the mobility in the deformed and non-deformed case, respectively, and $\mathcal{I}$ the 3×3 identity tensor, one may introduce a tensor $\mathcal{D}$ such that

$$\Delta \mathcal{M} = \mu_0 \mathcal{I} - \mathcal{M} = \mu_0 \mathcal{D}.$$

For symmetry reasons, $\mathcal{D}$ has only six independent components which, for the sake of simplicity, are thought of as a 6×1 vector $\mathbf{d}$. The components of the latter in terms of the deformation could in principle be derived by direct calculation; however, a much simpler way is that of determining these relationships experimentally, and recasting the result in terms of the stress components. More precisely, one obtains $\mathbf{d} = \mathcal{P}\mathbf{s}$, where the 6×1 vector $\mathbf{s}$ is made of the independent components of the stress tensor, and $\mathcal{P}$ is the 6×6 piezoeresistive tensor. The form of the latter is particularly simple in crystals with cubic symmetry like silicon [1]. Moreover, for a purely ohmic transport due to a uniform electron concentration n it turns out $\Delta \boldsymbol{\rho}/\rho_0 = -\Delta \boldsymbol{\sigma}/\sigma_0 = \mathbf{d}$, where $\boldsymbol{\rho}$, $\boldsymbol{\sigma}$ are made of the independent components of the resistivity and conductivity vectors, respectively, and $\sigma_0 = 1/\rho_0 = q\mu_0 n$. The calculation of the stress components is made easier by the simplicity of the structures at hand; typically, one takes a rectangular diaphragm in the xy plane, with built-in boundary conditions, whose thickness h is large compared to the deflection w. The latter is then found by solving

$$C_x \frac{\partial^4 w}{\partial x^4} + 2H \frac{\partial^4 w}{\partial x^2 \partial y^2} + C_y \frac{\partial^4 w}{\partial y^4} = p,$$

where p is the pressure (the coefficients are different due to anisotropy). The stress components are derived from w in the customary way [2].

3. Implementation and Results

The capability of handling stress components in a simplified manner has been added to the drift-diffusion version of the in-house developed, three-dimensional device simulator HFIELDS-3D. A number of tests have been carried out on the code, aiming at detecting the variations in the electrical behavior of identical devices placed at different position on a diaphragm. As a first example, the structure presented in [3] has been simulated, i.e., a pressure sensor.

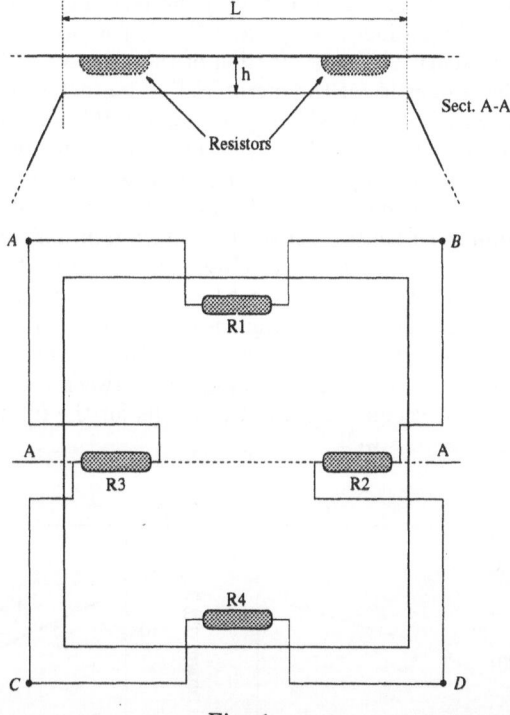

Fig. 1

The device is made of a square diaphragm, obtained by anisotropically etching a silicon wafer (Fig. 1). Three different thicknesses h of the diaphragm have been considered, namely 10, 20 and 30 μm, as well as three different side lengths L (100, 200 and 300 μm). Two etching modes were taken into account: in the first one, the sides L are oriented along the principal axes of the crystal while, in the second, they form a 45° angle with the latters. Four 20×4×2 μm^3 resistors are diffused near the sides' midpoint and connected to form a bridge. Their longer side belongs to the [110] plane in the first etching mode, and to the [100] plane in the second one. Let R_0 be the common value of the resistance when the diaphragm is not deformed; the relative unbalance of the bridge when a variation occurs in the resistances, at a given bias V_{AD}, is expressed by

$$\frac{V_{BC}}{V_{AD}} = \frac{\Delta V_{BC}}{V_{AD}} = \frac{-\Delta R_1 + \Delta R_2 + \Delta R_3 - \Delta R_4}{4R_0}.$$

The relative variations in the resistances have been calculated by HFIELDS-3D. The results can be summarized by defining the pressure sensitivities α_i of the individual resistors as

$$\frac{\Delta R_i}{R_0} = \alpha_i\, p, \qquad i = 1,\dots,4$$

and using the expression of the relative unbalance to obtain the total pressure sensitivity β :

$$\frac{V_{BC}}{V_{AD}} = \beta\, p, \qquad \beta = \frac{1}{4}\left(-\alpha_1 + \alpha_2 + \alpha_3 - \alpha_4\right).$$

The values of β corresponding to different conditions are reported in Figs. 2, 3 and 4 (expressed in 10^{-10} cm^2/dyne units). From Fig. 2 (3) it is seen that doubling the diaphragm length (thickness) results in increasing (decreasing) the pressure sensitivity by about a factor 4. This is consistent with the fact that in the linear regime and in the simpler, one-dimensional case, the deflection is proportional to L^2/h^2. The larger differences in sensitivity as a function of the resistors' orientation are a direct consequence of the properties of the piezoresistive tensor. In fact, its elements give rise to individual sensitivities α_i which, in the case [110], have all the same sign and similar magnitudes; in the case [100], on the contrary, the magnitudes are still similar while the sign of α_1 and α_4 is opposite to that of α_1 and α_4. Fig. 4, finally, illustrates the dependence of β upon the position of the resistor: d indicates the distance of the bridge resistors from the diaphragm edge ($L = 100\,\mu$m, in this case). As expected, the sensitivity decreases as the resistors move toward the center; this is due to both the smaller stress experienced by each element (for a given pressure) and the decreasing unbalance of the bridge. The sensible elements have then to be placed as close to the edge as possible: the accurate evaluation of the stress components in such regions, however, needs the actual profile of the transition between the thin and thick silicon layers to be taken into account. This, in turn, calls for the full numerical simulation of the mechanical system as well.

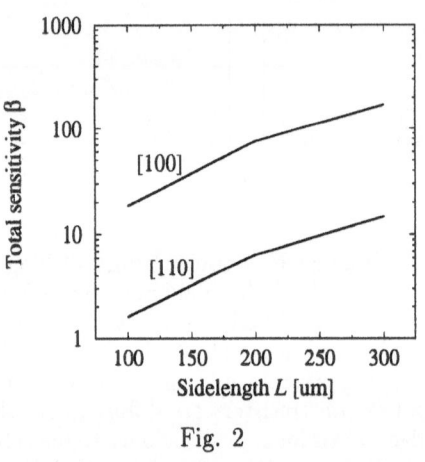

Fig. 2

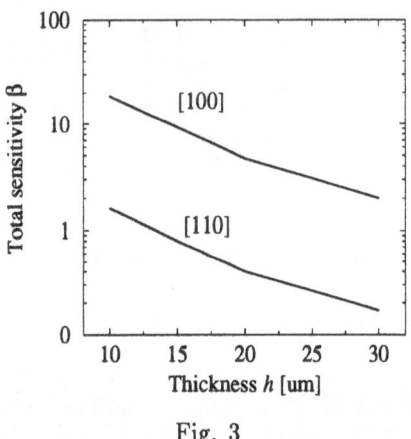

Fig. 3

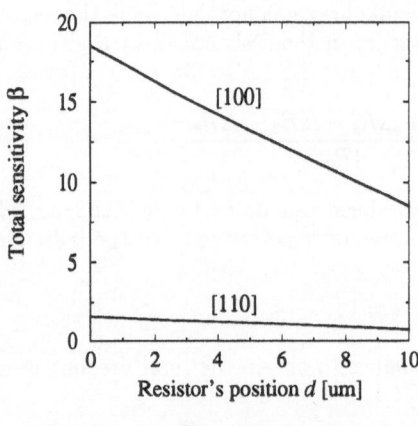

Fig. 4

References

[1] Y. Kanda, *A Graphical Representation of the Piezoresistance Coefficients in Silicon*, IEEE Trans. ED, Vol. ED-29, No. 1, Jan. 1982, p. 64.

[2] S. Timoshenko, S. Woinowsky-Krieger, *Theory of Plates and Shells*, McGraw-Hill, New York, 1959.

[3] K.W. Lee, K. D. Wise, *SENSIM: A Simulation Program for Solid-State Pressure Sensors* IEEE Trans. ED, Vol. ED-29, No. 1, Jan. 1982, p. 34.

Models for the Chemo-Physical Reactions at the Sensitive Layer of Semiconductor Gas Sensors and their Application in the Simulation of these Devices

Z. Gergintschew and D. Schipanski

Institut für Festkörperelektronik, Technische Universität Ilmenau
Weimarer-Straße 32, D-98684 Ilmenau, GERMANY

Abstract

In this paper we consider models for some adsorption effects, the resulting change of the work function at metal-oxide-semiconductor sensitive layers of gas sensors and their implementation in a two dimensional device simulator, which solves the Poisson equation. An example for the application of this simulation program is given with a Suspended Gate Field Effect Transistor.

1. Introduction

The importance of the development tools like the numerical device simulation in the sensor technics increases with the introduction of microelectronic sensor devices and manufacture methods. In order to regard the modelling of sensor devices as complete, apart from the common electrical description, the interaction of the sensor with the detected enviroment has to be taken into account in the simulation. That's why, we consider in our paper a model for the adsorption effects and the resulting work function change at the metal-oxide-semiconductor gas sensitive layers of microsensors.

2. Theoretical description of the adsorption effects for O_2

One of the main reactions of the metal-oxide-semiconductor sensitive layers (we consider here only n-type materials like ZnO, SnO_2) is the interaction with oxygen. Neglecting the reaction with instationar metal ions we have at the oxide surface the following effects:

$$O_2(g) \rightleftharpoons O_2 \ \}physiosorption$$
$$\left. \begin{array}{l} O_2 + e \rightleftharpoons O_2^- \\ O_2^- + e \rightleftharpoons 2O^- \end{array} \right\} chemiosorption$$

To work out the models for these reactions we have used 'The Electron Theorie of Catalysis on Semiconductors' by Volkenstein [1]. He distinguishes between the 'weak' ($^\circ$) and the strong ($^-$) chemiosorption in dependance on the wave function interaction between the binding electron of the adsorbed particle and the oxide lattice. For the further considering we assume a monocrystal oxide and an adsorption without dissociation. After complicated operations with the equation for the steady-state adsorption equilibrium (like Volkenstein [1] and Geistlinger [2]) we have for the fractional surface coverage θ with O_2^- the following expression:

$$\theta = \frac{N}{N_0} = \frac{\beta P_{O_2}}{\beta P_{O_2} + 1} \tag{1}$$

with:

$$\beta = \frac{s_0}{\nu(\sqrt{2\pi M k T})\exp\left(\frac{-Q^\circ}{kT}\right)} \cdot \frac{1}{f^\circ\left(1 + \exp\left(\frac{E_f - E_c}{kT}\right)\right)} \tag{2}$$

Here are: N the density of the occupied surface states; N_0 the density of the maximal available surface states; P_{O_2} the oxygen partial pressure; s_0 adhesion coefficient; M the mass of the adsorbed particle; k the Bolzmann constant; T the temperature; ν the phonon frequence of the adsorbed particle; Q° the weak chemosorption energy; E_c the energy of the conduction band edge; E_f the Fermi energy. E_s is the energy difference between the strong and the weak chemiosorbed particles. f° and f^- are the occupation probabilities for the weak and strong chemiosorption (eq.3):

$$f^\circ = \frac{1}{0.5\exp\left(\frac{E_f - E_s}{kT}\right) + 1}; \quad f^- = \frac{1}{2\exp\left(\frac{E_s - E_f}{kT}\right) + 1} \tag{3}$$

In eq.2 the fractional coverage depends additionally on the energy difference $(E_f - E_c)_s$ at the sensitive layer surface, that means on its doping, on the external electrical fields and on the number of the already adsorbed particles. The electron exchange between the adsorbed molecules and the oxide (mainly due to strong chemiosorption) leads to a charging of the surface:

$$Q_s = e\, N_0\, \theta^- = e\, N_0\, f^-\, \theta \tag{4}$$

e is the elementary electronic charge.
The changing of the work function $\Delta\varphi \cdot e$ (eq.5) of the sensitive layer, which the sensor effect of many devices is based on, is defined by the band bending $\Delta V_s \cdot e = \Delta(E_f - E_c)_s$ due to Q_s and by the electron affinity changing $\Delta\chi$ at the surface caused by dipol effect of the polarised adsorbed particles [3].

$$\Delta\varphi = \Delta V_s + \frac{\Delta\chi}{e} = \Delta V_s + \frac{\mu_{ad} N_0 \theta}{\varepsilon_s \varepsilon_0} \tag{5}$$

$\varepsilon_s \varepsilon_0$ is the permittivity at the surface of the sensitive layer and μ_{ad} the dipol moment of one adsorbed particle.

3. Implementation of the models in the simulator CADI-CHEM

For the calculation of the band bending at the sensitive layer surface and in this way of the work function changing in dependance on the gas adsorption, doping and external fields, it is necessary to solve the Poisson equation (eq.6). For the interface sensitive layer/atmosphere the following equation system is valid:

$$\frac{d^2\varphi}{dz^2} = -\frac{1}{\varepsilon}[Q_s(E_f, P_{O_2}) + eV_o^{\cdot} + eV_o^{\cdot\cdot} + N_A^+ - N_D^- - en(z) + ep(z)] \qquad (6)$$

$$Q_s = -eN_o f^- 0(E_f, P_{O_2}) \qquad (7)$$

Here are: φ the electrical potential; $V_o^{\cdot}, V_o^{\cdot\cdot}$ the density of the singly and doubly ionised oxygen vacancies; N_A, N_D the acceptor and donator doping density; n, p the electron and hole density of the sensitive layer. This equation system is solved by the 2-dimensional device simulator CADI-CHEM, which solves additionally the Poisson equation in the Si-area and the Laplace equation in the insulator areas with a Finite Difference Method.

4. Application on a SGFET gas sensor

The simulator CADI-CHEM is applied for the modelling of a Suspended Gate Field Effect Transistor (SGFET) [4]. As shown in Fig.1, it is a MISFET with an air gap between the gate electrode and the gate insulator (100 nm SiO_2, 100 nm Si_3N_4). A ZnO sensitive layer is located over the air gap. The changing of its work function due to gas adsorption leads to a changing of the transistor threshold voltage.

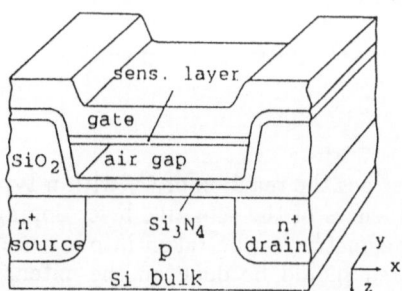

Figure 1: Schematic plot of a SGFET gas sensor

The gate area of a SGFET was calculated with the program CADI-CHEM. Fig.2 shows the energy band diagram of this stucture with and without gas adsorption.

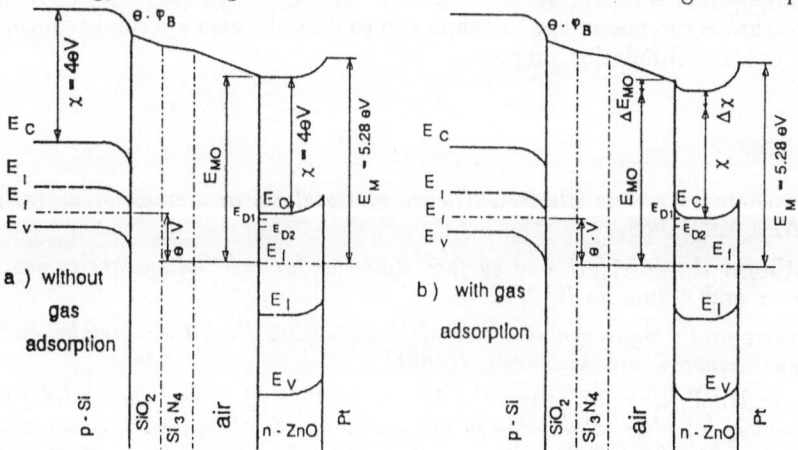

Figure 2: Energy band diagram of the gate structure of a SGFET with ZnO as sensitive layer

In this case the changing of the work function $\Delta\varphi$ is caused mainly by the band bending. Its modeled dependance on O_2 partial pressure is shown in Fig.3. Further on we use the calculated $\Delta\varphi$ for the simulation of the drain current dependance of a SGFET sensor on the O_2 partial pressure (Fig.3) with the 2-dimensional device simulator TOSCA [5] which solves the basic semiconductor equations with the Finite Element Method.

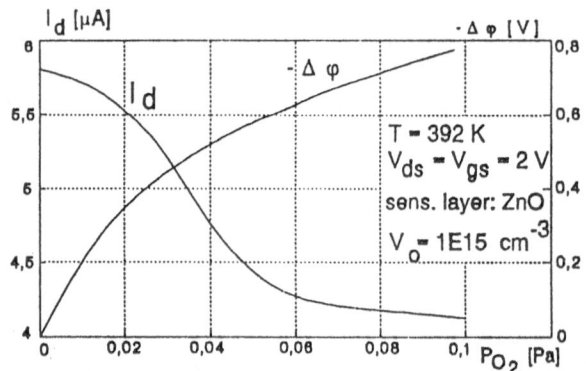

Figure 3: Threshold voltage and drain current dependance of a n-channel SGFET on the O_2 partial pressure

5. Summary

The introduced model describes the reaction of O_2 with n-type metal-oxide-semiconductors. This way the influences of the sensitive layer doping, the external voltage, the temperature and the geometry can be taken into account in the sensitive effect simulation. Further research should be done on the extension of these models by the reactions with reduction gases like H_2 or NH_3. For that it is necessary to determine a lot of unknown model parameters. The description of the chemo-physical reactions on the sensitive layers by the Electron Theory and the band model allows the implementation of these effects in the basic semiconductor equation system and so the extension of common device simulators with these models. An important condition for that is the possibility to define and to describe also a second semiconductor material in these simulation programs.

References

[1] Volkenstein,Th.: *The Electron Theory of Catalysis on Semiconductors, McMillan,New York,1963*.

[2] Geistlinger,H.: *Interface and surface statistics for a Schottky-barrier gas sensor, Sensors and Actuators B, 7(1992)*

[3] Lundstroem,I.: *Hydrogen sensitive MOS-structures Part 1: Principles and Applications, Sensors and Actuators, 1(1981)*

[4] Lortenz,H.; Peschke,M.; Riess,H.; Janata,J.; Eisele,I.: *New suspended gate FET technology for physical deposition of chemically sensitive layers, Sensors and Actuators, A21-A23 (1990)*

[5] Gajewski,B.; Heinemann,B.: *Two Dimensional Semiconductor Analysis Package, Handbuch, IAAS Berlin, (1991)*

SIMULATION OF SEMICONDUCTOR DEVICES AND PROCESSES Vol. 5
Edited by S. Selberherr, H. Stippel, E. Strasser – September 1993

Two-Dimensional Numerical Simulations of High Efficiency Silicon Solar Cells

G. Heiser, A. G. Aberle[†], S. R. Wenham[†], and M. A. Green[†]

School of Computer Science & Engineering, University of New South Wales
P.O.Box 1, Kensington, New South Wales, 2033, AUSTRALIA
[†]Centre for Photovoltaic Devices & Systems, University of New South Wales
P.O.Box 1, Kensington, New South Wales, 2033, AUSTRALIA

Abstract

This paper presents for the first time the use of two-dimensional (2D) device simulation for optimising design parameters of high-efficiency silicon solar cells of practical dimensions. We examine the influence of 2D effects on the operating conditions of these devices and give results for the optimal spacing of the front metal fingers. Numerical difficulties inherent in the simulations are being discussed.

1. Introduction

The *passivated emitter, rear locally-diffused* (PERL) silicon solar cell (Fig. 1) has recently been developed at the University of New South Wales (UNSW) and demonstrates record-high efficiencies of 23.2 % under non-concentrated ("one-sun") illumination [1]. It is esti-mated that PERL technology can yield efficiencies of up to 25 % if all design parameters are fully optimised. However, this requires improved understanding of the operating conditions of the devices. In this paper, state-of-the-art two-dimensional (2D) numerical simulation tools are for the first time applied to gain detailed insight into the operating conditions of practical high-efficiency Si solar cells and to optimise various design parameters.

2. Two-dimensional Effects in High-efficiency Silicon Solar Cells

One important design parameter of Si solar cells is the *front contact spacing*, i.e. the distance between the thin metal fingers contacting the emitter layer (the n^+ layer in a n^+p structure). Since the metal contacts are opaque to light, a small finger spacing reduces cell efficiency due to excessively large shading losses. For wide spacings, however, the limited conductivity of the thin emitter layer leads to large resistive losses. Furthermore, the voltage drop resulting from the emitter resistance leads to an increased operating voltage half way between two metal fingers. This voltage profile results in increased electron concentrations in the central base region and a 2D electron flow in the base (Fig. 2).

In order to determine the optimum finger spacing by simulation, accurate modelling of the resistive losses in the emitter is essential. 1D simulations, which have been successfully used to optimise other design parameters (see e.g. [2]), cannot provide this accuracy, as they overestimate the emitter current due to the assumption of a vertical electron flow in the base (p layer). As a result of the 2D electron flow in the base the spacing of the front fingers can be increased well above the limits predicted by 1D models, and cell efficiency increases due to reduced shading losses as well as reduced metal contact recombination.

3. Difficulties and Limitations

High-efficiency Si solar cells are notoriously difficult to simulate. The main reason for this lies in the huge device dimensions (typical cells are 250 μm deep with a front contact spacing of 800 μm) and the large diffusion lengths (1–2 mm). Making use of symmetry, a simulation domain of 400 μm × 250 μm is therefore required. On the other hand, some dimensions of the cells are more typical for VLSI devices: the emitter depth is 1 μm, while the contact fingers are 3 μm wide. For a simplified cell with a planar front surface, a fully metallised back, and a p$^+$ "back surface field" (Fig. 2), there are 3 regions of the device which require, in at least one dimension, grid lines spaced 0.2 μm or less: the emitter (where roughly half of the incident light is absorbed), the back contact (where recombination is high), and the shade boundary under the front contact (where the generation rates are discontinuous). This easily leads to grid sizes which require massive supercomputer use.

The recently developed ETH device simulation package [3] allows the user to control grid densities well enough so that we could limit grid sizes to some 3,000-5,000 points (Fig. 3), which enabled us to perform the simulations on a Sun SPARC-2 workstation. However, the price to be paid for the small grid sizes is the poor condition of the linear systems that have to be solved in the course of the simulation. We found occasionally that some linear systems could not be solved by a direct solver, particularly on the Cray, which has a smaller mantissa length than machines using IEEE arithmetic. Iterative solvers have generally failed to converge on our problems. This is a serious restriction, as we are planning to investigate problems that require 3D simulations (cf. Sect. 5), which will only be possible if we can use iterative solvers. We are presently investigating the sources of these numerical problems and hope to solve them by the use of better adjusted grids.

n_{ie} model	oS	S	dA	BW	BW*
$n_i/10^{10}\,cm^{-3}$	1.548	1.247	1.493	1.09	1.00
P/P_{BW^*}	0.804	0.885	0.934	0.994	1.000

Table 1: Power output as a function of n_i

Another difficulty results from the extreme sensitivity of the cells to material properties such as minority carrier lifetimes and intrinsic carrier density, n_i. Table 1 shows the dependence of the simulated power output of a cell on the model used for n_{ie} (effective n_i). The first four values in the table correspond to the models built into Simul [3], while the last one corresponds to model BW modified to yield $n_i = 1.00 \times 10^{10}\,cm^{-3}$, which is the value currently accepted in the photovoltaic community [4]. The results show that these models must be carefully tuned if the simulation results are to be compared with experimental data. In our simulations we used the BW* model and adjusted carrier lifetimes to 2 ms for good agreement with experimental data at a reference point (a finger spacing of 800μm).

4. Results

Fig. 4 shows the calculated electron flow in the base at the maximum power point of the solar cell of Fig. 2 (contact spacing of 800 μm) under long wavelength (1000 nm) illumination. The light intensity has been adjusted to yield the same short-circuit current as for standard white light of one-sun intensity (AM1.5 spectrum). As Si has an indirect bandgap of about 1.1 eV, light of such a wavelength is only very weakly absorbed and carrier generation is almost uniform throughout the device. The plot shows that the electron flow in the base has indeed a significant lateral component. The effect is particularly pronounced for electrons generated near the back contact: These electrons move a considerable distance along the rear surface until they diffuse upwards to be collected by the emitter region. Consequently, a significant fraction of the electron current that otherwise would have to be transported by the emitter flows through the base of the cell. This effect reduces the current density and hence the ohmic losses in the emitter. However, due to the enhanced electron path length in the base, this 2D effect is only beneficial for solar cell efficiency if the electron diffusion length is larger than about half the front finger spacing.

Fig. 5 shows the corresponding electron density and the direction of the electron flow under AM1.5 illumination. The combination of weak absorption at long wavelengths and stronger absorption at shorter wavelengths shifts the area of maximum electron concentration closer to the front surface and results in an electron flow which, in most of the device, is more horizontal than vertical.

Fig. 6 shows that the optimum front finger spacing under AM1.5 illumination is about 1.1–1.2 mm, compared to 0.8–0.9 mm as predicted by 1D models. Interestingly, the 2D model predicts only a weak decrease of efficiency for wide finger spacings: According to Fig. 6, the power output drops only by 1.3 % if the finger spacing is increased from 1.1 to 2 mm, while the 1D model predicts an efficiency loss more than twice as big.

5. Conclusions and Future Work

The 2D simulations presented in this paper considerably improve the general understanding of the operating conditions of high-efficiency Si solar cells. Improved values for an important design parameter, the front finger spacing, have been obtained. We expect that these and forthcoming simulation results will help to further improve the efficiency of laboratory cells at UNSW. Presently we are working on the optimisation of the back contacts of PERL cells, which will involve 3D simulations.

6. Acknowledgments

This work was partially funded by a grant from the Australian Research Council (ARC). The Centre for Photovoltaic Devices and Systems is supported by the ARC's Special Research Centres Scheme and Pacific Power. A.G.A. gratefully acknowledges the support of a Feodor Lynen Fellowship provided by the Alexander von Humboldt Foundation. Finally, we would like to thank Wolfgang Fichtner from ETH Zürich for granting access to the latest versions of Simul and Kevin Kells and Ulrich Krumbein of his group for their help with various problems encountered in the simulations.

References

[1] M. A. Green. Recent advances in silicon solar cell performance. In *Proc. 10th European Communities Photovoltaic Solar Energy Conf.*, p. 250, Lisbon, 1991.

[2] A. Aberle, W. Warta, J. Knobloch, and B. Voss. Surface passivation of high efficiency silicon solar cells. In *Proc. 21st IEEE Photovoltaic Specialists Conf.*, p. 233, Orlando, 1990.

[3] Integrated Systems Lab. *Simul Manual*. ETH Zürich, Switzerland, 1992.

[4] A. B. Sproul and M. A. Green. Improved value for the silicon intrinsic carrier concentration from 275 to 375 K. *J. App. Phys.*, 70:846–54, 1991.

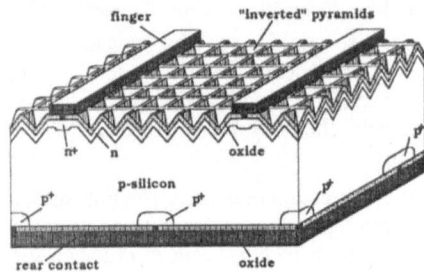

Figure 1: The UNSW PERL Si solar cell.

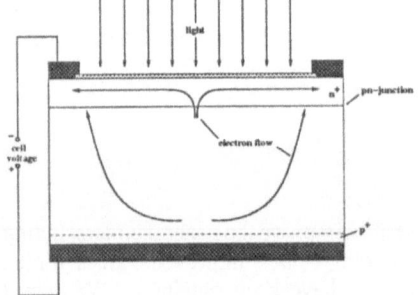

Figure 2: Schematic view of a simplified solar cell and the electron current flow within the device.

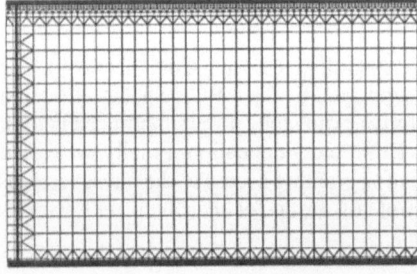

Figure 3: Simulation grid for the above cell. One emitter contact is at the top left corner of the picture, while the right edge coincides with the symmetry plane between two contacts

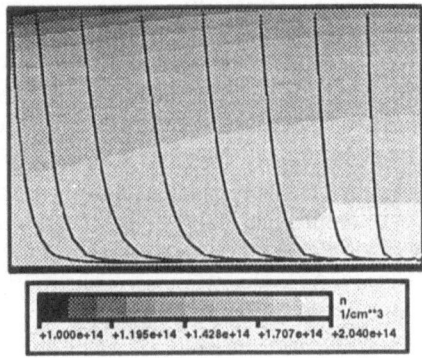

Figure 4: Electron density and electron flow lines under 1000 nm illumination.

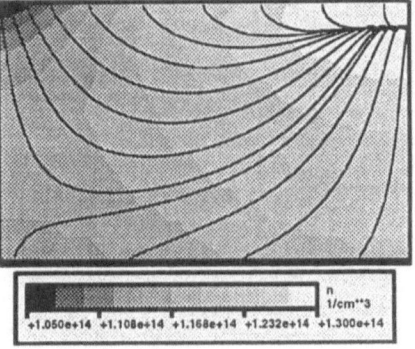

Figure 5: Electron density and electron flow lines under AM1.5 illumination.

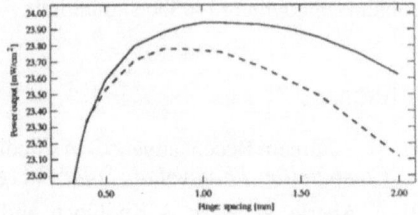

Figure 6: Output power according to 1D (broken line) and 2D (solid line) simulations under AM1.5 illumination.

Simulation in High Efficiency Solar Cell Research

S. Sterk and S. W. Glunz

Fraunhofer-Institut für Solare Energiesysteme
Oltmannstr. 22, D-79100 Freiburg, GERMANY

Abstract

The practical application of numerical device simulation in high efficiency silicon
solar cell research is presented. Aspects of the design development and the characterization are discussed.

1. Introduction

Two main features of high efficiency silicon solar cells [1], see Fig. 1, are high
minority carrier lifetime τ_B in the bulk
and low recombination velocities S_{ox} of
the minority carriers at the oxidized surfaces, especially at the rear side. Since
the recombination velocity S_m at the metallized surface regions is much higher,
the rear side is only locally contacted.
We present an optimization study for the
design of the rear point contact pattern,
and advanced characterization techniques
for the starting material, the interfaces
and the diffused regions.

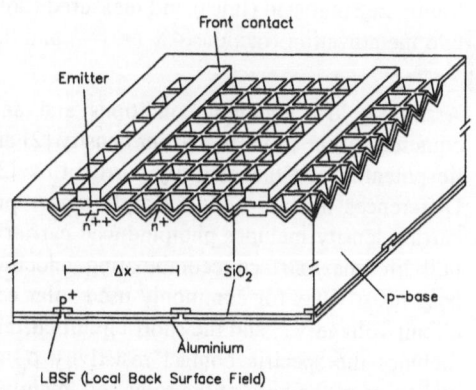

Figure 1: LBSF Solar Cell.

2. Optimization of the Rear Contact Design

Because the diffusion length L of the carriers is several times the base thickness, the total
carrier loss depends strongly on the recombination at the rear surface, mainly occurring at
the silicon-metal interface of the local rear contacts. In this paper we present an experimental and theoretical optimization study for the rear side point contact design (point
spacing and size). The results are presented in Fig. 2.

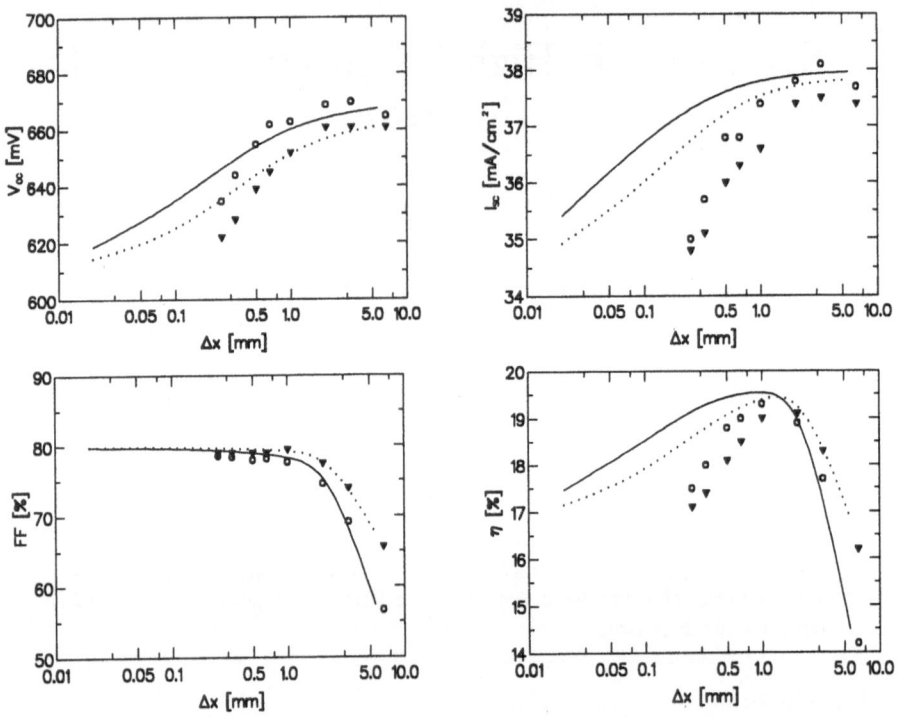

Figure 2: Simulated (lines) and measured (dots) solar cell parameters V_{oc}, I_{sc}, FF and η for two metallization fractions 4% (▼, ·····) and 0.5% (○, —) versus the contact point spacing.

Assuming low injection conditions and an ideal emitter, we solve the 3D diffusion equation for the minority carrier density [2] and the 3D Poisson equation for the electrostatic potential and the majority carrier flux [3] in the base separately. We use the Finite Differences method with a simple tensor product grid. The simulation of the minority carrier density includes photoinduced carrier generation (solar spectrum AM1.5), realistic bulk lifetimes, surface recombination velocities at the oxidized and metallized surfaces and base resistivities for commonly used solar cell materials, resulting in values for the open circuit voltage V_{oc} and the short circuit current I_{sc}. The model for the majority carrier flux includes the specific contact resistivity ρ_c and the bulk resistivity ρ_b. It computes the resistance of the base of the solar cell, mainly due to the current crowding around the point contacts, which decreases the fill factor FF of the I-V-curve of the solar cell under illumination. The efficiency η of the solar cell is given by the product of V_{oc}, I_{sc} and FF, devided by the power of the incident sun light.

We have simulated and processed solar cells with different rear contact design. The spacing Δx of the quadratic contact point pattern varies from 0.25 mm up to 6.6 mm, the metallization fraction f_m of the rear side was 4% (triangles, dotted lines) and 0.5% (circles, solid lines). The cells have been processed identically on 1.0 Ωcm p-type material, the rear contact was formed by aluminum alloying. The simulation parameters were L = 650 µm, S_{ox} = 10 cm/s, S_m= ∞, ρ_b = 1.0 Ωcm, ρ_c = 3x10^{-4} Ωcm^2.

Fig. 2 shows the increase of V_{oc} and I_{sc} with increasing point spacing, due to the decrea-sing influence of the recombination at the contacts. For $\Delta x \rightarrow 0$, the simulated V_{oc} and I_{sc} values adopt the 1D calculated limit for $S_{eff} = (1-f_m)S_{ox} + f_m S_m$ at the rear side. Because the injection level at the rear side decreases for smaller point spacing, the surface recombi-nation velocity at the oxide increases [4]. This effect is not incorporated in the present simulation model, thus the measured I_{sc} values for the small point spacing are lower than the calculated ones.

The decrease of the fill factor with larger point spacing and smaller point size is due to the increasing series resistance. The contrary behavior of V_{oc}, I_{sc} and FF results in an maximum for η. The width, the position Δx and the height of this maximum depend strongly on the material parameters and the metallization fraction.

3. Characterization

Recombination parameters at different steps of solar cell processing are conveniently deter-mined by contactless nondestructive measurements (e.g. IR-absorption [5], microwave re-flection, photoconductivity decay). Up to now simplifying analytical models have been used for extracting parameters, but numerical models (Finite Differences method) are beco-ming more and more important. This is due to their flexibility in handling more complex structures (doping profiles, inhomogeneities).

We concentrate on lifetime experiments using sine amplitude modulated light (in contrast to light pulses) to generate carriers. For a simple model with a homogeneous bulk lifetime τ_b and two surface recombination velocities S_{front} and S_{back} it is easy to extract analytically the recombination parameters from the phase shift and the frequency dependent amplitude of the integrated carrier density, measured by a transmitted IR-beam or MW-reflection.

In Fig. 3 the electron density in a 200 µm thick, uniformly doped p-type wafer, calculated by a time resolved numerical method, for three modulation periods of two different frequencies is shown. Obviously for the higher frequency (100 kHz) the phase shift is higher and the amplitude of the electron density is smaller. Using light of short penetration

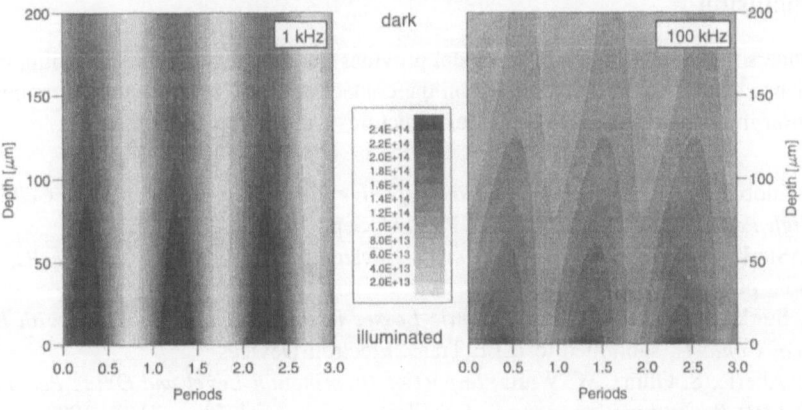

Figure 3: Density of electrons in a uniformly doped p-type wafer (200µm) illu-minated with modulated light of short penetration depth ($\alpha^{-1} = 1\mu m$)

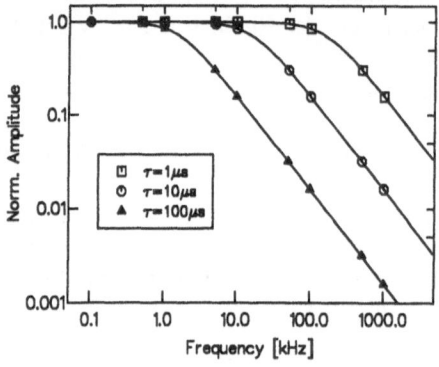
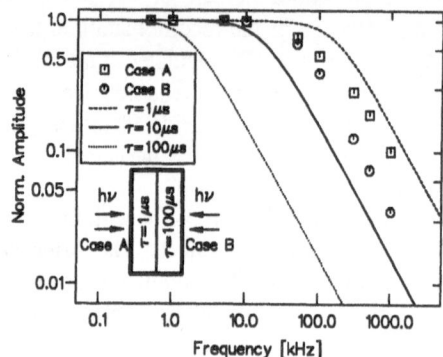

Figure 4: Comparision of analytical (solid lines) and numerical calculation.

Figure 5: Two layer structure with different lifetimes.

depth (α^{-1}=1µm), the carrier modulation for the high frequency takes place mainly under the illuminated surface, due to the fact that the modulation is faster than the carrier transfer by diffusion. Thus, the surface and bulk recombination can be separated.

The calculated numerical values for this simple structure are in perfect agreement with the analytical curves, which is demonstrated in Fig. 4 for three different τ_b values.

However more complex structures such as a two-layer system with two different τ_b values (e.g., thin film solar cells, where an epitaxial layer with a high lifetime is grown on a low quality substrate) cannot be described easily by an analytical model. The numerical calculation of this structure for both illumination directions in Fig. 5 shows that no simple model for an average τ_b can be used. For both cases the shape of the amplitude follows neither the behaviour of the geometrical average $\tau = 10$µs nor of the arithmetical average $\tau = 55$µs. Thus, for this structure only the numerical calculation is useful to extract the recombination parameters from the experimental values.

4. Conclusion

3D device simulation with a simple model provides good qualitative understanding of the dependence of the solar cell efficiency on the contact design. The advantage of numerical evaluation in characterization of complex structures has also been outlined.

[1] J. Knobloch, A. Aberle, B. Voss, *Cost Effective Processes for Silicon Solar Cells with High Performance*, Proc. 9th E.C. PVSEC, Sept. 1989, p.777
[2] S. Sterk, *Optimization of the rear contact design of high efficiency solar cells*, to be published
[3] S. Sterk, S. Glunz, W. Warta, *Ohmic Losses in the Base of Solar Cells with Local Rear Contacts*, submitted to IEEE Trans. Electron Devices
[4] A. Aberle, S. Glunz, W. Warta, *Impact of Illumination Level and Oxide Parameters on SRH Recombination at the Si-SiO₂ Interface*, J. Appl. Phys. 71(9), 1992, p. 4422
[5] F. Sanii, R. J. Schwartz, *The Measurement of Bulk and Surface Recombination by Means of Modulated Free Carrier Absorption*, Proc. 20th IEEE PVSC, 1988, p. 575

Modeling of Breakdown in SOI MOSFETs

G. A. Armstrong and W. D. French

Institute of Advanced Microelectronics, Queen's University
Belfast, BT9 5AH, UNITED KINGDOM

Abstract

This paper describes how a modified version of the two-dimensional device simulator MINIMOS4 has been used to simulate the dependence of the breakdown voltage in an SOI transistor on key device parameters. The strong influence of the gain of the parasitic lateral bipolar transistor is discussed.

1. Introduction

Silicon-on-insulator (SOI) MOSFETs suffer from low operating voltages as a result of an enhanced breakdown due to the lateral bipolar transistor[1]. Breakdown occurs when the product of the bipolar current gain β and avalanche multiplication rate M-1 tends to unity. As the drain voltage is increased, accumulation of positive charge in the film causes the threshold voltage to reduce, progressively shifting the characteristics in the direction of lower gate voltage eventually leading to hysteresis[2]. By simulating these effects, it is possible to define a very precise holding (or breakdown) voltage as that value of drain voltage at which the transistor will just be able to turn off, when swept from positive to negative gate voltage, as shown in Fig.1 for $V_D = 5.3V$. This paper describes how simulation has been used to both quantify and explain the dependence of breakdown voltage on key parameters such as gate length, oxide thickness and SOI film doping.

2. Simulation details

The simulations of breakdown have been carried out using a modified version of MINIMOS4[3], which has been adapted to include bandgap narrowing and a non-local 'lucky electron' impact ionisation model[4]. This model, which also incorporates an energy dependent electron mean free path requires no fitting parameters and yet provides accurate estimates of substrate current for sub-micron bulk silicon transistors, both with and without a lightly doped drain[5]. By evaluating the individual components of current flow, it is possible to use the simulator to derive an effective value for the equivalent bipolar current gain[6], an extremely useful parameter in interpreting the contribution of minority carrier injection at the source junction in lowering the breakdown voltage. A fixed value of carrier lifetime of 0.1 microseconds was used in all simulations.

An initial validation of the accuracy of the simulator is illustrated in Fig.2, which compares measured and simulated bipolar holding voltage as a function of gate length. A reduction in holding voltage occurs because of an increase in bipolar current gain due to reduced base width (gate length). The effect of reduced gate length on bipolar current gain, (calculated at $V_D = V_h$), is shown in Fig.3, along with the corresponding *reduction* in lateral electric field. This reduction in field at $V_D = V_h$ is a consequence of the condition that the product of current gain and ionisation rate tends to unity at the onset of snapback, so that snapback is triggered at a lower field for shorter gate lengths.

The simulated dependence of holding voltage on SOI film thickness is shown in Fig.4, for two gate lengths. As the film thickness is reduced, current crowding causes a reduction in bipolar current gain, as shown in Fig.5. Despite this reduction in gain, the corresponding increase in the lateral electric field however has a dominant effect on the impact ionisation rate, due to its non-linear dependence on electric field.

Fig.6 shows the effect of reducing gate length for three different film thicknesses. In each case the film doping was chosen to give a threshold voltage of 0.6V. Clearly for short gate lengths, the three graphs tend to converge. The reduction in the relative change in holding voltage as the gate length is reduced from $2\mu m$ to $0.5\mu m$ is a result of body charging effects. In the thick film device, the film is partially depleted, so that all charge accumulates within the film and as the gate length reduces, the increase in the electric field has its full effect. However, for the thinnest film of 400Å the film is fully depleted, so that the removal of the potential barrier at the source results in a removal of charge from the film. Therefore, the increasing lateral drain electric field has a more limited effect due to a reduction in body charging.

The simulated dependence of holding voltage on film doping is shown in Fig.7, for an SOI transistor, both with and without a lightly doped drain. Superimposed on this plot is the corresponding measured variation (including an LDD), taken from the published literature[7]. The increase in holding voltage for higher doping is caused by a combination of lower current gain and higher lateral electric field at the drain junction. Negative values of doping correspond to an accumulation mode transistor. Fig.8 shows the expected decrease in bipolar current gain with increased film doping, in agreement with standard bipolar theory. At low doping, high bipolar gain dominates, whilst for high doping, reduction in bipolar gain is compensated by increase in impact ionisation due to higher electric field, so that eventually the holding voltage becomes independent of film doping when only partial as opposed to full depletion of the SOI film occurs.

The pattern shown for the ultra-thin film transistor of Fig.7 differs when the same parameters are used but the film thickness is increased. Fig.9 compares the dependence of thick and thin film transistors, with no lightly doped drain, on film doping. Clearly for the thicker film a clear optimum doping exists to maximise the holding voltage, defined by the boundary between full and partial depletion. An explanation of this dependence may be inferred from Fig.10. Clearly the current gain is more sensitive to film doping for the transistor with the thicker film. In this case, as the film doping is increased from 0 to $3x10^{16}cm^{-3}$, the current gain is reduced by almost two orders of magnitude and this reduction predominates over the increase in electric field, resulting in an increase in holding voltage. For larger values of doping, the current gain reduces much less rapidly, whilst the electric field continues to increase, so that the holding voltage is lowered. In the thin film case however, because the current gain is much less dependent upon the film doping it is compensated much more by the electric field.

3. Conclusions

Simulation has shown that both reduced gate length and film thickness, result in lower breakdown voltage in an SOI transistor. In the former case, both the current gain and the lateral electric field are increased. In the latter case, however, the dependence of holding voltage on film thickness depends on the relative effects of lower bipolar gain and higher electric field with reduction in film thickness. In fully depleted sub-micron gate transistors the high film doping associated with enhancement mode transistors offers the possibility of low threshold voltage, steep subthreshold slope and high holding voltage.

References
1. J.P.Colinge, *Silicon-on-Insulator Technology*, Kluwer Academic Publishers, 1991.
2. G.A.Armstrong et al, IEEE Trans Electron Devices ED-38,1991, pp.328-336.
3. S.Selberherr, IEEE Trans Electron Devices ED-36, 1989, pp.1464-1472.
4. T.Thurgate and N.Chan, IEEE Trans Electron Devices ED-32, 1985, pp.400.
5. G.A.Armstrong and W.D.French, Solid State Electronics, 35, 1992,pp.1761-1770.
6. J.Choi and J.G.Fossum, IEEE Trans Electron Devices, ED-38, 1991, pp.1384-1392.
7. Y.Yamaguchi et al, IEDM Tech. Dig., 1990, pp.591-594.

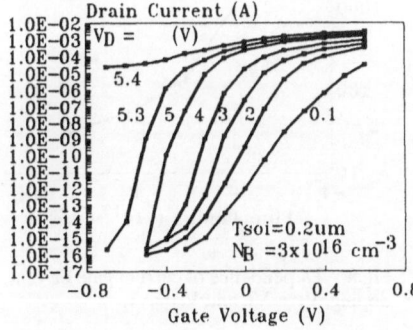

Fig.1 Variation of drain current with gate voltage for a 1μm SOI transistor, $T_{OX} = 27$nm.

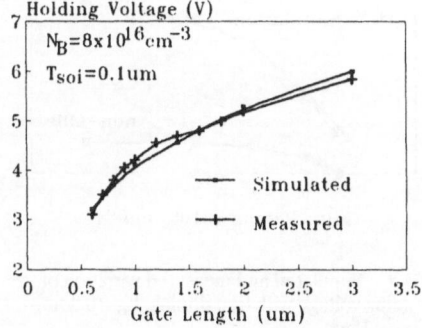

Fig.2 Dependence of holding voltage on gate length for SOI transistors, $T_{OX} = 20$nm.

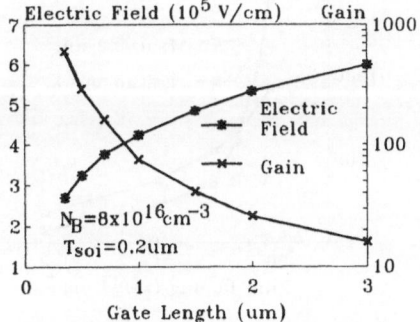

Fig.3 Variation of lateral electric field and current gain, for $V_D = V_h$, $V_G = V_t$.

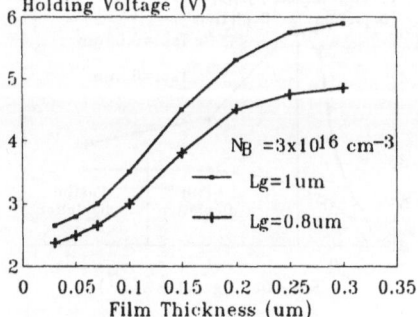

Fig.4 Dependence of holding voltage on film thickness and gate length.

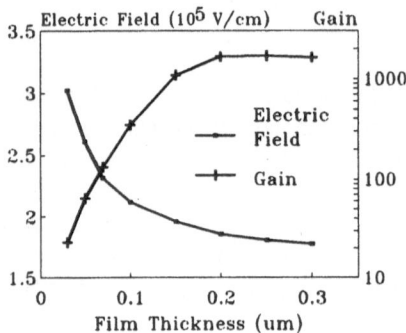

Fig.5 Variation of lateral electric field and current gain at $V_D = 2.5V$ and $V_G = Vt$, for a $1\mu m$ SOI transistor.

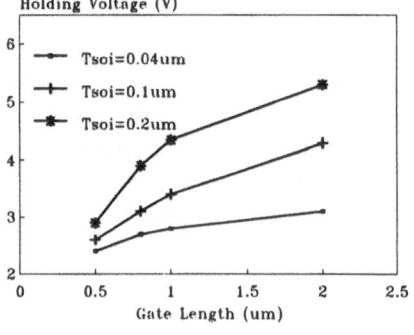

Fig.6 Variation of holding voltage with gate length and film thickness.

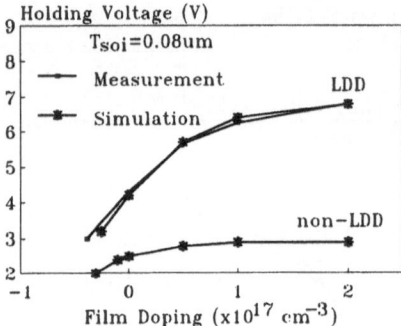

Fig.7 Simulated and measured variation of holding voltage with film doping, $Lg = 0.8\mu m$, $T_{OX} = 15nm$.

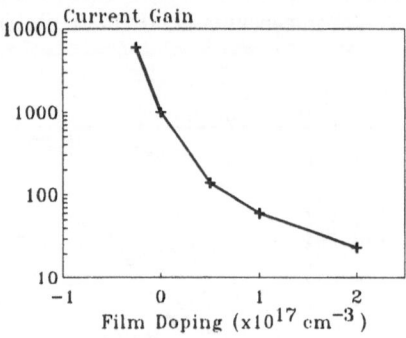

Fig.8 Dependence of current gain on film doping, $V_D = Vh$ and $V_G = Vt$.

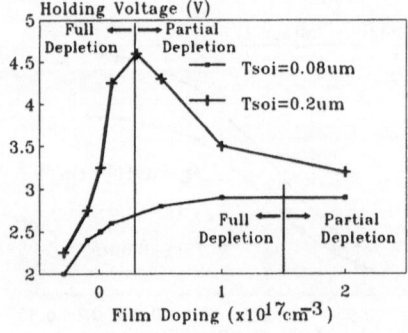

Fig.9 Dependence of holding voltage on film doping.

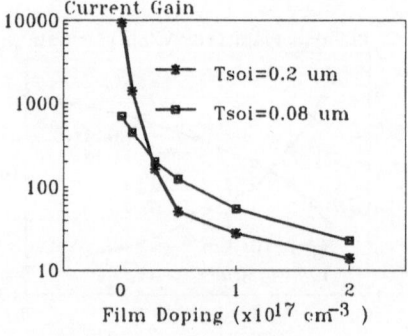

Fig.10 Variation of current gain with film doping, at $V_D = Vh$ and $V_G = Vt$.

3D Simulation of MOS Transistors with Inversion Condition in Two Directions

G. Punz

PSE, Siemens AG Österreich
Gudrunstraße 11, A-1101 Wien, AUSTRIA

Abstract

The 3D device simulator MINIMOS has been extended for simulation of complex oxide structures, allowing inversion condition in two dimensions. The results for a test structure with a threefold bent gate oxide show enhanced channel formation in regions where the inversion condition is fulfilled in two directions.

1. Introduction

Full 3D simulation is a necessity in order to describe the effects of e.g. narrow channel width or field implants for realistic MOSFET structures. Compared to the heavy numerical and computational effort of general–purpose simulators [1], [2] specialized codes for MOSFET's are more efficient. In MINIMOS [3] the simulation is performed in steps, seeking first a full solution for a 2D cut plane which serves as an initial guess for the 3D problem, and introducing a quasi–2D mode (during which only Poisson's equation is solved fully threedimensionally, whereas the solution of the carrier continuity equations are estimated assuming vanishing currents in the third dimension) to benefit further from the computationally cheaper 2D solution. However, this restricts the allowed 3D geometries to be smoothly generable from the 2D geometry and allows only modest variation of the solution with respect to the third coordinate.

In this paper simulations of 3D MOS structures with complex geometry features with respect to the width direction are presented. MINIMOS has been extended to allow flexible oxide body specification by polygons (see sketch of a quarter transistor in fig. 1). A heuristic scheme for the setup of an appropriate 3D initial solution from the 2D solution is introduced. The applicability of our method is demonstrated for an extreme test example.

2. Method

The first step is to adapt the grid generation algorithm – also for 2D – for a redefinition of the simulation area where carrier continuity equations are solved by shifting the corresponding milestone up to the overall maximum of the lower oxide contour. (Like for 2D nonplanarities, MINIMOS masks all points assigned to boxes fully in oxide for the continuity equations).

Next the 3D initial distributions have to be specified. In plain MINIMOS the initial potential (ψ) and carrier distributions (n, p) are taken from the 2D simulation plane and continued with constant values along the width direction. This is sufficient for oxide bodies with a monotonous variation from the thin gate oxide to the field oxide. With our complex structures, however, this is not well defined for semiconductor regions higher than the oxide contour in the 2D simulation plane (region A in fig. 2). The possibility to solely extend the distributions from the well defined region B upwards suffices in the case of only moderate parameters h as compared to the inversion layer depth d, but leaves too many carriers in subregion C of B and is not convergent for $d \sim h$. Therefore a heuristic transformation is applied to extend the distributions in z–direction according to the upper and lower oxide contours $f^u(x, z), f^l(x, z)$ (z_c is a cutoff for the channel width estimated from the oxide thickness, for meaning of other symbols see figures 1 and 2):

$$\psi(x, y, z) = \psi(x, y_{ref}, z_0)$$
$$n(x, y, z) = n\,(x, y_{ref}, z_0) \cdot \Theta(z_c - z) \qquad p(x, y, z) = p\,(x, y_{ref}, z_0) \cdot \Theta(z_c - z) \qquad (1)$$

$$y_{ref} = \begin{cases} \left(y - f^l(x, z)\right) \cdot \dfrac{y_{max} - f^l(x, z_0)}{y_{max} - f^l(x, z)} + f^l(x, z_0) & \text{for } y > f^l(x, z) \\[2ex] \left(y - f^u(x, z)\right) \cdot \dfrac{f^l(x, z_0) - f^u(x, z_0)}{f^l(x, z) - f^u(x, z)} + f^u(x, z_0) & \text{for } f^u(x, z) < y < f^l(x, z) \quad (2) \\[2ex] \left(y - y_{min}\right) \cdot \dfrac{f^l(x, z_0) - y_{min}}{f^l(x, z) - y_{min}} + y_{min} & \text{for } y < f^u(x, z) \end{cases}$$

For the carrier concentrations only the first line of eq. (2) is relevant.

This scheme does not claim to be based on sound physical arguments, but is extremely practical for two reasons: (i) it is well defined for the whole MOS transistor and is applicable for quite general oxide contours (ii) looking at yz–cuts within the channel region, the resulting initial potential and carrier distributions match the oxide boundaries well and are sufficiently close to to their final solutions, which is crucial for reaching convergence. (For the regions near source and drain more sophisticated initial potential distributions might be found.)

Considering the 3D solution hierarchy for MINIMOS we notice that a quasi–2D mode is feasible for h small compared to d, but fails – for the same, abovementioned reasons – in the general case of highly nonplanar oxide bodies. In this case Poisson's and continuity equations for both carrier types are solved fully 3D from the very start.

For most flexible applications we combine gate oxide nonplanarities within the 2D simulation plane with complex oxide geometries within the cut normal to it in the middle of the channel. A user interface allowing oxide specification by two (upper, lower) polygons $\left(x_i, f_x^{u,l}(x_i)\right)$ and $\left(z_i, f_z^{u,l}(z_i)\right)$ is already implemented in our version of MINIMOS. Two algorithms have to be specified: (i) for 3D doping profile generation from 2D cuts and (ii) for matching and extending oxide contours from 2D cuts to full 3D oxide surfaces. For the second task e.g. we use

$$f^l(x, z) = f_x^l(x) + \left(f_z^l(z_{max}) - f_x^l(x)\right) \cdot \frac{f_z^l(z) - f_z^l(z_0)}{f_z^l(z_{max}) - f_z^l(z_0)} \qquad (3)$$

for the lower contour, assuming the oxide boundary to be lowest at z_{max} (and an analogous formula for the the upper contour). Ideally – using reliable full 3D process simulation – the effort of constructing oxide contours from 2D cuts should be obsolete.

3. Results

To demonstrate the capability of the method a rather drastic geometry is chosen as a test example (1 μm transistor at $U_G = 1V, U_D = 3V$ with a threefold bent gate oxide along the width direction combined with a nonplanarity in length direction, a yz–cut in the middle of the channel is seen in fig. 3). Figs. 4 and 5 show the solutions for potential and minority carrier concentrations in the region where the horizontal and vertical inversion interact (a cut plane intersecting on the source side of the gate is shown). The resulting enhancement of the channel is clearly seen. Table 1 lists the drain current I_D as a function of the "horizontal" gate oxide thickness t_h.

$t_h[nm]$	80	100	120	140	160
$I_D[mA]$	1.79	1.59	1.43	1.31	1.15

Table 1

CPU time is about one hour on an HP 900-750 for a 59·66·60 grid.

4. Conclusion

An enhancement of the 3D device simulator MINIMOS for the efficient simulation of MOS transistors with complex oxide structures has been implemented, allowing routine analysis and assessment of their usefulness in VLSI design. Effects stemming from inversion conditions in two directions may now be investigated in detail.

References

[1] P. Ciampolini, A. Pierantoni, G. Baccarani: "Efficient 3–D Simulation of Complex Structures", IEEE Trans. CAD, Vol. 10, Nr. 9, Sep. 1991, P. 1141

[2] G. Heiser, C. Pomerell, J. Weis, W. Fichtner: "Three-Dimensional Numerical Semiconductor Device Simulation: Algorithms, Architectures, Results", IEEE Trans. CAD, Vol. 10, Nr. 10, Oct. 1991, P. 1218

[3] M. Thurner, P. Lindorfer, S. Selberherr: "Numerical treatment of Nonrectangular Field–Oxide for 3D MOSFET Simulation", Proc. of SISDEP 3, 1988, p. 375

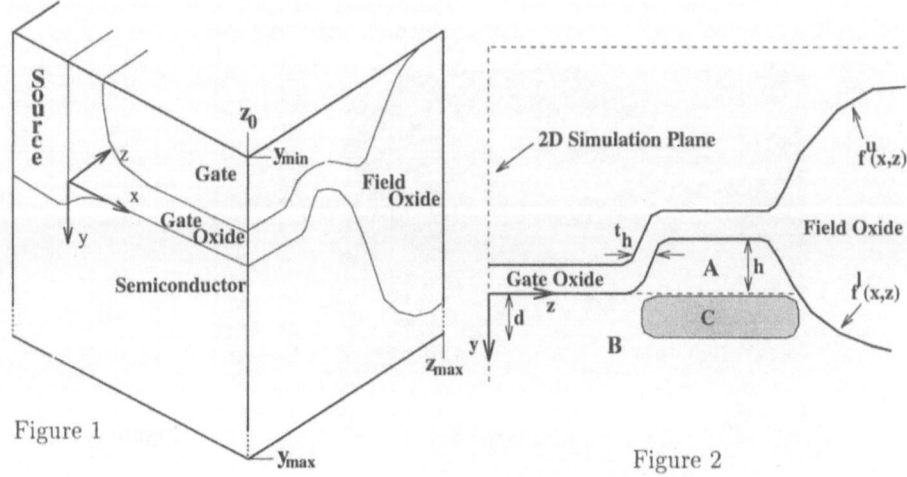

Figure 1

Figure 2

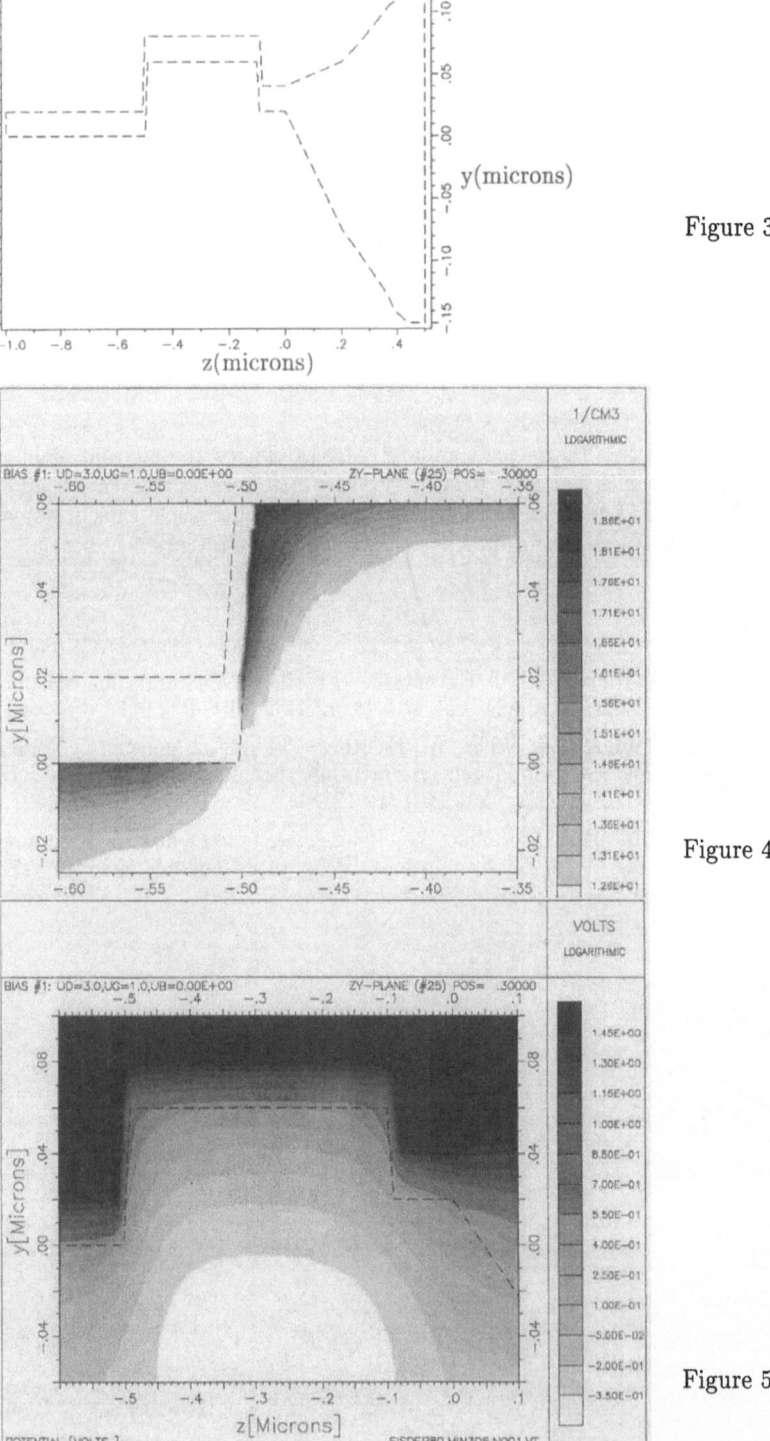

Figure 3

Figure 4

Figure 5

SIMULATION OF SEMICONDUCTOR DEVICES AND PROCESSES *Vol. 5*
Edited by S. Selberherr, H. Stippel, E. Strasser – September 1993

Quasi Two-Dimensional Numerical Simulation of SiGe/Si MOSFETs

T. Manku and A. Nathan

Electrical and Computer Engineering, University of Waterloo
Waterloo, Ontario, N2L 3G1, CANADA

Abstract

Results of the charge distribution and terminal behaviour of Si/SiGe/Si p-channel devices are presented for various gate voltages and device geometries. The results are based on a quasi two-dimensional numerical model where Poisson's equation is solved for the electric potential in the longitudinal direction with the subsequent lateral current transport obtained using an integral method.

1. Introduction

The differing lattice constants between the epitaxially grown SiGe alloy and the underlying Si substrate leads to an inherent strain in the alloy which significantly enhances the optical and electronic properties of the material [1]. The in-plane compressive stress in the $Si_{1-x}Ge_x$ layer results in a considerable reduction of the energy gap [2] making the alloy very attractive for the fabrication of high performance transistors and optoelectronic devices. This includes heterojunction bipolar transistors [3], MODFETs [4], and photodetectors for 1.3 μm wavelength [5].

More recently, there has been growing interest on SiGe MOSFETs for purposes of achieving higher performance p-channel transistors [6]. They are based on the Si/SiGe/Si structure where the active region is within the SiGe layer which is separated from the gate oxide by a thin cap (or spacer) layer. These devices have the advantage of higher hole mobilities [7] and furthermore, they do not suffer from mobility degradation due to surface scattering. However, one main problem that still remains with these devices is the gate-induced inversion layer at the oxide/Si interface which reduces the performance at moderate gate bias.

In this paper, we present modelling results of charge distribution and current-voltage characteristics of p-channel Si/SiGe/Si MOSFETs based on a quasi 2-D model. At very low gate voltages, conduction predominantly takes place in the SiGe layer where the hole mobility is inherently higher thus enhancing the device transconductance. However, with increasing gate bias, an inversion layer is induced at the oxide/Si interface; the conduction cross over point occurs at a lower gate voltage for larger cap layer thickness.

2. Modeling Equations

The device cross section and corresponding energy band diagram is shown in Fig. 1. The device comprises of a strained n-SiGe layer pseudomorphically grown on a (001) n-Si substrate followed by a n-Si cap (or spacer) layer passivated by a low temperature oxide. With a negative potential on the gate electrode, the valence band moves upwards causing an accumulation of holes at the Si/SiGe interface (see Fig. 1). With further increase in the gate voltage, a hole inversion layer is created at the oxide/Si interface. Assuming that the electron Fermi potential is uniform from the bulk to the oxide/Si interface, and the hole Fermi potential is displaced by the applied voltage V, the longitudinal potential distribution is obtained by solving Poisson's equation [8],

$$d^2\psi/dx^2 = -(q/\varepsilon_s)\,[(p_{no}e^{q(V-\psi)/kT} - 1) - (n_{no}e^{q\psi/kT} - 1)]. \qquad (1)$$

in one-dimension. Here, Boltzmann's statistics is assumed for the mobile charge density. In (1), ε_s denotes the semiconductor permittivity, and n_{no} and p_{no} are the equilibrium densities of electrons and holes, respectively. The majority carrier concentration, n_{no} is taken to be equal to $N_D \sim 10^{16}\,cm^{-3}$ for all regions. The minority carrier concentration, p_{no} is given by n_{ie}^2/n_{no}, where n_{ie} is the intrinsic carrier concentration which is position-dependent due to bandgap variations. The resulting p_{no} may lead to error in view of possible discontinuity in the conduction band. But the discontinuity is significantly smaller than ΔE_v and hence the relative error is expected to be small. The intrinsic carrier concentration in the SiGe layer is modelled as

$$n_{ie,SiGe} = (n_{ie,Si}/f_c f_v)\exp\,[q\Delta E_v/2kT], \qquad (2)$$

where $n_{ie,Si}$ is the effective intrinsic concentration in Si, f_c accounts for the strained induced energy shifts in the conduction band and f_v takes into account the strained induced distortions and sub-band splitting in the valence band. For 30% Ge fraction, their values are $\sqrt{(3/2)}$ and 2, respectively (see [9] and references therein).

Equation (1) is solved within the semiconductor regions subject to the following boundary conditions. Deep in the substrate, the electric potential is assumed zero. At the oxide/Si interface, the potential is prescribed by

$$\psi_s = V_G - t_{ox}\,(d\psi/dn), \qquad (3)$$

where V_G denotes the gate voltage, t_{ox} is thickness of the oxide, and n is normal to the interface. The discretization of (1) was based on a finite difference scheme. The permittivity was assumed to be uniform; the variation in the permittivity due to Ge content was small and its effect on the potential distribution is negligible. The oxide thickness was assumed to be 100 Å. The solution to eqns. (1)-(3) yield the electric potential which is tabularized for various voltages V, to be used in subsequent numerical computations of the charge distribution and terminal behaviour, viz.,

$$\mathcal{Q}_{p,Si} = q\,p_{no}\,\int_{\ell_{Si}} e^{q(V-\psi)/kT}\,dx \qquad (4)$$

$$\mathcal{Q}_{p,SiGe} = q\,p_{no}\,\int_{\ell_{SiGe}} e^{q(V-\psi)/kT}\,dx \qquad (5)$$

$$I_{DS} = (qW/L)\,\int_{V_{DS}}[\int_\ell \mu_p p_{no}\,e^{q(V-\psi)/kT}\,dx]\,dV. \qquad (6)$$

Here, ℓ_{Si} and ℓ_{SiGe} denote the thickness of the Si and SiGe layers, respectively, and ℓ being the total layer thickness, and μ_p is the hole mobility which is taken to be electric field dependent.

3. Results and Discussion

The hole charge concentration at the oxide/Si interface, $\mathcal{Q}_{p,Si}$ and Si/SiGe interface $\mathcal{Q}_{p,SiGe}$, computed for various gate voltages is illustrated in Fig. 2 for two spacer layer thicknesses. The charge $\mathcal{Q}_{p,SiGe}$ is seen to be larger than that of $\mathcal{Q}_{p,Si}$ only at small gate voltages, $|V_G|$. As the gate voltage is increased, the inversion layer charge starts to predominate; the behaviour being consistent with experimental observations reported in [6]. The conduction cross over point, at which the charge in the Si and SiGe layers become equal, occurs for low values of $|V_G|$ for thicker spacer layers (see Fig. 3). The thickness of capping layer used is 10.8 nm. The dependence of drain current on gate voltage is illustrated in Fig. 4 for the SiGe MOSFET and an equivalent surface inversion MOSFET serving as a control. Also shown are the measured results of Garone et al. [10]. The SiGe MOSFET clearly exhibits enhanced transconductance due to the higher hole mobility in the strained layer as well as due to reduced surface scattering. The discrepancy between simulations and measurement results [10] at low $|V_G|$, where conduction predominantly takes place at the Si/SiGe interface, can be attributed to the possible presence of dislocations at the interface which cause the mobility to decrease due to additional scattering. The simulated current-voltage characteristics or both devices is shown in Fig. 5 for different gate voltages. Despite the high concentration of surface layer charge, the contribution of the higher mobility inversion layer charge at the Si/SiGe interface is significant thus yielding enhanced terminal behaviour.

4. Conclusions

In this paper, we have presented the charge distribution at the oxide/Si and Si/SiGe interfaces for various gate voltages and cap layer thickness. A significantly enhanced performance has been obtained with the SiGe MOSFET due to the high hole mobility in the strained layer and due to reduced surface scattering.

References

[1]   T. P. Pearsall, , in CRC Critical Reviews in Solid State and Material Sciences, 15 (1989) 551-600.
[2]   R. People, Phys. Rev., 32 (1985) 1405-1408.
[3]   S. S. Iyer, G. L. Patton, J. M. C. Stork, B. S. Meyerson, D. L. Harame, IEEE Trans. Electron Devices, ED-36 (1989) 2043-2061.
[4]   R. People, J. C. Bean, D. V. Lang, A. M. Sergennt, H. L. Stormer, K. W. Wecht, R. T. Lynch, K. Balkwing, Appl. Phys. Lett., 45 (1984) 1231-1334.
[5]   H. Temkin, T. P. Pearsall, J. C. Bean, R. A. Logan, S. Luryi, Appl. Phys. Lett., 58 (1986).
[6]   P. M. Garone, V. Venkataraman, and J. C. Sturm, IEEE Electron Device Letts. 12 (1991) 230-232.
[7]   T. Manku and A. Nathan, IEEE Electron Device Letts. 12 (1991) 704-706.
[8]   S. M. Sze, Physics of Semiconductor Devices, 2nd Ed., Wiley:New York, 1981.
[9]   T. Manku and A. Nathan, J. Appl. Phys. 73 (1993) 1205-1213.
[10]   P. M. Garone, V. Venkataraman, and J. C. Sturm, IEEE Electron Device Letts. 13 (1992) 56-58.

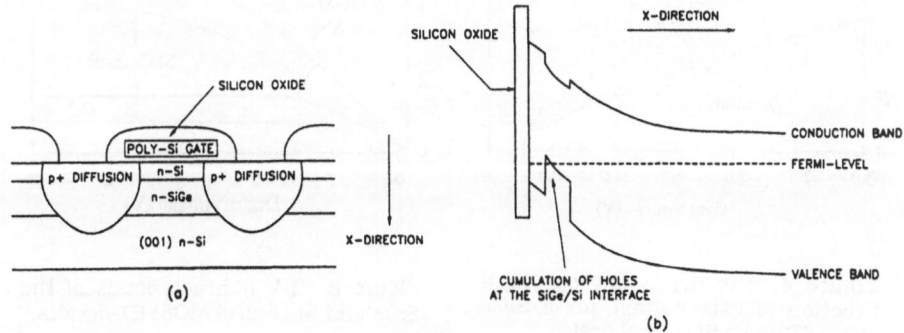

Figure 1: (a) Cross section of SiGe MOSFET and (b) its band structure.

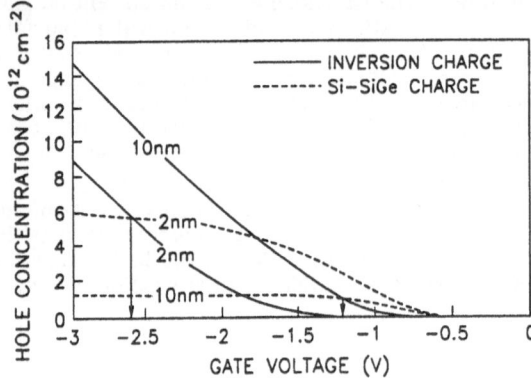

Figure 2: The oxide/Si and Si/SiGe inversion layer charge density as a function of gate voltage for different spacer layers. The vertical arrows denote cross over points, $Q_{p,Si} = Q_{p,SiGe}$.

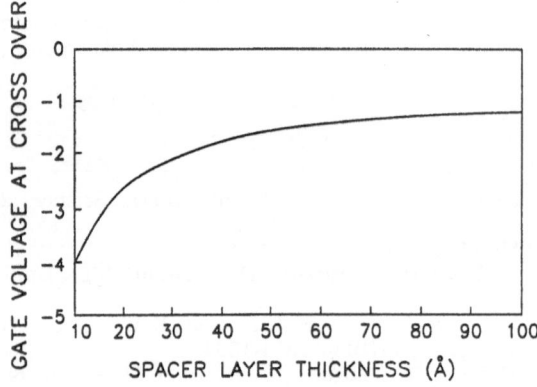

Figure 3: The gate voltage at which cross over occurs as a function of spacer layer thickness.

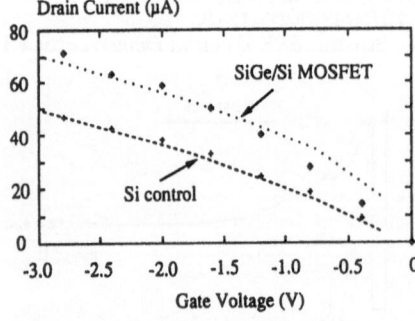

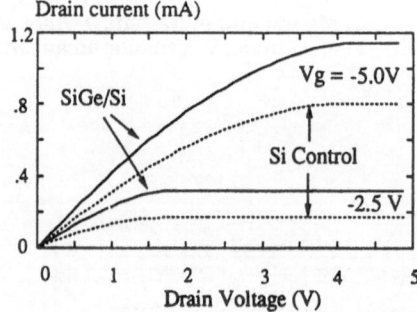

Figure 4: The drain current as a function of gate voltage for a SiGe MOSFET and a Si control device. $V_{DS} = 0.2\,V$.

Figure 5: I-V characteristics of the SiGe and Si-control MOSFET devices.

Large-Signal RF and DC Performance of p-Type Diamond FETs

M. W. Shin, T. A. Winsloow, G. L. Bilbro, and R. J. Trew

High Frequency Electronics Laboratory,
Department of Electrical & Computer Engineering, North Carolina State University
Raleigh, NC 27695, USA

Abstract

Large signal RF and DC simulation of diamond FETs using incomplete dopant ionization statistics is reported for the first time. The simulation was performed using a harmonic balance circuit simulator which uses the two-dimensional simulator, PISCES-IIB, to investigate the high temperature performance of FET devices. Simulations indicate the DC and RF performance of diamond MESFET is improved as temperature is increased in a temperature range of 573 K to 923 K. At 923 K the simulated diamond MESFET produces about 0.5 W/mm for an operating frequency of 3 GHz and 29% power-added efficiency (PAE) at 19 dBm input power. Due to the high activation energy of dopant in diamond the DC and RF performance of device is found out to be dominated by the carrier ionization process.

1. Introduction

The properties of semiconducting diamond make it an excellent material for high-frequency, high-power and high-temperature device operation. This is due to diamond's high saturation velocity, wide band-gap and high thermal conductivity [1]. Despite the excellent properties, however, a present disadvantage of diamond is the low ionization probability of carriers due to high activation energies. For this reason, the use of an incomplete dopant ionization model in simulation is important for determining the high temperature performance of diamond devices. For verification, the simulated DC performance of diamond IGFET is presented and compared with experimental results. Several physical models including the incomplete ionization model are employed in diamond device simulation. The temperature dependence of diamond MESFET DC and RF performance is investigated in the temperature range of 573 K to 923 K.

2. Simulation Experiment

The RF simulations are performed by a harmonic balance circuit simulator [2] which utilizes the accuracy of the two-dimensional numerical device simulator, PISCES-IIB, to simulate MESFETs of various geometries and materials [3]. The harmonic balance simulator extracts the FET gate and drain current and capacitance characteristics

MATERIAL PARAMETERS	VALUE
Activation energy of Dopant (eV)	0.34
Low field mobility $(cm^2 V^{-1} s^{-1})$	30
Energy gap (eV)	5.45
Permittivity	5.7
Saturation velocity of holes (cm/s)	2×10^7
Valence band density of states (300 K)	1.8×10^{19}
Life time of holes (s)	1×10^{-9}

Table 1: Material parameters used in the simulation.

and then performs large and small signal analysis. Several physical models are employed in the simulations. For example, the mobility of the carriers depends on both the electric field and temperature, and the carrier velocity saturates at high fields. The effect of temperature on free carrier activation is modeled using Fermi-Dirac statistics for incomplete ionization. Several parameters such as intrinsic carrier concentration, density of states, and band gap energy exhibit temperature dependence and are included in the simulation. The harmonic balance simulator utilizes a table based look-up approach (which describes the device) to increase the overall efficiency of the simulator when performing DC and RF simulations as well as optimizations and yield analysis. The semiconductor devices in this work are characterized using PISCES-IIB. DC-IV and contact capacitances are extracted and formed into a two-dimensional look-up table which spans the typical drain and gate operational voltage range. The look-up table is then automatically incorporated into the harmonic balance simulator. The harmonic balance simulator has demonstrated excellent accuracy in predicting the RF performance of a variety of industrial devices fabricated from GaAs [2].

3. DC simulation of diamond IGFET

For verification, the DC simulation of diamond IGFET is performed using the device structure fabricated by Hewett et. al. [4]. The IGFET was fabricated in a concentric ring structure. The diameter of the circular gate was accounted for to properly compare the measured and the simulated drain current. The channel depth and channel doping concentration are 0.18 μm and $1.1 \times 10^{17} cm^{-3}$, respectively. The oxide thickness on top of the channel is 0.1 μm and the workfunction of the gate electrode is 5.1eV. The material parameters used in the simulation are listed in Table I. Figure 1(a) demonstrates an excellent match between the simulated and measured IGFET DC characteristics. Incomplete ionization of acceptor doping was found to severely limit the peak channel current of the device at room temperature. The peak drain current(at 0 gate bias, and 50V_{ds}) is only 45 μA. The input capacitance is found to be approximately 0.3 pF. The gate displacement current was calculated to be approximately 40 mA. Thus, at RF frequencies, the gate displacement current dominates the device channel current by three orders of magnitude and the resulting theoretical RF power performance is very poor.

4. Analysis of diamond MESFET at high temperatures

The material parameters used for the analysis of the diamond MESFET are extracted from the DC simulation of the diamond IGFET previously discussed. The same material parameters are also found to accurately characterize polycrystalline IGFETs [5] with the exception of dopant activation energy. The structure of the simulated diamond MESFET is shown in Figure 1(b). The structure of the MESFET has been designed to be appropriate for device performance at high frequencies. The operation temperatures are 573 K, 773 K, and 923 K, and the resulting mobilities are calculated to be 600, 100, and 83 $cm^2/V - s$, respectively. Figures 2(a)-(c) show the simulated MESFET I-V curves. The drain current at 773 K is slightly higher than that at 573 K due to the increased carrier activation. The gate width is 1 mm and gate voltages vary from 0 to 20 V. The drain current at 923 K shows approximately a 20% increase compared to that at 773 K because of a higher activated carrier concentration, with similar mobilities at both temperatures. At higher temperature, there is an improvement in RF output power as well as gain and power-added efficiency (PAE) as shown in Figures 2(d)-(f). The result of RF simulations clearly show a consistent increase in RF performance with the increase in channel current as the temperature increases. The device was characterized at an operating frequency of 3 GHz with V_{gs} = 7V and at V_{ds} = 30V . The MESFET device produced approximately 0.5 W/mm of RF output power at 923 K. The PAE at this temperature has approximately 29% peak at an input power of 19 dBm. The linear gains are about 12.0, 13.5, and 15.0dB at 573 K, 773 K, and 923 K, respectively. The comparison of the DC and RF performance at different temperatures shows a distinct trade-off between decreased mobility and carrier saturation velocity and the increased activated carrier concentration with increasing temperature. Due to a high value of activation energy for ionization of acceptors (Boron) in diamond, the magnitude of channel current and thus the RF performance are shown to be dominated by the ionization process.

5. Conclusion

This study shows that the DC performance of diamond IGFET can be predicted by employing proper material parameters. The comparison of the DC and RF performance of diamond MESFET at different temperatures indicates a trade-off between decreased mobility and carrier saturation velocity and increased activated carrier concentration with increasing temperature. It is shown that the theoretical RF performance of diamond MESFETs improves with increasing temperature. This is due to a high value of activation energy of dopant ionization.

References

[1] J. E. Field, — Properties of Diamond, Academic Press, London, 1979.

[2] M. A. Khatibzadeh, R. J. Trew, — A large-Signal, Analytic model for the GaAs MESFET, IEEE Trans. Microwave Theory and Techniques, MTT-36, 231-238, 1988.

[3] PISCES-IIB, Stanford Electronics Lab., Stanford University, Stanford, CA.

[4] C. A. Hewett, C. R. Zeisse, R. Nguyen, and J. R. Zeider, — Fabrication of an Insulated Gate Diamond FET for High Temperature Application, First International High Temperature Electronics Conference, 168-173, 1991.

[5] M. W. Shin, G. L. Bilbro, and R. J. Trew, and D. L. Dreifus, A. J. Tessmer, — Variable Temperature Current-Voltage Characteristics of Polycrystalline Diamond FETs: Modeling and Experiment, submitted to Appl. Phys. Lett., 1993.

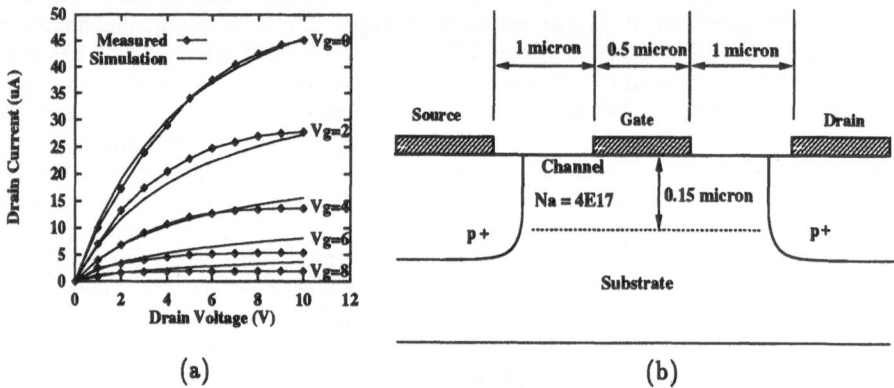

(a)   (b)

Figure 1: (a)Comparison between the measured [4] and theoretical I-V characteristics for a diamond IGFET. (b)Cross sectional sketch of the diamond MESFET.

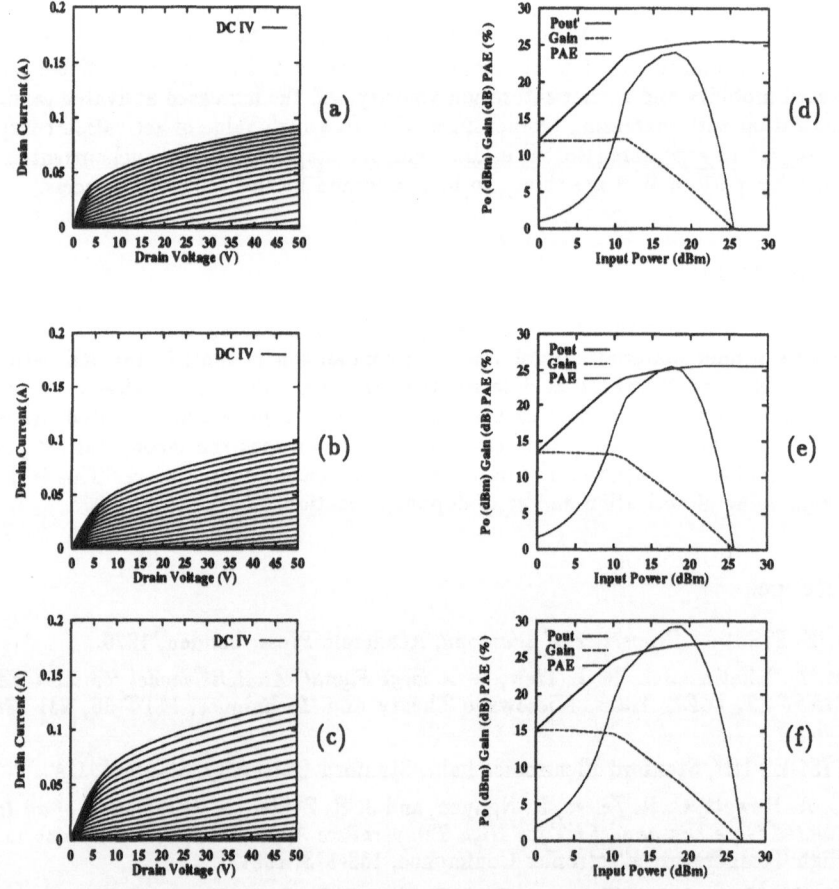

Figure 2: Simulated I-V characteristics for a diamond MESFET at (a) 573 K, (b) 773 K, and (c) 923 K. The gate voltage ranges from 0 to 20V in steps of 2V. Simulated RF (3 GHz) characteristics for a diamond MESFET at (d) 573 K, (e) 773 K, and (f) 923 K.

SIMULATION OF SEMICONDUCTOR DEVICES AND PROCESSES Vol. 5
Edited by S. Selberherr, H. Stippel, E. Strasser – September 1993

3D Grid Generation for Semiconductor Devices Using a Fully Flexible Refinement Approach

N. Hitschfeld and W. Fichtner[†]

Dpto Ciencias da la Computación, Universidad de Chile
CL-2120 Blanco Encalada Santiago, CHILE
[†]Integrated Systems Laboratory, ETH-Zürich
Gloriastraße 35, CH-8092 Zürich, SWITZERLAND

Abstract

We present a new algorithm for the generation of 3-dimensional (3-D) grids for the simulation of semiconductor devices. The fitting of the device geometry and the required mesh density is obtained by partitioning the elements at an optimal point at each refinement step. This allows the fitting of more general 3-D device geometries and the reduction of grid points in comparison with previous grid generators.

1. Introduction

Successful grid generators typically use iterative refinement of coarse elements to obtain a required tessellation. Iterative refinement can be implemented by either bisecting the element edges (*bisection* based approach) or dividing the element edges at an arbitrary position (*intersection* based approach). In 2-D, both approaches have been already used. Algorithms based on the *intersection* based approach generate better grids because they choose the *best* refinement point at each refinement step. To our knowledge, this *intersection* based approach still has not be used in 3-D because of its inherent complexity.

This paper presents the main aspects of a grid generator using an *intersection* based approach and compares it, as far as it has been implemented, with its ancestor Ω_{me}(Ω with mixed elements) [1].

2. Previous work

Currently, Ω_{me} is the most powerful 3-D mesh generator for semiconductor devices to generate grids for box method discretizations. Ω_{me} is implemented using mixed element trees, a generalization of modified octrees[2]. Instead of only bricks, 3-D elements of different shapes are used to fit the device geometry. The main algorithm looks as follows:

1. Generate a grid that fits the geometry of the modeled device exactly. This initial grid consists of bricks, rectangular prisms and rectangular pyramids. The grid is handled as a forest where each element is the root of a tree.

2. Refine due to variations in the impurity profile and certain geometrical parameters. In order to fit physical and other geometrical parameters, an irregular grid is generated by refining each element independently of the others.

3. Generate a proper Delaunay mesh. A finite element grid is obtained after tessellating all the irregular elements into tetrahedra, pyramids, prisms and bricks if the Voronoi cross sections inside of each irregular element can be properly computed.

The most serious problem of Ω_{me} originates from the algorithm that fits the original device geometry. This algorithm generates an initial (tensor product based) grid that is a complete partition of the device (i.e. grid elements have no green points). A complete partition is required in order to be able to use a bisection based approach. During the generation of the initial grid, small geometry features are propagated (by inserting planes) to the boundaries of the device (see Fig. 1). Therefore, the initial grid contains a high number of unnecessary elements with a very bad aspect ratio. In addition, the repetitive generation of new points due to intersections between the inserted lines (planes in 3-D) and some boundary or material interfaces makes it impossible to fit several device geometries (see Fig. 1).

3. New Algorithm

The same consecutive steps are used to generate a grid using an *intersection*-based approach but each step is focused in a different way.

The geometry is fitted by refining the grid elements at the best possible point (see Fig. 2). The best point—the one whose associated refinement generates sons with the smallest aspect ratio—is chosen from the available green points and intersection points. Elements are bisected for generated sons with bad aspect ratio.

After the geometry is completely fitted, the irregular macro grid is further refined until the density requirements are fulfilled. Elements are partitioned in the required direction according to the best located green point.

The grid is made 1-irregular before looking for proper tessellations. Subsequently, the algorithm checks the *splittable* condition, i.e., a condition that guarantees the existence of a proper tessellation. If an element is non splittable, proper points are inserted by looking at the neighbors.

Once all elements are splittable, each local tessellation is computed using an algorithm to compute Delaunay tessellations inside of a convex-hull. We are currently exploring several different possibilities for this algorithm.

4. Comparison of the new algorithm with its ancestor

The algorithms to fit the device geometry and to fulfil the density requirements have been already implemented [3].

Figure 3 shows two views of the silicon part of the same ECL bipolar transistor. It can be noticed that the new algorithm strongly reduces the number of unnecessary

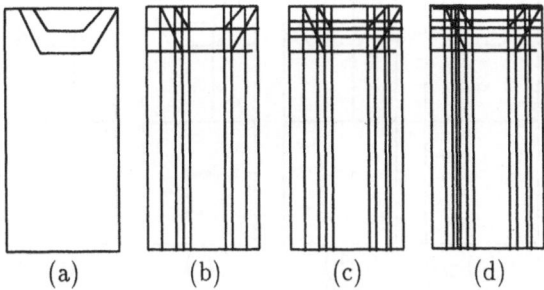

(a)   (b)   (c)   (d)

Figure 1: Fitting a 2-D device geometry using a tensor product grid. (a) Device geometry (b), (c), and (d) sequence of steps to fit the device geometry. After the step in (d) the geometry is still not optimally fitted

grid points and elements. In addition, it avoids the propagation of grid planes to the whole device and small changes in the geometry are fitted locally. Empirically, the new algorithm fits device geometries of complex devices much faster than the old one.

Because of the better fitting of the device geometry, the new algorithm also fulfils the required point density using fewer points (Fig. 4). The rest can be only empirically compared after the new algorithm to generate the final grid is implemented.

The main improvement of the new algorithm is the fitting of more realistic device geometries. In addition, its implementation shows theoretically a more predictable behavior because it does not use loops neither to fit the device geometry nor to look for proper tessellations.

Acknowledgments

This work was partially supported by ESPRIT-6075 (DESSIS) project in 1992 and by FONDECYT project No. 1930765 in 1993.

References

[1] N. Hitschfeld, P. Conti, and W. Fichtner, *Mixed elements trees: A generalization of modified octrees for the generation of meshes for the simulation of complex 3-d semiconductor devices*, Accepted for publication in IEEE Trans. on CAD/ICAS, 1993.

[2] M. A. Yerry and S. Shephard, *Automatic Three-dimensional Mesh Generation by the Modified-Octree Technique*, Int. J. Numer. Methods Eng. p. 1965–1990, vol. 20, 1984.

[3] N. Hitschfeld, *Grid Generation for Non-Rectangular Semiconductor Devices*, PhD thesis publish. by Hartung-Gorre Konstanz, Series in Microelectronics, Vol. 21, 1993.

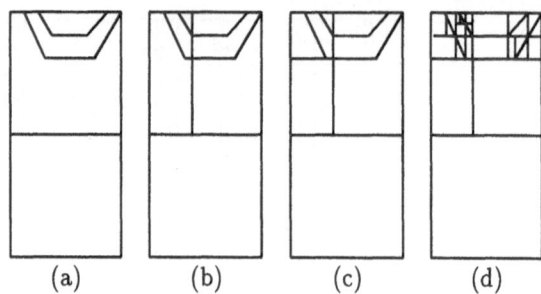

Figure 2: Fitting a 2-D device geometry using the *intersection* based approach. (a), (b), and (c) first steps to fit the device geometry (d) final initial grid.

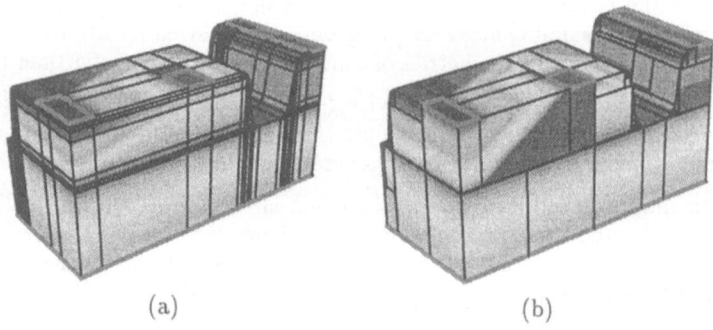

Figure 3: Fitting the device geometry for the bipolar transistor (a) tensor product grid: 2365 pts (b) *intersection* based approach: 881 pts

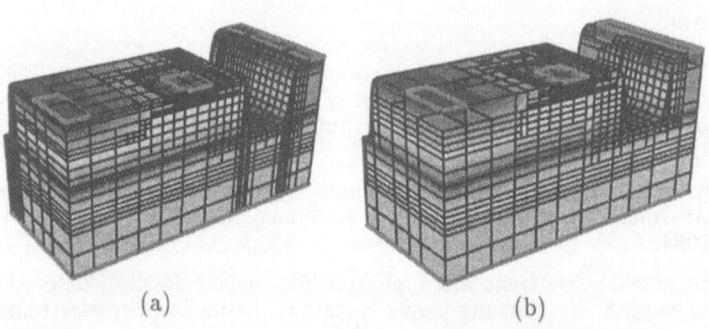

Figure 4: Achieving the desired mesh density for the bipolar transistor (a) *bisection* based approach: 10,376pts (b) *intersection* based approach: 7,097pts

SIMULATION OF SEMICONDUCTOR DEVICES AND PROCESSES Vol. 5   417
Edited by S. Selberherr, H. Stippel, E. Strasser – September 1993

Grid and Geometry Techniques for Multi-Layer Process Simulation

Z. H. Sahul, R. W. Dutton, and M. Noell[†]

Integrated Circuits Laboratory, Stanford University
Stanford, CA 94305, USA
[†]Motorola Inc.
3501 Ed Bluestein Blvd., Austin, TX 78721, USA

Abstract

A robust grid generation and evolution technique for process simulation is presented. A quadtree approach is used to decimate the geometry and final triangulation is performed using templates. Warping is used to improve the grid quality near boundaries. The quadtree depends only on the structure's spatial extent and not on the specific vertices. This allows independent manipulation of the geometry and grid by various process simulation tools.

1. Introduction

The goal of our work is to insure a high quality computation grid at all times in multi-layer process simulation. A grid module for process simulation should grid complex geometries, enable easy grid adaptation, and must be responsive to rapid topography changes due to etching, deposition, and oxidation simulations. An automatic quadtree based 2-D grid generator, called Forest, has been written to specifically address these issues. Forest stores and manipulates both geometry and grid data and can be used by grid based diffusion/oxidation simulators as well as string based etch/deposition simulators.

The geometry of the device is stored in a hierarchy of points, edges, boundaries and regions. This allows specification and manipulation of complex structures including those that contain voids. String based simulators use and update this data without directly manipulating the grid.

2. Quadtree Decimation

Initial grid generation (Figure 1) proceeds by surrounding the geometry with a root square and recursively decimating the root square until a final terminated lines quadrilateral mesh is obtained. Neither the root square nor the resulting quadtree depends on the specific vertices of the geometry; they only depend on the spatial extent of the structure. The level of decimation can be controlled by a number of factors including user specification, doping variation, or solution error.

3. Triangulation

Triangulation of the terminated lines quadrilateral mesh is performed by using templates (Figure 2). The triangles are optimized for aspect ratio and the quality of the triangulation is measured by considering the ratio of the area of a triangle to the sum of square of its sides [1]. Normalizing this aspect ratio to be 1.0 for an equilateral triangle, the triangulation aims to produce triangles whose aspect ratios are at least 0.5.

Triangulation of interior quadrilaterals is easy and produces triangles of high quality - all triangles are right or acute with high aspect ratios. Further treatment, however, is required for quadrilaterals with region boundaries. As shown in Figure 2b, boundary templates produce triangles of adequate aspect ratio if:

- The boundary does not intersect the quadrilateral very close to its corners.

- The boundary does not contain a vertex point that is very close to a quadrilateral corner or an edge.

Straightforward analysis yields a condition of closeness to be one third of the quadrilateral side length for producing triangles of aspect ratios greater than 0.5. Quadrilaterals that fail this criteria are preprocessed by a technique called warping (Figure 2c) [2]. The quadrilaterals are slightly deformed by moving the corner closest to an intersection point onto the intersection point. For vertex points close to a side, a similar concept is used to warp the side onto the vertex point. The resulting quadrilaterals are then triangulated, producing triangles with aspect ratios greater than 0.5. If obtuse triangles need to be eliminated or if the aspect ratio needs to be higher, then the quadrilaterals can be further divided and the whole process is repeated in a recursive fashion. This allows an user to trade-off grid quality with grid quantity. Another example of a Forest generated grid used in oxidation simulation is shown in Figure 3.

4. Advantages of Forest

One of the main advantages of this gridding technique for process simulation is illustrated in Figure 4 which shows a prototype interface between Forest and a string based etch/deposition simulator SPEEDIE [3]. The geometry of the structure is used to extract the top layer of the device and deposition simulation is performed. The grid is subsequently altered to conform to the new geometry. This entails only pruning or growth of the quadtree and triangulation locally over the regions where the boundary has changed. This results in minimal change to the existing grid. The technique can be applied to great advantage for coupled oxidation and diffusion simulation.

Acknowledgement: This work was supported by the Semiconductor Research Corporation under contract # 92-SP-101.

References

[1] R. Bank, *PLTMG User's Guide - Edition 5.0*, Technical Report. Dept. of Mathematics, U.C. San Diego, 1988

[2] M. Bern, et. al. *Provably Good Mesh Generation.* Proc. 31st IEEE Symp. Foundations of Computer Science (1990) 231-241.

[3] J. McVittie, et. al. *SPEEDIE User's Manual Version 2.0. Integrated Circuits Laboratory.* Stanford University. Feb. 1992

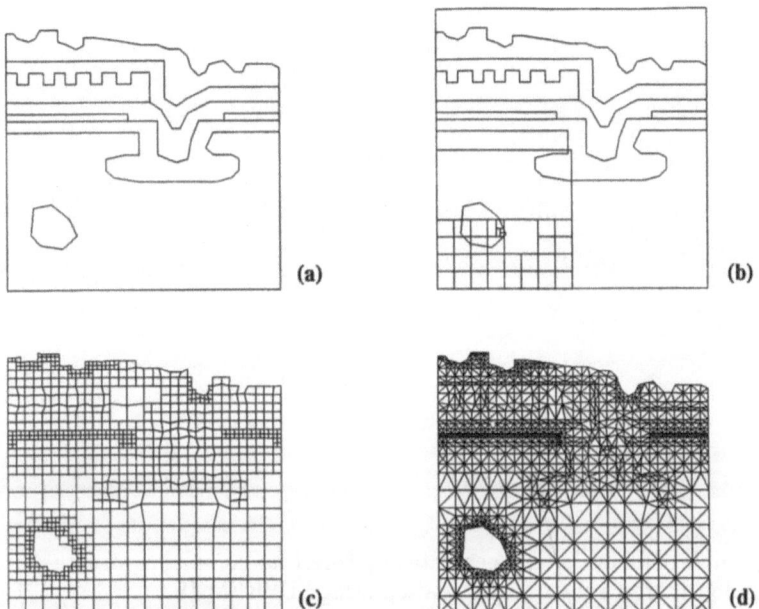

Figure 1: Grid generation proceeds by enclosing a geometry (a) by a root square and decimating the square (b) resulting in a terminated lines mesh (c). The terminated lines mesh in then triangulated (d).

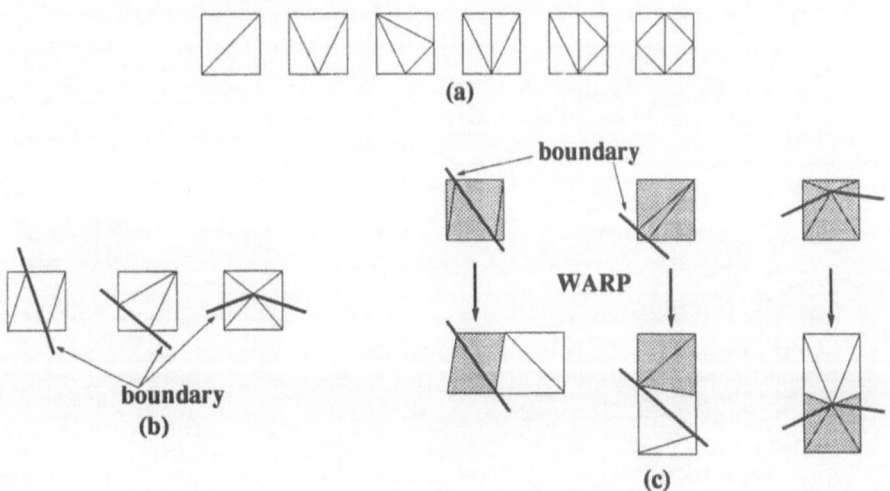

Figure 2: (a) Template used for triangulating interior quadrilaterals. (b) Template used for triangulating boundary quadrilaterals. (c) Warping used to improve triangle qualities of boundary quadrilaterals.

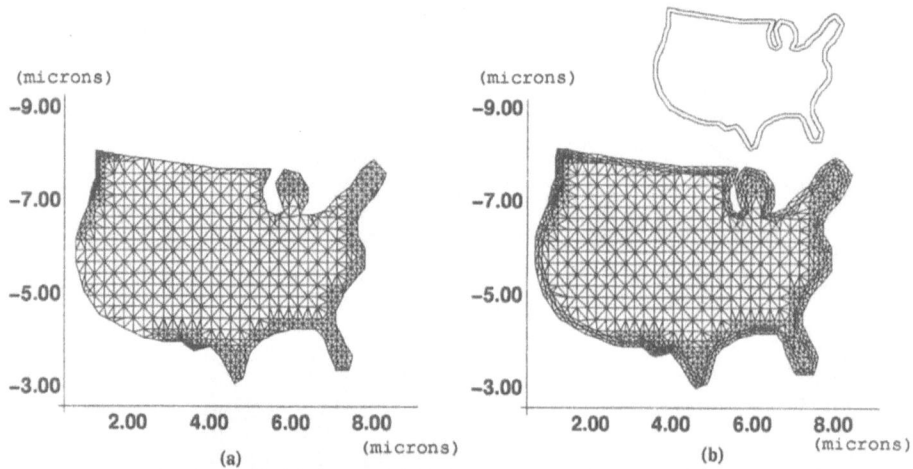

Figure 3: (a) Example of an initial grid generated by Forest for non-planar structures. (b) Geometry and grid after subsequent oxidation using SUPREM IV.

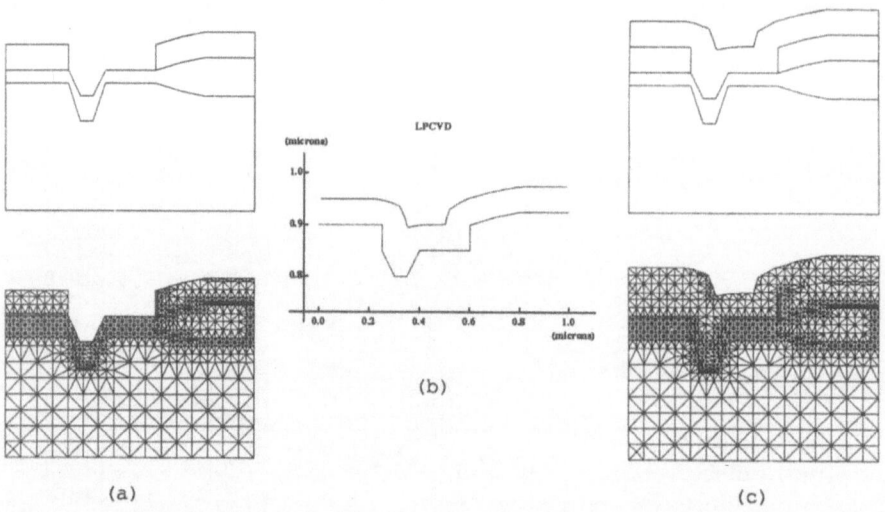

Figure 4: Prototype use of Forest and SPEEDIE to simulate deposition. (a) Grid and geometry in Forest. (b) Deposition simulation using SPEEDIE. (c) Conforming grid to new boundary.

Multigrid Becomes a Competitive Algorithm for some 3D Device Simulation Problems

J. Fuhrmann and K. Gärtner[†]

Institute for Applied Analysis and Stochastics
Hausvogteiplatz 5-7, D-10117 Berlin, GERMANY
[†]Interdisciplinary Project Center for Supercomputing, ETH-Zürich
ETH-Zentrum, CH-8092 Zürich, SWITZERLAND

Abstract

For situations where Gummel's decoupling scheme is applicable a multigrid algorithm for the continuity equations fully consistent with the usual Scharfetter-Gummel discretization can be used to solve the van Roosbroeck equations. The main problems for applying a multigrid algorithm are that discrete spaces are not nested in the usual sense for the refined grids because of the Scharfetter-Gummel discretization and that problem coefficients vary strongly. Transforming the equations to symmetric form and applying a block MILU decomposition based on the coarse-fine splitting of the discrete spaces with a perturbed Schur complement defines the prolongation and restriction operators. The transformation back to the original variables is possible. Coarse grid matrices are M-matrices.

Let $\mathcal{M}_0, \ldots \mathcal{M}_j$ be a sequence of Euclidean vector spaces with growing dimension. In order to define a standard multigrid algorithm to solve the continuity equation of the van-Roosbroeck system $A_j u = f$ on the j-th level there need to be the following components (in the terminology of [2]):
   - scalar products $((\cdot, \cdot))_k : \mathcal{M}_k \times \mathcal{M}_k \to R$
   - symmetric, positive definite with respect to $((\cdot, \cdot))_k$ operators $A_k : \mathcal{M}_k \to \mathcal{M}_k$,
   - interpolations $I_k : \mathcal{M}_{k-1} \to \mathcal{M}_k$
   - restrictions $P_k^0 : \mathcal{M}_k \to \mathcal{M}_{k-1}$.
   - smoothers $R_k : \mathcal{M}_k \to \mathcal{M}_k$

While the smoothers R_k are provided by one or more steps of a classical iteration method (Jacobi, Gauß-Seidel, ILU), the design of other components in cases of strongly varying coefficients or missing standard finite element background is unclear. Here, we try a 'semi-algebraic' method as described in previous stages in [4, 5, 6]. Rather similar ideas of constructing multigrid or multilevel preconditioners have been used in [1, 3, 9, 10].

In what follows, we abbreviate the level-k-indices, to mean a fine grid corresponds to space $\mathcal{M}_k$, a coarse grid then corresponds to $\mathcal{M}_{k-1}$.

On a three-dimensional grid with quadrilateral cells generated by standard refinement from a coarser one, we have the splitting of the grid vertex set $V(A) = V_C \cup V_F \cup V_E \cup V_N$ into sets of coarse grid cell midpoints, coarse grid cell face midpoints, coarse grid cell edge midpoints and coarse grid node points, respectively. We get the matrix partitioning

$$
A = \left(\begin{pmatrix} A_C & B_{CF} & 0 \\ B_{FC} & A_F & B_{FE} \\ 0 & B_{EF} & A_E \\ (0 & 0 & B_{NE}) \end{pmatrix} \begin{pmatrix} 0 \\ 0 \\ B_{EN} \\ A_N \end{pmatrix} \right) = \begin{pmatrix} A_{11} & A_{12} \\ A_{21} & A_{22} \end{pmatrix},
$$

where the off diagonal blocks are nonpositive and the diagonal blocks $A_C = A_{CF} + M_C$, $A_F = A_{FC} + A_{FE} + M_F$, $A_E = A_{EF} + A_{EN} + M_E$ and $A_N = A_{NE} + M_N$, are positive diagonal matrices which consist of sums of off diagonal row entries and a nonnegative "mass" term. The assumptions made on A imply at least one entry of the "mass" M_* is positive.

Let $\tilde{A}_F = A_{FE} + M_F$, $\tilde{A}_E = A_{EN} + M_E$, and choose

$$
F = \begin{pmatrix} A_C & B_{CF} & 0 \\ 0 & \tilde{A}_F & B_{FE} \\ 0 & 0 & \tilde{A}_E \end{pmatrix}, \qquad G = A_{12}, \qquad U = \begin{pmatrix} F & G \\ 0 & I \end{pmatrix}.
$$

U can be seen as a transformation matrix to an approximate harmonic basis [7]. Then for $\cdot^T$ being the transposition with respect to the $((\cdot, \cdot))$-scalar product we have

$$
A = U^T \begin{pmatrix} F^{-T} A_{11} F^{-1} & \Delta \\ \Delta^T & S \end{pmatrix} U, \tag{1}
$$

with $\Delta = F^{-T}(A_{12} - A_{11} F^{-1} G)$ and $S = \hat{S} + \Delta^T (F^{-T} A_{11} F^{-1})^{-1} \Delta$, and $\hat{S} = A_{22} - A_{21}(A_{11})^{-1} A_{12}$ is the Schur complement. To create a block diagonal preconditioner for A in the new basis, one takes the decomposition (1) and omits the off diagonal blocks Δ. Omitting the error correction in the fine grid part, too, yields a coarse grid correction by projecting the error vector onto the fine grid space in the new basis. It has the form

$$
B = U^{-1} \begin{pmatrix} 0 & 0 \\ 0 & S^{-1} \end{pmatrix} U^{-T} = I_k S^{-1} P_k^0
$$

with

$$
I_k = \begin{pmatrix} -F^{-1} G \\ I \end{pmatrix} = \left(\begin{pmatrix} -A_C^{-1} B_{CF} \tilde{A}_F^{-1} B_{FE} \tilde{A}_E^{-1} B_{EN} \\ \tilde{A}_F^{-1} B_{FE} \tilde{A}_E^{-1} B_{EN} \\ -\tilde{A}_E^{-1} B_{EN} \\ I \end{pmatrix} \right)
$$

and P_k^0 being its $((\cdot, \cdot))$-adjoint. S is the Galerkin coarse grid operator corresponding to the given choice of the intergrid transfer operators:

$$
\begin{aligned}
S = P_k^0 A I_k &= (A_N - B_{NE} \tilde{A}_E^{-1} B_{EN}) + B_{NE} \tilde{A}_E^{-1} (A_{EF} - B_{EF} \tilde{A}_F^{-1} B_{FE}) \tilde{A}_E^{-1} B_{EN} \\
&\quad + B_{NE} \tilde{A}_E^{-1} B_{EF} \tilde{A}_F^{-1} (A_{FC} - B_{FC} A_C^{-1} B_{CF}) \tilde{A}_F^{-1} B_{FE} \tilde{A}_E^{-1} B_{EN}.
\end{aligned}
$$

Some geometrical considerations and numerical experiments suggest that in the sense of spectral equivalences, it should hold that $S \approx 4(A_N - B_{NE} \tilde{A}_E^{-1} B_{EN}) =: A_{k-1}$, when the coefficients are not too strongly varying. This suggests replacing S by A_{k-1} in B. At the other hand, A_{k-1} is the Schur complement of the positive definite matrix

$$
4 \begin{pmatrix} A_{EN} + M_E & B_{EN} \\ B_{NE} & A_N \end{pmatrix}
$$

and inherits the $((\cdot, \cdot))$-symmetry, the M-property, and the seven-diagonal structure of A, so the process described above can be continued recursively. To ensure the preservation of the M-property in a floating point representation, one has to choose a matrix data structure where, on all the levels, instead of the main diagonal entries of A, the difference between these entries and the sum of the remaining entries of the same column is stored.

It can be shown [4] that for $((\cdot, \cdot))$-symmetric, positive definite operators, the convergence of a multigrid method with components defined this way depends on a number of reasonable factors:
 - the spectral equivalence of A_{11} and F, and of A_{11} and its diagonal;
 - the cosines of the angles between coarse grid and fine grid spaces in the A-energy scalar product;
 - the spectral equivalence of S and A_{k-1};
 - a smoothing property for R_k, which for Jacobi and Gauß-Seidel smoothers is valid for any symmetric M-matrix [11]

The whole multigrid operator described above is selfadjoint in the $((\cdot, \cdot))$-scalar product provided the smoothers are selfadjoint. Without smoothing, the MG-operator has a special recursively defined MILU decomposition interpretation. One can use it as a preconditioner for conjugated gradients in this scalar product. If one considers the Scharfetter-Gummel discretization of carrier transport equations in semiconductors, the discrete operators are selfadjoint with respect to a scalar product using a weight $e^{\pm\psi}$ where ψ is the electrostatic potential.

Here, we compare the algorithm above with a classical iterative one — ILU(1) preconditioning using Chebyshev polynomials and CG with weighted inner product.

The pictures show results for a photo diode with multiple differently doped horizontal layers. The aim is to deplete the whole diode and to compute the recombination current. The kink at 3.5V in the I-U-curve is what the designers are looking for. The Gummel iteration has been stopped at 10^{-7} U_T to fulfill the current balance better then 10^{-4}. A second example shows the results for a MOSFET at the 1MBit DRAM design level.

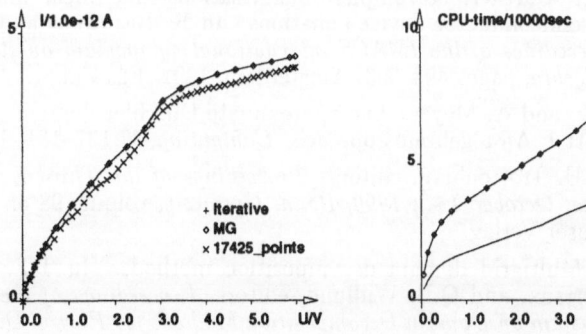

MEDEA, photo diode, Multigrid versus iterative method, 3 grids:
$130977 = 49 \times 33 \times 81$, $17425 = 25 \times 17 \times 41$, $2457 = 13 \times 9 \times 21$ points, CONVEX C220

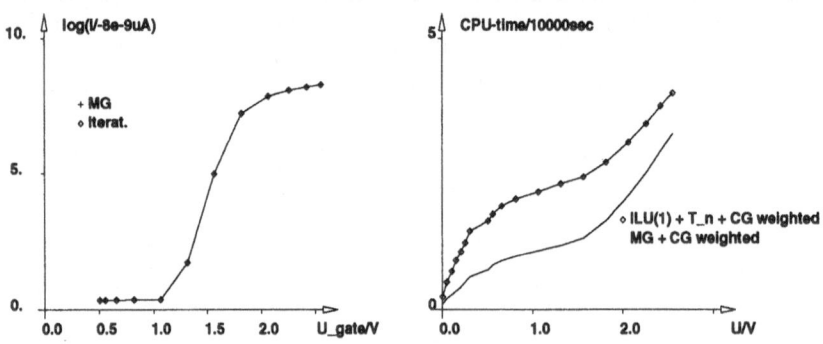

MEDEA, 'half-MOSFET', 1MBit DRAM design level, $U_{bulk} = -2V$, $U_{drain} = 0.1V$,
$156849 = 97 \times 33 \times 49$, $20825 = 49 \times 17 \times 25$, $2925 = 25 \times 9 \times 13$, $455 = 13 \times 5 \times 7$,
$84 = 7 \times 3 \times 4$ points, DEC ALPHA (3000/500, 64MB)

References

[1] O. Axelsson and P. S. Vassilevski. Algebraic multilevel preconditioning methods, III. preprint, Cath. Univ. Dept. of Math., Nijmegen, 1990. Report 9045.

[2] J.H. Bramble, J.E. Pasciak, and J. Xu. The analysis of multigrid algorithms with nonnested spaces or noniherited quadratic forms. *Math. of Comp.*, 56:1–34, 1991.

[3] W. Dahmen and L. Elsner. Algebraic multigrid methods and the Schur complement. In W. Hackbusch, editor, *Robust Multigrid-Methods*, volume 23 of *Notes on numerical fluid mechanics*, pages 58–69. Vieweg, Braunschweig, 1989.

[4] J. Fuhrmann. On the convergence of algebraically defined multigrid methods. preprint no.3, Institut für Angewandte Analysis und Stochastik Berlin, 1992.

[5] J. Fuhrmann and K. Gärtner. A multigrid method for the solution of a convection - diffusion equation with rapidly varying coefficients. In [8].

[6] J. Fuhrmann and K. Gärtner. Incomplete factorizations and linear multigrid algorithms for the semiconductor device equations. In R. Beauwens and P. de Groen, editors, *Proccedings of the IMACS international symposium on iterative methods in linear algebra*, pages 493–503, Amsterdam, 1992. Elsevier.

[7] G. Haase, U. Langer, and A. Meyer. The approximate Dirichlet domain decomposition method. part I: An algebraic approach. *Computing*, 47:137–151, 1991.

[8] W. Hackbusch and U. Trottenberg, editors. *Proceedings of the Third European Multigrid Conference, October 1 - 4,1990, Bonn, Germany*, volume 98 of *ISNM*, Basel, 1991. Birkhäuser Verlag.

[9] Yu.A. Kuznetsov. Multigrid domain decomposition methods. In T.F. Chan, R. Glowinski, J. Periaux, and O.B. Widlund, editors, *Proceedings of the Third International Symposium on Domain Decomposition Methods for Partial Differential Equations, Houston, Texas, March 20-22,1989*, pages 290–313, Philadelphia, 1990. SIAM.

[10] C. Popa. ILU decomposition for coarse grid correction step on algebraic multigrid. In Hackbusch and Trottenberg [8].

[11] J.W. Ruge and K. Stüben. Algebraic multigrid. In S. McCormick, editor, *Multigrid methods*, volume 4 of *Frontiers in Applied Mathematics*, chapter 4, pages 73–130. SIAM, Philadelphia, 1987.

A 2D Analytical Model of Current-Flow in Lateral Bipolar Transistor Structures

D. Freund, A. Kloes, and A. Kostka

Solid State Electronics Laboratory, Technical University of Darmstadt
Schloßgartenstraße 8, D-64289 Darmstadt, GERMANY

Abstract

In this paper we present a new theoretical approach for analytical models of lateral bipolar transistors (CLBT, LBT) by using conformal mapping techniques. It is shown that this method leads to closed form solutions for the currents in the devices under realistic boundary conditions. Model equations for CLBT's and LBT's are derived and will be compared to numerical simulation results as well as to measurements.

1. Introduction

In the design of analog integrated circuits lateral bipolar transistors offer greater design flexibility despite their behaviour is normally restricted with respect to optimized vertical BJT's. Typical lateral BJT's are shown in Figure 1: they appear as lateral pnp-transistors (LBT) in standard bipolar or BiCMOS-processes or as compatible lateral transistors (CLBT) in CMOS-technology.

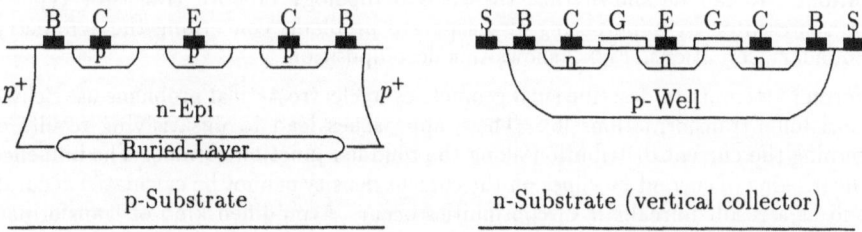

Figure 1: Cross-sections of pnp-LBT and p-well-CLBT

In spite of their widespread use, literature on compact, analytical modelling of lateral BJT's is not too abundant. Nearly all modelling approaches use a Gummel-Poon-Analysis of the device behaviour [1, 2]. But the fundamental assumptions of the Gummel-Poon-Analysis ignore the two-dimensional current-flow in lateral devices and lead to inaccurate results. Other models are numerical and do not offer the possibility to investigate the influence of device-parameters in a handy way [3, 4]. As a result, numerical models cannot properly be used when the layout- and geometry-dependence of electrical parameters has to be investigated.

In this paper we present an approach to describe the current-flow in lateral devices, especially LBT's and CLBT's, that focusses on a two-dimensional,but still analytical description.

2. 2D-calculation of carrier densities and currents by conformal mapping

To calculate the minority-carrier distribution and the collector currents in lateral transistor structures, it can be shown that the Laplacian equation

$$\Delta n = 0 \tag{1}$$

is valid if recombination effects and internal electric fields are neglected. It is possible to solve (1) by using a conformal representation $z = F(w)$, that maps the geometrical structure onto the real axis of the complex plane. A complex minority carrier density function N can then be calculated, whose real part reflects the carrier distribution and whose imaginary part can be regarded as a flux function [6]. It can be shown that the current flowing between two points A and B [1] can then be written as:

$$I' = Im(N(w_B)) - Im(N(w_A)). \tag{2}$$

This shows that in order to calculate the currents in a 2D-device problem it is necessary to first find a mapping function that transforms the device structure into a geometry in the complex plane, where the solution of (1) is known and then to apply (2).

3. Conformal mapping of lateral structures

Fig. 1 suggests a structural affinity of LBT and CLBT. The main difference is the presence of the buried-layer which mathematically imposes different boundary conditions. It can be shown that the general topology of both transistor types can be represented by superposing two separate problems concerning the symmetry of boundary conditions. Fig. 2 shows this decomposition.

Former attempts to describe such geometries in electrostatical problems use Schwarz-Christoffel-transformations [6]. These approaches lead to unsatisfying results concerning the current distribution along the rounded junction corners. The influence of the fringing of current flowlines on the current density cannot be calculated accurately and as a result unrealistic discontinuities occur. A modified kind of transformation was developed that is much better suited to describe junctions corners in integrated devices. The resulting mapping function is:

$$F(w) = i\frac{vg}{\pi}(ln\frac{1+\eta}{1-\eta} + 2\sqrt{\varrho}\ arctan(\sqrt{\varrho}\ \eta)) + i(1-v)\frac{2g}{\pi}ln(\sqrt{w-q} + \sqrt{w-1}) \tag{3}$$

---

[1]Throughout this 2D-analysis only line current densities will be regarded.

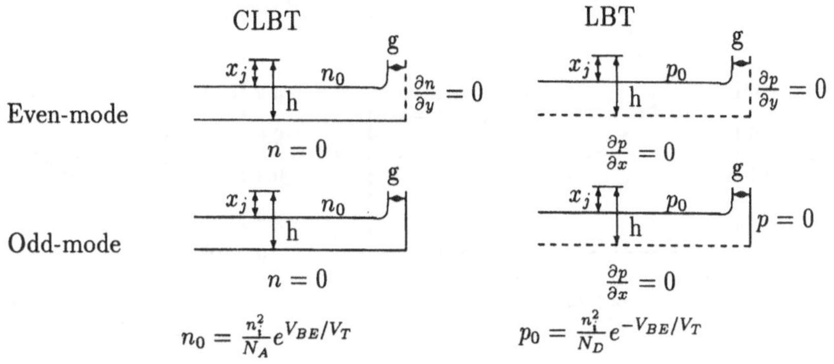

Figure 2: Odd- and even-mode decomposition of LBT and CLBT

$$\eta = \frac{\sqrt{w-q}}{\sqrt{w+1}} \quad (4)$$

$$\varrho = \frac{p+1}{q-p} \quad (5)$$

$$p = \frac{qh^2 - v^2g^2}{h^2 + v^2g^2} \quad (6)$$

$$q = \frac{\varepsilon^4 + 6\varepsilon^2 + 1}{(\varepsilon^2 - 1)^2} \quad (7)$$

$$\varepsilon = e^{\frac{x_j\pi}{2(1-\nu)g}} \quad (8)$$

$$\nu = \frac{1}{4}\sqrt{\frac{g}{x_j}} \quad (9)$$

$\eta, \varrho, p, q, \varepsilon, \nu$ are parameters which are needed to map those points of the structure that are relevant for the calculation of currents following the procedure outlined in section 2.

By using this mapping function the 2D-structure is transformed onto the real axis of the complex plane and (1) can be solved in a manner analogous to field plate problems in electrostatical calculations [6].

4. Modelling the collector currents in CLBT's

Under the conditions of homogenous base doping, flat-band-operation and $V_{CB} = 0$, the lateral and vertical collector currents are derived from (3) - (9) and applying (2):

$$I'_L = eD_N\frac{n_0}{2\pi}(arcosh(a) - ln(2a - 2)) \quad (10)$$

$$I'_V = eD_N\frac{n_0}{2\pi}(arcosh(c) + ln(2c + 2)) \quad (11)$$

with

$$a = \frac{1 + 2q - p}{1 + p} \quad (12)$$

$$b = \frac{1 - 2q + p}{1 - p} \quad (13)$$

$$c = 1 - \frac{2\psi^2(b+1)}{(\psi^2 - 1)(a+1)} \quad (14)$$

$$\psi = tanh(\frac{\pi(g + \frac{W_E}{2})}{2\sqrt{\varrho\nu g}}) \quad (15)$$

To include the effect of internal drift fields it is possible to introduce a drift factor m, that reflects the effect of inhomogenous base doping profiles. Substituting the parameter h by $h/\sqrt{m}$ and multiplying (11) with $\sqrt{m}$ allows to calculate the currents in a CLBT with a drift base.

Figure 4 shows a comparison of numerical simulation results with the presented model concerning the prediction of current-splitting in a CLBT. Measurements at CLBT's,that were fabricated in a $3\mu m$-standard CMOS-technology are in good accordance with the model.

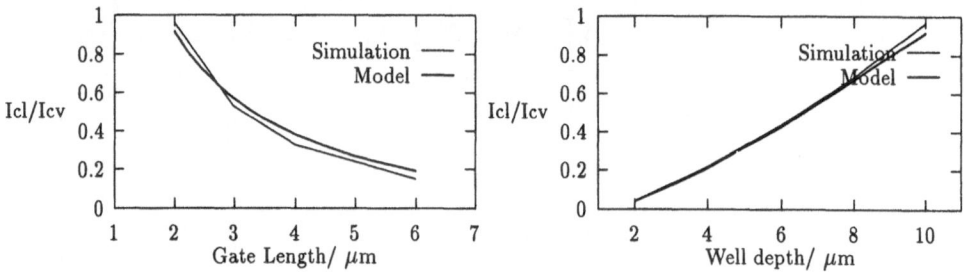

Figure 3: Comparison of simulation results and theory

5. Modelling the collector current in an LBT

Due to the influence of the buried-layer, which is absent in a CLBT, approximately all vertical currents in an LBT are suppressed. Preceeding in the same way as discussed in the above sections, the current in an LBT can be calculated as:

$$I'_L = eD_P \frac{p_0}{2\pi} arcosh(\frac{q+p+2}{q-p})$$
(16)

with p, q according to the mapping definitions in (6) and (7).

6. Conclusions

The theoretical approach presented in this paper leads to a physics-based model for the current-flow in lateral BJT's under shorted collector conditions. Variations of collector potentials can be included by modification of well depth and effective gate length in dependance of the extension of space-charge-regions into the base [5]. The model describes the correspondence of process-, geometry- and electrical parameters of the device in a rather exact, but still analytical way, enabling the circuit designer to calculate currents in the device as well as its scaling behaviour.

References

[1] H.H.Berger,U.Dreckmann: *The lateral pnp-transistor - a practical investigation of the DC-characteristics*, IEEE Trans. Electron Devices, vol. ED-27, pp.1038 - 1046, July '79

[2] K.S.Seo, C.K.Kim: *On the geometrical factor of lateral pnp-transistors*, IEEE-Trans. Electron devices, vol. ED-27, pp. 295 - 297, Jan. '80

[3] D.E.Fulkerson: *A two-dimensional model for the calculation of common-emitter current-gains of lateral pnp-transistors*, Solid-State-Electronics, vol.11, pp.821 - 826, 1968

[4] J.D.Last,D.W.Lucas,G.W.Sumerling: *A numerical analysis of the DC-performance of small geometry lateral transistors*, Solid-State Electronics, vol. 17, pp. 1111 - 1118, 1974

[5] D.Freund,A.Kloes,A.Kostka: *A 2D-Model for current-splitting in CMOS-compatible lateral bipolar transistors*, accepted for publication at ESSDERC '93

[6] Weber: *Electromagnetic fields*, Volume 1, Wiley,1950

Consistent Treatment of Carrier Emission and Capture Kinetics in Electrothermal and Energy Transport Models

G. Wachutka

Physical Electronics Laboratory, ETH-Zürich
ETH-Hönggerberg, CH-8093 Zürich, SWITZERLAND

Abstract

The quasi-static approximation usually employed to model the reaction kinetics of electrons, holes, and impurities in semiconductor devices is critically re-examined. It is shown that in the case of high trap concentrations, hot carriers, low temperature operating conditions or wide-gap devices the commonly used balance equations for particle and energy flow must be supplemented by additional terms in order to correctly include the emission and capture kinetics of the free carriers.

1. Introduction

In semiconductor device modeling the reaction kinetics of carrier emission and capture by the action of impurities is usually described within the so-called quasi-static approximation. For shallow impurities, local thermal and chemical equilibrium between donors and electrons or acceptors and holes, respectively, is assumed. So, for instance, incomplete ionization is modeled with the implicit supposition that the lattice temperature T_L and the electrochemical potential of the donors μ_D are equal to the electron temperature T_n and the electron quasi-Fermi level $-q\Phi_n$, respectively, and with the analogous assumption for acceptors and holes ($T_L = T_p$ and $\mu_A = -q\Phi_p$).

Trapping and generation-recombination processes through deep centers are described by a steady-state quasi-equilibrium Gibbs distribution

$$f_{tp}^{(o)} = (1 + g_{tp}\, exp\,((E_{tp} - \mu_{tp})/kT_L))^{-1} \tag{1}$$

leading to the well-known Shockley-Read-Hall net reaction rate $(G - R)_{SRH}$. Here the underlying assumption is that the relaxation time τ_{tp} of the transient occupation number $f_{tp}(t)$ is much faster than the time scale relevant to the electric and thermal device operation.

Considering the operating conditions of lifetime-tailored power devices with a high concentration of complex recombination centers, highly integrated circuits with hot carriers, low temperature devices or novel wide-gap silicon carbide or diamond devices, the above-mentioned assumptions seem questionable and, therefore, have been subjected to critical reexamination.

2. Consistent Completion of the Governing Equations

Applying thermodynamic methods, the composite system of (possibly hot) electrons and holes, host lattice, donors, acceptors, and deep centers can be treated in a consistent way. It turns out that, indeed, the basic equations constituting the drift-diffusion-model, the electrothermal model [1] or the hydrodynamic [2] or energy transport models [3],[4],[5] have to be supplemented and completed by additional terms.

On the right-hand side of the current continuity equations (6) and (7), the time-derivatives of the ionized donor and acceptor concentrations, $\partial N_D^+/\partial t$ and $\partial N_A^-/\partial t$, respectively, have to be added. Even within the quasi-static approximation, where we assume instantaneous ionization and neutralization of the shallow impurities on the electrical and thermal time scale of interest, these terms must not be neglected, because N_D^+ and N_A^- may quasi-stationarily vary with time through the temporal evolution of temperature and quasi-Fermi levels according to

$$\frac{\partial N_D^+}{\partial t} = \left(\frac{\partial N_D^+}{\partial \Phi_n}\right)_{T_L} \frac{\partial \Phi_n}{\partial t} + \left(\frac{\partial N_D^+}{\partial T_L}\right)_{\Phi_n} \frac{\partial T_L}{\partial t} \tag{2}$$

and the analogous equation for $\partial N_A^-/\partial t$. Omitting these contributions from the particle balances may lead to a serious violation of charge conservation, as has been demonstrated in [6]. By the same token, the charge of the ionized deep centers must not be neglected on the right-hand side of Poisson's equation (8) when low temperatures or high trap densities are considered. Moreover, as it has been discussed in [7], recombination via donors and acceptors may become significant in VLSI bipolar devices. To model this effect, the shallow impurity net generation-recombination rates $(G - R)_A$ and $(G - R)_D$ (see eq. (2) in [7]) have to be substituted for $\partial N_D^+/\partial t$ and $\partial N_A^-/\partial t$, respectively, the two rates adding to that of recombination via the deep centers.

Furthermore, in each of the balance equations (6), (7), (9), (10) and (12) the Shockley-Read-Hall net generation-recombination rate $(G - R)_{SRH}$ has to be replaced by the individual non-steady-state electron and hole reaction rates $(\partial n/\partial t)_{tp}$ and $(\partial p/\partial t)_{tp}$, respectively, if the relaxation time τ_{tp} of the transient occupation number $f_{tp}(t)$ becomes comparable or longer than the electrical or thermal rise and fall times. In this case, the system of dynamic equations has to be augmented by eq. (14) in order to determine $f_{tp}(t)$ consistently. Here it is important to recognize that, dependent on capture cross sections [8], injection levels and temperature, the trap relaxation time

$$\tau_{tp} = (e_n + e_p + c_n n + c_p p)^{-1} \tag{3}$$

varies over many orders of magnitude (0.1ps - 1ms). Evidently it depends on the specific situation considered (device structure and operating conditions) whether or not the steady-state occupation number $f_{tp}^{(o)}$ (cf. eq.(1)) is an acceptable approximation of the true value $f_{tp}(t)$.

With the view to energy transport modeling, the right-hand sides of the electron and hole energy balance equations (9) and (10) must be supplemented by the terms

$$(w_{tp} - w_n)(\partial n/\partial t)_{tp} + (w_D - w_n)\partial N_D^+/\partial t \tag{4}$$

and

$$-(w_{tp} + w_p)(\partial p/\partial t)_{tp} + (w_A - w_p)\partial N_A^-/\partial t \tag{5}$$

respectively, which account for the energy lost or gained by the hot carrier subsystems due to carrier emission or capture by the impurity subsystems. For deep centers and/or heated carrier distributions, the respective energy transfer per particle $w_{tp} - w_n$ etc. can easily attain half or more of the band gap energy. Therefore the additional contributions to the heat exchange rate in the electron and hole temperature equations (9) and (10) may become substantial not only for fast transient processes, but also under quasi-static conditions in device regions where trap-assisted generation-recombination dominates.

The additional terms in (9) and (10) originate from the fact that under non-equilibrium conditions, the electrochemical potentials of the impurities, μ_D, μ_A and μ_{tp}, and the carrier quasi-Fermi levels, $-q\Phi_n$ and $-q\Phi_p$, differ each from the other, as well as the respective temperatures. Thus, for instance, we may regard the expressions $(w_D - w_n)\,\partial N_D^+/\partial t$ and $(w_A - w_p)\,\partial N_A^-/\partial t$ as corrections of the invalidated assumption of local thermal and chemical equilibrium between donors and electrons or acceptors and holes, respectively (which can be attained under global isothermal conditions solely).

Finally, our analysis also brings up some consequences for the formulation of the electric and thermal boundary conditions. In particular, it shows that inconsistencies arise in the widely used model of an "ideal ohmic contact", because the postulates of local thermodynamic equilibrium and charge neutrality are in contradiction to charge conservation and the energy balance equations.

3. Summary

The completed set of basic equations for particle and energy transport in semiconductor devices reads as follows:

particle balance equations

$$\frac{\partial n}{\partial t} = \frac{1}{q}\,div\,\vec{J}_n + \frac{\partial N_D^+}{\partial t} + \left(\frac{\partial n}{\partial t}\right)_{tp} + (G - R)_{else} \tag{6}$$

$$\frac{\partial p}{\partial t} = -\frac{1}{q}\,div\,\vec{J}_p + \frac{\partial N_A^-}{\partial t} + \left(\frac{\partial p}{\partial t}\right)_{tp} + (G - R)_{else} \tag{7}$$

($(\partial n/\partial t)_{tp}$ and $(\partial p/\partial t)_{tp}$: electron and hole reaction rate with deep centers)

Poisson's equation

$$div(\,\epsilon\,\nabla\psi) = q\,(n - p + N_A^- - N_D^+ - (z^{emp}(1 - f_{tp}) + z^{occ}f_{tp})N_{tp}) \tag{8}$$

(f_{tp} : trap occupation number; z^{emp} and z^{occ}: electric charge number of empty and occupied trap, respectively; N_{tp} : total trap concentration).

energy balance equations for hot carriers

$$c_n\frac{\partial T_n}{\partial t} = div\,(\kappa_n\nabla T_n + L_n\vec{J}_n) + H_n^{(old)} + (w_{tp} - w_n)\left(\frac{\partial n}{\partial t}\right)_{tp} + (w_D - w_n)\frac{\partial N_D^+}{\partial t} \tag{9}$$

$$c_p\frac{\partial T_p}{\partial t} = div\,(\kappa_p\nabla T_p + L_p\vec{J}_p) + H_p^{(old)} - (w_{tp} + w_p)\left(\frac{\partial p}{\partial t}\right)_{tp} + (w_A - w_p)\frac{\partial N_A^-}{\partial t} \tag{10}$$

where

$$w_\alpha := \mu_\alpha - T_\alpha \left(\frac{\partial \mu_\alpha}{\partial T_\alpha} \right)_{f_\alpha} \tag{11}$$

(μ_α : electrochemical potentials of electrons, holes, donors, acceptors and traps, respectively)

heat flow equation for lattice

$$c_L \frac{\partial T_L}{\partial t} = div\,(\kappa_L \nabla T_L) + H_L \tag{12}$$

with

$$H_L = q \left[T_p \left(\frac{\partial \phi_p}{\partial T_p} \right)_{n,p} - T_n \left(\frac{\partial \phi_n}{\partial T_n} \right)_{n,p} + \phi_n - \phi_p \right] (G - R)_{else}$$

$$+ \frac{3}{2} k\,n \frac{T_n - T_L}{\tau_{nL}} + \frac{3}{2} k\,p \frac{T_p - T_L}{\tau_{pL}} \tag{13}$$

rate equation for trap occupation number

$$\frac{\partial f_{tp}}{\partial t} = -(e_n + e_p + c_n n + c_p p) f_{tp} + e_p + c_n n \tag{14}$$

(e_α, c_α : emission and capture coefficients of electrons and holes, respectively)

References

[1] G. Wachutka, *Rigorous Thermodynamic Treatment of Heat Generation and Conduction in Semiconductor Device Modeling*, IEEE Trans. on CAD, Vol. CAD-9, pp 1141–1149, 1990.

[2] M. Rudan, F. Odeh, *Multi-Dimensional Discretization Scheme for the Hydrodynamic Model of Semiconductor Devices*, COMPEL, Vol. 5, pp 149–183, 1986.

[3] C. McAndrew, E. Heasell, K. Singhal, *Modelling Thermal Effects in Energy Models of Carrier Transport in Semiconductors*, Semicond. Sci. Technol., vol. 3, pp 758–765, 1988.

[4] G. Wachutka, *Unified Framework for Thermal, Electrical, Magnetic, and Optical Semiconductor Device Modeling*, COMPEL, Vol. 10, pp 311–321, 1991.

[5] D. Chen *et al*, *An Improved Energy Transport Model Including Non-parabolicity and non-Maxwellian Distribution Effects*, IEEE Electron Dev. Lett., Vol. 13, pp 26–28, 1992.

[6] F. Pfirsch, M. Ruff, *Note on Charge Conservation in the Transient Semiconductor Equations*, submitted for publication.

[7] H. P. D. Lanyon, *Shallow Level Recombination Current Dominance in Transistor Betas*, IEEE Electron Dev. Lett., Vol. 14, pp 49–50, 1993.

[8] K. W. Boer, *Survey of Semiconductor Physics*, Van Nostrand Reinhold, New York, Vol. I, Ch. 43, 1990.

SIMULATION OF SEMICONDUCTOR DEVICES AND PROCESSES Vol. 5   433
Edited by S. Selberherr, H. Stippel, E. Strasser – September 1993

Modeling of Electron-Hole Scattering in Semiconductor Power Device Simulation

E. Velmre, A. Koel, and F. Masszi

Scanner Lab, Electronics Department, Institute of Technology, Uppsala University
P.O.Box 534, S-75121 Uppsala, SWEDEN

Abstract

This paper gives an overview of different approaches of including electron-hole scattering (EHS) in physical semiconductor simulation models and compare the influence on the simulation results.

In the majority of semiconductor simulation codes, the physical model is based on the numerical solution of the *Van Roosbroeck* equations [1], consisting of the *Poisson* equation, the continuity equations, and the drift-diffusion transport equations. The latter originate from the solution of the *Boltzmann* kinetic equation in case of negligible electron-hole scattering (abbreviated as EHS in the following). Thus, the carrier transport in semiconductors is conventionally described by two equations in the form of

$$j_n = q\mu_n nE + qD_n \frac{\partial n}{\partial x} \, , \qquad j_p = q\mu_p pE - qD_p \frac{\partial p}{\partial x} \tag{1}$$

The conventional way to account for the EHS is to modify the carrier mobilities in Eq.(1) using the following reciprocal mobility summation rule (*Mathiessen*'s rule),

$$\frac{1}{\mu_n} = \frac{1}{\mu_{no}} + \frac{1}{\mu_{np}} \, , \qquad \frac{1}{\mu_p} = \frac{1}{\mu_{po}} + \frac{1}{\mu_{pn}} \tag{2}$$

where μ_{np} is the mobility component due to electron scattering on holes, μ_{pn} is the mobility component due to hole scattering on electrons, and the subscript o denotes the mobility components independent of EHS. This approach was proposed by *Fletcher* [2] and since then it has widely been used. However, by a conventional use of Eq.(2) it is assumed, that $\mu_{np}=\mu_{pn}$, which is correct only in a particular case of equal concentrations $n=p$.

For practical device analysis, the EHS related mobility components had to be evaluated. The first theoretical expression for μ_{np}, when $n=p$, was proposed by *Fletcher* and the first measurements were made by *Davies* [3] on a Ge P$^+$–L–N$^+$ structure. By extraction of μ_{np} from a measured voltage drop on the device *Davies* made the assumption, that

$$\frac{1}{\mu_n+\mu_p} = \frac{1}{\mu_{no}+\mu_{po}} + \frac{1}{\mu_{np}} \tag{3}$$

This summation rule differs principally from the *Fletcher* summation of Eq.(2), which turns out to be incorrect in case of any momentum exchange between the electron and hole subsystems. *Davies* also presented a theoretical expression for the calculation of μ_{np}.

Models from [4,5,6] and [7,8,9] are all of *Fletcher*- or *Davies*-type, respectively. Models from [10,11] are empirical and are mainly based on experimental data by *Dannhäuser* [12] and *Krausse* [13]. The *Davies*-type models give somewhat lower mobilities in comparison to the *Fletcher*-type models. The *Dorkel–Leturcq* model [14], employed in the commercially available simulator MEDICI [29], is using *Choo*'s [4] μ_{np} expression and gives at the case $n=p$ lower mobility values than the model used in the program DYNAMIT (developed at the Institute of Electronics, Tallinn Technical University). This is the main reason of elevated forward voltage drops.

Another way for accounting EHS was proposed by the Armenian scientist

Avakyants and his co-workers in 1963 [15], starting from phenomenological plasma theory. From the *Avakyants*-type carrier motion equations the following current equations can be derived [16]

$$j_n = q\mu_{nl}nE + k_BT\mu_{n2}\frac{\partial n}{\partial x} + k_BT\mu_{n3}\frac{\partial p}{\partial x}$$ (4)

$$j_p = q\mu_{pl}pE - k_BT\mu_{p2}\frac{\partial p}{\partial x} - k_BT\mu_{p3}\frac{\partial n}{\partial x}$$ (5)

where $\mu_{nl} = \mu_{n2}-\mu_{p3}$, $\mu_{pl} = \mu_{p2}-\mu_{n3}$ and $n\cdot\mu_{p3} = p\cdot\mu_{n3}$ (6)

These current equations differ from the *Van Roosbroeck* equations by two principal features. To begin with, there are extra cross terms representing the carrier drag effect [17–19]. Thereafter, the *Einstein* relationship does not hold for the electron and hole drift mobilities μ_{nl}, μ_{pl} and for the relevant diffusion coefficients in the second and the third terms. Later on it was shown [20], that the validity of *Einstein*'s relationship can be restored by writing the current equations in a matrix form.

The *Avakyants*-type current equations with explicitly written drag terms appeared probably for the first time in 1972 [8] and in 1976 [9]. However, the same equations can be obtained by solving the *Boltzmann* kinetic equation using *Kohler*'s variational principle [21 - 24]. These more accurate, modified current equations, Eqs.(4-6), have been introduced into the numerical simulation practice in 1983 [16, 25] by the simulation code called DYNAMIT and since then have been used for power device simulation, eg. [26–28].

The purpose of this paper is to show how these different approaches influence the simulation results in case of high current densities (J>100 A/cm2). We have chosen a semiconductor power diode and have made simulations for the following cases:

Case #1: Full EHS model with the modified transport equations, Eqs.(4-6) These simulations were made by using the code DYNAMIT. The mobilities are specified as follows:

$$\mu_{nl} = \mu_{n2}-\mu_{n3}, \quad \mu_{n2} = \mu_{nB}+\mu_{n3}, \quad \mu_{pl} = \mu_{p2}-\mu_{p3}, \quad \mu_{p2} = \mu_{pB}+\mu_{p3}$$ (7)

$$\mu_{n3} = \mu_{nB}\mu_{po}n\cdot J^{eh}, \quad \mu_{p3} = \mu_{pB}\mu_{no}p\cdot J^{eh}, \quad \mu_{nB} = B\cdot\mu_{no}, \quad \mu_{pB} = B\cdot\mu_{po}$$ (8)

$$B = \frac{1}{1+\mu_{no}p\cdot J^{eh}+\mu_{po}n\cdot J^{eh}}$$ (9)

where J^{eh} is the EHS function [30]. In our DYNAMIT simulations, the following empirical formula [10] has been used

$$J^{eh} = 1.3513\cdot10^{-20}\frac{1 + 3.591\cdot10^{-18}\frac{n+p}{2}[cm^{-3}]}{1 + 2.857\cdot10^{-17}\frac{n+p}{2}[cm^{-3}]} \quad [V\cdot sec\cdot cm]$$ (10)

The function J^{eh} can be represented using the mobilities μ_{np} or μ_{pn} as follows[3]

$$J^{eh} = \frac{1}{p\cdot\mu_{np}} = \frac{1}{n\cdot\mu_{pn}}$$ (11)

Case #2: Like in case #1, but neglecting the cross-terms in the modified transport equations, Eqs.(4-5). The mobilities are specified in the followings:

$$\mu_{n3} = \mu_{p3} = 0, \quad \mu_{nl} = \mu_{n2} = \mu_{nB}, \quad \mu_{pl} = \mu_{p2} = \mu_{pB}$$ (12)

Case #3: Like in case #2, but using the *Fletcher*-type summation of mobilities. Electrons and holes as scattering centers are treated as non-drifting impurities. As a consequence, cross-terms disappear ($\mu_{n3} = \mu_{p3} = 0$) and the following formulas are valid:

---

[3]   This relationship was originally established by *Avakyants* and *Lazarev* [31], who used the notation α instead of our J^{eh}.

$$\mu_{n1} = \mu_{n2} = \mu_{nB} \Bigg|_{\mu_{po}=0} = \frac{\mu_{no}}{1+\mu_{no}p \cdot J^{eh}} = \frac{1}{\frac{1}{\mu_{no}} + \frac{1}{\mu_{np}}} \tag{13}$$

$$\mu_{p1} = \mu_{p2} = \mu_{pB} \Bigg|_{\mu_{no}=0} = \frac{\mu_{po}}{1+\mu_{po}n \cdot J^{eh}} = \frac{1}{\frac{1}{\mu_{po}} + \frac{1}{\mu_{pn}}} \tag{14}$$

Case #4 : With no EHS taken into account at all, calculated by DYNAMIT. The scattering function $J^{eh} = 0$, and

$$\mu_{n3} = \mu_{p3} = 0, \quad \mu_{n1} = \mu_{n2} = \mu_{no}, \quad \mu_{p1} = \mu_{p2} = \mu_{po} \tag{15}$$

Case #5: EHS effect included only in the mobilities. These calculations were made by MEDICI [29], using it's *Dorkel-Leturcq*-type mobility model. All other physical models and their parameters were identical to those used in the DYNAMIT simulations, also

$$\mu_{n3} = \mu_{p3} = 0, \quad \mu_{n1} = \mu_{n2} = \mu_n, \quad \mu_{p1} = \mu_{p2} = \mu_p \tag{16}$$

Next, we demonstrate the differences between these 5 cases in 3 figures.

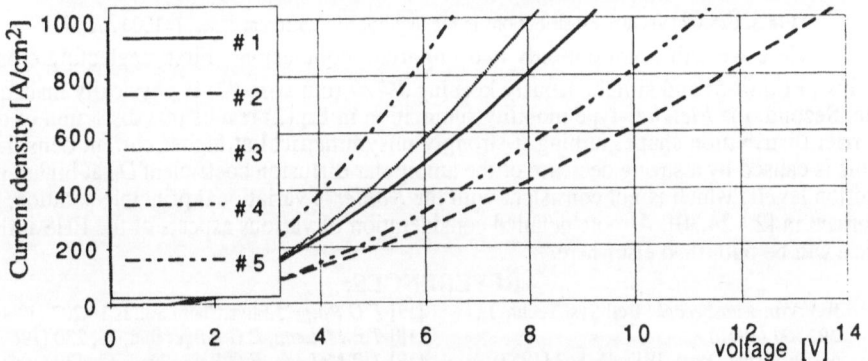

Fig.1. Calculated IV characteristics in the mentioned 5 different cases

The simulation results show a rather big difference in I-V characteristics and the structure internal variable distributions depending on the EHS accounting way. At higher current densities the forward voltage drop is predominantly determined by the electric field integral over the thick base region, where $n \approx p$ and the drift transport of carriers is dominating. The local electric field in the base is then approximately given by

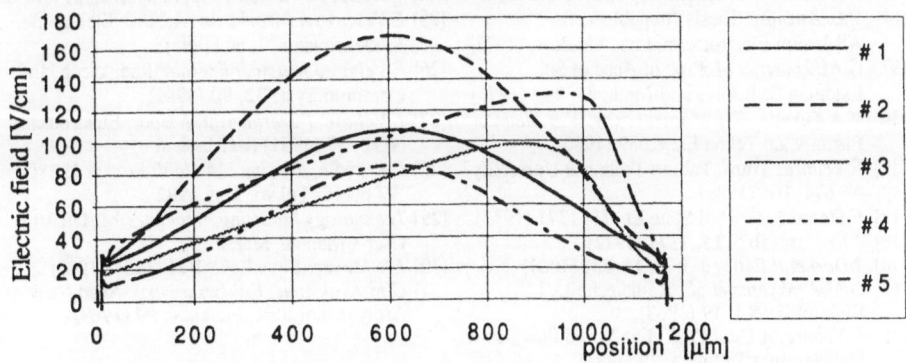

Fig.2. Calculated electric field distributions in the mentioned 5 different cases, J=1000 A/cm²

$$E = \frac{j}{qn(\mu_{nl}+\mu_{pl})} \tag{17}$$

where $j=j_n+j_p$. Thus, the structure voltage drop at the given total current density j is directly depending how the selected EHS accounting way is affecting both the carrier distribution $n(x)$ and the sum of the drift mobilities $\mu_{nl}+\mu_{pl}=f[n(x)]$.

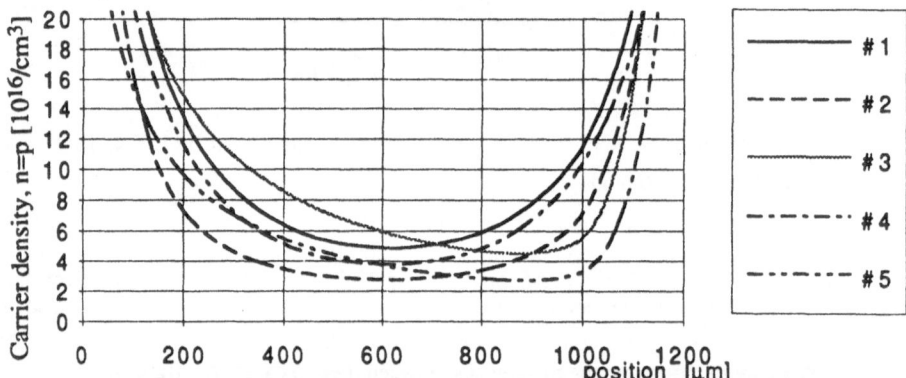

Fig.3. Calculated carrier distributions in the mentioned 5 different cases, J=1000 A/cm^2

Here we will point out only two important conclusions. First, neglecting cross-terms in Eq.(4-5) and simultaneously keeping $J^{eh}\neq0$ (our case #2) is physically inadequate. Second, the *Fletcher*-type mobility summation in Eq.(2) results in a distortion of the carrier distribution shape, turning it strongly unsymmetrical at higher current densities. This is caused by a strong decrease of the ambipolar diffusion coefficient D_a at higher injection levels, which is not consistent with the *Kohler*'s variational principle solution, as noticed in [22-24,30]. A more detailed consideration of various aspects of the EHS influence will be published elsewhere.

REFERENCES:

[1] W.V.Van Roosbroeck: Bell Syst.Techn. J., 29, 560 (1950)
[2] N.H.Fletcher: Proc. IRE, 45, 862 (1957)
[3] L.W.Davies: Nature, 194, 762 (1962)
[4] S.C.Choo: IEEE Trans. El. Dev, ED-19, 954 (1972)
[5] N.G.Nilsson: El. Letters, 8, 580 (1972)
[6] M.S. Adler, G.E. Possin: IEEE Trans. El. Dev., ED-28, 1053 (1981)
[7] R.A.Kokosa: Proc. IEEE, 55, 1398 (1967)
[8] V.L. Kuzmin: Thesis (unpublished), All-Union Electrotechn. Inst., Moscow, (1972)
[9] G.Ashkinazi et al.:Proc.of Acad.of Sc., Estonian SSR,Phys.and Math, 25, 299 (1976)
[10] V.A.Kuzmin, T.T.Mnatsakanov, V.B.Shuman: Pisma v Zh.Tekhn.Fiz,6,689 (1980)
[11] E.Velmre: Trans. Tallinn Technical University, N° 674, 166 (1988)
[12] F.Dannhäuser: Sol.State El., 15, 1371 (1972)
[13] J.Krausse: ibid, 15, 1377 (1972)
[14] J.Dorkel,P.Leturcq: ibid, 24, 821 (1981)
[15] G.M.Avakyants et al.: Radiotechnika i Elektronika,8,1919 (1963)
[16] E.Velmre, A.Udal:Proc.High Speed Pow.Semic. Dev.Seminar,Tallinn, p.39 (1984)

[17] E.G.Paige:J.Phys.Chem.Solids,16,207 (1960)
[18] T.P McLean, E.G.Paige: ibid, 16, 220 (1960)
[19] T.P.McLean, E.G.Paige: ibid, 18, 139 (1961)
[20] T.Mnatsakanov, I.L.Rostovtsev, N.I.Philatov: Fiz.Tekh.Polupr,18,1293 (1984)
[21] M.Kohler: Z. für Physik,124,772 (1948)
[22] T.Mnatsakanov:Phys.Stat.Sol,B143,225(1987)
[23] E.Velmre: Transactions of Tallinn Technical University, N°674, 166 (1988)
[24] V.Kane,A.Swanson:J.Appl.Phys,72,5294(1992)
[25] E.Velmre, A.Pirozhenko, A.Udal: Elektronnoc Modelirovanie, 7, 66 (1985)
[26] E.Velmre,J.Nurste,B.Freydin: Radioelectr. and Commun.Syst, 32, 90 (1989)
[27] E.Velmre, P.Dermenzhi, A.Udal: Elektrotechn, N°10, 37 (1991) (in Russian)
[28] B.Freydin, E.Velmre, A.Udal: Proc.of ISPSD '92 p.118, Tokyo, May 1992
[29] Technology Modeling Associates:MEDICI 1.0 User's Manual, March 1992
[30] J.R Meyer: Phys.Rev., B21, 1554 (1980)
[31] G.M.Avakyants, E.V.Lazarev:Izv.Akad.Nauk Armyanskoi SSR, Fizika, 4, 89 (1969)

Modeling of Localized Lifetime Tailoring in Silicon Devices

P. Hazdra and J. Vobecký

Department of Microelectronics, Czech Technical University of Prague
Technická 2, CS-16627 Praha 6, CZECH REPUBLIC

Abstract

A new accurate method for simulation of the device that is subjected to low-dose high-energy ion irradiation (hydrogen, helium) is compared with the ordinary simulation technique which utilizes a structured lifetime profile as an input into SRH recombination model. The novel method, involving ion-implantation process simulator, expert system, and multilevel recombination model is described. Spatial distribution of the minority carrier lifetime of irradiated device is drawn for different injection levels. Simulated trade-off between forward voltage drop and reverse recovery time is presented for both the hydrogen and helium irradiation.

1. Introduction

Low-dose high-energy ion irradiation is emerging as a powerful tool for local lifetime control in silicon power devices. Light ions, e.g. hydrogen or helium, with energies of several MeV are used to produce vacancy-related point defects mostly appearing towards the end of the ion range. Since these defects act as recombination centers and carrier traps, the reduction of minority carrier lifetime takes place. Consequently, localized lifetime profile may be optimized, choosing appropriate ion type, irradiation energy and dose, and annealing temperature in order to reach a superior trade-off curve of forward voltage drop versus turn-off time or reverse recovery time. Up to this time, an optimization of irradiation-induced lifetime profiles has been based mostly on the principle of trial-and-error experimentation due to unreliable results of device simulations. This is because the simulators are using only a structured lifetime profile [1, 2], a simple recombination model which consider only a single ideal recombination/generation centre, and there is no connection between the parameters of irradiation and lifetime. However, it is possible to fit some way the profile of the lifetimes τ_{n0} and τ_{p0} for SRH model in order to get a better agreement with experimental results [6], but it is a rather hopeless activity, since the

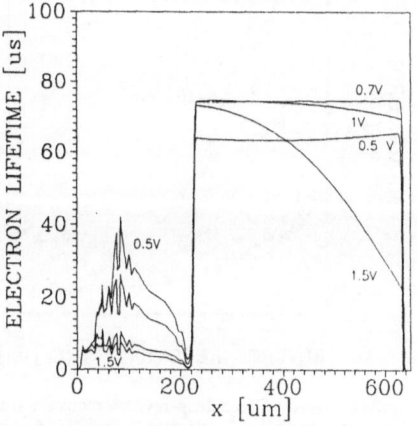

Fig.1: Spatial distribution of electron lifetime $\tau_n = \Delta n/R_n$ in diode 0.5, 0.7, 1, and 1.5V dc (Irradiated by hydrogen, 5MeV/5x10^{10}cm^{-2})

actual lifetime profile is very complicated, hard to be accurately measured with desirable spatial resolution, and is bias dependent. It is documented by Fig.1 which shows the simulated lifetime profile calculated by our new method for a long P^+PNN^+ power diode to be used throughout the remainder of the paper. The irradiated device volume is located within $0 < x < 215\mu m$ (the defect maximum is at $x \doteq 215~\mu m$). During the transient simulation, e.g. reverse recovery process, the profile is even more complicated, especially in the area of desaturated p-n junction. All these circumstances motivated us to developed an exact simulation system to be further described.

2. The simulation system and its application

The simulation system (see the Flow Chart) is based on the integration of simulation procedures that simulate the processes of primary and secondary defect generation during irradiation and their influence on electrical properties of irradiated device. Simulation of primary and secondary defect reactions is replaced by an expert system based on an extensive set of DLTS measurements performed on irradiated silicon [3, 4] which takes into account technological parameters such as target type (CZ, FZ, resistivity), type of ion projectile, dose, dose rate, irradiation temperature, temperature and time of annealing. It assigns the appropriate deep level to each defect created assuming the fact that the secondary defects are mostly vacancy-related and their number increases linearly with the dose within the interval under study [3, 4]. The parameters of deep levels (emission and capture rates, charge state and concentration profile) enter into 1-D device simulator [7] which includes improved models of thermal generation / recombination based on SRH statistics and the complete solution of trap-dynamic equations necessary for detailed description of charge flow through complex structure of traps/recombination centres appearing within silicon bandgap after irradiation. An output of the device simulator provides important linkage of the process and electrical characteristics that have been up to this time available only via experiment. Figs.2 and 3 show the simulated trade-off between voltage

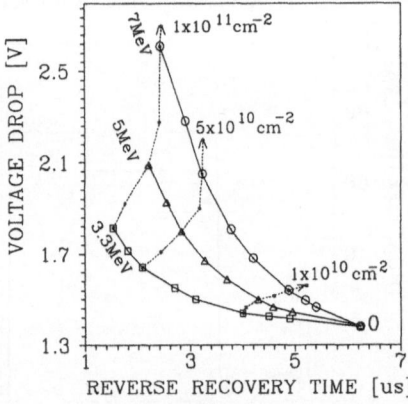
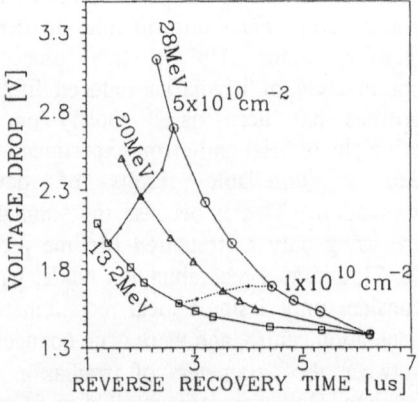

Fig.2: Forward voltage drop-reverse recovery time trade-off for hydrogen (DUT from Figs.1, 3,4,5), t_{rr}: -20V/5Ω/25μH from ON-state

Fig.3: Forward voltage drop-reverse recovery time trade-off for helium (DUT from Figs.1, 2, 4, 5) t_{rr}: -20V/5Ω/25μH from ON-state (200A.cm^{-2})

SIMULATION PROCESS FLOW CHART

PRIMARY
DEFECT GENERATION
■

Ion Irradiadion Process
Simulation
MC code TRIM 90

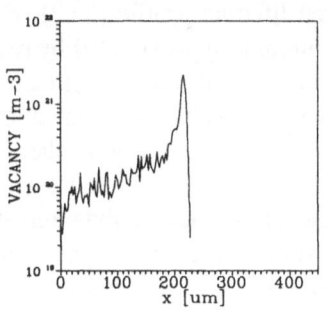

SECONDARY
DEFECT GENERATION
■

Expert System
Based on Experiment

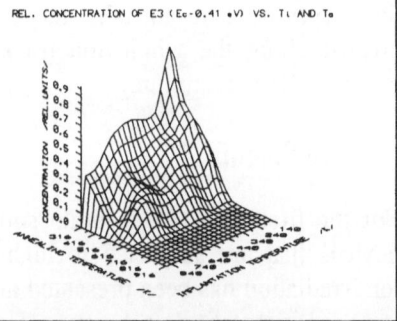

DEVICE SIMULATION
■

1D Simulator
with Multilevel G-R Models
Including Full Trap Dynamics

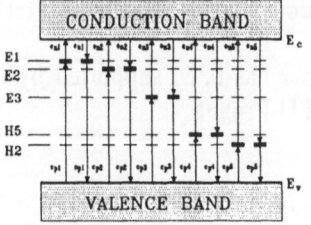

He irradiated CZ Si
model of thermal generation/recombination

SIMULATED
DEVICE PERFORMANCE
■

Reverse Recovery
of a Power Diode

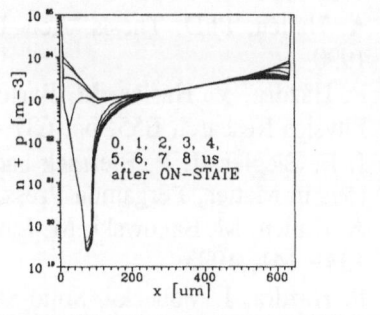

drop and diode reverse recovery time with the irradiation dose and energy as parameters. Figs.4 and 5 provide a comparison of the new method (no.1) with the ordinary simulation based only on a simple-structured lifetime profile (no.2). For the case no.2, the maximum is approximated by rectangular lifetime profile of τ_{n0} and τ_{p0} magnitudes in SRH model, where FWHM values calculated by TRIM [5] are used as a width. In Fig.4, the curve no.1 is in excellent qualitative agreement with measurements [3], while curve no.2 exhibits unrealistic decrease of the the the voltage drop for higher energies. In Fig.5, the curve no.2 gives much more pesimistic trade-off between forward voltage drop and reverse recovery time in comparison with no.1. This is because both the curves no.2 do not take into account the defects created along the whole ion track, i.e. before the concentration maximum.

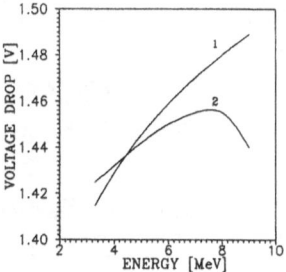

Fig.4: Forward voltage drop vs. irradiation energy of hydrogen (exact simulation curve no.1, simple-structured lifetime curve no.2) for the current density of 200 A.cm^{-2}.

3. Conclusion

For the first time, an accurate simulation of power devices that are subjected to high-energy low-dose ion irradiation has been presented and compared with standard simulation technique. The system proved useful for exact prediction of electrical parameters in conjunction with the irradiation technology.

This paper has been supported by the grant no.8068 from CTU Prague.

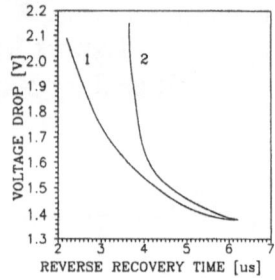

Fig.5: Trade-off between forward voltage drop and reverse recovery time for hydrogen irradiation with energy of 5MeV (exact simulation no.1, structured lifetime no.2).

References:
[1] V. A. K. Temple and F. W. Holroyd, IEEE Trans. Electron Devices, Vol. ED-30, no.7, pp.782 - 790, 1983
[2] M. Hátle, J. Vobecký, Proceedings of the MADEP'91, Florence, pp. 402 - 406, 1991
[3] A. Hallén, PhD. Thesis, Acta Universitatis Upsaliensis, Uppsala, Sweden, 1990
[4] P. Hazdra, V. Hašlar, M. Bartoš, Nuclear Instruments and Methods in Physics Research B55, pp. 637 - 641, 1991
[5] J. F. Ziegler, J. P. Biersack and U. Littmark, The Stopping and Range of Ions in Matter, Pergamon Press, New York, 1985
[6] A. Hallén, M. Bakowski, M. Lundqvist, Solid-State Electronics, Vol. 36, pp. 133 - 141, 1993
[7] P. Hazdra, J. Vobecký, Solid-State Electronics, submitted

Analytical Model of the Metal-Semiconductor Contact for Device Simulation

A. Schenk and S. Müller

Integrated Systems Laboratory, ETH-Zürich
Gloriastraße 35, CH-8092 Zürich, SWITZERLAND

Abstract

We report on the implementation and first numerical results of a new analytical
model of the metal-semiconductor contact in a drift-diffusion device simulator.
The model covers the entire range from Schottky to Ohmic contacts and fits
well with experimental I(V)-characteristics of intermediately doped silicon.

1. Introduction

Usually, in device simulation the physical system "metal-semiconductor (MS) interface"
is treated in form of idealized boundary conditions. Neutrality and equilibrium are assumed for Ohmic contacts and thermionic emission for rectifying (Schottky) contacts. A
model of the non-ideal contact is not only of general interest, but also desirable for certain applications, e.g. the combined Schottky-pn-structure in power diodes (MPS diodes)
or the Schottky Injection Field Effect Transistor (SINFET). Obviously, such a model cannot reflect the entire complicated physics involving barrier tunneling, inelastic scattering,
recombination, trapping and trap-assisted tunneling, potential fluctuations, lateral barrier
height fluctuations, roughness, band-state mixing, carrier heating, image forces, and some
other effects. Since barrier tunneling is commonly accepted to produce Ohmic behavior,
the concept of thermionic field emission (TFE) is successful in explaining the transition
from Schottky to Ohmic contacts as the doping level is increased. Schroeder [1] used a
simplified version of the WKB transmittance of a parabolic barrier (neglecting quantum
reflection) and derived an analytical expression of the emission current j_e suitable for a
boundary condition in device simulation. We believe that because of the importance of
barrier tunneling for the properties of MS contacts with arbitrary doping the substitution
of the WKB approximation by a better approach should be a reasonable improvement, despite the mentioned variety of other physical effects. Details of the new model including the
lengthy formulas are published elsewhere [2]. Here we concentrate on the implementation
of the model in a general drift-diffusion simulator and report on first numerical results. The
essentials of the model are outlined in the following section.

2. Theory

Idealizing assumptions are: parabolic potential barrier (constant doping in the barrier region, Schottky approximation, no image effect, no interfacial layer, etc.), 1D approximation
for the transmission probability, and unique effective mass in the semiconductor.

The WKB approximation is by-passed by interpolating analytically between the asymptotic forms of the eigenfunctions (parabolic cylinder functions) by means of Airy functions. The maximum error at the classical turning points, where the WKB solutions diverge, is shown to be less than 0.2 %. To enable analytical integration the maximum peak of the Airy function is fitted to a Gaussian with an universal attenuation parameter for all doping concentrations. In that way good agreement is achieved with the true transmission probability up to an energy E_{max} well above the maximum of the barrier. For still higher energies the simpler WKB approximation is sufficient to account for quantum reflection there. A fully analytical model is derived if the arguments of the Gaussians are developed with respect to the energy at the maximum of the spectral current density. This maximum is solution of a transcendental equatio. and may be approximated by an expression similar to that given by Crowell and Rideout [3]. To avoid expensive numerical integration including Fermi integrals, we use Boltzmann statistics above and total degeneracy below the Fermi energy, respectively. The final expression then contains error functions as the most complicated ingredients.

Fig. 1 compares $j(V)$-characteristics of the MS contact calculated with the new analytical model against the results of an "exact" reference model, where the correct transmission probability in terms of parabolic cylinder functions (Conley et al. [4]) was used in a numerical integration (not changing the statistics model). Curves labeled "Schroeder" are the corresponding characteristics, if his simplified WKB transmittance [1] is used.

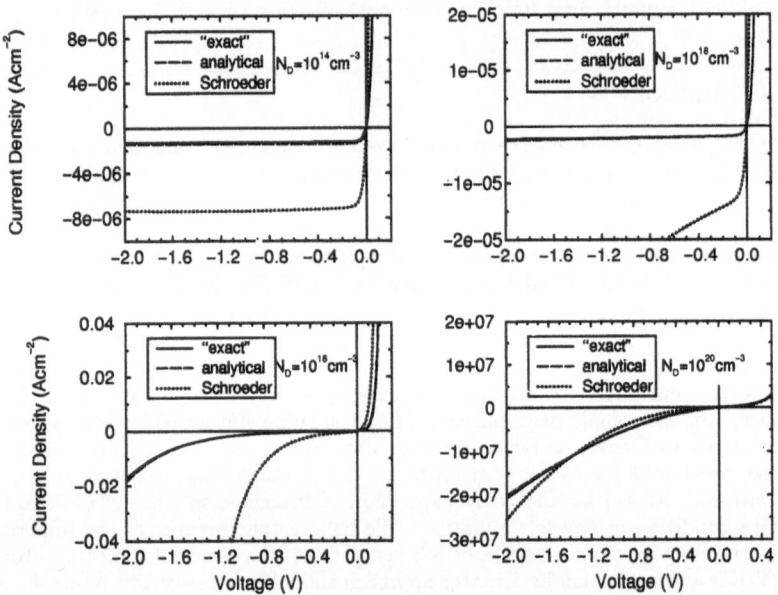

Figure 1: Calculated $j(V)$-characteristics of an Al on n-Si contact with barrier height $= 0.7\,eV$ and $m_c = 0.258\,m_0$ for different models.

3. Implementation

The implementation of the above model in a drift-diffusion device simulator requires the definition of boundary conditions for the electrostatic potential ψ and the quasi Fermi potentials φ_n and φ_p. For feasibility we assume equilibrium, i.e. $\varphi_n = \varphi_p =: \varphi$. This variable is determined by numerically balancing the drift-diffusion current and the TFE current as determined by the analytical model.

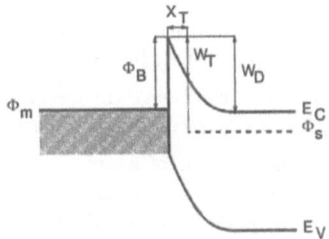

Figure 2: Schematic band diagram

The simplifying assumption of equilibrium is caused by the problem that the TFE current depends on the variable V_{app}, the potential drop over the barrier, which is a non-local variable and hence not available in a device simulator. This is why we approximate it by $V_{app} = \Phi_m - \Phi_s = V_{contact} - \varphi$ (see Fig. 2).

To be able to derive the boundary condition for ψ we have to identify the position in the barrier until which the current is determined by the TFE current and from which it can be treated as a pure drift-diffusion current. In Fig. 2 this point is shown at the depth X_T under the contact. It can be determined from the TFE model by the condition that tunneling remains negligible at lower energies. From the parabolic barrier assumption we can then derive a corresponding energy W_T and using this arrive at the following formula for ψ at the point X_T: $\psi = \varphi + \Phi_{bi} + (W_T - W_D)/q$, Φ_{bi} is the built-in potential. Unfortunately, $W_T - W_D$ depends again on V_{app} and is hence not available. Using the same approximation as above we arrive at: $\psi = \varphi + \Phi_{bi} + (W_T - W_D)_{eq}/q$, $W_{T,eq}$ and $W_{D,eq}$ are the equilibrium values of the energies W_T and W_D shown in Fig. 2. Note, that the expression for ψ reduces to the common boundary conditions in the two extreme cases of a pure Ohmic ($W_{T,eq} = W_{D,eq}$) and the Schottky case ($W_{T,eq} = 0$).

4. Examples

Fig. 3 compares simulation results with the drift-diffusion simulator SIMUL [5] against experimental data of a Kelvin structure (Ti on n-Si with ⟨100⟩-orientation, barrier height – $0.50\,eV$, $N_D = (1.8 - 2.2) \times 10^{18}\,cm^{-3}$). Such a contact represents an intermediate case be-

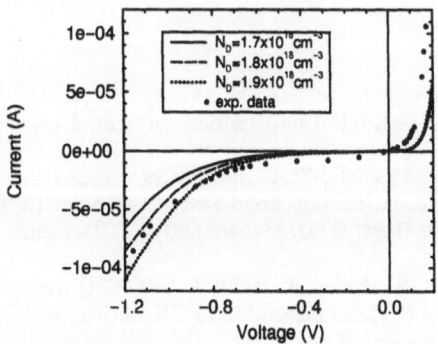

Figure 3: Comparison of simulations with a measured $I(V)$-characteristic of Ti/n-Si with $N_D = (1.8 - 2.2) \times 10^{18}\,cm^{-3}$ and an area of $3.4 \times 10^{-7}\,cm^2$ (dots). Simulated curves are based on the parameter set: $m_c = 0.19\,m_0$, $\Phi_B = 0.50\,eV$, $m_M = m_0$, and $E_{F,M} = 11.7\,eV$.

tween Ohmic and Schottky, and the data are not influenced by an unknown bulk resistance. The reverse bias branch can be well fitted with the transverse effective mass $m_t = 0.19\,m_0$ ($\langle 100\rangle$-orientation!) for doping concentrations in the range $N_D = (1.8 - 1.9) \times 10^{18}\,cm^{-3}$.

Note that no ideality factor was used to remove the deviations at low reverse and forward biases, which are presumably caused by recombination inside the barrier region.

As another example we show the behavior of a *nin* structure (e.g. the Schottky part of a combined Schottky-pn-structure) with a variation of the surface doping concentration. The structure under consideration is $10\,\mu m$ long with a bulk value of $10^{14}\,cm^{-3}$. The one contact is Ohmic with a surface concentration of $N_D = 10^{20}\,cm^{-3}$ and the other is varied in steps from $10^{18}\,cm^{-3}$ to $2 \times 10^{19}\,cm^{-3}$ as shown in Fig. 4a.

The simulated $j(V)$-characteristics in Fig. 4b show the transition from a Schottky diode like behavior to a resistive behavior.

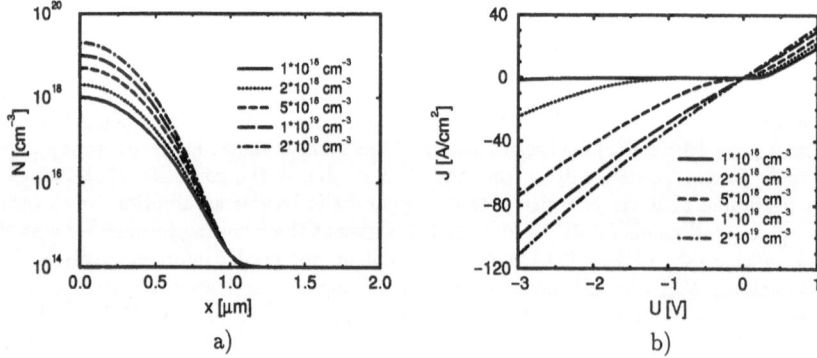

Figure 4: *nin* structures for varying surface doping. a) Doping versus spatial coordinate for the varying well. b) $j(V)$-characteristics.

Acknowledgements

We are grateful to Dr. D. Schroeder (TU Hamburg-Harburg) for many valuable discussions. This work has been financially supported by the Swiss Research Project LESIT.

References

[1] D. Schroeder, Proc. 4th Int. Conf. on Sim. of Sem. Dev. and Proc., Zürich 1991, p. 313
[2] A. Schenk, submitted to Solid-State Electronics
[3] C. R. Crowell and V. L. Rideout, Solid-State Electronics **12**, 89 (1969)
[4] J. W. Conley, C. B. Duke, G. D. Mahan, and J. J. Tiemann, Phys. Rev. **150** (2), 466 (1966)
[5] SIMUL 1.1 manual, S. Müller, K. Kells, J. Litsios, U. Krumbein, A. Schenk, and W. Fichtner, Integrated Systems Laboratory, ETH Zurich, Switzerland, 1993

Nonlinear Contact Resistance and Inhomogeneous Current Distribution at Ohmic Contacts

D. Schröder, T. Ostermann, and O. Kalz

Technische Elektronik, TU Hamburg-Harburg
Eißendorfer Straße 38, D-21073 Hamburg, GERMANY

Abstract

Results of simulations are presented that make use of a recently proposed model for non-ideal ohmic contacts. The model considers both tunneling and thermionic emission currents across the contact. The nonlinearity of the contact resistance is discussed. The two-dimensional current distribution under the contact arising from doping variations is investigated. It is shown that slight doping variations can result in strong current inhomogenities.

1. Introduction

In [1], [2], a model of non-ideal metal-semiconductor contacts for semiconductor device simulation has been proposed. The model considers both tunneling and thermionic emission currents across the contact and allows the simulation of contacts on very low to very highly doped material with a single model. In this paper, we investigate the properties of the model for the case of nearly-ohmic contacts by applying it in actual device simulations.

The paper is organized as follows. In section 2, some details of the implementation are given. In section 3, we compute the current-voltage characteristics of the contact for various doping concentrations and compare them to the usual model of ideal contacts. The nonlinearity of the contact resistance is discussed. In section 4, we investigate the current distribution under the contact in a two-dimensional simulation. In particular, the current distributions at an ideal and a non-ideal contact are compared. Finally, we investigate the change of the current distribution with respect to a slight variation of the doping concentration under the contact.

2. Implementation of the contact model

We implemented a simplified version of the model for nearly ohmic contacts into the device simulator PARDESIM [3], [4]. The model assumptions are 1) charge neutrality on the boundary, 2) tunneling and thermionic emission current of the majority carriers according to [1], and 3) Fermi level continuity across the contact for the minority carriers [5]. These assumptions are used as boundary conditions for the Poisson equation and for the electron and hole continuity equations, respectively. Assumption 1) corresponds to the special case that the tunneling length comprises the total depletion region of the interface [2]. This is only the case if the depletion region is very thin, i.e. if the doping is high. Thus, the model is valid for ohmic contacts. Assumption 2) leads to a finite quasi-Fermi energy step across the contact, and thus to a nonzero specific contact resistance [1]. The well-known ideal contact model differs only in assumption 2) by assuming instead continuous Fermi energy across the contact also for the majority carriers.

Simulations have been carried out for Al contacts on n-Si (0.7 eV barrier height). Since the semiconductor is mostly in degeneration in the considered doping range, we used for the simulations a simplified version of the heavy-doping transport model as described in [6]. Briefly, this model accounts for heavy-doping effects by using a doping-dependent apparent bandgap narrowing.

3. Nonlinearity of the contact resistance

First, we tested the model with a one-dimensional simulation of a resistor, consisting of a bar of homogeneous n-semiconductor with length $L = 6\mu m$. The resistor has an ideal ohmic contact at $x = L$, and a non-ideal contact at $x = 0$ [4]. Fig. 1 shows the current-voltage relationship of the resistor for various doping concentrations.

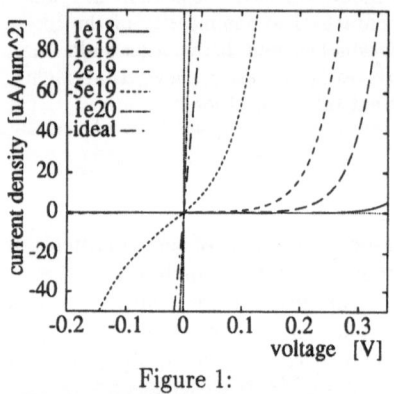
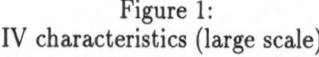
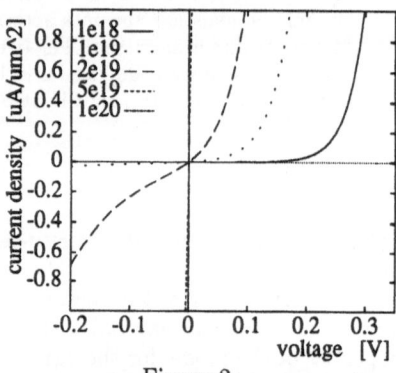

Figure 1:
IV characteristics (large scale)

Figure 2:
IV characteristics (small scale)

We see a linear behaviour for a doping of 10^{20} cm^{-3}. At lower doping concentrations the curves become nonlinear, thus indicating the non-ideal behaviour of the contact in these cases. The physical reason is that the tunneling probability is strongly sensitive to variations of the electron Fermi level drop across the contact [1]. We found that the applied bias dropped almost entirely across the non-ideal contact, indicating that its resistance is much larger than that of the semiconductor bulk.

For comparison, the line for a 10^{18} cm^{-3} doped resistor with two *ideal* ohmic contacts is displayed in Fig. 1, too. We note large differences between the devices with the ideal and non-ideal contacts. The slope of the line in the ideal contact case is inversely proportional to the resistance of the semiconductor bulk, since the ideal contact has zero resistance.

Fig. 2 shows the same plot as Fig. 1, but on a much smaller current density scale. It is interesting to note that the character of the curves depends on the scale. While the $5 \cdot 10^{19}$ cm^{-3} curve in Fig. 1 appears nonlinear, it looks like a nearly ideal contact in Fig. 2. The $2 \cdot 10^{19}$ cm^{-3} curve in Fig. 1 shows a rectifying behaviour, while on the smaller scale it looks like a nonlinear resistor. From this observation we conclude that the ideal or non-ideal appearance of a contact depends on the magnitude of the current density flowing through the contact, which might not be determined by the contact alone but also by depletion regions etc. inside the device. Thus it depends on the particular operating conditions if the use of an ideal contact model, a current independent contact resistance, or the non-ideal model is appropriate.

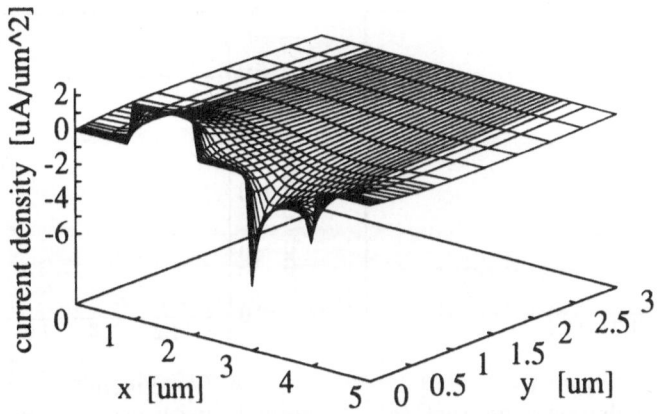

Figure 3: y-component of electron current density

4. Current distribution under the contact

In a second example, we investigated the current density distribution obtained in a two-dimensional simulation [4]. The simulated device was a $5\mu m \times 3\mu m$ piece of silicon with two planar contacts of $1\mu m$ length and a constant doping of 10^{19} cm^{-3}. Fig. 3 shows the distribution of the y-component of the electron current density at a bias of 0.2 V. The non-ideal contact ($x = 1..2\mu m$) has a smooth current distribution, while the ideal contact model at $x = 3..4\mu m$ exhibits sharp current density peaks at the ends of the contact. The reason for the latter effect is that the ideal contact model demands constant Fermi levels and potential and thus effectively "shortcuts" the semiconductor. Hence, the electrons leave the metal predominantly at the ends of the contact. The contact resistance of the non-ideal contact, on the contrary, ensures a uniform distribution of the current density. We found that the uniform distribution in this example is nearly unaffected by the magnitude of the bias or the electric field near the contact. Thus, the use of a position independent contact resistance is justified as long as the doping under the contact does not change.

In order to investigate the influence of doping variations on the contact resistance, we introduced a slight increase of the doping concentration in the region near the non-ideal contact. Fig. 4 shows that the doping in the contact plane now varies from $4 \cdot 10^{19}$ cm^{-3} to $\approx 2 \cdot 10^{19}$ cm^{-3}. In Fig. 5 we can see that this slight doping variation causes a significant reduction of the current density under that portion of the contact where the doping is lower. While the doping drops by a factor of two, the current density drops by a factor of 10. In other words, the effective contact size reduces to ca. 75% of the original metallization size. As apparent from Fig. 1, this effect is due to the drastic sensitivity of the contact resistance to the doping concentration. Thus, the slight doping variation of Fig. 4 leads to a strong inhomogenity of the specific contact resistance, and a correspondingly inhomogeneous current distribution. A simple estimation of the contact current density from the total current and the contact area might underestimate the peak current density in such cases. For simulations where the current distribution near the contacts is important, it can be expected that the non-ideal model would give more satisfactory results.

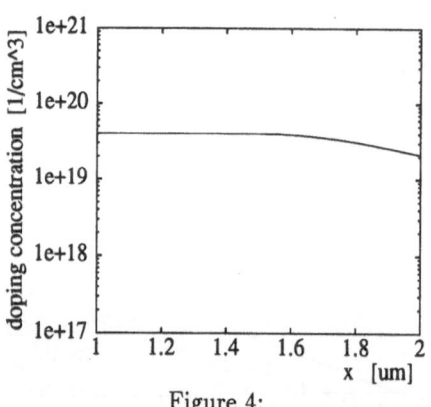
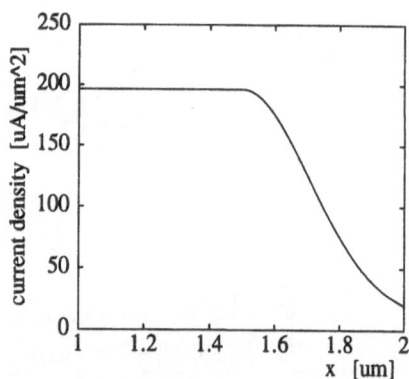

Figure 4:
Doping distribution under contact

Figure 5:
Normal current density under contact

References

[1] D. Schroeder. An analytical model of non-ideal ohmic and Schottky contacts for device simulation. In W. Fichtner and D. Aemmer, editors, *Proc. 4th Int. Conf. on Simulation of Semiconductor Devices and Processes, Sept. 12-14, 1991, Zurich.* Hartung-Gorre, Konstanz, 1991.

[2] D. Schroeder. A boundary condition for the poisson equation at non-ideal metal-semiconductor contacts. In *Proc. 8th Int. Conf. on Numerical Analysis of Semiconductor Devices and Integrated Circuits (NASECODE VIII), Dun Laoghaire, May 18-22, 1992, Vienna.* Boole Press.

[3] O. Kalz, D. Schroeder. PARDESIM - A parallel device simulator on a transputer based MIMD machine. *Proc. 5th Int. Conf. on Simulation of Semiconductor Devices and Processes, Sept. 7-9, 1993, Vienna* (this issue).

[4] T. Ostermann. *Implementierung und Erprobung eines Modells für nichtideale Metallkontakte in einem Devicesimulator* Diplomarbeit TU Hamburg-Harburg, 1993.

[5] E.H. Rhoderick and R.H. Williams. *Metal-semiconductor contacts.* Clarendon Press, 1988.

[6] D. Schroeder, T. Ostermann, O. Kalz. Comparison of transport models for the simulation of degenerate semiconductors. To be published.

SIMULATION OF SEMICONDUCTOR DEVICES AND PROCESSES Vol. 5   449
Edited by S. Selberherr, H. Stippel, E. Strasser – September 1993

Simulations of Carrier-Blocking Effects on Cutoff Frequency Characteristics for AlGaAs/GaAs HBTs with Insulating and Semi-Insulating External Collectors

K. Horio and A. Nakatani

Faculty of Systems Engineering, Shibaura Institute of Technology
307 Fakasaku, Omiya 330, JAPAN

Abstract

Two-dimensional simulation of AlGaAs/GaAs HBTs with a perfectly insulating external collector is performed. It is shown that the cutoff frequency can degrade heavily due to its carrier blocking effect that leads to an increase in base delay time. In relation to this effect, a design criterion for collector-up HBTs will also be discussed.

1. Introduction

Recently, AlGaAs/GaAs heterojunction bipolar transistors (HBTs) have received great interest for application to high-speed and high-frequency devices. To reduce the parasitic base-collector capacitance and to improve the high- frequency performance, semi-insulating external collectors are often introduced [1]. Also, collector-up HBTs without external collectors are fabricated and examined [2]. In a previous work, we simulated AlGaAs/GaAs HBTs with a semi-insulating external collector [3], and found that the cutoff frequency f_T could degrade when electrons become injected into the semi-insulating layer [4]. In this work, we have simulated AlGaAs/GaAs HBTs with a perfectly insulating external collector and studied how its current blocking affects f_T characteristics. In relation to this, a collector-up IIBT is also simulated, and a design criterion for it will be discussed.

2. Physical model

Device structure simulated in this study are shown in Figure 1. (a) and (b) are emitter-up HBTs with semi-insulating (SI) and perfectly insulating (I) external collectors, respectively. We assume that the SI-layer is achieved by introducing a deep acceptor into the n^--layer [3]. (Its density N_T must be higher than the n^--layer doping density N_{C1}.) In the I-layer, current flow is prohibited. Figure 1(c) is a collector-up structure where the external emitter is assumed to be ideally insulating. The Poisson's equation and continuity equations for electrons and holes are solved numerically in two dimension. Here, we concentrate our attention on how the f_T characteristics are affected by the collector structures.

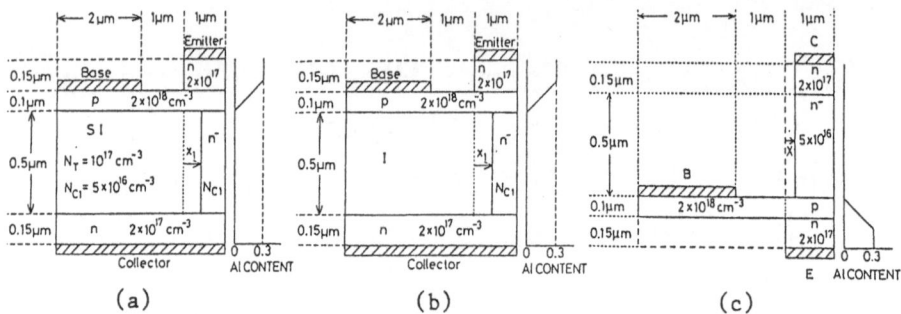

Figure 1: Device structures simulated here. (a) and (b) are usual emitter-up HBTs with semi-insulating (SI) and ideally insulating ($^\intercal$) external collectors, respectively. (c) is a collector-up HBT with an ideally insulating external emitter.

3. f_T for emitter-up HBTs and carrier-blocking effect

Figure 2 shows f_T versus collector current density I_C curves of emitter-up HBTs with different x_1 in Figures 1(a) and (b). Positive x_1 means that the SI- or I-layer extends into the intrinsic collector region, while negative x_1 means that it is away from the intrinsic collector region. $N_T = 0$ in Figure 2 corresponds to a case with an n^- external collector. It is seen that by introducing a SI- or I-layer, f_T is improved, as expected, when $x_1 < 0$. But, when $x_1 \geq 0$, f_T is degraded and the degradation is much more remarkable for a case with an I-layer. The latter point is an unexpected result. It is understood that when $x_1 > 0$, effective channel width becomes narrow in the collector region and so a high injection condition appears earlier, leading to a lower f_T. However, this situation is similar for the two cases with SI- and I-layers. So, another mechanism for degrading f_T should exist for the case with an I-layer.

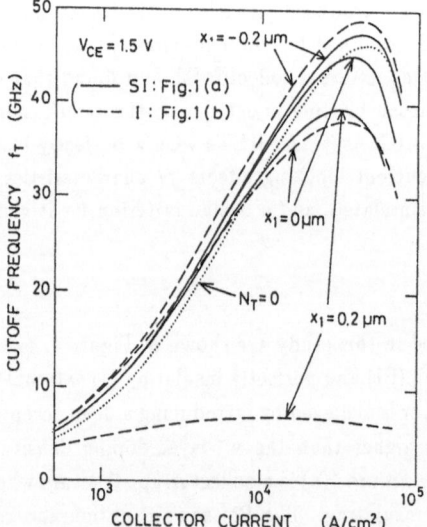

Figure 2: Cutoff frequency f_T versus collector current density I_C curves of emitter-up HBTs, with x_1 as a parameter. $N_T = 0$ corresponds to a case with a usual n^- external collector.

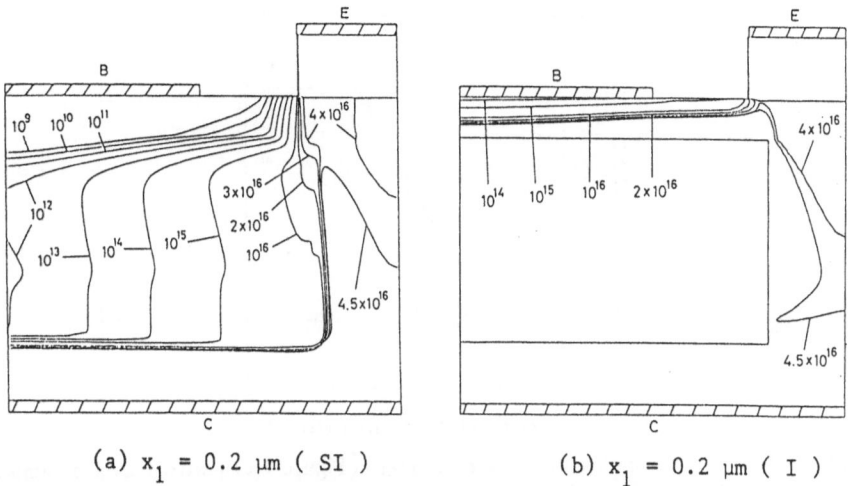

(a) $x_1 = 0.2$ μm (SI)   (b) $x_1 = 0.2$ μm (I)

Figure 3: Electron density profiles of emitter-up HBTs with different external collectors. $V_{CE} = 1.5$ V and $I_C = 6.5 \times 10^4$ A/cm^2.

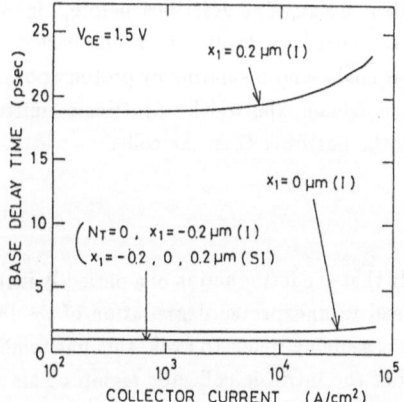

Figure 4: Base delay time versus I_C curves of emitter-up HBTs with different collector structures, corresponding to Figure 2. No remarkable differences are seen for three cases with semi-insulating (SI) external collectors.

Figure 3 shows electron density profiles in a high current region. It is seen that for the case with an I-layer, relatively high densities of electrons exist in the external base region. It is interpreted that these electrons are blocked by the I-layer. In this case, the effective base delay time τ_B becomes very long as shown in Figure 4, because τ_B is given by $(\delta Q_{nB}/\delta I_C)_{V_{CE}}$ where Q_{nB} is electron charges in the base region. Therefore, the remarkable degradation of f_T for the case with an I-layer ($x_1 > 0$) is due to the increase in the base delay time.

4. f_T for collector-up HBTs

The above carrier-blocking phenomenon may become a problem in a collector-up HBT. Figure 5 shows $f_T - I_C$ curves of collector-up HBTs as a parameter of x in Figure 1(c). When $x > 0$, that is, when the collector width becomes narrower than the emitter width,

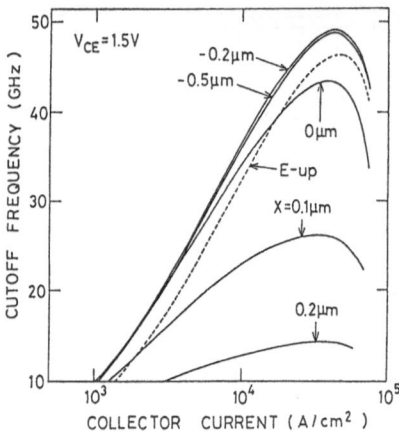

Figure 5: f_T versus I_C curves of colletor-up HBTs (Figure 1(c)), with x as a parameter. E-up corresponds to a case of $N_T = 0$ in Figure 2.

f_T is degraded heavily. This is because, as described before, high densities of electrons are blocked and remain in the external base, leading to a long base delay time. In real devices, the external emitter region is made semi-insulating by proton or oxygen ion-implantation [2]. So, to avoid the above phenomenon, the width of intrinsic emitter region determined by the ion-implantation should be narrower than the collector width.

5. Conclusion

We have shown theoretically that the introduction of a perfectly insulating external collector in emitter-up HBTs may lead to unexpected degradation of f_T due to its carrier-blocking and the resulting increased base delay time. To avoid this phenomenon, the insulating layer should be slightly away from the intrinsic collector region. This means that in collector-up HBTs, the width of effective intrinsic emitter region must be made narrower than the collector width.

References

[1] J. J. Liou, "Modeling the cutoff frequency of $Al_x Ga_{1-x} As$/GaAs heterojunction bipolar transistors with proton-implanted collector region", *Solid-State Electron*, Vol.33, pp.1329-1331, 1990

[2] S. Yamahata, Y. Matsuoka, and T. Ishibashi, "High-f_{max} collector-up AlGaAs/GaAs heterojunction bipolar transistors with a heavily carbon-doped base fabricated using oxgen-ion implantation", *IEEE Electron Device Lett.*, Vol.14, pp.173-175, 1993

[3] K. Horio, A. Oguchi, and H. Yanai, "Two-dimensional simulation of AlGaAs/GaAs HBTs with various collector structures", *Proceedings of SISDEP'91*, pp.81-90, 1991

[4] K. Horio, A. Oguchi, and H. Yanai, "Two-dimensional analysis of high injection effects in AlGaAs/GaAs HBTs with semi-insulating external collectors", *Solid-State Electron.*, Vol.34, pp.1393-1400, 1991

Non-Stationary Transport HBT Modeling Under Non-Isothermal Conditions

A. Benvenuti, G. Ghione, and C. U. Naldi

Dipartimento di Elettronica, Politecnico di Torino
Corso Duca degli Abruzzi 24, I-10129 Torino, ITALY

Abstract

Because of the interaction between thermal and hot carriers effects, neither isothermal nor conventional macro-thermal models are adequate for state-of-the-art power heterojunction bipolar transistors (HBTs); instead, a non-isothermal hot carrier transport model, such as the *thermal-fully hydrodynamic* model, is required. We apply such a detailed thermal model to the simulation of an AlGaAs/GaAs HBT, comparing the results with those provided by simplified models, and highlighting how deeply both non-stationary transport and self-heating affect the predicted device performance.

1. Introduction and modeling approach

The dynamical behaviour of each of the subsystems composing the semiconductor (electron, hole and phonon populations) can be conveniently and quite generally described by means of a Boltzmann transport equation, relating the time evolution of the corresponding distribution function to the collision contribution.

Applying the method of moments in $\underline{k}$ space to such Boltzmann equations, three coupled sets of partial differential equations are obtained in the single electron and hole gases approximation. Retaining the moments of order $0 \div 2$ for the electron gas subsystem, conservation equations for the electron concentration, average velocity and average kinetic energy, respectively, are obtained [1]. In particular, we restricted ourselves to the steady-state, 1D case, and closed the system by expressing the electron gas heat flux with the Fourier law [2].

As to the lattice, the moment of order 2 suffices for the description of its dynamics:

$$\frac{\partial(\rho c_l T_l)}{\partial t} + \nabla \cdot \underline{Q}_l = \left. \frac{\mathrm{d}(\rho c_l T_l)}{\mathrm{d}t} \right|_{coll}. \tag{1}$$

Being physically related to scattering events, the collision terms can be split into an intraband contribution, described with the relaxation time approximation, plus an interband contribution, depending on generation-recombination mechanisms [3].

The charge, momentum and energy conservation equations for the electrons, a similar set for the holes, the Poisson equation and the lattice heat equation together have been named *thermal-fully hydrodynamic* (T-FH) model [1]. The T-FH formulation is completed by imposing a set of coupled, mixed boundary conditions for the electron, hole and lattice temperatures T_n, T_p and T_l, accounting for the 3D heat spreading through the metallization and the substrate by means of a geometrical transformation.

If the so-called convective terms [4] are neglected, the *thermal-energy balance* (T-EB) model [5] is recovered as a particular case, and the moments of order $0 \div 1$ can be joined into a single second-order current continuity equation.

Both T-FH and T-EB may be considered as *detailed* thermal models, since they allow for a separate description of the energy stored in (and carried by) the electron, hole and phonon subsystems. Just a few detailed thermal simulations have been reported up to now; the T-FH model has been proposed in [6, 3], and applied to HBT simulation in [1], while the simpler T-EB model has been proposed in [7] and applied to BJT simulation in [5].

Most thermal models are based on a *macroscopic* approach, whereby the thermal contributions arising from electrons, holes and phonons are considered as a whole. In fact, if optical generation-recombination mechanisms are neglected, applying a global energy balance principle for the collision terms [8, 9], a lumped heat equation is obtained, which describes the time evolution of the total energy density stored in the semiconductor. Assuming $T_n \simeq T_p \simeq T_l$, and neglecting the Peltier term $5k_B T_l / 2q \nabla \cdot (\underline{J}_n + \underline{J}_p)$, the widely-used *macro-thermal* model [10, 11] is recovered:

$$\frac{\partial(\rho c T_l)}{\partial t} - \nabla \cdot (\kappa \nabla T_l) = \underline{J}_n \cdot \underline{\mathcal{E}}_n + \underline{J}_p \cdot \underline{\mathcal{E}}_p + E_g R, \qquad (2)$$

where $c = c_l + 3k_B(n+p)/(2\rho)$ and $\kappa = \kappa_l + \kappa_n + \kappa_p$ are the lumped specific heat and thermal conductivity.

To preserve the main advantage of the macro-thermal approach, that is its simplicity, the differential equations expressing conservation of carrier energy must be eliminated from the electrical model as well. Usually, this is implicitly accomplished resorting to the drift-diffusion formalism of carrier transport, *i.e.* neglecting the carrier energy fluxes and assuming local quasi-field dependent mobilities (T-DD model).

An appropriate modeling of the T_l-dependence of some key parameters is critical in obtaining accurate thermal simulations: the relaxation times can be conveniently expressed [2] in terms of the low field mobility μ_0 and saturation velocity v_{sat}, for whose T_l dependence both experimental data and analytical expressions are available, while the Varshni formula can be applied for the thermal band gap shrinkage [2].

We introduced both surface and bulk trap-related recombination, described with the conventional SRH model, direct band-to-band recombination, and avalanche genera-tion: for the latter we tested average velocity [6] and average energy [3] dependent impact ionization coefficients, as well as the conventional field dependence. Also, we modified the model proposed in [12] to describe hot carrier effects on the Auger recombination coefficients.

All of the mentioned thermal models (T-FH, T-EB, T-DD) and their isothermal coun-terparts (FH, EB, DD) have been implemented in the framework of a highly flexible device simulation code [1] based on the 1-liner technique, exploiting generalized con-tinuation, grid adaption and automatic jacobian calculation, and allowing for an arbitrary number of differential/algebraic equations. Several discretization schemes (including Scharfetter-Gummel, pure 1-sided and optimal upwinding [4]) may be ap-plied, while all equations are solved self-consistently applying the full-Newton method.

2. Results and discussion

We simulated a 10-emitter fingers AlGaAs/GaAs HBT with a base-emitter junction linearly graded over 20 nm; since hot holes effects are expected to play a minor role in the electro-thermal behaviour of such Npn devices, we adopted the drift-diffusion model to describe hole transport.

The dramatic impact of self-heating on the peak cutoff-frequency is apparent in Fig.1; furthermore, neglecting non-stationary transport, the DD model (and even more the T-DD model) gives a much lower f_T in high injection conditions, where the collector delay dominates.

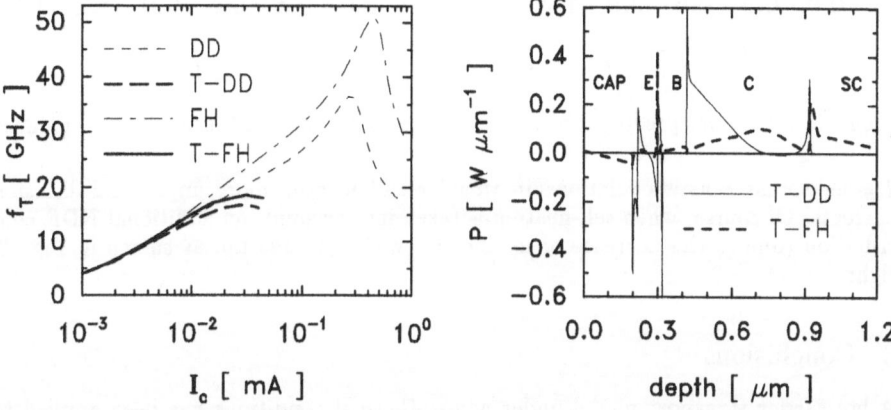

Figure 1: Cutoff frequency for the different models.

Figure 2: Lattice thermal sources profiles for the T-FH and T-DD models.

Conventional macro-thermal models do not correctly locate the lattice heat sources, for the same reason that they cannot describe non-stationary transport, i.e. beacause they neglect the energy flux related to the electron and hole populations. Due to such fluxes, the lattice heat source profile is much smoother than suggested by the Joule terms ($\underline{J}_n \cdot \underline{\mathcal{E}}_n + \underline{J}_p \cdot \underline{\mathcal{E}}_p$), with a maximum substantially displaced from the base-collector to the collector-subcollector junction (Fig.2). In contrast with the results of Liou and co-workers [11], no $T_l < T_a$, where T_a is the ambient temperature, is observed in the emitter, even though a substantial lattice cooling takes place.

The performances of high-power HBTs can be severely limited by the onset of a negative output differential resistance (NDR). Although this effect, which is frequently observed in measured output characteristics, is commonly ascribed to device heating, our simulations (see Fig. 3) point out that NDR might not be entirely due to thermal effects. In fact, only the isothermal DD model results in flat output I-V characteristics, while, even for $T_l = 300$ K, both non-stationary transport models (EB and FH) predict a decreasing I_c versus V_{ce}. This appears to be due to the drift current being opposed by a higher thermal (i.e., due to a diffusivity gradient, related to a T_n gradient) diffusion current in the collector region.

Even if the DD model is modified in order to take into account the carrier energy fluxes (and thus non-stationary transport), I_c remains nearly flat in the saturation region, though at a higher value than for DD, as long as the diffusivity/mobility ratio is assumed to be proportional to T_l. The estimate of the electron temperature T_{eff}, provided by the DD model on the basis of the static field-temperature relationship, can be very inaccurate, especially near the b-c junction; therefore, assuming a

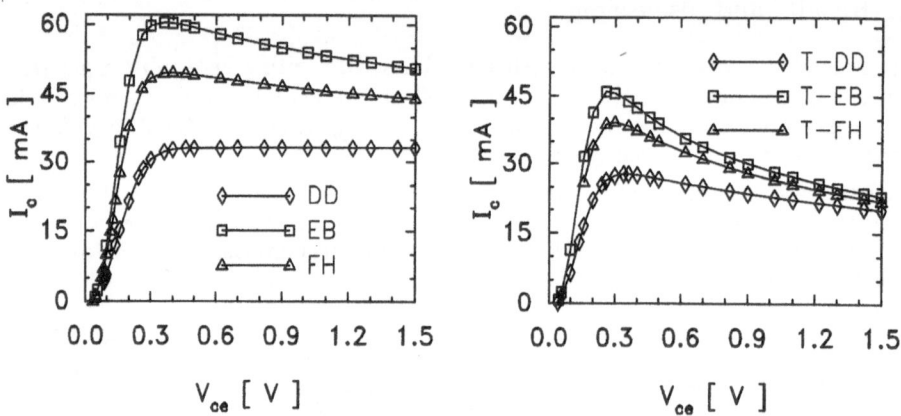

Figure 3: I-V curves predicted by the isothermal (left) and the thermal (right) models.

T_{eff}-dependent Einstein relationship would result in even more unphysical diffusion currents. Of course, when self-heating is taken into account, an additional NDR contribution (due to the decrease of μ_0 and v_{sat} with T_l) adds up, as shown in Fig. 3, right.

3. Conclusions

A hot-carrier transport model under non-isothermal conditions has been applied to the simulation of multi-emitter power HBTs. Comparisons with previous models confirm that the interplay between non-stationary and thermal effects is essential in power HBT modeling. Further work will aim at gaining a deeper insight on the electrical and thermal mechanisms limiting the performances of these devices.

Acknowledgements. We are indebted to SIEMENS AG for providing information concerning HBT devices, and to the AT&T Bell Labs for software support. The work has been partly funded by EEC through the MANPOWER ESPRIT project.

References

[1] A. Benvenuti et al., *IEDM Tech. Dig.-92*, pp. 737-740. 1992

[2] A. Benvenuti et al., *Proc. 4th Int. Work. GaAs Tel.*, pp. 139-156. 1993

[3] J. W. Roberts, S. G. Chamberlain, *COMPEL*, Vol. 9, pp. 1-22. 1990

[4] A. Benvenuti et al., *Proc. NUPAD-92*, pp. 155-160. 1992

[5] S. Szeto, R. Reif, *Solid-St. Electron.*, Vol. 32, pp. 307-315. 1989

[6] C. T. Wang, *Solid-St. Electron.*, Vol. 28, pp. 783-788. 1985

[7] G. Amaratunga, W. Ying-Jian, *Solid-St. Electron.*, Vol. 33, pp. 1343-1346. 1990

[8] P. Ciampolini et al., *IEDM Tech. Dig.-92*, pp. 733-736. 1992

[9] C. C. Mc Andrew et al., *Sem. Sci. Technol.*, Vol. 3, pp. 758-765. 1988

[10] A. Chryssafis, W. Love, *Solid-St. Electron.*, Vol. 22, pp. 249-256. 1979

[11] L. L. Liou et al., *IEEE Trans. Electron Devices*, Vol. ED-40, pp. 35-43. 1993

[12] W. Quade et al., *IEEE Trans. CAD*, Vol. 10, pp. 1287-1294. 1991

An Enhanced Two Dimensional Hydrodynamic Energy Model for Transient Time Simulation of Complex Heterostructure Field Effect Transistors

K. Sherif, G. Salmer, and O. L. El-Sayed[†]

IEMN, UMR CNRS 9929, Département Hyperfréquences et Semiconducteurs,
Université des Sciences et Technologies de Lille
Bât. P4, F-59655 Villeneuve d'Ascq Cédex, FRANCE
[†]Electronics and Telecommunications Department,
Faculty of Engineering, Cairo University
EG-11517 Cairo, EGYPT

Abstract

In this work we present an enhanced two dimensional hydrodynamic energy model used for the simulation of complex heterostructure field effect transistors. We highlight its capabilities and potential performance using the state of art computational technologies. The major modifications are presented and discussed; then typical obtained results are presented.

1. Introduction

The interest in complex heterostructure field effect transistors has grown in the past years. The attention was particulary drawn towards pseudomorphic channel AlGaAs\InGaAs because of their particularly unprecedented performance in microwave power applications. Yet their performance falls at 60 GHz mainly due to non-optimised structures.

The optimisation of such devices passes by the investigation of different structural parameters and studying their effect upon the device performance.

From the numerous parameters emerges the necessity of a reliable simulation tool that should satisfy two major criterea : Accuracy and Rapidity. The two dimensional hydrodynamic energy models satisfy these criterea.

Based upon a simpler model used for conventional HEMT simulation, our model was modified to permit the simulation of the more complex pseudomorphic HFETs.

In the following section we will be presenting the features of our enhanced model and the computational technics employed to increase its efficiency. Then we will present typical results obtained.

2. The enhanced model

The enhanced 2D model is based upon the three transport equations driven through the integration of Boltzmann Transport Equation for heavily doped multivalley semiconductors [1]. This model was then modified by including the effect of minority carriers (holes) into Poisson's equation, adding the holes current continuity equation and the holes current density equation.

Carrier generation by impact ionisation was included in the two charge conservation equations. Quantization effects were accounted for by assuming that the electrons start with an energy equivalent to that of the first sub-band and not the bottom of conduction band as was done before.

Another modification made was that we included the screen effect. This effect was noticed experimentaly, and was introduced based upon an emperical formula $\mu(Ns)$.

The final set of equations composing our model then reads :

Poisson Equation :

$$\nabla^2 V = \frac{q}{\epsilon}(n - Nd - p) \tag{1}$$

Electrons current continuity :

$$\frac{\partial n}{\partial t} = -\nabla n \vec{v}_n + \alpha_n \frac{J_n}{q} + \alpha_p \frac{J_p}{q} \tag{2}$$

Holes current continuity :

$$\frac{\partial p}{\partial t} = \nabla p \vec{v}_p + \alpha_p \frac{J_p}{q} + \alpha_n \frac{J_n}{q} \tag{3}$$

Average electron energy balance :

$$\frac{\partial \varepsilon}{\partial t} + \vec{v} \cdot \nabla \varepsilon = \frac{\vec{v}^2}{\mu(\varepsilon)} - KT(\varepsilon)\nabla \cdot \vec{v} - \frac{\varepsilon + \delta\varepsilon}{\tau(\varepsilon)} \tag{4}$$

And the current density J calculated by :

$$J_n = \mu(\varepsilon)nE - \nabla \cdot (\mu(\varepsilon)nKT) \tag{5}$$

$$J_p = -\mu(E)pE \tag{6}$$

The dependance of the different physical parameters upon energy or field is driven from the steady-state Monte-Carlo simulations. The ionisation coefficient α, is taken as a function of the average total energy for electrons.

The whole system of equations is discretised using a finite difference scheme with a variable mesh size; linearised with respect to time using a dynamic time step that is controlled by the simulation program. The system of equations is decoupled based upon the relaxation time approximation. Poisson's equation is solved using the MDS technique [2] , the energy equation is linearised and solved, as well as the continuity equations, using an iterative scheme.

The decoupling of equations permitted the use of parallel execution techniques to solve the energy balance equation, and the two current continuity equations in the same elapsed time.

The whole system of equations was written, in the discretised form, in a vectorial formulation. This allowed the use of the inherently fast vector processors resulting in a significant reduction in the execution time.

3. Computational performance

The simulations were done using an IBM-3090-600E equiped with parallel vector processors of which we used at least 4. A typical execution time recorded was 1 cpu minute for 1 psec of simulation time, keeping in mind that we need 3-4 picosecond to reach the steady state. This time however increased as the complexity of the structure increases; for a 0.1 μ gate,δ-doped, double plan, pseudomorphic transistor we were at 10 cpu min for 1 psec real time. The simulation of dual-gate conventional HFET came up with a time performance of about 40 cpu min for 1 psec real time. It is important to note that the mesh size varies typically from 91x42 uptill 120x49, which is considered relatively large mesh.

4. Examples of obtained results

Figure 1 shows the effect of 2DEG screening, where we notice considerable reduction in the maximum intrinsic current gain cut-off frequency F_{ci}. The obtained values are in good agreement with those obtained experimentaly (* points).

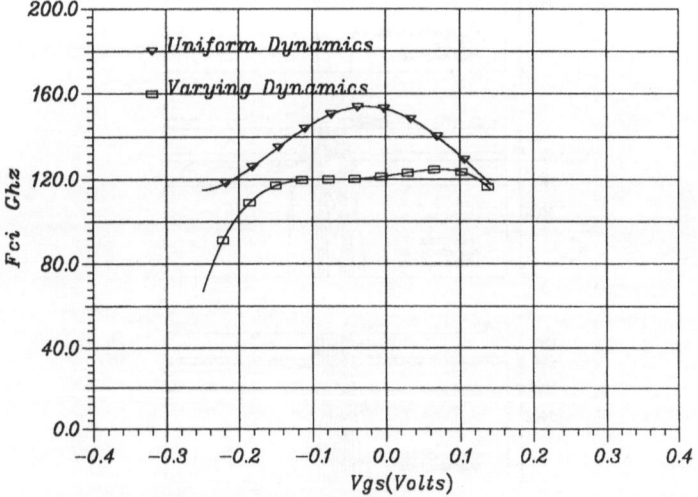

Figure 1: Effect of 2DEG screening on F_{ci} in .3μ gate, δ-doped HFET

Our model is quite capable of simulating the HFET under breakdown conditions. In figure 2 we can see such an effect occuring in a 0.3μ gate uniformely doped HFET.

To our knowledge this is the only model that is capable of simulating the avalanche breakdown in HFETs in transient time

Another feature of this model is that it gives a good insight of the device physics through isolines of different quantities. Figure 3 shows such isolines of electrons density, energy and potential contours.

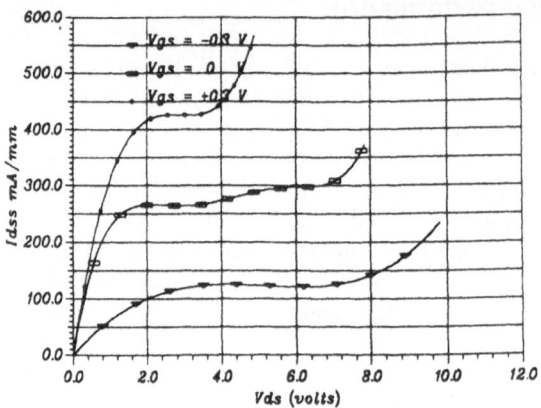

Figure 2: I-V characterstics of .3μ gate HFET including avalanche multiplication

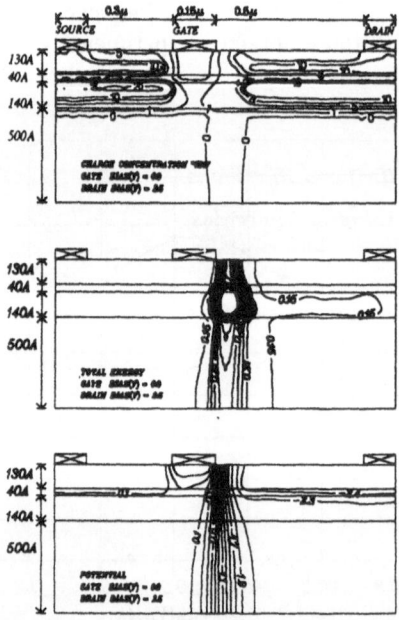

Figure 3: Isolines of physical quantities in a 0.15μ gate, δ-doped pseudomorphic HFET with AlGaAs buffer

5. Conclusion

We have developped an enhanced two dimensional hydrodynamic energy model that serves as a powerfull tool in the simulation and optimisation of largerly submicronic gate pseudomorphic δ - doped HFET. The model is capable of predicting transient time as well as steady state DC characteristics.

References

[1] Tarek Shawky et al, *Optimisation of MODFETS...*, IEEE, TED, Jan. 1991.

[2] M.Saadoon, *M.Sc Theses*, Cairo University, 1983.

Evaluation of Effective Device Parameters by Comparison of Measured and Simulated C-V Characteristics for Conventional and Pseudomorphic HEMTs

R. Deutschmann, C. Fischer[†], C. Sala[†], and S. Selberherr[†]

Corporate Research and Development, Siemens AG
Otto-Hahn-Ring 6, D-81739 München, GERMANY
[†]Institute for Microelectronics, TU Vienna
Gußhausstraße 27-29, A-1040 Wien, AUSTRIA

Abstract

Measurements of the gate C-V characteristics of several conventional and pseudomorphic high electron mobility transistors (HEMT) on wafer and the comparison with simulations are presented. In order to study the influence of important technological parameters on the capacitance, the Schrödinger and Poisson
equations were solved self-consistently in the structure, using the thickness of
the doped layer d_A, the doping density N_D and the built-in voltage V_b as fit
parameters. Measurement and simulation were found to be in good agreement
and the fit parameters can be shown to be the effective device parameters. We
demonstrate how to apply this technique for monitoring the spatial variation
of d_A, N_D and V_b over the wafer, a result of particular importance for the development of the manufacturing process and for calibrating the design of the
device.

1. Introduction

It is well known that the shape of the gate capacitance C_g versus gate voltage V_g
relationship of heterojunction devices is strongly influenced by the properties of the
quasi two-dimensional electron gas (Q2DEG) which forms the active region of the device [1, 2]. Another important contribution to the capacitance is given by the change
of the ionized donor concentration in the doped layer, as well as by the onset of a
so-called parasitic channel in the wide bandgap material. The one-dimensional self-
consistent Schrödinger-Poisson solver (SPS) computes the charge distribution and
the layer capacitances for arbitrary conventional or pseudomorphic heterojunction
structures. Only physical parameters like the effective electron masses, the dielectric constants of the different materials and the conduction band-edge discontinuities
forming the heterojunction were used as input parameters in addition to the thickness
of the different layers and the doping densities. SPS is also able to take into account
the three conduction band valleys (Γ, X, L), the local exchange-correlation potential

and deep donor levels according to the model of Schubert and Ploog [3]. The physical parameters used for the modelling are described in the literature (Table 1, [1, 4, 5, 6]). For the heterojunction band-edge discontinuity we used a $\Delta E_c = 0.65 \, \Delta E_g$ rule [6]. The solution of the Schrödinger equation follows a description of P.C. Chow [7].

Measurements were performed on wafer on conventional $Al_{0.23}Ga_{0.77}As/GaAs$ and pseudomorphic $Al_{0.23}Ga_{0.77}As/In_{0.2}Ga_{0.8}As/GaAs$ HEMTs using an HP 4275 A impedance analyzer. The source and the drain contacts were set to the same potential for the measurements and the measurement frequency was 1MHz.

The nominal thickness for the supply layer was designed to $d_A = 40$ nm (Table 1). This thickness d_A is sometimes affected by etching process steps while forming the gate on top of the supply layer. A reduction of the effective d_A influences all electric properties of the device, including the gate capacitance.

Layer	x, y	d (nm)	$N_{D,A}$ (cm^{-3})	E_g (eV)	m^*/m_0	$\varepsilon/\varepsilon_0$
$Al_xGa_{1-x}As$-supply	0.23	40	1.5E18 (D)	1.711	0.084	12.4
$Al_xGa_{1-x}As$-spacer	0.23	2	0	1.711	0.084	12.4
$In_yGa_{1-y}As$-channel	0.2	12	0	1.140	0.058	13.2
GaAs-buffer	—	>1000	2.4E14 (A)	1.424	0.067	13.1

Table 1: Nominal process parameters for a pseudomorphic HEMT and physical parameters used for the simulations. Same for conventional HEMTs, but without $In_yGa_{1-y}As$-channel layer.

2. Results

The gate C-V characteristics measured on sub-μm gate-length transistors show significant contributions of pad and fringe capacitances (Fig. 1a). They agree very well

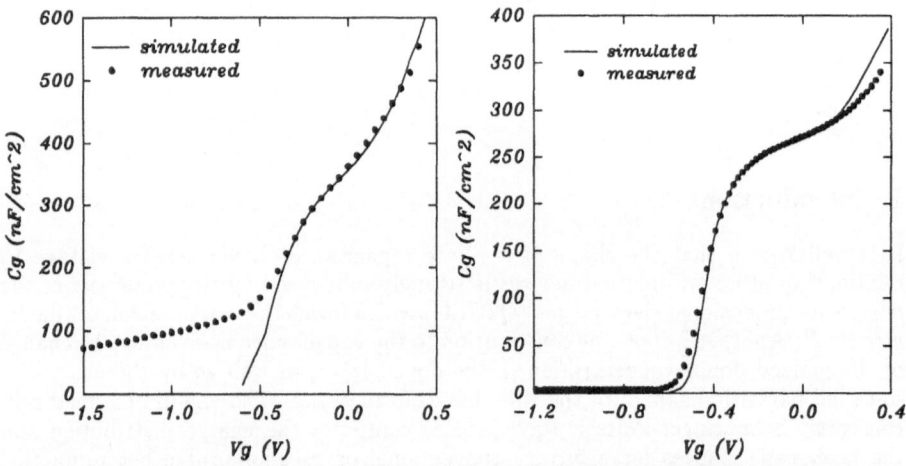

a) conventional HEMT with a sub-μm gate-length structure

b) pseudomorphic HEMT with a 100 μm gate-length structure ("fat FET").

Figure 1: Examples for the comparison between simulated and measured gate C-V characteristics of conventional and pseudomorphic HEMTs.

with the SPS simulation above the threshold voltage. Pad capacitances were assumed to be constant over the observed voltage range and could be subtracted from the measurement signal. The remaining contribution to the capacitances at negative voltages was assumed to be mainly due to fringe capacitances. If measurements are performed on so-called fat FETs (gate-length 100 μm) the pad and fringe capacitances do not contribute significantly to the data (Fig. 1b).

Moreover, the C-V characteristics measured on the fat FETs agree very well with the SPS simulation over the whole voltage range. If measurements were performed on sub-μm gate-length transistors ($L_g = 250$ nm, $W_g = 180$ μm) across a wafer, a systematic displacement between the curves measured in the outer and in the inner regions of the wafer was found (Fig. 2a). The origin of this displacement was assumed to be an irregularity in one of the manufacturing process steps. By comparison of the measured data with SPS simulations (Fig. 2b) we could show that the displacement of the measured C-V characteristics resulted from a systematic change of the effective thickness d_A of the doped AlGaAs layer in the pseudomorphic HEMTs (Table 2). The doping concentration N_D as well as the built-in voltage V_b were found to be fairly constant over the wafer.

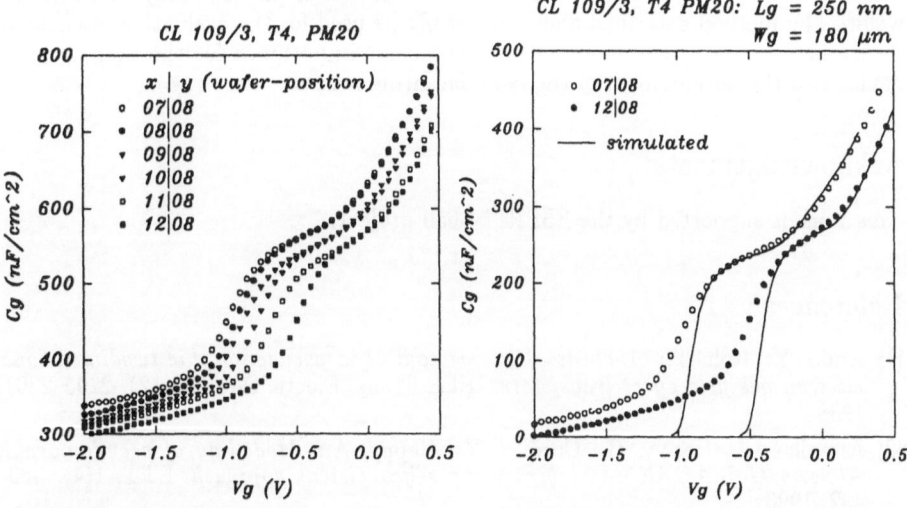

a) Measured capacitances from the outer (right curve) to the inner (left curve) region of the wafer

b) Comparison between selected curves and simulations

Figure 2: Measurement and simulation of the gate capacitances of pseudomorphic HEMTs.

This result could be corroborated e.g. by SEM photographs which showed that the thickness of the doped layer d_A was indeed diminished by several etching steps. In addition, it explains systematic shifts in other HEMT electrical properties (e.g. the transfer

Device #	d_A (nm)	N_D (cm^{-3})	V_b (V)
07|08	37	1.5E18	1.0
12|08	32	1.5E18	1.1

Table 2: Parameters evaluated from the simulations shown in Fig. 2b

characteristic) we found when measurements were performed across the wafer. Consequently, the evaluated fit parameters were taken to be the effective device parameters. Furthermore, the comparison of measured and simulated C-V characteristics and the resultant effective device parameters led to an improvement of the manufacturing process.

3. Conclusion

We have developed a one-dimensional self-consistent Schrödinger-Poisson solver (SPS) which is capable of simulating the charge control behaviour and the associated capacitances for arbitrary heterojunction structures. The gate-to-source capacitances of several conventional and pseudomorphic HEMTs were measured and compared with C-V characteristics simulated by SPS. A very good agreement between measurement and simulation of the C-V characteristics was found for so-called fat FET structures over the whole voltage range, as well as for sub-μm gate-length transistors for voltages higher than the threshold voltage. When measurements were performed across a wafer a systematic displacement of the C-V characteristics was found. It could be shown by comparison of measured data with SPS simulations that this displacement was mainly due to a variation of the effective thickness of the doped layer d_A over the wafer. The method described above can easily be used for the evaluation of effective device parameters. This is a result of particular importance for the design of the device and the development of the manufacturing process.

Acknowledgements

This work is supported by the ESPRIT 6050 project.

References

[1] Ando, Y., Itoh, T.: *Analysis of charge control in pseudomorphic twodimensional electron gas field-effect transistors.* IEEE Trans. Electr. Dev., 35(12), 2295–2301, 1988

[2] Alamkan, J., Happy, H., Cordier, Y., Cappy, A.: *Modelling of pseudomorphic AlGaAs/GaInAs/AlGaAs layers using selfconsistent approach.* ETT, 1(4), 429–432, 1990

[3] Schubert, E.F., Ploog, K.: *Shallow and deep donors in direct-gap n-type $Al_x Ga_{1-x} As$: Si grown by molecular-beam epitaxy.* Phys. Rev. B, 30(12), 7021–7029, 1984

[4] Adachi, S.: *GaAs, AlAs, and $Al_x Ga_{1-x} As$: Material parameters for use in research and device applications.* J. Appl. Phys., 58(3), R1–R29, 1985

[5] Zhao, K., Kuhn, K.J.: *Dislocation scattering in n-type modulation doped $Al_{0.3} Ga_{0.7} As/In_x Ga_{1-x} As/ Al_{0.3} Ga_{0.7} As$ quantum wells.* IEEE Trans. Electr. Dev., 38(12), 2582–2589, 1991

[6] Giugni, S., Tansley, T.L.: *Comment on the compositional dependence of bandgap in AlGaAs and band-edge discontinuities in AlGaAs-GaAs heterojunctions.* Semicond. Sci. Technol., 7, 1113–1116, 1992

[7] Chow, P.C.: *Computer solutions to the Schrödinger equation.* Am. J. Phys., 40, 730–734, 1972

Helena: A Physical Modeling for the DC, AC, Noise and Non Linear HEMT Performance

H. Happy, F. Kapche-Tagne, F. Danneville, J. Alamkan, G. Dambrine, and
A. Cappy

Institut d'Electronique et de Microélectronique du Nord,
Université des Sciences et Technologies de Lille
F-59655 Villeneuve d'Ascq Cédex, FRANCE

Abstract

A friendly software for the modeling of HEMTs called HELENA for Hemt ELEctrical properties and Noise Analysis is presented. Using this software, the DC, AC, noise and non linear performance of any kind of HEMT realized on either GaAs or InP substrates can be obtained. HELENA is very fast and gives results in a good agreement with experiments.

1. Physical model used in HELENA

The physical HEMT modeling used in HELENA is based on the quasi-two dimensional (Q2D) approach [1, 2, 3, 4]. However, significant improvements have been introduced as compared with the previously published works.

- Concerning the charge control law determination, a new model including quantum effects has been developed for HEMT layers [5]. This new model is faster than the self-consistent resolution of Schrödinger and Poisson's equation although it gives results in good agreement with the more rigorous model and with experiments. As a consequence, it is well suited for the Q2D modeling of HEMTs.

- For the AC and noise performance, the device is considered as a non uniform active line, and the electrical properties are calculated using the method described in [6]. A key feature of this approach is to provide all the small signal parameters (Gm, Gd, Cgs, Cgd, Ri, τ, Rgd, Cds), the S-parameters as well as any small signal performance at any frequency of operation. The noise performance is calculated using the impedance field method [7] associated together with the correlation matrix approach [8]. The main advantage of this method is its validity in the millimeter wave range, contrary to the modelings based on the quasi-static approach. The details of the method used are well described in [9].

The latest improvements included in the physical modeling concern the reverse gate leakage current (GLC) of the schottky barrier, and the calculation of the non linear device parameters. The gate leakage current effect is important because it introduces a parasitic conductance at the device input and also adds shot noise sources that influence the noise performance [10].

As an example, figure (1) shows the minimum noise figure Fmin versus frequency, for differents values of the GLC in the case of a 0.4 x 100 μm^2 PM HEMT. As expected, the GLC strongly degrades the minimum noise figure, especially at low frequencies.

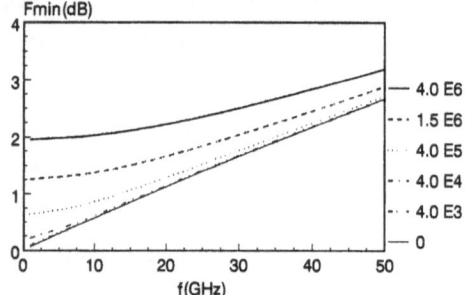

Figure 1. Influence of gate current on Fmin

When the DC and AC performance is calculated, the non linear device behaviour is deduced. The Ids(Vds, Vgs) and Q(Vds, Vgs) relationships are fitted using usual modeling (Curtice, Tajima, Materka ...). The non linear modeling based on the active line method is also available. These non linear parameters can be directly used in the circuit simulators.

2. Software description

The physical model previously described is included in the friendly software called **HELENA.** All the physical and technological parameters are easily introduced and file-managed thanks to pull-down menus and data illustrations. The results are displayed in a convenient form using linear or logarithmic scales as well as polar or Smith chart when it is necessary. The flow chart of the software is shown in figure (2).

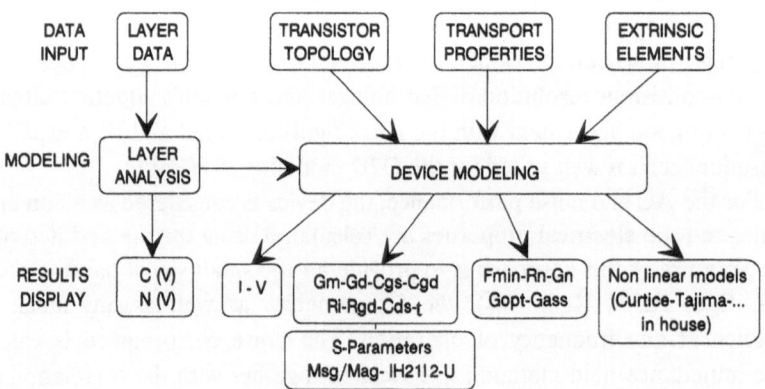

Figure 2. HELENA flow chart

It is divided in three different blocks: data input, modeling of HEMT performance and results display.

Two main routines are used for the device performance calculation: in the first step the layer analysis is made, and in the second step, DC, AC, noise and non linear performance of the device are calculated.

The device library contains the following structures:

- GaAlAs/GaInAs/GaAlAs system on GaAs.

  Pseudomorphic - Conventional HEMT - Epitaxial or implanted MESFET.
- AlInAs/GaInAs/AlInAs system on InP: Lattice-matched - Pseudomorphic.

HELENA is now running on PC with MSDOS operating system. HELENA is very fast (about 3 mn are necessary to make a DC, AC and Noise analysis of a device with a 486/33MHz personnal computer). HELENA is then an interesting tool for the device optimization as well as for the C.A.D. of microwave and millimeter wave circuits.

3. Comparison with measurements

The validity of HELENA is obviously an important problem and a number of comparisons between theoretical results and experimental measurements have been carried out. The structure used for this comparison is a 0.20 x 48 μm^2 gate device realized by THOMSON-LCR on a PICOGIGA delta-doped pseudomorphic layer. To make a comparison, it should be noted that no fitting parameters have been introduced for the intrinsic device simulation while the value of the parasitic elements, needed for the extrinsic performance calculation, have been deduced from measurements. Figures (3:a-d) shows the comparison between theoretical and experimental results for the intrinsic transconductance Gm, the intrinsic resistance Ri, the S11 parameters in the frequency band 1-36 GHz, and the minimum noise figure Fmin. The experimental noise results have been obtained using a new on-wafer measurements [11]. As is shown, HELENA gives results in a good agreement with measurements.

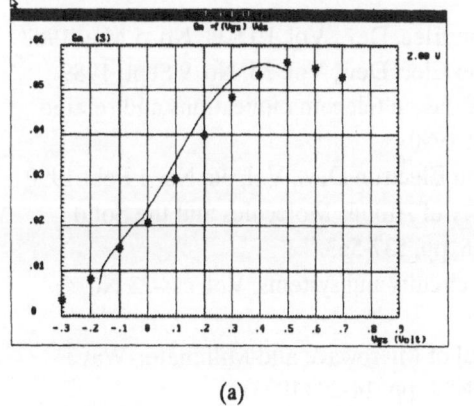
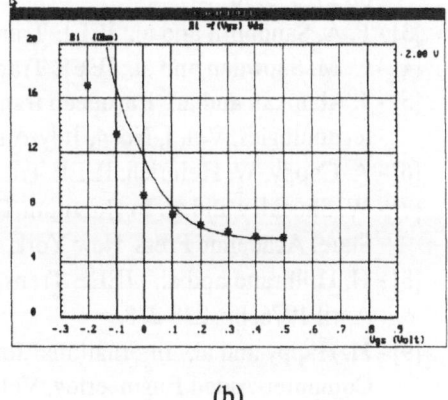

(a)   (b)

Figure 3-a-b. Comparison between the theoretical (solid line) and experimental (points) intrinsic elements: (a) Transconductance versus Vgs; Vds = 2V. (b) Intrinsic resistance versus Vgs; Vds = 2V.

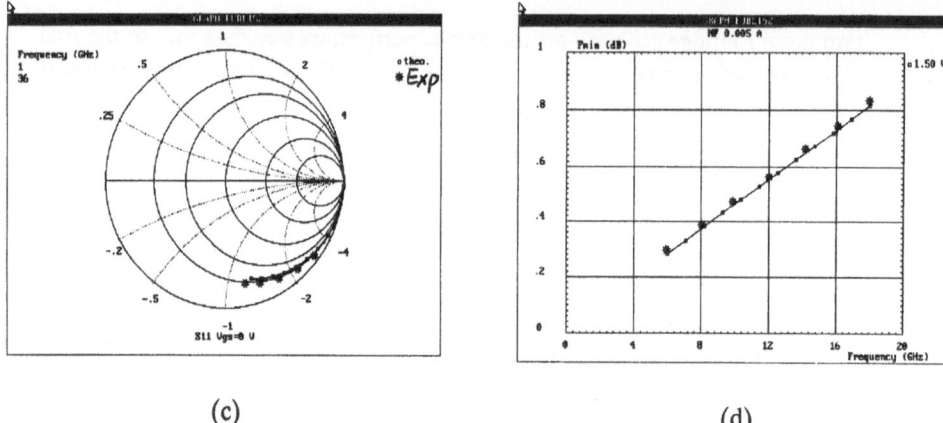

(c)   (d)

Figure 3-c-d. Comparison between the theoretical (solid line) and experimental (points):
(c) S11 parameters: $Vgs = 0$; $Vds = 2V$. (d) Minimum noise figure $Ids = 5mA$; $Vds = 1.5V$.

4. Conclusion

A friendly software called HELENA has been presented. Based on a quasi-2D model, HELENA provides all the electrical performances of many kind of HEMT with a short computing time and a good accuracy. It is then well suited for the study of systematic influence of device parameters. In order to easily integrated HELENA in the microwave circuits environnement, the work station version of the software will be available in the future.

References

[1] - B. Carnez and al.- J. Appl. Phys., Vol. 51, Jan. 1980, pp 784-790
[2] - A. Cappy and al. - IEEE Trans. on Elec. Dev., Vol.32, No.12 Feb. 1985,
      pp 2787-2796
[3] - P. A. Sandborn and al.: IEEE Trans. on Elec. Dev., Vol. ED-34, No. 5 May 1987
[4] - C. M. Snowden and al.: IEEE Trans. on Elec. Dev., Vol. 36, No. 9 Sept. 1989
[5] - J. Alamkan and al.- European transactions on telecommunications and related
      technologies, Vol. I, No. 4, July-August 1990
[6] - A. Cappy, W. Heinrich: IEEE Trans. on Electron Dev., Vol. 36, No. 1 Feb. 1990
[7] - W. Shockley and al. In Quantum Theory of Atoms, Molecules and the Solid
      State. Academic Press, New York, 1966, pp. 537-563.
[8] - H. Hillbrand and al. - IEEE Trans. on circuits and systems, Vol. cas-23 No. 4
      April 1976, pp. 235-238.
[9] - H. Happy and al.- International Journal of Microwave and Millimeter-Wave
      Computer-Aided Engineering, Vol. 3, N° 1, pp. 14-28 (1993)
[10] - F. Danneville and al.: MTT Symposium - Atlanta - June 1993.
[11] - G. Dambrine and al. - Trans. IEEE-MTT, March 1993.

SIMULATION OF SEMICONDUCTOR DEVICES AND PROCESSES Vol. 5
Edited by S. Selberherr, H. Stippel, E. Strasser – September 1993

Lifetime Calculations of MOSFET's Using Depth-Dependent Non-Local Impact Ionization

M. J. van Dort, J. W. Slotboom, G. Streutker, and P. H. Woerlee

Philips Research Laboratories
P.O.Box 80.000, NL-5600 JA Eindhoven, THE NETHERLANDS

Abstract

In this paper, we present a simple but accurate engineering tool for optimizing the lifetime of MOSFET's against hot carrier degradation. The method consists of simulation of the substrate currents in conjunction with the use of an empirical relation between the transistor lifetime and the MOSFET currents. Accurate calculations of the substrate currents are only possible if depth-dependent impact ionization is used in combination with the energy-balance equation.

1. Introduction

Hot carrier degradation of MOSFET's has been investigated extensively. Experiments have shown the existence of empirical relationships between the lifetime against hot-carrier degradation and the currents in a MOSFET. For n-channel MOSFET's at the worst-case bias conditions, the lifetime is simply determined by the ratio of the substrate and the drain currents [1, 2, 3, 4]. This relationship is of course technology dependent, but does not depend on the process generation. The simplest way to predict the lifetime of an n-channel MOSFET is to combine the calculated currents with the empirical relation for the lifetime. The accuracy of the resultant lifetime is then determined by the accuracy of the simulation of the drain and substrate currents.

We have tested the method on a set of devices belonging to different fully-optimised process generations (from 2.0 down to 0.17 μm). The leveling-off of the maximum allowable power supply voltage with decreasing design rule is correctly simulated.

2. Substrate currents

It is of course possible to tune the impact-ionization coeffients to match the calculated with the measured substrate currents of a certain MOS generation, but if we want accurate results for the substrate currents over a wide range of MOS generations with a single model, two physical mechanisms have to be taken into account. Firstly, the mean free path of an electron near the Si-SiO$_2$ interface is significantly smaller than in the bulk of the substrate, and impact ionization in this region is therefore not as efficient as far away from the interface. Impact ionization near the Si-SiO$_2$ interface was experimentally investigated in surface CCD, where the currents flow very close to the interface (Fig. 1, [5]). In an MOS transistor, the position of the maximum generation rate depends on the bias conditions, but is generally close to the interface. Secondly, for deep-submicron MOSFET's the lateral electric field peaks are very steep and narrow, and due to non-local effects the carriers are not very efficiently heated up. For a realistic treatment of the substrate currents, we therefore have to introduce a depth-dependent impact ionization model for non-local carrier heating.

For the modeling of the high-energy tail of the electron distribution, we have used the model proposed in [6]. In Ref. [6], it was shown that the solution of the energy-balance equation calculated in post processing is sufficiently accurate to account for the non-local effects of the carrier heating in advanced MOSFET's and in bipolar transistors.

After the determination of the electron temperature T_e, the impact ionization rate is calculated using the bulk [7] and the surface [5] ionization parameters to obtain $\alpha_B(x,y)$ and $\alpha_S(x,y)$, respectively. The effective impact ionization rate α_{eff} is then approximated by

$$\alpha_{\text{eff}}(T_e) = F(y)\alpha_S(T_e) + (1 - F(y))\,\alpha_B(T_e), \tag{1}$$

with $F(y) = 2\exp(-y^*)/(1 + \exp(-2y^*))$ and $y^* = (y/\zeta)^2$ (Fig. 2). This is the same function as used in MINIMOS-4 [8] to restrict the surface-roughness mobility to the region near the interface (at $y = 0$).

We have implemented this model for impact ionization in MINIMOS [8] and in MEDICI [9]. The value of the characteristic length ζ has been carefully calibrated from devices of fully scaled NMOS processes with design rules of 0.17, 0.25, 0.35, 0.50 μm (conventional S/D structures) and 0.7 μm (LDD devices). Best fit is obtained for $\zeta = 25$ nm. The measured and simulated I_{sub}– V_g characteristics are shown in Fig. 3 for the deep-submicron generations. Fig. 4 shows the same characteristics for the 0.7-μm generation, where the position of the maximum avalanche rate is at much larger depth than for the deep-submicron generations.

3. Lifetime calculations

The lifetime τ of a MOSFET is usually defined as the time in which the transconductance degrades 10 percent with the gate biased at the maximum substrate current. For these bias conditions, stress experiments show a unique relation between τ and I_{sub}/I_d [1, 2, 3]. Figure 5 depicts τ versus I_{sub}/I_d for MOSFET's with the nominal gate length of different MOS generations. If we use this empirical relation, we can predict the expected lifetime of the transistors. Specifying the minimum lifetime of a MOSFET to be 10 years, the maximum power supply voltage V_{dd} can also be determined. Figure 6 shows the simulated and measured [10] power supply voltage versus the design rule of the MOS generation. For deep-submicron MOS generations, V_{dd} remains constant at 2.5 V [10]. This is caused by the effects of non-local carrier heating. The power supply voltage can remain at fairly high values without affecting the lifetime of the transistor. The agreement between simulation and experiment is excellent.

In conclusion, accurate simulations of the substrate currents over a wide range of MOS generations are only possible with a single model if depth-dependent impact ionization is used together with the energy-balance equation. We have used the same depth dependence as is used for modeling of the surface mobility in MINIMOS-4. Using the simulated drain and substrate currents, the lifetime of MOSFET's can be predicted from empirical lifetime models. Non local carrier heating is correctly taken into account as is illustrated by the simulation of the constant power supply voltage in the deep sub-micron regime. All the calculations are completely done in post processing after a simple MINIMOS or MEDICI run and consume negligible CPU time.

The authors thank J.M.F Peters for the implementation of the model into MEDICI.

References

[1] C. Hu *et. al.*, IEEE Tr. Elec. Dev. ED-32, p. 375 (1985).

[2] R. Bellens *et. al.*, El. Dev. Lett. 10 p. 553 (1989).

[3] Hazema *et. al.*, Techn. Digest IEDM, p. 569 (1990).

[4] R. Woltjer *et. al.*, Techn. Digest IEDM, p. 535 (1992).

[5] J. Slotboom *et. al.*, Techn. Digest IEDM, p. 494 (1987).

[6] J. Slotboom *et. al.*, Techn. Digest IEDM, p. 127 (1991).

[7] R.v.Overstraeten and H. deMan, Solid-St Electr. 13, p. 583 (1969).

[8] S. Selberherr, IEEE Tr. Elec. Dev. ED-36, p. 1464, 1989.

[9] TMA – MEDICI, Version 1.10, modified version (1993)

[10] P. Woerlee *et. al.*, Techn. Digest IEDM, p. 537 (1991).

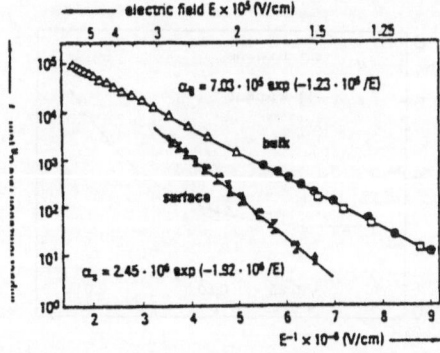

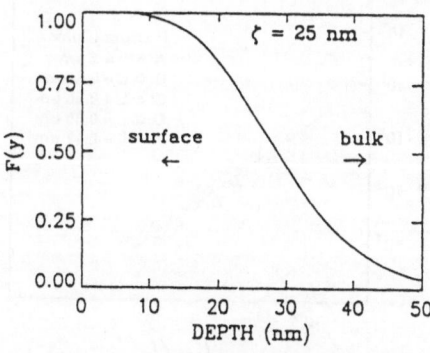

Figure 1: Measured impact ionization rates in the bulk [7] and near the Si-SiO$_2$ interface [5].

Figure 2: Function used for depth-dependence impact ionization. This function is also used in the MINIMOS-4 mobility model [8].

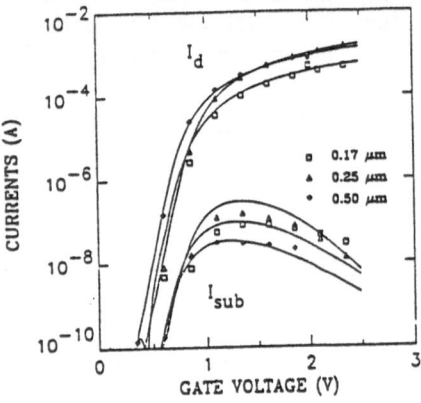

Figure 3: Measured and simulated substrate currents for various MOS generations (design rules 0.17, 0.25 and 0.50 μm). The simulations are done with $\zeta = 25$ nm. For these devices, surface impact-ionization is important.

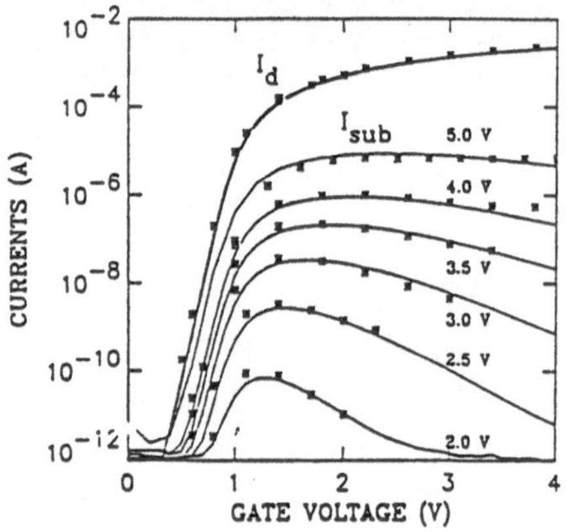

Figure 4: Measured and simulated substrate currents for the 0.70-μm generation (LDD). The simulations are done with $\zeta = 25$ nm. For this MOS generation, bulk impact-ionization is dominating.

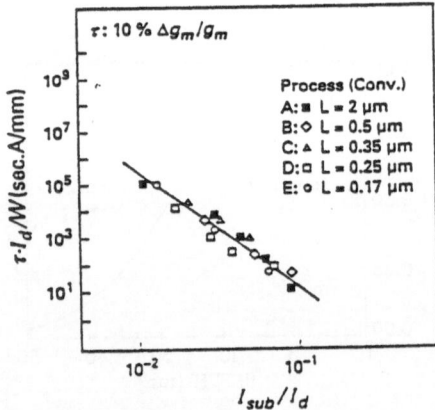

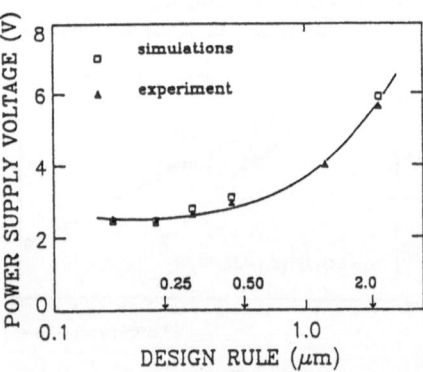

Figure 5: Lifetime plot of τI_d versus I_{sub}/I_d for devices with the minimum gate length of various MOS generations with a conventional S/D. The gate is biased at maximum substrate current. The lifetime criterium is a 10 percent change in the transconductance g_m The data are taken from Ref. [10].

Figure 6: The measured and simulated maximum power supply voltage versus the design rule of the MOS generation. Due to non-local carrier heating V_{dd} remains constant in the deep sub-μm regime. Date are taken from Ref. [10].

SIMULATION OF SEMICONDUCTOR DEVICES AND PROCESSES Vol. 5   473
Edited by S. Selberherr, H. Stippel, E. Strasser – September 1993

Inverse Modeling of Impact Ionization Rate Formula Through Comparison Between Simulation and Experimental Results of MOS Device Characteristics

S. Imanaga, K. Hane, and Y. Hayafuji

Yokohama Technology Center, Sony Corporation
134 Goudo, Hodogaya, Yokohama, Kanagawa 240, JAPAN

Abstract

This paper appraises the degree of agreement between simulated and experimental results of drain current versus drain voltage $(I_d - V_d)$. It also derives the impact ionization rate formula inversely by compared the simulated and experimental dependence of the substrate current(I_{sub}) on the gate voltage(V_g). We found that : (1) for $I_d - V_d$ characteristics, the agreement in the linear region was off, but overall agreement was fairly good, and (2) the simulated $I_{sub} - V_g$ characteristics were in fairly good agreement with the experimental characteristics when the modified Keldish formula $P_{ii} = P_0((E - 1.12)/1.12)^n$ with n of 7 and P_0 of $2.8 \times 10^{11} s^{-1}$ was used as the formula for the impact ionization rate.

1. Introduction

Rigorous models of the transport characteristics of electrons in bulk Si, such as the real band model [1], have been developped. Direct comparison of device characteristics of MOS device between Monte Carlo simulation results and experimental results, however, is rare. We think that in order to assess the status of the conventional full Monte Carlo simulation, including highly doped regions, it is necessary to compare Monte Carlo simulation results with experimental results even if some of the models are not sufficient rigorous. Our first aim is to appraise the degree of agreement between simulated and experimental results of drain current versus drain voltage $(I_d - V_d)$. Our second aim is to determine an impact ionization rate formula inversely through a comparison between simulated and experimental results of the dependence of the substrate current(I_{sub}) on the gate voltage(V_g).

2. Simulation method

As a band model, a spherical band is employed, and nonparabolicity $(\alpha = 0.35eV^{-1})$ is taken into account. The scattering mechanisms included[2] are intervalley scattering, acoustic phonon scattering, ionized impurity scattering (Brooks-Herring formula), surface roughness scattering, and impact ionization. We employed Park's model[3] of surface roughness scattering. This is a partial diffusive scattering model in which a critical parameter P_s is introduced to identify the boundary between a specular and diffusive scattering event. The value of P_s is taken to be 0.77 by Park. Various formulas have been proposed for the dependence of the impact ionization rate on the electron energy ($P_{ii} - E$ dependence). Among those formulas, those proposed by Fischetti [1] and Thoma [4] base their claim on validity on the close agreement of simulated and experimental results of the impact ionization coefficient for bulk Si. However, Fischetti and Thoma have not compared the simulated and experimentally derived substrate current of MOS devices produced by impact ionization. We attemted, therefore, to determine $P_{ii} - E$ dependence inversely by comparing the simulated and experimentally derived $I_{sub} - V_g$ dependence. We assume that Eq. 1 is the formula for $P_{ii} - E$ dependence,

$$P_{ii} = P_0((E - E_{th})/E_{th})^n, \tag{1}$$

where E_{th} is the threshold energy for impact ionization and is assumed to be 1.12 eV. Then, we attempted to determine the value of n and P_0 to give the best fit to the experimental results. Ohmic contact for source and drain is modeled conventionally, namely, a layer of cells beneath the electrode is heavily doped and is maintained neutral every simulation time step. Electron-electron scattering and the degeneracy effect are not taken into account in the present simulation. The impurity profile of the device is obtained as a result of the process simulation using real process steps. The simulated MOS device has an LDD structure and its effective channel length is 0.6 μm. The maximum carrier concentration of source/drain and LDD are $2.4 \times 10^{20} cm^{-3}$ and $1.5 \times 10^{19} cm^{-3}$, respectively. The maximum doping concentration of p-type substrate is $1.6 \times 10^{17} cm^{-3}$ near the surface.

3. Simulation results and discussion

Figure 1 shows the impurity profile of the simulated device. Figure 2 shows the simulated and experimental results of $I_d - V_d$ characteristics. The agreement in the linear region and that in saturation region with V_g of 5 V are off, but overall agreement is fairly good. Figure 3 compares the experimental and simulated $I_{sub} - V_g$ characteristics. Here, we assume that the substrate current is the electron charge times the simulated number of electrons produced by impact ionization divided by simulation time after the system reaches the steady state. The squares in Fig.3 show $I_{sub} - V_g$ characteristics when Eq. 1, with n of 2 (the standard Keldish formula) and P_0 of 7.5×10^{12}, which is Tang and Hess's model [5], is employed as the formula for the impact ionization rate. In this case, the dependence of I_{sub} on V_g is not in agreement with the experimental results. Moreover, the absolute values of I_{sub} are two orders larger than the experimental values. The crosses in Fig.3 show $I_{sub} - V_g$ characteristics when Eq. 1 with n of 7 and P_0 of 2.8×10^{11} is used as the formula for the impact ionization rate. In this case, it is found that the simulated dependences of I_{sub} on V_g for both V_d of 4 V and V_d of 5 V are in fairly good agreement with the experimental results. The functional form of $I_{sub} - V_g$ characteristics is determined by the subtle balance between the increase of drain current and the decrease of the ratio of the number of high energy electrons in the channel which have enough energy to create electron-hole pairs by impact ionization, as the gate voltage increases. Figure 4 shows the dependence of the impact ionization rate on the electron energy proposed by various groups. Among those shown in Fig.4, the present result is most closely related to that of Taniguchi et.al [6] . Figure 5 shows the distribution of averaged electron energy in the device. We can see that the peak of the electron energy increases and becomes sharper as the gate voltage decreases. Figure 6 shows the comparison of energy distributions between (a) at the drain-LDD edge and (b) at the source-LDD edge in the channel. At the position (b), the energy distribution is sharp and the peak of the distribution is about 0.15 eV. At position (a), the energy distribution broadens a great deal and is tailing towards higher energy. The peak of the distribution is about 1.14 eV. Figure 7 compares the energy distributions under the conditions of V_g of 1 V and V_g of 5 V. Figure 8 shows the distribution of electron positions in the device. It shows the increase of electron number in the channel as the gate voltage increases. We can see that at low gate voltages, pinch off occurs, and the depleted region with a high electric field near the LDD edge of the drain broadens gradually as the gate voltage decreases.

4. Conclusion

We simulated the device characteristics of a real MOS device by the self consistent full Monte Carlo method including high doping regions and compared the simulation results with the experimental results. We found that (1) as for $I_d - V_d$ characteristics, the agreement in the linear region was off, but overall agreement was fairly good. (2) The simulated $I_{sub} - V_g$ characteristics were in fairly good agreement with the experimental characteristics when Eq. 1 with n of 7 and P_0 of 2.8×10^{11} s^{-1} was used as the formula for the impact ionization rate.

References

[1] M. V. Fischetti and S. E. Laux Phys. Rev. B, vol.30, no. 14, 9721(1988).

[2] C. Jacoboni, R. Minder, and G. Majni, J. Phys. Chem. Solids 36, 1129(1975).

[3] Y. -J. Park, T. -W. Tang, and D. H. Havon, IEEE Trans. Electron Devices, ED-30, no.9, 1110(1983).

[4] R. Thoma, H. J. Peifer, W. L. Engl, W. Q. Brunetti, and C. Jacoboni, J. Appl. Phys. 69(4), 2300(1991).

[5] J. Y. Tang and K. Hess, J.Appl.Phys. 54, 5139(1983).

[6] T. Kunikiyo, Y. Kamakura, M. Yamaji, H. Mizuno, M. Takenaka, K. Taniguchi and C. Hamaguchi, VPAD, 40(1993).

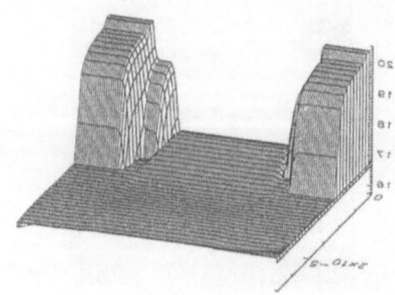

Fig.1 Impurity profile of the simulated MOS device

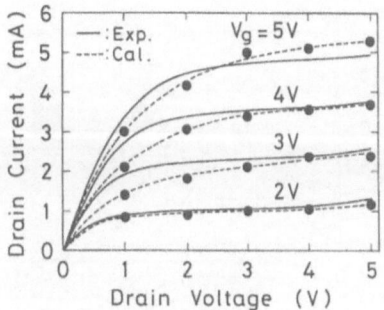

Fig.2 Comparison of Id-Vd characteristics between simulation and experimental results

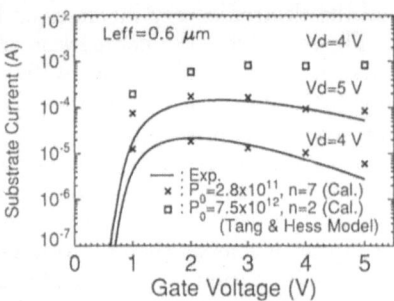

Fig.3 Comparison of Isub-Vg characteristics
between simulation and experimental results

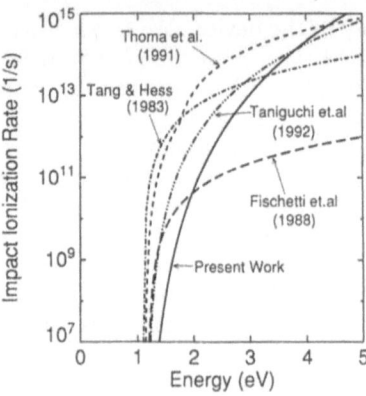

Fig.4 Dependence of the impact ionization rate
on the electron energy proposed by various groups.

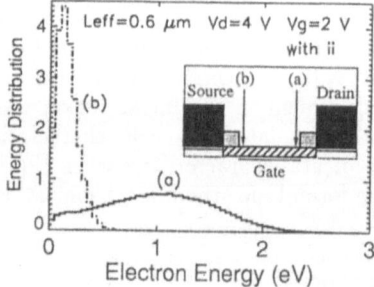

Fig.6 Energy distributions (a) in the channel near the drain-
LDD edge and (b) in the channel near the source-LDD edge

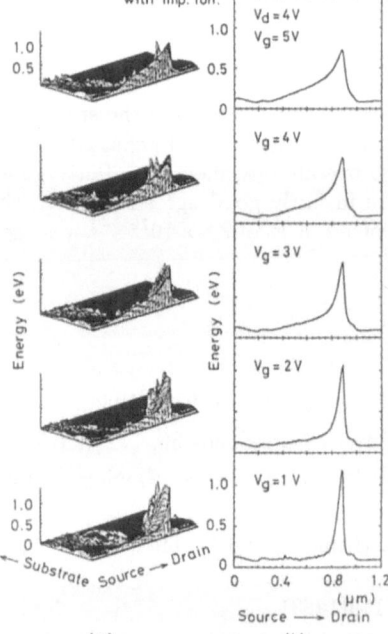

Fig.5 Distribution of averaged electron energy in the device

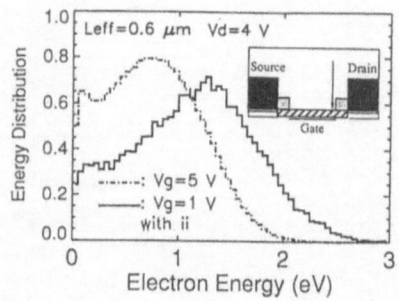

Fig.7 Comparison of the energy distributions under
the conditions of Vg of 1 V and Vg of 5 V.

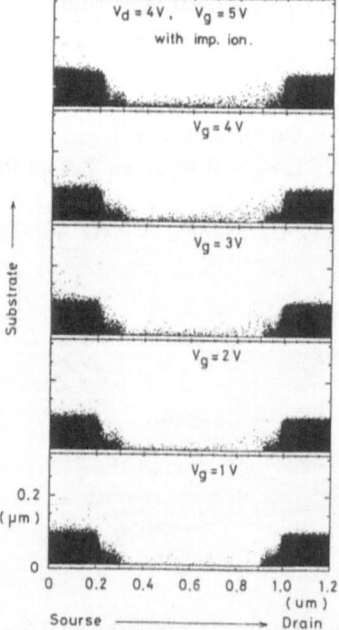

Fig.8 Distribution of electron positions in the device

The MicroMOS 3D Monte Carlo Simulation Program – a Tool for Verifying the MINIMOS Mobility Models

K. Tarnay[‡†], F. Masszi[†], A. Poppe[‡], P. Verhás[‡], T. Kocsis[‡], and Zs. Kohári[‡]

[‡]Deptartment of Electron Devices, Technical University of Budapest
H-1521 Budapest, HUNGARY

[†]Scanner Laboratory, Electronics Department,
Institute of Technology, Uppsala University
Box 534, S-75121 Uppsala, SWEDEN

Abstract,

A molecular dynamics Monte Carlo simulation method was developed for examining the behaviour of submicron MOSFET devices. A brief description of the simulation principles and physical backgrounds is presented. Special attention was paid to use first physical principles. An important advantage of the method is that the Coulomb scattering are taken into account inherently. A MOS structure of 0.25 μm channel length and 0.25 μm channel width has been analyzed on an Alpha-chip DEC 7000 computer.

1. Introduction

MiCroMOS is a *three-dimensional Monte Carlo semiconductor device simulation program* developed primarily for studying the behaviour of sub-halfmicron Si MOSFET devices.[2]. The *molecular dynamics method* [1] seems to be the best suitable technique[3]. The program development is concentrated on applying *first physical principles* in the active device region, without any fitting factors. [4]

2. The device structure

In the channel region, and in limited parts of the source, drain and bulk all carriers are examined individually. Outside this region classical approximations are

---

[1] This research has been sponsored by the Digital Equipment Co. External European Research Projects HG-001 and SW-003, and by the Swedish and Hungarian governments' scientific research funds.

[2] The following considerations have lead us to apply this concept:
  - the classical *drift-diffusion method* or the *hydrodynamic method* - both based on some statistical considerations for the distribution functions - are no more valid, since in such a structure the number of carriers is only in the order of thousands.
  - for the relatively small number of particles, *the momentum and space-trajectories of each individual carrier can be followed* within acceptable CPU time.

[3] The more sophisticated Monte Carlo methods (using charge clouds, superparticles, etc.) offer far more effective numerical solution tools, whereas the physics of the simulated system is obscured.

[4] At present, the following effects are neglected: generation - recombination, impact ionization surface quantization, magnetic field, presence of split-off holes.

applied[5]. The dimensions of the examined device can be seen on Fig. 1. In the nondepleted parts of the source, drain and bulk charge neutrality is forced by introducing the required number of carriers.

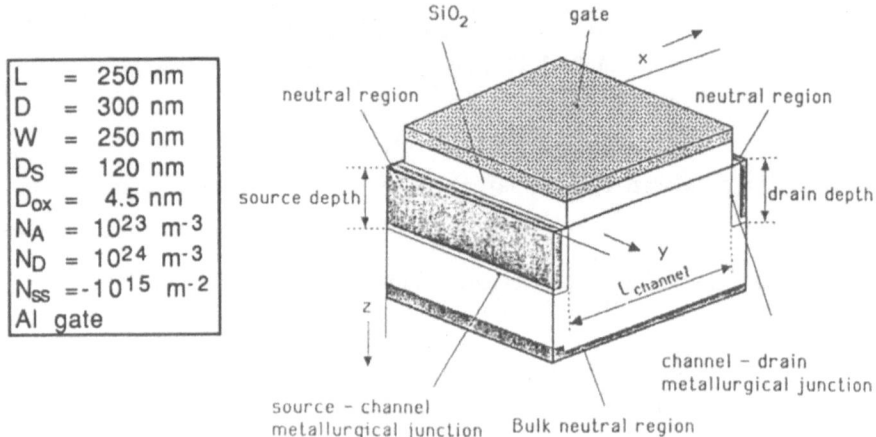

L	=	250 nm
D	=	300 nm
W	=	250 nm
D_S	=	120 nm
D_{ox}	=	4.5 nm
N_A	=	10^{23} m^{-3}
N_D	=	10^{24} m^{-3}
N_{ss}	=	-10^{15} m^{-2}
Al gate		

Fig. 1

3. Simulation principle[6]

The flow chart of the simulation can be seen on Fig. 2.

3.1 Potential calculation

A unique feature of the program, that the field and the potential distributions are not determined by solving the Poisson equation. Instead, the field and potential are separated into two parts, one originating from the *charges inside the active region* (and from the Si-SiO₂ interface charges) and calculated *analytically*, the other from the *charges outside the active region and external voltages* (representing the boundary conditions) and determined by solving the Laplace equation *numerically*. The potential distribution is given on Fig. 3.

3.2 The dispersion relation

For electrons, there are *six ellipsoid shaped constant energy surfaces*. For <100> oriented Si their principal axes lies on the positive or negative coordinate axes of the k - space and their centre is located at 0.85 k_{Max} [3]. The constant *effective mass concept* is used, and only diagonal elements of the effective mass tensor differ from zero. Different transversal and longitudinal effective masses are considered. The relationship between the k-vector, velocity and momentum for the ith ellipsoid electrons is given by

$$p = m_i^{-1} \, v = h(k - k_{oi})$$

The effective masses of heavy and light holes correspond to two concentric constant energy spheres (ie. the warped shape of the valence band is approximated by spheres). Caused by the spherical symmetry, the diagonal elements are equal, ie. two scalar effective mass can be considered for the light and heavy holes.

---

5   In this sense, the program is a hybrid Monte Carlo method
6   For a detailed discussion see Ref. [2]

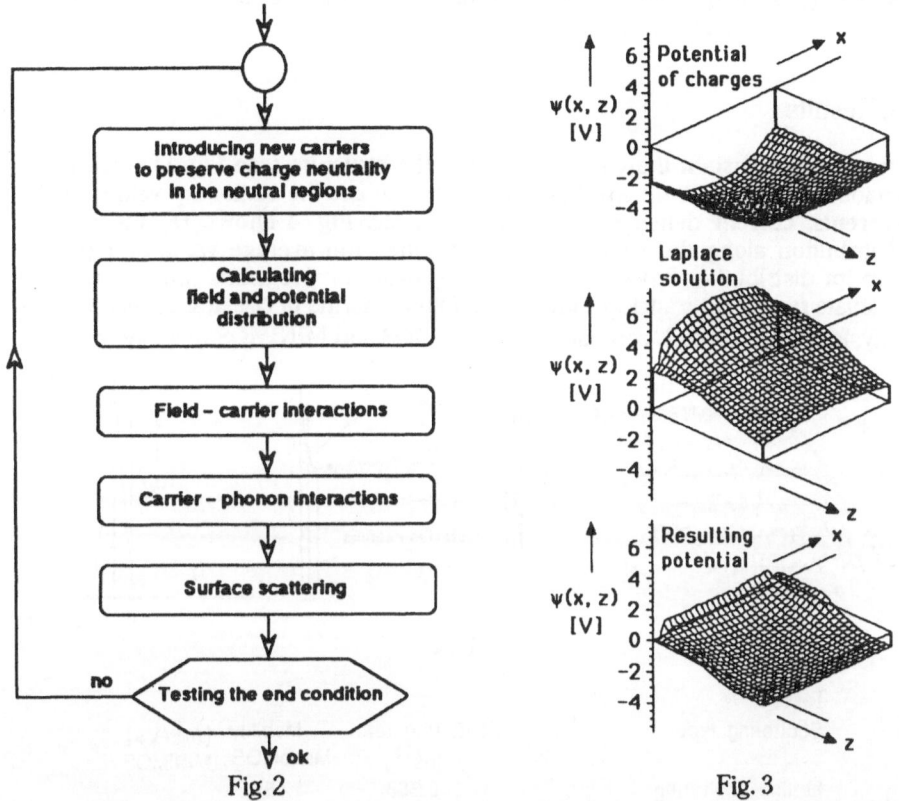

Fig. 2   Fig. 3

3.3 Carrier Dynamics

The carrier motion is described by the classical *Newtonian law of motion*, using the reciprocal effective mass tensor. The force acting to carriers is proportional to the electric field, which consists of two parts. One of them can be analytically calculated from the positions of point charges, and the second one is derived from the potential distribution calculated by the Laplace solver. The first integral the of motion equation yields the *velocity*, and the second one the *position vector* of carrier[7]. This method enables exact simulation of carrier trajectories and thus the exact evaluation of all Coulomb scattering processes.

3.4 Scattering processes

Our approach for the field and potential calculation results in the exact simulation of all *Coulomb scattering* (ionized impurity scattering, scattering on charged interface states and carrier-carrier scattering)[8]. For the *intervalley scattering* of the electrons a *thermodynamic approach* is applied: an electron corresponds to that ellipsoid, where it has minimal energy. For *surface scattering* the *Coulomb*

---

7 Applying a time increment Δt small enough to assume a constant force during this time, the integrations can be carried out by first order numerical quadrature formulae. Since the field is a strongly varying function of the position, using low order formula with small time step Δt yields better accuracy, than a higher order one with a larger time step.

8 An *empirical factor* (the only one in the program) must be used in the postprocessing phase to make a decision that the change of the direction of the particle movement is large enough to consider this as a Coulomb scattering or not.

scattering on charged interface states, further the *elastic* and *specular surface scatte-*
ring[9] are taken into consideration.

4. Results

In each time instant the circumstances in the structure (see Fig. 1) seem to be
chaotic, there-fore *time averaging* is needed to get the stationary values of the
currents, current densities, concentrations, etc. Fig. 4 shows (a) the current
distribution along the x-axis vs. time, (b) the time average vs. x, and (c) the
current distribution function. Special *postprocessor programs* must be used to
evaluate results (to transform the results into the terms of classical semiconductor
physics)[10] [5]. Table 1. compares the MicroMOS and MINIMOS mobility results.

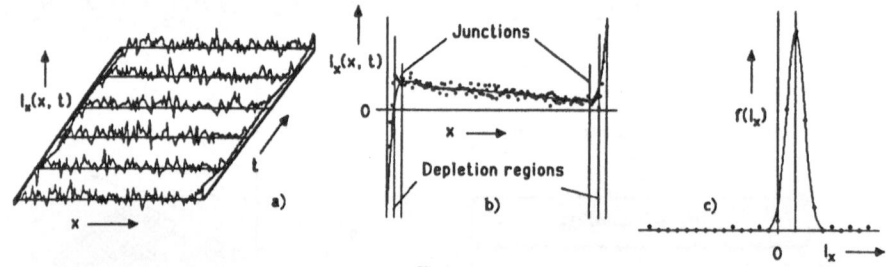

Fig. 4

Table 1.

Scattering type	Scattering rate [ps⁻¹]	Mobility [m²/Vs]	
		MicroMOS	MINIMOS
Lattice scattering	0.368	1.489	
Intervalley scattering	0.543	0.902	
Impurity scattering	1.665	0.329	
Electron - electron scattering	7.150	0.077	
Electron - hole scattering	0.223	2.461	
Interface state scattering	0.481	1.140	
Elastic surface scattering	0.332	1.648	
Specular surface scattering	0.407	1.348	
Resulting mobility		0.048	0.034

References

[1] R.W. Hockney and J.W. Eastwood: Computer Simulation Using Particles. Adam Hilger,
Bristol and Philadelphia, 1988
[2] K. Tarnay et al.: A 3D Monte Carlo Semiconductor Device Simulator for Submicron Silicon
MOS Transistors. Accepted paper for the Electrosoft'93, Southampton, 6 - 8 July 1993
[3] C. Jacoboni and P. Lugli: The Monte Carlo Method for Semiconductor Device Simulation,
Springer Verlag, Wien New York, 1989, page 133
[4] K. Fuchs: Proc. Cambr. Phil. Soc. Vol. 34 (1938)
[5] K.Tarnay - F.Masszi - A.Poppe - P.Verhás - T.Kocsis: Evaluation and Results Interpretation of
3D Monte Carlo Simulation of Submicron MOS Transistor. Hungarian Telecommunication.
Accepted for publication in the July 1993 issue.

---

[9] The ratio of the elastic and specular surface scattering are determined by the *Fuchs parameter*
[4] and by random numbers.

[10] For the estimation of the *drain current* components and their *RMS noise;* regression analysis
of the *number of carriers, electrons entering to drain, the average time spent in the active region,*
etc.; *distribution of electrons* over different ellipsoids; the *electron temperature* vs. time spent in
the structure, or vs. x ; *statistics of scattering rates;* estimation of *mobility components,* etc.

Accurate Determination of Silicon Inversion Layer Mobility by the Monte Carlo Method

F. Gámiz, J. A. López-Villanueva, J. Banqueri, J. A. Jiménez-Tejada, and
P. Cartujo

Departamento de Electrónica y Tecnología de Computadores,
Facultad de Ciencias, Universidad de Granada
Campus Universitario Fuentenueva, E-18071 Granada, SPAIN

Abstract

An accurate determination of the mobility in an n-Si (100) inversion layer has
been performed by the one-electron Monte-Carlo method. The calculation is
based on precise models of scattering mechanisms, with special attention to
coulomb scattering, for which different effects have been considered.

1. Introduction

In current MOS (Metal-Oxide-Semiconductor) technology, detailed modelling of the
electron mobility in the channel of an NMOS transistor is highly desirable in order to
accurately predict its behaviour in circuit applications. The Monte Carlo method is a
powerful tool for calculating mobility; nevertheless, to obtain accurate results in simulation the electron transport properties, and in particular, the scattering mechanisms,
need to be properly modelled.

We have performed an accurate determination of the n-Si(100) inversion layer mobility
at 300 K by the one-electron Monte Carlo method. The electron mobility in silicon
inversion layers had been previously determined using this procedure [1, 2], but the
calculation that we present in this paper is based on more precise models of scattering
mechanisms. Some details and results of our simulation are given below.

2. Method

The electron inversion layer has been treated as a two-dimensional electron gas con-
tained in energy subbands [3]. The minima of the electric subbands arising from
the equivalent minima of the silicon conduction band and the envelope functions
are calculated by self- consistently solving the Schrödinger and Poisson equations
[4]. Electrons can move parallel to the interface, undergoing phonon, coulomb, and
surface-roughness scattering [5].

Both intervalley and intravalley phonon scattering have been considered, allowing
the electron to move in the lowest six subbands. Both zero- and first-coupling order
interactions have been taken into account [5, 6]. The phonon-limited mobility thus

obtained coincides with the phonon-limited mobility for the silicon bulk in the zero transverse-electric- field limit.

For Coulomb scattering, we have developed a model which, starting from previous formulations [3, 7, 8], incorporates significant improvements: it simultaneously takes into consideration the effects of i) screening by mobile carriers, ii) correlation of the oxide charges (which may be important at high concentrations), iii) distribution of the charged centres into the oxide, iv) distribution of the electrons in the inversion layer, and v) image effects caused by the difference in the dielectric constants of Si and SiO_2. In our simulation we have evaluated the local perturbation caused by the spatial fluctuations of the point-charge distributions in the potential acting on electrons in the channel. The perturbation of the potential is affected by the charges in the oxide or at the interface, by the induced electronic charge, and by the image charges of both of them. Correlation of the oxide charge distribution also affects it by decreasing the scattering, since the greater the degree of the distribution uniformity, the lower the magnitude of the fluctuations.

This theory can be used to calculate the effective mobility of the electrons in the inversion layer, and to study its dependence on different physical parameters such as: a) effective electric field (as defined in Ref.9), b) temperature, c) concentration and position of charge in the oxide, interface, and silicon, and d) correlation of the charged centres in the oxide. In this paper we show the effect of the amount and position of charge in the oxide and the influence of each scattering mechanism on the electron mobility.

3. Results

We have obtained the electron mobility in an N-type silicon inversion layer, for a sample with $N_A = 10^{16}$ cm^{-3} and $N_{ox} = 0.35 \times 10^{11}$ cm^{-2} assuming that: a) only phonon scattering exists, b) electrons are also scattered by surface-roughness, and c) all scattering mechanisms are present. Figure 1 shows the influence of the different scattering mechanisms on the mobility. It is apparent that at low fields coulomb scattering is the most important mechanism, while at high fields surface-roughness scattering is the most important.

One of the most influential limiting agents on the mobility is, according to the results in Figure 1, coulomb scattering produced by electric charges in the oxide and at the Si-SiO_2 interface. Although a great deal of work on modelling this mechanism has been done by other researchers [1, 2, 3, 8], there remain certain aspects requiring more work. For these reasons, we have centred our attention on this agent. In Figures 2 and 3, several curves of the effective mobility are represented versus the effective transverse field. The influence of the oxide- charge concentration is shown in Figure 2, where it can be seen that the greater the concentration, the smaller the electron mobility. These curves are in good agreement with experimental results [10]. In Figure 3, the influence of the depth of the oxide-charge is analyzed. It can be observed that the scattering of electrons by oxide charges quickly decreases when the charge is kept away from the interface. For charges placed at 100 Åor more from the interface, coulomb scattering is negligible. It is also apparent that all curves are almost superposed at high electric fields, which indicates that coulomb scattering loses its importance with respect to the other mechanisms, mainly surface roughness scattering, as the electric field grows.

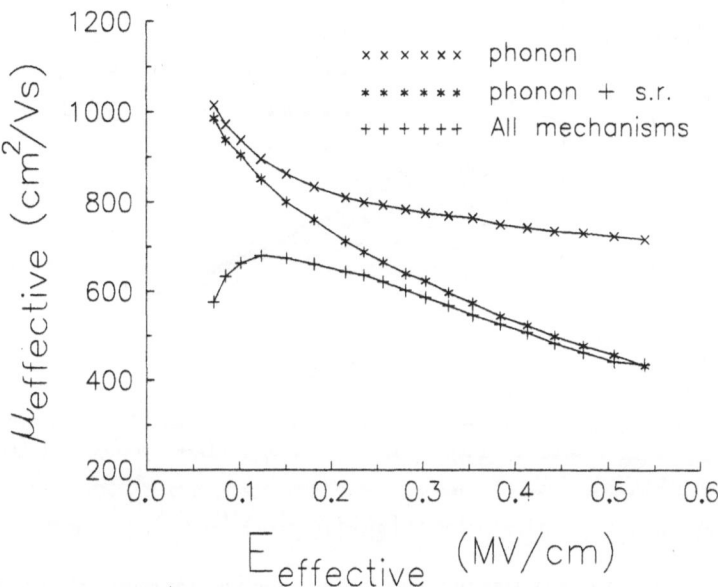

Figure 1: Plot of the effective mobility vs the effective transverse electric field showing the relative importance of the different scattering mechanisms.

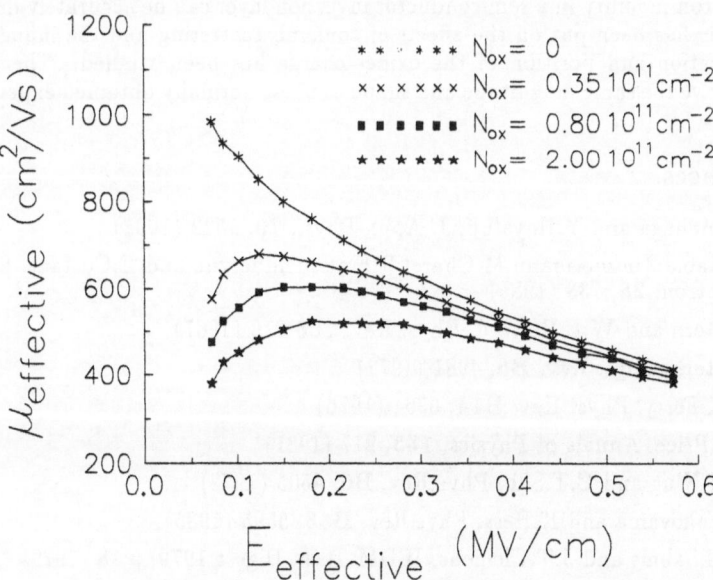

Figure 2: Plot of the effective mobility vs the effective transverse electric field for different concentrations of the oxide charge. (The position of the oxide charge is right at the interface Si-SiO$_2$)

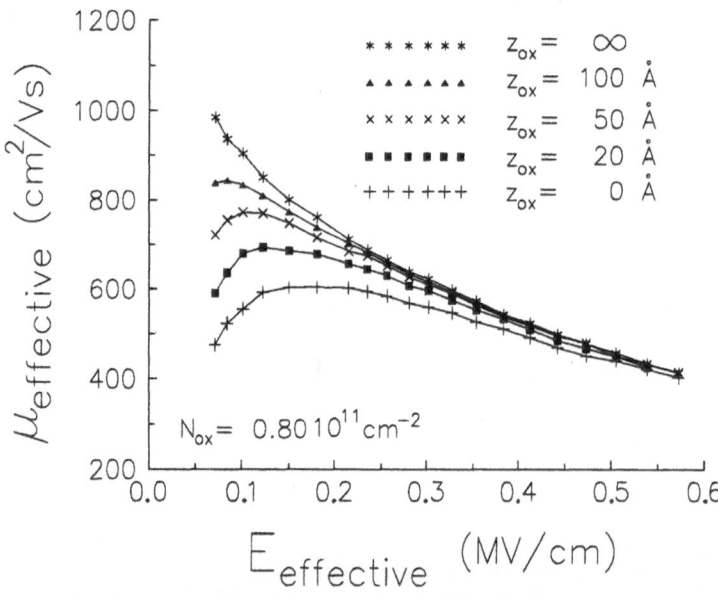

Figure 3: Plot of the effective mobility vs the effective transverse electric field for a fixed density of the oxide charge as a function of its depth inthe oxide.

4. Conclusions

In conclusion, it has been shown that by using precise models of scattering mechanisms the electron mobility in a semiconductor inversion layer can be accurately determined. Emphasis has been put on the effects of coulomb scattering and the influence of the concentration and position of the oxide- charge has been studied. The calculated curves have the same magnitude and shape as those normally obtained experimentally.

References

[1] S.Imanaga and Y.Hayafuji, J. Appl. Phys., **70**, 1522 (1991)

[2] C.Hao,J.Zimmermann,M.Charef,R.Frauquembergue and E.Costant, Solid State Electron. **28**, 733 (1985)

[3] F.Stern and W.E.Howard, Phys. Rev. **163** 816 (1967)

[4] F.Stern, Phys. Rev. **B5**, 4981 (1972)

[5] D.K.Ferry, Phys. Rev. **B14**, 5364 (1976)

[6] P.J.Price, Annals of Physics, **133**, 217 (1981)

[7] T.H.Ning and C.T.Sah, Phys.Rev. **B6**, 4605 (1972)

[8] K.Yokoyama and K.Hess, Phys.Rev. **B33**, 5595 (1986)

[9] A.G.Sabnis and J.T.Clemens, IEDM Tech. Digest 1979, p.18 (1979)

[10] S.Manzini, J. Appl. Phys. **57** 411 (1985)

SIMULATION OF SEMICONDUCTOR DEVICES AND PROCESSES Vol. 5   485
Edited by S. Selberherr, H. Stippel, E. Strasser – September 1993

Importance of Hole Generation on Modeling and Simulation of Schottky and MESFET Structures

D. Donoval, J. Racko, and C. M. Snowden[†]

Microelectronics Department, Slovak Technical University
Ilkovicova 3, SQ-81219 Bratislava, SLOVAKIA
[†]Microwave and Terahertz Technology Group, University of Leeds
Leeds, LS2 9JT, UNITED KINGDOM

Abstract

A new semiclassical model for the modelling and simulation of the electrical properties of rectifying metal-semiconductor structures has been developed. The contribution of hole current to the total current through the interface is significant for reverse biased Schottky structures and cannot be neglected in the model.

1. Introduction

The need for an accurate simulation of the electrical characteristics of MESFET, HEMT and other related structures which use the Schottky contact as a gate electrode to control the output characteristics requires a rigorous model. The simulated results are strongly dependent on the proper definition of discretization schemes and boundary conditions. Although there exists many different approaches for expressing the boundary conditions at the Schottky interface [1,2,3], the agreement between simulated and experimental results is still generally not satisfactory, especially in the reverse direction of applied voltages [4]. Most of the models express the concentrations of electrons and holes at the interface, but in many cases, for example the simulation of MESFET structures, the minority carriers are neglected [5]. The aim of this paper is to point out the importance of generation-recombination processes within the analysed structure. In some cases the hole current through the interface exceeds the electron current and the minority carriers cannot be neglected.

2. Description of the simulator

For the simulation of electrical characteristics of Schottky structures we have used the recently developed simulator DEVSIM [6]. It is based on a revised theory of current flow through the Schottky contact [7]. The final expression of current flow takes into account not only the thermionic-emission/drift-diffusion current but also the generation-recombination current within the space charge region and the injection of holes into the quasi-neutral epitaxial region. It can be rewritten in the following form

$$J = \frac{qv_{rn}^s}{\frac{v_{rn}^s}{v_{dn}}+1}\left\{n_0\left[\exp\left(\frac{V_a}{V_t}\right)-\frac{v_{rn}^m}{v_{rn}^s}\right]+n_{gr}\right\} - \frac{qv_{rp}^s}{\frac{v_{rp}^s}{v_{dp}}+1}\left\{p_0\left[\exp\left(-\frac{V_a}{V_t}\right)-\frac{v_{rp}^m}{v_{rp}^s}\right]+p_{gr}\right\} \qquad (1)$$

where

$$n_{gr} = \frac{1}{V_t}\int_0^L \frac{\exp(-\psi(x)/V_t)}{\mu_n(x)}\cdot\left[\int_0^x U(x')\,dx'\right]dx \qquad (2)$$

$$p_{gr} = \frac{1}{V_t}\int_0^L \frac{\exp(\psi(x)/V_t)}{\mu_p(x)}\cdot\left[\int_0^x U(x')\,dx'\right]dx \qquad (3)$$

relate contribution of generated/recombinated carriers within the analysed structure to their actual concentrations. The equilibrium free carrier concentrations at the interface are

$$n_0 = N_c\exp\left(-\frac{\phi_b}{kT}\right), \qquad p_0 = N_v\exp\left(-\frac{\phi_b - E_g}{kT}\right) \qquad (4)$$

where ϕ_b is the effective Schottky barrier height, v_{dn} and v_{dp} have the dimensionality of velocity [8], v_{rn}^m, v_{rn}^s, v_{rp}^m and v_{rp}^s are the recombination velocities through which the influence of electric field strength and current density at the interface is introduced [1,9]. All other symbols have their usual meaning.

When the generation-recombination rate is set to zero and the contribution of holes is neglected, the derived expression for current has the same form as derived by Crowell and Sze [8].

3. Experimental work

For simulation of electrical characteristics we have used the doping profile which can be seen in Fig. 1. Also the samples with $MoSi_2$-Si Schottky structures with nearly ideal I-V characteristics were prepared experimentally and $\phi_b = 0.64$ eV was determined from forward I-V measurement. The distributions of free carriers at two different applied voltages in reverse direction are presented in Fig. 1a. The free holes generated within the space charge region are attracted towards the interface and the increase in their concentration with increase of generation rate is evident. Here the hole current in the vicinity of the interface may exceed the electron current (Fig. 1b). Fig. 2 shows the simulated reverse I-V characteristics for different lifetimes of free carriers. For low barrier height $\phi_{bo}=0.657$ eV, which corresponds to the experimentally measured value of $\phi_b=0.64$ eV for $MoSi_2$ - Si structure (Fig. 2a), the thermionic emission electron current is very high and the contribution of the generation current is masked, when the lifetimes of free carriers is less than 10^{-8} s.

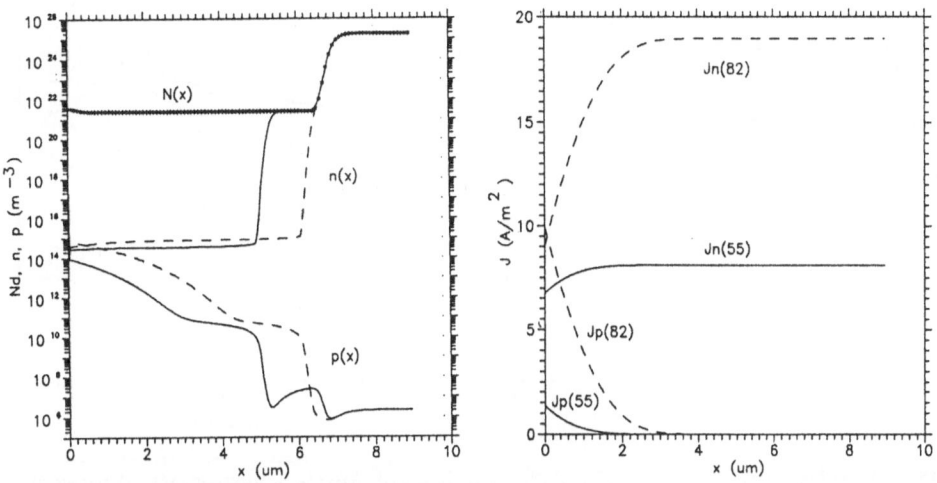

Fig.1 a) Doping profile $***$ and concentrations of free carriers and b) current densities at $V_r = 55$ V (——) and $V_r = 82$V (——)

The electron current remains constant for all lifetimes and the increase of total current can be attributed to the increase of hole current. For high barrier height $\phi_{bo} = 0.85$ eV the above mentioned effect dominates the I-V characteristics even for relatively long lifetimes of free carriers. From the previous analysis it is clear that the generated holes create the hole current which can exceed the electron current at certain reverse bias. Therefore their contribution cannot be neglected in the model. This effect has a particular significance for all structures with reverse biased Schottky contact on III-V compounds where the barrier height is high and lifetime of free carriers is short.

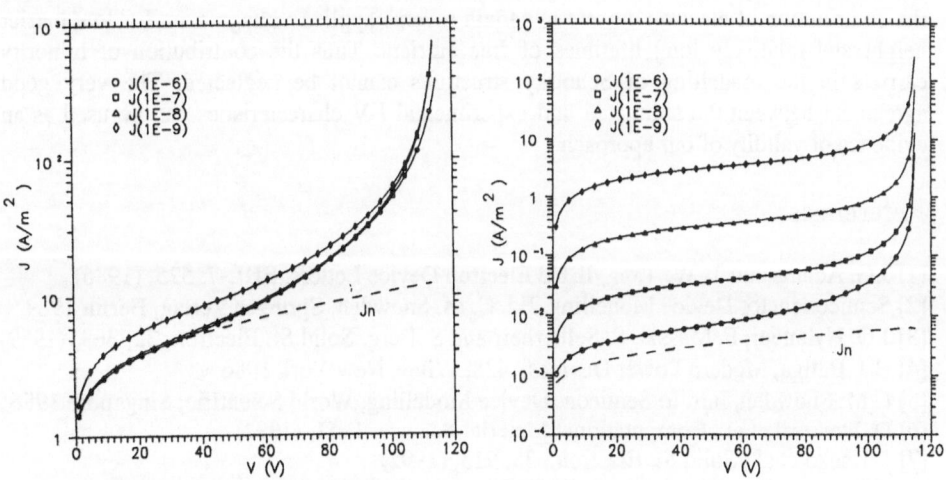

Fig. 2 Simulated reverse I-V characteristics for different lifetimes of free carriers for
a) $\phi_b = 0.657$ eV and b) $\phi_b = 0.85$ eV

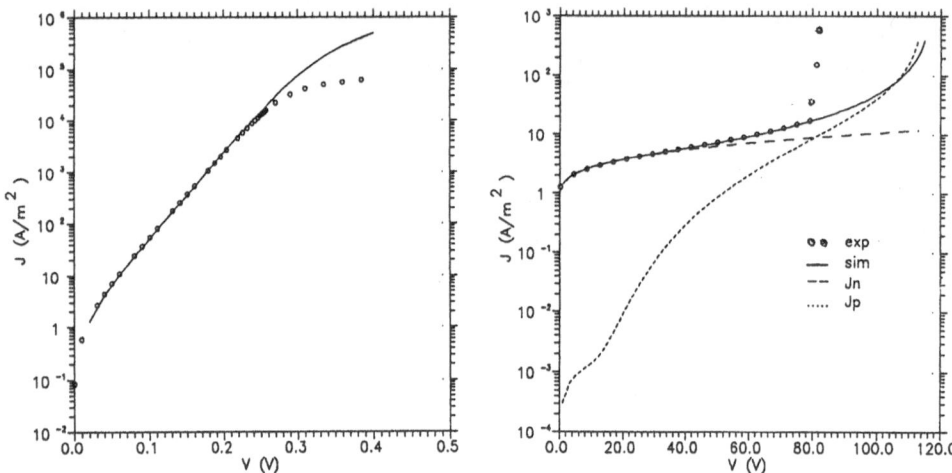

Fig.3 Simulated and experimental I-V characteristics of $MoSi_2$ - Si structure a) forward, b) reverse

We also compare the results of simulation with experimentally measured I-V characteristics (Fig.3). The very good correlation between simulated and experimental characteristics in the reverse direction is based on the introduction of hole current. The lower breakdown of experimental sample is determined by the breakdown of a guarded pn-junction at the contact periphery and appears earlier as the theoretical limit due to higher field in the cylindrical junction [10].

4. Conclusion

The influence of generation-recombination effects on the total current through the Schottky contact has been presented. The contribution of hole current to the total current can exceed the contribution of electron current at certain reverse applied voltage even for low barrier height and relatively long lifetimes of free carriers. Thus the contribution of minority carriers in the modelling of Schottky structures cannot be neglected. The very good agreement between the simulated and experimental I-V characteristics can be used as an evidence of validity of our approach.

References

[1] J.G. Adams and T. W. Tang, IEEE Electron Device Letters, EDL-7, 525, (1986)
[2] Semiconductor Device Modelling, Ed. C. M. Snowden, Springer Verlag, Berlin 1989
[3] J.O. Nylander, F. Masszi, S. Selberherr and S. Berg, Solid St. Electron, 32, 363, (1989)
[4] B.J. Baliga, Modern Power Devices, 428, Wiley, New York 1986
[5] C.M. Snowden, Intr. to Semicon. Device Modelling, World Scientific, Singapore 1986
[6] D. Donoval et al., Computational Material Science, 1, 51, (1992)
[7] J. Racko et al., Solid St. Electron., 35, 913, (1992)
[8] C.R. Crowell and S.M. Sze, Solid St. Electron., 9, 1035, (1966)
[9] R.B. Darling, Solid St. Electron., 31, 1031, (1988)
[10] S.M. Sze, Physics of Semiconductor Devices, Wiley, New York 1981

SIMULATION OF SEMICONDUCTOR DEVICES AND PROCESSES Vol. 5
Edited by S. Selberherr, H. Stippel, E. Strasser – September 1993

489

Thermionic Current in Direct-Indirect Energy-Gap $GaAs/Al_xGa_{1-x}As$ Interfaces

D. Tammaro, K. Hess[†], and F. Capasso[‡]

Dipartimento di Elettronica, Politecnico di Tornio
Corso Duca degli Abruzzi 24, I-10129 Torino, ITALY
[†]Beckman Institute for Advanced Science and Technology and Coordinated Science
Laboratory, University of Illinois at Urbana-Champaign
Urbana, IL 61801, USA
[‡]AT&T
600 Mountain Avenue, Murray Hill, NJ 07974, USA

Abstract

Experimental and theoretical studies, on the decrease of the Richardson constant
for the thermionic emission in Al-rich ($x \geq 0.45$) heterojunctions by more than 3
orders of magnitude reveal that transport in the (100) crystallographic direction,
across these interfaces is still an open research field. We present a phenomenological model based on envelope wavefunctions which involves two important transport
mechanisms: *zero-phonon transitions* due to $\Gamma - X$ mixing and *phonon-assisted
transitions*. The model makes use of tunneling calculations and transmission coefficients, evaluated for the above two mechanisms. These coefficients are different
from the step function used in the classical theory.

1. Introduction

Thermionic current across single [1, 2] and double [3], Al-rich ($x \geq 0.45$) heterojunctions
has been measured by several experimental researchers both in steady state [4, 5] and
in dynamic regimes [6, 7]. Their data showed a dramatic decrease of the Richardson
constant for thermionic emission (Fig. 1). Theoretical studies have been therefore applied
in order to discover phenomena which are responsable for this decrease [8] - [11]. This
effect has been attributed to the transition of the alloy from a direct to indirect energy-gap

material as the Al mole fraction (x) is increased. In this paper we provide a fully quantum-mechanical model for the thermionic current valid for every value of the Aluminum fraction x. Transport across the heterojunction is associated with the Γ minimum band edge if the AlAs fraction is less than 0.45. As x exceeds this value the AlGaAs energy gap becomes indirect and the Γ electrons in GaAs are transmitted to AlGaAs via electronic states associated with the X minimum.

2. The model

The classical thermionic current expression can be derived from Bethe's model [12]

$$J = A^* T^2 exp\left(-\frac{\Delta E_b}{kT}\right)$$ (1)

where the ΔE_b is the barrier height, kT the thermal energy and $A^* = 8A/cm^2 K^2$ following Ref.[9]. However this model can not explain the decrease by more than 3 orders of magnitude in the indirect range of the $GaAs/Al_xGa_{1-x}As$ interface and gives overestimated currents. The importance of the completely quantum-mechanics multivalley transport is crucial for Al-rich heterojunctions. The $\Gamma - X$ transition can occur by two different processes [8]: the transfer via the two X minima (X_z) aligned in the normal (100) direction with the Γ minimum (Γ point at k=0) or the transfer through the four lateral X minima (X_x, X_y). In a previous paper [14], we described a complete model for thermionic emission in steady state as well as for transient response and we compare it with a large number of experimental results. In such a model the transmission coefficient must account for the $\Gamma - X$ transfer via both the zero-phonon and the phonon-assisted mechanisms. The former are elastic coherent processes in which $\Gamma-$ electrons (Fig. 2) e.g. from $GaAs$ transfer to $Al_xGa_{1-x}As$ via the two X minima (X_z) aligned in the normal (100) direction with the Γ minimum. The latter, via the four lateral X minima (X_z, X_y), require the assistance of electron-phonon scattering events in order to conserve momentum in the lateral direction. Using the thermally enhanced tunneling current expression given by Duke [13]

$$J = \frac{em_1^* kT}{2\pi^2 \hbar^3} \int t(E) \cdot S(E) dE$$ (2)

where $t(E)$ is the transmission coefficient and $S(E)$ is the supply function

$$S(E) = \ln \frac{1 + exp[(E_f - E)/kT]}{1 + exp[(E_f - E - qV)/kT]}$$ (3)

we describe the transmission coefficient as [15]

$$t_{\Gamma X}(E) \simeq A_{\Gamma X}(E - \Delta E_{\Gamma X})^{3/2} \tag{4}$$

where $A_{\Gamma X}$ is for the two cases:

zero-phonon transitions

$$A_{\Gamma X} = \frac{16\alpha^2}{3h^2} \frac{\frac{M_X^{(1)} m_X^{(2)}}{(M_X^{(2)} m_\Gamma^{(1)})^{3/2}}}{\Delta E_X \Delta E_{\Gamma X}^{1/2} \left[\Delta E_{\Gamma X} + \frac{m_\Gamma^{(1)}}{m_\Gamma^{(2)}}(\Delta E_\Gamma - \Delta E_{\Gamma X}) \right]} \tag{5}$$

phonon-assisted transitions

$$A_{\Gamma X_z} \approx \frac{2\sqrt{2}D_{\Gamma X}^2}{9\pi\hbar^2 \rho\omega_{\Gamma X}} \left(N_{\Gamma X} + \frac{1}{2} \pm \frac{1}{2}\right) \frac{m_X^{(2)} \dot{M}_X^{(1)} m_\Gamma^{(2)}}{(M_X^{(2)} m_\Gamma^{(1)})^{\frac{1}{2}}} \frac{\Delta E_{\Gamma X}^{\frac{1}{2}}}{\Delta E_\Gamma \cdot \Delta E_X}$$

$$\times \left\{ \frac{1}{(\Delta E_\Gamma m_\Gamma^{(2)})^{\frac{1}{2}}} \left[1 + \left(\frac{\Delta E_X M_X^{(2)}}{\Delta E_\Gamma m_\Gamma^{(2)} M_X^{(1)}} \right)^{\frac{1}{2}} + \frac{\Delta E_X M_X^{(2)}}{\Delta E_\Gamma m_\Gamma^{(2)} M_X^{(1)}} \right] \right.$$

$$\left. + \frac{1}{(\Delta E_X M_X^{(2)})^{\frac{1}{2}}} \left[1 + \left(\frac{\Delta E_\Gamma m_\Gamma^{2(2)}}{\Delta E_X m_\Gamma^{(2)} M_X^{(1)}} \right)^{\frac{1}{2}} + \frac{\Delta E_X M_X^{(2)}}{2\Delta E_X m_\Gamma^{(2)} M_X^{(1)}} \right] \right\} \tag{6}$$

$A_{\Gamma X_z}$ and $A_{\Gamma X_y}$ are perfectly equivalent one another and have similar expressions [14]. The total $A_{\Gamma X}$ can be therefore written as

$$A_{\Gamma X}^T = A_{\Gamma X} + A_{\Gamma X_z} + 2A_{\Gamma X_z} \tag{7}$$

3. Conclusions

Our calculations clearly show that both *zero-phonon* and *phonon-assisted* contributions are needed in order to correctly evaluate the thermionic current and emission rates in the direct-indirect range of composition of the $GaAs/Al_xGa_{1-x}As$ interface system.

Acknowledgement

K.Hess and D.Tammaro have been supported by the Joint Services Electronics program.

References

[1] P.M.Solomon, S.L.Wright and C.Lanza *Superlattice Microstruct.*, 2, p.521, 1986 , and References therein.

[2] F.Capasso, F.Beltram, J.F.Walker and R.J.Malik *IEEE Electr. Dev. Lett.* 9,p.377,1988

[3] A.R.Bonnefoi, D.H.Chow and T.C.McGill *J.Vac.Sci.Technol.* B4, p. 988,1986

[4] C.S.Kyono,V.P.Kesan,D.P.Neikirk,C.M.Maziar and B.G.Streetman *Appl. Phys. Lett.* 54, p.546, 1989

[5] M.Rossmanith,J.Leo,E.Boeckehoff, K.Ploog and von Klitzing *Proc. Int.Symp. on the Phys.of Sem.* 1990 p.1142

[6] F.Beltram, F.Capasso, J.F.Walker and R.J.Malik *Appl. Phys. Lett.* 53,p.377,1988

[7] J.A.Lott,J.F.Klem and H.T.Weaver *Appl. Phys. Lett.* 55,p.1226, 1989

[8] P.J.Price *Surface Sci. 196,p.394-98, 1988*

[9] A.A.Grinberg *Phys. Rev.B.* 33,p.7256, 1986

[10] M.Rossmanith, K. Syassen,E. Böckenhoff,K.Ploog and K. von Klitzing *Phys. Rev. B* 44,p.3168, 1991

[11] H.C.Liu *Appl.Phys. Lett.* 51, p.1019, 1987

[12] H.A.Bethe *MIT Radiat. Lab. Rep.* 43-12, 1942

[13] C.B.Duke *Academic, New York* 1969, p.32-59 and References therein.

[14] D.Tammaro, K.Hess and F.Capasso to be published in *J.Appl.Phys.* vol. 73, 15 June 1993

[15] Z.S.Gribnikov and O.E.Raichev *Sov. Phys. Semicond.* 23,p.1344, 1989

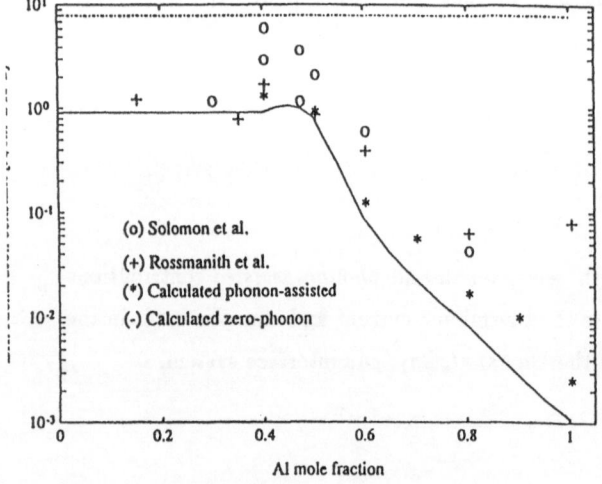

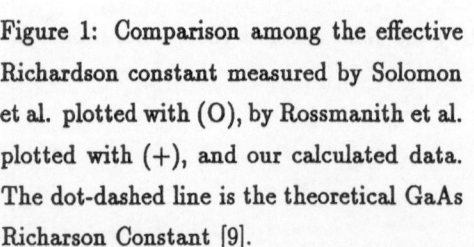

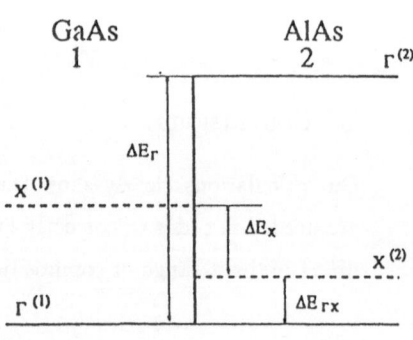

Figure 1: Comparison among the effective Richardson constant measured by Solomon et al. plotted with (O), by Rossmanith et al. plotted with (+), and our calculated data. The dot-dashed line is the theoretical GaAs Richarson Constant [9].

Figure 2: Two valleys energy band diagra for the GaAs/AlAs interface. The figu shows the relative energy difference betwee Γ and X valleys.

SIMULATION OF SEMICONDUCTOR DEVICES AND PROCESSES Vol. 5
Edited by S. Selberherr, H. Stippel, E. Strasser – September 1993   493

Preliminary Results of Quantum Directional Coupler Simulation Using a Beam Propagation Method

A. D. Sadovnikov, A. Sarangan, and W.-P. Huang

Electrical and Computer Engineering Deptartment, University of Waterloo
Waterloo, Ontario, N2L 3G1, CANADA

Abstract

A simulation method for electron wave propagation in a quantum directional coupler is presented. As an example of the method, the switching behavior of quantum directional coupler is simulated.

1. Introduction

Recent advances in technology have enabled the fabrication of nano-scale devices, whose quantum effects can be put to use. The Quantum Directional Coupler (QDC) is an example of such a device which has been extensively investigated in the last three years [1-3]. A cross-section of a possible QDC structure is shown in Fig. 1. Space charge layers under the Schottky contacts create the required potential wells to confine the electrons in x-direction. The heterojunction between the undoped $Al_{0.3}Ga_{0.7}As$ and $GaAs$ layers confines the electrons in the y-direction. This creates two parallel quantum wires of two-dimensional electron gas (2DEG) with very high mobility at low temperature. If the separation between the wires is small enough, then the electrons propagating in these wires will interact with each other. For example, if initially the electron wave is in one wire, then after traveling some distance it will be transferred into the second wire. This can be viewed as the symmetric and antisymmetric modes beating with each other, causing the electrons to slosh between the two wires. The transfer length depends on the barrier height and upon the separation between the two wires. Therefore, by changing the voltages, one can switch the electron current from one wire to the other.

In previous papers [1-3], the two-dimensional Schrodinger's and Poisson's equations in the QDC structure were solved to calculate the electron eigenfunctions, and conclusions about the characteristics of the QDC were made. The purpose of this paper is to demonstrate the applicability of the beam propagation method (BPM), that has been used to simulate optical waveguides [4], to QDC simulation.

2. Physical model

To apply BPM one should assume that there is a preferred direction of wave propagation, along which the changes of electron wave-function are slow. In a QDC this is the z-direction. We shall also assume that the QDC supports only two eigenfunctions: first symmetric and first antisymmetric modes [2], with nearly equal wave numbers k_z^s and k_z^a along the z-direction. Therefore the envelope of the electron wavefunction Ψ can be described as [2]

$$\Psi = \psi^s exp(-jk_z^s z) + \psi^a exp(-jk_z^a z) = \psi exp(-jk_z^s), \tag{1}$$

where ψ^s and ψ^a are z-independent wave functions, corresponding to the lowest symmetric and asymmetric modes, and $\psi = \psi^s + \psi^a exp[-j(k_z^a - k_z^s)z]$. If $k_z^a \approx k_z^s = k_z$ then ψ varies slow along the z-axis. Using this property of ψ we can use a *paraxial* approximation, $|k_z^2| \gg |\partial^2\psi/\partial z^2|$, to simplify Schrodinger's equation. In this study we shall assume that the electrons in the quantum wells are fully confined in the y-direction. Therefore it is sufficient to solve Schodinger's equation in 2D instead of in 3D. Substituting (1) into Schrodinger's equation, one can derive the following reduced equation for ψ [5]:

$$2jk_z\partial\psi/\partial z = [2m^*(E - E_c)/\hbar^2 - k_z^2]\psi + \partial^2\psi/\partial x^2. \tag{2}$$

Here m^* is the effective mass of electrons in $GaAs$, E is the electron energy, $E_c = -qV + E_g/2$ is the edge of the conduction band, V is the electrostatic potential, and E_g is the semiconductor bandgap.

3. Numerical method

The simulation method is as follows:

1. For given gate voltages and semiconductor layer parameters we solve Poisson's equation

$$\nabla \cdot (\epsilon_s \epsilon_0 \nabla V) = -q(p - n + N_d^+ - N_a^-), \tag{3}$$

where the hole and electron concentrations p and n are calculated using a constant Fermi level approximation. This equation is solved for the "classical" case, where we neglect electrons in the localized states. After linearization and discretization on a non-uniform rectangular grid, we solve a system of linear algebraic equations with a five-diagonal symmetric matrix using the Incomplete Cholesky-Conjugate Gradient method.

2. For a given electron energy E and an initial electron wavefunction, we calculate k_z using the variational principle (as suggested in [6] for optical wave-guides):

$$k_z^2 = \frac{\int 2m^*(E - E_c)/\hbar^2 |\psi|^2 dx - \int |\partial\psi/\partial x|^2 dx}{\int |\psi|^2 dx}. \tag{4}$$

3. Next we solve (2) on the uniform x-grid, for one step in the z-direction starting from $z = 0$. We have compared several different first- and second-order integration methods, and have found that the Crank-Nicholson scheme gives the the best results.

4. If the contact geometry changes with z, we repeat steps 1-3 until we reach the end of the simulation region. For the particular QDC considered later, we need only repeat the third step.

All calculations were done on IBM PC AT/486 computer. The typical calculation time was 2 to 5 minutes for 40×70 spatial grid nodes in Poisson's equation, 200 nodes, and 100 to 200 z-steps in Schrodinger's equation.

4. Results and discussion

A QDC structure similar to the one investigated in [3] is considered in our simulation (see Fig. 1). The Schottky barrier height is 1.0 eV, and the lattice temperature T = 4.2 K. The length of the structure is 100 nm, which can be regarded as a reasonable value for ballistic electron motion length for such temperature [2]. A low voltage $V_{G1} = V_{G5} = 0.2$ V is applied to the gates G1 and G5, creating deep space charge regions which prevent electrons from moving out of the simulation region. A high voltage $V_{G2} = V_{G4} = 0.75$ V is applied to the gates G2 and G4, creating two quantum wires underneath them. The potential on gate G3 is varied to control the barrier height and the distance between these wires. To estimate the currents flowing in the wires we calculate the quantity $k_z \int \psi \psi^* dx$, where the integral is taken over the left or right sides with respect to the QDC centerline.

The initial $\psi(x)$ distribution simulates the injection of electrons into the left wire. From Fig. 2 we see that changing V_{G3} from 0 to 0.3 V can easily change the potential barrier between the wires from quite a large value to almost zero. This results in a different transmission probability, and in turn changes the length required for electrons to penetrate from the left wire to the right one (see Fig. 3). Calculating the currents for the given QDC dimension in z-direction for different V_{G3} values, we obtain *current-voltage* characteristics (see Fig. 4). A few remarks about the method in general and these results are appropriate.

1. The *paraxial* approximation is not important for our method. In fact, one can derive an equation slightly more complicated than (2) using a *wide angle* approximation. However we have found that all these approximations give only small differences in the final results (at least for our particular device).

2. Calculations made in [1] show that the occupancy levels of highest eigenstates decrease drastically with decreasing temperature. Therefore we suppose that for very low temperatures, which are required to obtain the reasonable values of collision-free electron length, our method will be accurate.

3. The results of Fig. 3 and 4 depend on the chosen value of electron energy and initial $\psi(x)$ distribution, which have been chosen arbitrarily in the present study. Therefore one should be cautious about drawing specific conclusions from those figures, but rather consider them only as an illustration of the possibilities of the proposed simulation method.

4. Future QDC structures will be non-uniform in z-direction [2,3] as our investigated QDC is. Moreover, to account for impurity de-ionization, electron concentration on the localized states, and other Fermi level related effects, one should solve a three-dimensional Schrodinger's equation instead of the two-dimensional one (2). However with suitable modifications, our method can handle these changes, and we hope that it still will be faster than a standard approach [1-3].

References

1. T. Kerkhoven, A.T. Galick, U. Ravaioli, J.H.Arends, Y. Saad, J. Appl. Phys., Vol. 68, p. 3461-3469, N.7, 1990.

2. N. Dagli, G. Snider, J. Waldman and E. Hu, J. Appl. Phys, Vol. 69, p. 1047-1051, N. 2, 1991.

3. M. Macucci, U. Ravaioli, T. Kerkhoven, Superlattices and Microstructures, Vol. 12, p. 509-512, N.4, 1992.

4. W.-P. Huang, C. Xu, S.-T. Chu, and S. K. Chaudhuri, J. of Lightwave Technology, Vol. 10, p. 295-305, N. 3, 1992.

5. A. Sarangan and W.-P. Huang, submitted for publication in IEEE Transactions of Electron Devices.

6. W.-P. Huang, H. A. Haus, J. of Lightwave Technology, Vol. 9, p. 56-61, N. 1, 1991.

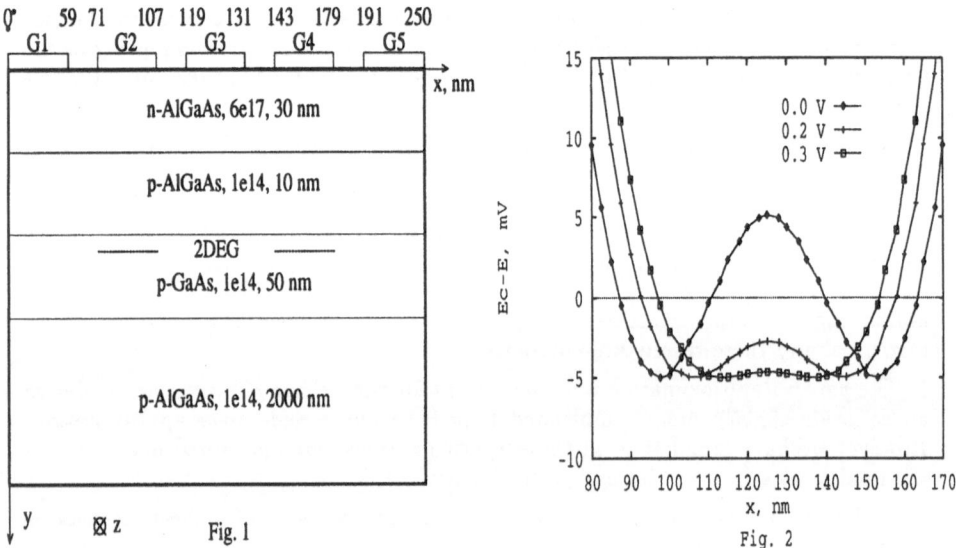

Fig. 1. A cross-section of a quantum directional coupler. Lateral dimensions are in nanometers.

Fig. 2. $E_c - E$ distribution for y = 41 nm for various V_{G3} values.

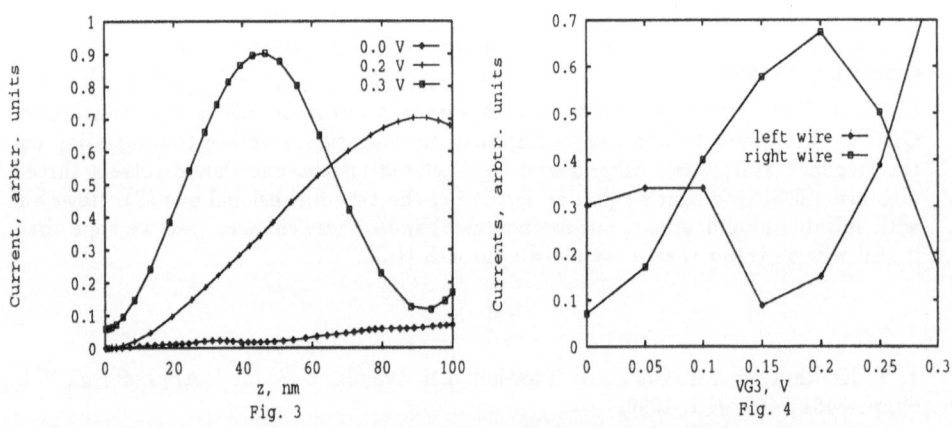

Fig. 3. Right wire current dependence on z-distance for various V_{G3} values.

Fig. 4. Dependence of the output currents in the right and left wires on V_{G3}.

SIMULATION OF SEMICONDUCTOR DEVICES AND PROCESSES Vol. 5
Edited by S. Selberherr, H. Stippel, E. Strasser – September 1993

Numerical Simulation of Auger-Induced Hot Electron Transport in InGaAsP/InP Double Heterojunction Laser Diodes: Hydrodynamics versus Drift and Diffusion

K.-W. Chai, Z.-M. Li[†], J. G. Simmons, and S. P. McAlister[†]

Centre for Electrophotonic Materials and Devices, McMaster University
Hamilton, Ontario, L8S 4L7, CANADA
[†]Institute for Microstructural Sciences, National Research Council
Ottawa, K1A 0R9, CANADA

Abstract

A strategy for the numerical simulation of Auger-induced hot electron transport related to the electron leakage problem encountered in the design and analysis of InGaAsP/InP laser diodes is presented. The theoretical structure used in conventional device simulation is extended to include the otherwise neglected interactions between the Auger hot electrons and the low-energy carriers in the device. The transport behavior of the Auger hot electrons is examined from both the hydrodynamic and the drift-diffusion perspectives.

1. Introduction

Electron leakage in InGaAsP/InP laser diodes originates from the overflow of the Auger-induced energetic electrons from the active region over the heterobarriers into the adjacent cladding layers. This causes significant loss of electrons that may otherwise undergo radiative recombination for light emission. This loss is reflected in the increase in the threshold current density [1] - [3]. The Auger hot electron concentration is typically two to three orders of magnitude smaller than that of the low-energy electrons in the active region [4]. Therefore, without detailed knowledge of the high-energy tail of the electron distribution function, the transport characteristics of these energetic electrons cannot be modeled using an average description of all the electrons in the device as a single system. As is well known, this average description forms the basis of conventional device simulation, in which the physical origins of the charged carriers are often ignored because of the global averaging process involved in the derivation of the semiconductor equations [5]. This renders existing device simulation programs not directly applicable to the analysis of the electron leakage problem in double heterojunction laser diodes.

To alleviate such difficulties, we present an expedient approach by extending the structure used in conventional device simulation to capture the otherwise neglected interactions between the Auger hot electrons and the low-energy carriers in the device.

The extension is achieved by decomposing the conventional electron current continuity equation into two components, with one for the Auger hot electrons, and the other for the low-energy electrons. This allows one to track down explicitly the transport behavior of the Auger hot electrons within the conventional device simulation environment. Coupling these two current continuity equations is an appropriately derived set of carrier statistics terms that account for the Auger, the Shockley-Read-Hall, and the spontaneous recombination processes. We present the simulation results for a one dimensional N-p-P InGaAsP/InP laser diode with composition corresponding to 1.3 μm emission wavelength. Hydrodynamic equations formulated for heterostructures [6] are used to model the dynamics of the Auger hot electrons, whereas drift-diffusion transport is assumed for the low-energy electrons and holes. Fermi-Dirac statistics is used in the simulation.

2. Problem Formulation

In the studies of the electron leakage problem in double heterostructure laser diodes, the electron system should be considered as consisting of an low energy electron gas interacting with a population of Auger-induced energetic electrons. A macroscopic description of the conservation of these two categories of electrons leads to the following continuity equations for the Auger hot electron current density $\vec{J}_{n'}$ and the cool electron current density $\vec{J}_{n_c}$, respectively,

$$-\frac{1}{q}\nabla \cdot \vec{J}_{n'} = G_{aug}\big|_{n'} - R_{spon}\big|_{n'} - R_{srh}\big|_{n'} - \frac{n'}{\tau} \tag{1}$$

$$-\frac{1}{q}\nabla \cdot \vec{J}_{n_c} = - R_{aug}\big|_{n_c} - R_{spon}\big|_{n_c} - R_{srh}\big|_{n_c} + \frac{n'}{\tau} \tag{2}$$

where the energy relaxation time τ appearing in (1) models the relaxation process of the Auger hot electrons due to intraband scatterings. The meaning of the generation and recombination terms on the right of (1) and (2) can be seen from their self-explanatory subscripts. These generation-recombination terms are given by

$$G_{aug}\big|_{n'} = C_n n (pn - n_i^2). \tag{3}$$

$$R_{aug}\big|_{n_c} = (2C_n n + C_p p)(pn - n_i^2). \tag{4}$$

$$R_{spon}\big|_{n'} = Bn'p \tag{5}$$

$$R_{spon}\big|_{n_c} = B[n_c p - n_i^2] \tag{6}$$

$$R_{srh}\big|_{n'} = \frac{\tau_p n_1 n'/\tau_n + pn'}{\tau_n(p + p_1) + \tau_p(n + n_1)} \tag{7}$$

$$R_{srh}\big|_{n_c} = \frac{-\tau_p n_1 n'/\tau_n + pn_c - p_1 n_1}{\tau_n(p + p_1) + \tau_p(n + n_1)} \tag{8}$$

where the electron concentration n is given by the sum of the Auger hot electron concentration n' and the cool electron concentration n_c. The parameters appearing in the above generation-recombination terms have their usual physical meanings. Note that the factor of two in (4) accounts for the loss of two low energy electrons in each CHCC events in order to create one Auger hot electron.

In this work, the following conservation equation for the Auger hot electron energy $W_{n'}$ is used

$$\nabla \cdot \vec{S}_{n'} - \nabla E_c \cdot \frac{\vec{J}_{n'}}{q} - G_{aug}|_{n'} E_g = -n \frac{W_{n'} - W_{n'}^0}{\tau} - \left[\frac{n'}{\tau} + R_{spon}|_{n'} + R_{srh}|_{n'} \right] W_{n'}. \quad (9)$$

A recombination energy equal to the energy gap E_g of the material is assumed. The hot electron energy flux $\vec{S}_{n'}$ and $\vec{J}_{n'}$ are treated using the formulation given by Azoff [6]. These equations are solved together with the Poisson and the hole current continuity equations for self-consistent solutions.

3. Simulation Results and Discussion

Fig.1 shows a schematic diagram of the device considered in this work. Fig.2 shows the energy distribution of the Auger hot electrons in the device. The injection current density is 7.4 kA/cm^2. The energy gradient throughout the device causes enhanced

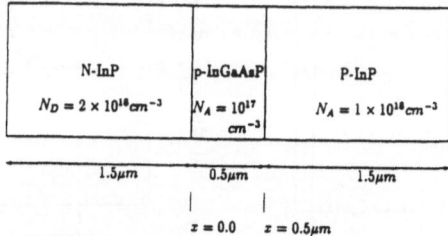

Figure 1: Schematic diagram of
the InGaAsP/InP laser diode

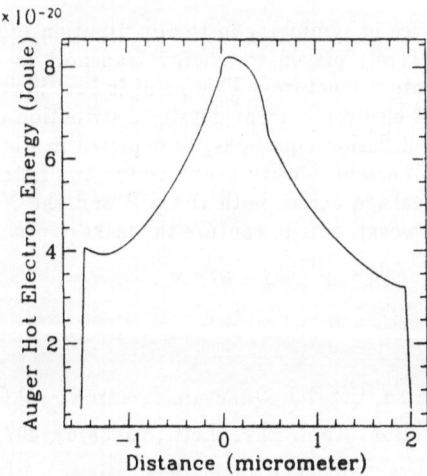

Figure 2: Auger hot electron
energy distribution

diffusion of the hot electrons into the P-InP cladding layers. This is illustrated in Fig.3, where profiles of the Auger hot electron concentration obtained from the hydrodynamic and the drift-diffusion equations are compared. The drift-diffusion result shows a highly asymmetrical distribution of the Auger hot electrons. The Auger hot electron concentration in the N-InP layer is much higher than that in the P-InP region. This is consistent with the fact that, unlike the p-P heterojunction, the built-in field across the N-p heterointerface favours the overflow of hot electrons into the N-InP layer. However, with the effects of energy transport included, the hydrodyna-

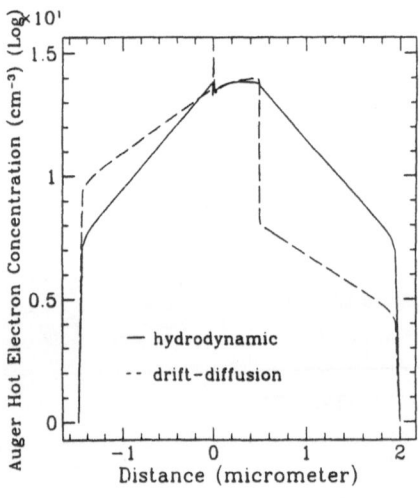

Figure 3: Auger hot electron concentration

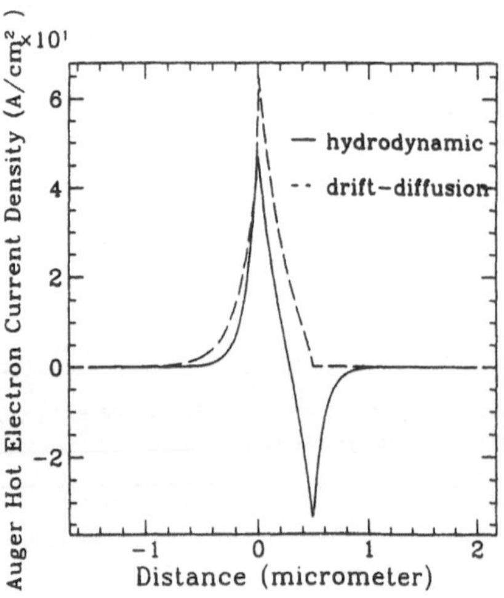

Figure 4: Auger hot electron current density

mic result shows a high degree of symmetry in the distribution of the hot electrons. This indicates the dominant role played by energy transport in the simulation of electron leakage in double heterostructures. This point is best illustrated by a direct comparison of the Auger hot electron current density distribution obtained from the hydrodynamic and the drift-diffusion equations, as depicted in Fig.4. It can be seen that the Auger hot electron current density peaks at the two heterointerfaces, indicating Auger hot electron leakage across both the p-P and the N-p heterobarriers. The drift-diffusion result, however, fails to capture the leakage process across the p-P heterointerface.

References

[1] M. Asada and Y.Suematsu, IEEE J. Quantum Electron., vol.QE-19, 917, 1983.

[2] N.K. Dutta and R.J. Nelson, Appl. Phys. Lett., vol.38(6), 407, 1981.

[3] H.C. Casey, J. Appl. Phys., vol.56(7), 1959, 1984.

[4] K.D. Chik, J. Appl. Phys., vol.63(9), 4688, 1988.

[5] K. Blotekjaer, IEEE Trans. Electron Devices, vol.ED-17, 38, 1970.

[6] E.M. Azoff, Solid-State Electron., vol.30(9), 913, 1987

SIMULATION OF SEMICONDUCTOR DEVICES AND PROCESSES Vol. 5
Edited by S. Selberherr, H. Stippel, E. Strasser – September 1993

501

Author Index

Wolfgang Joppich
Slobodan Mijalkovic

Multigrid Methods for Process Simulation

(Computational Microelectronics)
1993. 126 figures. Approx. 340 pages.
Cloth öS 1386,–, DM 198,–
ISBN 3-211-82404-9

Prices are subject to change without notice

This book is the first one that combines both research in multigrid methods and a particular application field here - process simulation. It is the declared intention of this book to convince by practically demonstrating the power of the multigrid principle and to establish an example of fruitful interdisciplinary interaction. The introduction to multigrid is therefore strictly directed towards the goal to provide the algorithmical overview one needs to compose optimal multigrid algorithms for evolution problems of process simulation and similar applications. The necessary explanation how and why multigrid works is derived from the roots. So the book preassumes no advanced familiarity with numerical analysis. Additionally a complete strategy to implement different algorithmical components on an adaptive multilevel grid structure is presented. The outlined principle of grid definement and adaption is based on the control of errors and is reliable as well as general. Last but not least the described strategies are applied to "real life" problems of process simulation.

Consequenly this book is an important contribution to the interdisciplinary challenge of improving numerical techniques for diffusion problems of process simulation.

Springer-Verlag Wien New York

Sachsenplatz 4-6, P.O. Box 89, A-1201 Wien · Heidelberger Platz 3, D-14197 Berlin
175 Fifth Avenue, New York, NY 10010, USA · 37-3, Hongo 3-chome, Bunkyo-ku, Tokyo, 113, Japan

Narain D. Arora

MOSFET Models for VLSI Circuit Simulation
Theory and Practice

(Computational Microelectronics)
1993. Approx. 260 figures. Approx. 600 pages.
Cloth öS 2086,–, DM 298,–, US $ 198.00
ISBN 3-211-82395-6

Prices are subject to change without notice

The book covers the MOS transistor models and their parameters required for VLSI simulation of MOS integrated circuits. It gives the first detailed presentation of model parameter determination for MOS models. Various models are developed ranging from simple to more sophisticated models that take into account new physical effects observed in submicron devices used in today's MOS VLSI technology. The assumptions used to arrive at the models are emphasized so that the accuracy of the model in describing the device characteristics are clearly understood. Understanding these models is essential when designing circuits for the state of the art MOS IC's. Threshold voltage being the single most important MOSFET parameter, a full chapter is devoted to the development of the device threshold voltage model. Due to the importance of designing reliable circuits, the device reliability models as applied for circuit simulations are also covered. Since the device parameters vary due to inherent processing variations, how to arrive at worst case design parameters are covered.
Presentation of the material is such that even an undergraduate student not well familiar with semiconductor device physics can understand the intricacies of MOSFET modeling. The book serves as a technical source in the area of MOSFET modeling for state of the art MOSFET technology for both practicing device and circuit engineers and engineering students interested in the said area.

Springer-Verlag Wien New York

Sachsenplatz 4-6, P.O. Box 89, A-1201 Wien · Heidelberger Platz 3, D-14197 Berlin
175 Fifth Avenue, New York, NY 10010, USA · 37-3, Hongo 3-chome, Bunkyo-ku, Tokyo, 113, Japan